Astronomy:
Principles and Practice

Fourth Edition

T0256404

Astronomy:
Principles and Practice

Fourth Edition

A E Roy, PhD, FRAS, FRSE, FBIS

*Professor Emeritus in Astronomy, Honorary Lecturer
and Senior Research Fellow, University of Glasgow*

D Clarke, PhD, MInstP, FRAS

*Honorary Research Fellow in Physics and Astronomy,
University of Glasgow*

CRC Press
Taylor & Francis Group
Boca Raton London New York

CRC Press is an imprint of the
Taylor & Francis Group, an **informa** business

First Edition published 1977
Reprinted 1978
Second Edition 1982
Third Edition 1991

CRC Press
Taylor & Francis Group
6000 Broken Sound Parkway NW, Suite 300
Boca Raton, FL 33487-2742

CRC Press is an imprint of Taylor & Francis Group, an Informa business

No claim to original U.S. Government works

British Library Cataloguing-in-Publication Data

A catalogue record for this book is available from the British Library.

ISBN 9780750309172

Library of Congress Cataloging-in-Publication Data are available

Visit the Taylor & Francis Web site at
http://www.taylorandfrancis.com

and the CRC Press Web site at
http://www.crcpress.com

To our students
who have learned almost as much from us
as we have from them

Contents

Foreword to fourth edition

The remarkable advances in astronomy, space research and related technology in the past twelve years since the third edition of this text was published have provided a need for a fourth edition. Again it gives us an opportunity to improve some presentations, to expand the material devoted to new astronomies and techniques and to correct a very few residual errors.

Much of the philosophy used in preparing the book's first edition and laid out in the foreword to that edition, published in 1977, has been retained. In almost a quarter of a century, nothing in our teaching experience has weakened our belief in the need for a comprehensive and systematic treatment of astronomy from first principles of the kind provided in this book. Indeed, this belief has been strengthened by the welcome favourable reactions to the book's successive editions and use of them by many of our teaching colleagues throughout the world. It is not a coffee table 'Oh, the wonder of it all!' text full of glossy pictures, taking the reader on a tour of the Universe in much the same way that a book on the world's greatest architectural feats or one on engineering exploits does. Such books are doubtless of value in creating an interest in and appreciation of the subject they depict but they are not the texts required by the serious student of architecture or engineering who wishes to pursue a career in the subject or who desires to understand the principles and practices the architect or engineer needs to know. The science of astronomy is no exception.

The exercises, examples and problems remain the best way of showing the student whether a real understanding has been achieved of the principles involved and an ability to use them with confidence and accuracy. In the diagrams for these examples and problems, we have deliberately not redrawn the celestial spheres using computer technology to remove their slight distortion from a true perspective. Such spheres are no more unrealistic than the usual diagrams used in explaining why eclipses of the Sun and Moon occur, in which the relative sizes of the Sun, Moon and Earth and the distances between them are necessarily depicted inaccurately. Hand-drawn celestial spheres for estimation of angular sizes are adequate, giving with practice an accuracy of plus or minus ten degrees; they are also good enough for showing which quadrant the celestial object is in, so providing a useful check on calculations using spherical trigonometry.

Since the first edition, the value of the obliquity of the ecliptic has been reducing steadily and, at the beginning of the third millennium, the value of ε is now $23° 26'$ rounded to the nearest arc minute and no longer $23° 27'$. The revised value has been adopted in this edition. To a cruder approximation, however, the value still remains as $23\frac{1}{2}°$.

As already mentioned at the beginning of this foreword, great advances in technology have been made since the publication of the third edition, no more so than in the recording of optical data. Many electronic detectors have appeared, made their brief mark and then been discarded. The almost universal embracing of CCD cameras under computer control as a means of performing efficient observations, however, suggests that their contribution to the recording of data is as significant a step forward as was the establishment of the photographic process in the middle part of the 19th century. Although the descriptions of instrumentation retain some of the technologies that in the professional arena are being abandoned, the contents of Part 3 on Observational Techniques certainly reflect the current importance of the CCD. For how long this detector will dominate the scene of optical information recording remains an interesting question.

Over the last ten years, the internet has developed with myriads of sites giving information on astronomical events—what can be seen in the sky—telescope facilities, forthcoming space probes, etc. It would be impossible to provide a comprehensive list of sites that are worthwhile to visit and, in any case, many of them have only transient value. A short list is, however, supplied in an appendix and where it is useful to do so, reference to a particular site is given in the text using the superscript 'W' followed by a chapter reference number corresponding to the listing in the appendix. We leave it to the student to 'surf' and 'trawl' but hopefully not at the expense of getting to grips with a deeper understanding of the subject by applying brain and pen to the information and exercises in this book.

Archie E Roy
David Clarke

Foreword to third edition

The need for a third edition of this book has given us a welcome opportunity to improve some presentations, to expand the material in new astronomies (chapter 22) and correct a few residual errors. We now have removed the logarithmic displays of reduction because of the widespread use of pocket calculators, a practice that removes most of the drudgery of spherical trigonometrical calculations. Nothing that we have learned from our teaching experience over the past few years has weakened our belief in the need for a comprehensive and systematic treatment of astronomy from first principles of the kind provided by this book.

Archie E Roy
David Clarke

Foreword to second edition

Very little need be added to the foreword of the first edition. Astronomy has continued to develop in fruitful and exciting ways both in its more traditional branches and in space research. There is a need more than ever for a comprehensive and systematic treatment of astronomy from first principles for those students who want to obtain a thorough grounding in physical and mathematical topics so often omitted from other textbooks. Because of this, because of the gratifying reception of the first edition of this book and also because our further teaching experience has not fundamentally changed our opinions regarding what the contents of a textbook on the principles and practice of astronomy should be, we have confined ourselves largely to correcting a number of errors, removing some defects and improving some presentations. In the worked examples, we have retained the logarithmic displays of reduction since even now there are many places in the world where pocket calculators are still not available. To help those people who use calculators, however, we have expanded on the 'Number' inserting additional steps.

Archie E Roy
David Clarke

Foreword to first edition

The present text has grown out of university and extra-mural courses in astronomy given by both authors over a number of years. In particular, much of the material here is presented in lectures and practical work given in the first-year course in the Department of Astronomy at Glasgow University. At this stage of their degree course, most of the students study a range of science disciplines in order to obtain a broad base of general science before taking more specialized courses in later years.

The course at Glasgow also attracts some arts students who wish to gain some knowledge of the workings of science and the scientific method. We feel that astronomy is perhaps the best discipline within science for doing this. Many of the major breakthroughs in the subject (for example, the understanding of the nature of stellar spectral lines) can best be appreciated by putting them into their historical perspective. They allow an inspection of how our knowledge of a subject develops—facts may stare us in the face but until the right person comes along at the right time to take perhaps just one new simple step, making all the facts fall into a series of connected relationships, they remain just a confusing jumble. The book in conjunction with a companion volume—*Astronomy: The Structure of the Universe*—has, therefore, been written to fulfil the need of a *preliminary science course* or a *liberal arts course* at the first-year university level. Indeed, those polytechnics that now provide courses in astronomy may find the present text suitable. We also hope that the serious amateur will find helpful discussions of topics in which he/she is interested.

The contents of this volume prepare the student for the presentation to be found in the companion volume of the facts about the Universe and their interpretation. In a sense, this book presents the basic software and hardware of the subject, so providing some of the simple mathematical tools and discussing some of the simple physical processes which are either involved in the astronomer's tools of trade or concerned in the mechanisms associated with astronomical bodies.

Our experience has shown that where a serious attempt is to be made to teach the basic principles of astronomical methods by which our present knowledge of the Universe has grown, the student must be prepared to step beyond the easy 'Oh-the-wonder-of-it-all' reading of purely descriptive matter and to come to grips with the methods in current use.

The best way of doing this and appreciating the techniques and the difficulties involved is to observe, to measure, to apply formulas, to solve problems. For these purposes, a certain level of mathematics is included though its use is not, we feel, to excess. The basic mathematical structure is kept to a level suitable for a first-year student with a knowledge of high school mathematics and physics. Indeed, the intelligent student without such a mathematical background will find it possible to read around the mathematical sections and still gain a measure of understanding.

The book has been divided into four parts. Briefly these perform the following functions:

Part 1: Sets the scene.
Part 2: Provides the software—the ideas and mathematics of positional and dynamical astronomy.
Part 3: Describes the hardware—the physics of radiation and the astronomer's tools.
Part 4: Provides exercises and practical work.

Problems are also given in the relevant places to test thoroughly the student's appreciation of the principles involved. Many of the problems are worked out in detail in the text as examples: all the problems have answers provided. Although logarithms have been used throughout the tabulated examples, the authors appreciate that many students now possess pocket calculators; because these machines use different languages, the authors believe it best to leave the modification of the examples to the students.

Most of the problems are original to the authors—these and others presented have appeared in examinations set in the past by the Department of Astronomy of Glasgow University.

Exercises suitable for laboratory or outdoor work are also provided in Part 4 and indeed have been developed and tested in recent years in the laboratories of the Department of Astronomy, Glasgow University. A guide is given to practical laboratory and observational work which can be carried out according to the availability of apparatus.

With respect to references, a small, carefully selected list of books is included but no papers from research journals are listed since our experience suggests that a first-year student has not yet developed either the ability to profit by reading in great detail or the necessary critical power. The references given seem to us to enlarge usefully on particular topics.

The book contains more material than would normally fill a first-year course, allowing teachers to select portions well adapted to their particular needs and resources.

No attempt has been made in the book to discuss every aspect of astronomy. With a subject which includes so great a diversity of method, and range in condition of and assembly of matter, this must be left to the canopy of an astronomical dictionary or encyclopedia. Thus, we hope that the professional astronomer will excuse us if his particular pet theme has been excluded or if we appear to have oversimplified any treatment which perhaps deserves better and deeper attention. We have tried wherever possible to treat the included topics of the subject at the same level. The authors have been very conscious of a difficulty in treating the subject systematically. It has been found that some overlap of material has been inevitable; for instance, some seeming repetition results where the nature of the observed source is described and elsewhere the technology of observation makes a reference to the same source.

It will be appreciated that with recent spectacular advances in space research, as well as in the more traditional branches of astronomy, much new knowledge has accumulated and we have made a particular effort to include this so that the text is as up-to-date as possible without adding material which may have only transient value.

Where new concepts are defined or introduced for the first time, they are given in a bold typeface. Figures and equations are given the number of the chapter in which they appear, followed by a number denoting the order in that chapter.

Astronomers use various systems of units, depending upon the astronomical topic under discussion. Without being slaves to pedantry, we have tried, wherever possible, to keep to a consistent set of units (SI) and symbols and to adopt those preferred by the International Astronomical Union. In an appendix, a useful set of constants and conversion factors has been included.

Astronomy has long been called the Queen of Sciences because of the fascination and influence it has exerted on the minds of men. If this book, together with its companion volume—*Astronomy: The Structure of the Universe*—manages to engender some degree of that fascination as well as to provide the main facts of the subject, the authors will be well content.

Archie E Roy
David Clarke

Acknowledgments

First edition

Teaching is a constant process of reiteration and regeneration. The response that a teacher has from his students provides a form of feedback, effecting an improvement to the content and method of presentation of succeeding courses. Without this intelligent response and students, courses would remain dull and static. Our first acknowledgment must be to our former students who have helped in giving shape to the contents of this book.

The presentation of many other topics in the book has undoubtedly benefited from innumerable discussions with past and present members of the staff of the Department of Astronomy, Glasgow University. In particular, we would like to thank Professor Peter A Sweet not only for helpful advice but also for providing some source material.

We would like to acknowledge and thank the Oxford University Press and Professor A Thom for permission to include in our text figure 7.17 redrawn from figure 12.2 of his work, *Megalithic Sites in Britain*, and the Senate of Glasgow University for permission to use a number of examination problems set by the Department of Astronomy.

The authors also acknowledge with sincere gratitude their debt to Mrs Margaret I Morris, MA, FRAS, of the Department of Astronomy, Glasgow University—not only did she type the first draft of the manuscript but she also read it critically, suggesting a large number of improvements in the mode of presentation that the authors have been happy to incorporate. Our thanks are also due to Mrs L Williamson, of the same Department, for much additional typing.

Inevitably, however, some errors of fact and misprints will remain and the authors will be glad to hear of them.

Second edition

We are sincerely grateful to those who have taken the trouble to notify us about some errors which have now been corrected.

Third edition

The third edition has given us the chance to improve and expand the index and we are pleased to express our thanks to Mr Douglas Taylor for his help. Once again we are greatly indebted to Mrs Margaret Morris for preparing the typescript of the new material and modifications.

Again we thank those who have drawn our attention to a few residual errors which have now been removed.

Fourth edition

We wish to thank our colleague, Dr Graham Woan, for his advice on radio matters and for permission to include a description in chapter 24 of the Radio Astronomy Exercise which he developed.

Archie E Roy
David Clarke

PART 1

INTRODUCTION

Chapters 1–6

PROGRAMME: This first part presents the simple observations which can be made by eye and discusses how such observations were interpreted by previous civilizations. It describes the nature of the observables by which we gain knowledge of the Universe and comments on the role the astronomer has to play. The effects of the Earth's atmosphere on such observations are described. Comment is made on how the present scientific age has advanced the range of astronomical instruments. The concepts and terms related to the basic observations of the brightnesses of objects, their position in the sky and the times of measurement are introduced. Simple advice on observing the night sky is provided.

Chapter 1

Naked eye observations

1.1 Introduction

The etymology of the word 'Astronomy' implies that it was the discipline involved in 'the arranging of the stars'. Today we might say that astronomy is our attempt to study and understand celestial phenomena, part of the never-ending urge to discover order in nature. We do not know who were the first astronomers—what we do know is that the science of astronomy was well advanced in parts of Europe by the middle of the third millennium BC and that the Chinese people had astronomical schools as early as 2000 BC. In all ages, from the burgeoning of man's intelligence, there have been people fascinated by the heavens and their changing aspect and these people, as far as their cultural environment has allowed them, have tried to formulate cosmologies. We are no different today.

Nowadays, the word 'Astrophysics' is also used to describe the study of the celestial bodies. In fact, many astronomers use both terms quite generally and it is not infrequent to find Departments of Astronomy and Astrophysics within educational establishments. The question may well be asked 'What is the difference between *Astronomy* and *Astrophysics*?' Very loosely, Astronomy might be defined as the subject of the '*where and when*' related to the description of a celestial body with the '*why and how*' being covered more by Astrophysics. Rather than trying to provide a hard and fast rule for the terminology, we will simply use *Astronomy* to cover all aspects of the description of the skies and the Universe.

If our current theories of the Universe are nearer the truth, it is probably not that our intelligence has increased in the past six millennia. It is more likely that the main factor has been the discovery and development of the 'scientific method', which has led to our present civilization based on the flood of technological advantages provided by this method. This has enabled scientists in far greater numbers than ever before to devote their lives to the study of the heavens, assisted by telescopes, computers, space vehicles and a multitude of other equipment. Their attempts to interpret and understand the wealth of new information provided by these new instruments have been aided by allied sciences such as physics, chemistry, geology, mathematics and so on.

We must remember, however, that for more than nine-tenths of the last five thousand years of our study of the heavens, we have had to rely on the unaided eye. The Mediterranean people who set the constellations in the sky, the Babylonians, Egyptians and Greeks, the Arabian astronomers who flourished during the Dark Ages of Post-Roman Europe, the Chinese, the Mayan and other early American astronomers, all built their theories of the Universe on naked eye observations. And so we begin by following in their footsteps and seeing what they saw as they observed over a few minutes (see section 1.2), over a few hours (see section 1.3), over a month (see section 1.4) or over at least a year (see section 1.5). In this way, we will find it easier to understand why their cosmological theories were formulated in their particular ways.

1.2   Instantaneous phenomena

1.2.1   Day

During the day a variety of phenomena may be seen. In a particular direction lies the Sun, so bright it is impossible (and dangerous) to look directly at it. In general, the sky background is blue. The Moon may also be visible, having a distinct shape though certainly not circular. If the Sun has just set or if dawn is not far away, there is sufficient daylight to see clearly. We call this condition **twilight**.

On the horizon opposite to the twilight glow, a dark purple band is sometimes seen. This area corresponds to a zone on the sky which is cut off from the direct sunlight by the Earth and is receiving very little light by scattering from the atoms and molecules in the atmosphere. It corresponds, in fact, to the shadow of the Earth in the sky. Its presence tells us of the extreme purity and low humidity of the local atmosphere. Needless to say, it is very rarely seen in Britain.

To the ancients, clouds, wind, rain, hail and other atmospheric phenomena were inadequately distinguished from what we term celestial events. Our civilization includes them in **meteorology**, a science quite distinct from astronomy, so that we need not consider them further, except to remark that astronomers' observations have, until recently, been dependent entirely upon good weather conditions being available. With the development of radio telescopes and the fact that other equipment can be placed in artificial satellites and operated above the Earth's atmosphere, this dependence is no longer complete.

1.2.2   Night

If seeing conditions are favourable, a view of the night sky provides a far wider variety of celestial phenomena. If the Moon is visible, its brightness will dominate that of all other objects. Its shape will be crescent or gibbous or even circular. At the last condition, its apparent diameter is very close to that of the Sun. To anyone with reasonable eyesight, its surface will not be evenly bright. Areas darker than their surroundings will be noticed, so that the fancy of primitive man could see a 'Man in the Moon', a 'Beautiful Lady' or a 'Rabbit', sketched out by these features.

In addition to the Moon, some two to three thousand tiny, twinkling points of light—the stars—are seen, ranging in brightness from ones easily visible just after sunset to ones just visible when the Moon is below the horizon and the sky background is darkest. Careful comparison of one bright star with another shows that stars have different colours; for example, in the star pattern of *Orion*, one of the many constellations, *Betelgeuse* is a red star in contrast to the blue of *Rigel*. The apparent distribution of stars across the vault of heaven seems random.

With the eyes becoming accustomed to the darkness, a faint band of light, the Milky Way, catches the observer's attention. Modern astronomers, with the aid of telescopes, know that this luminous region stretching from horizon to horizon across the sky in a great circle is made up of a myriad of stars too faint to be resolved with the naked eye. To the ancient observer, its presence inspired all kinds of speculations, none of them verifiable.

One or two of the tiny points of light may draw a closer scrutiny. They shine steadily, in contrast to the twinkling of the stars and they are among the brightest of the star-like objects. There must be some reason why they are different. If our observer is going to watch for a few hours, attention will be returned to these objects.

1.3   A few hours

1.3.1   Day

The heavens are never static. The slowly-moving shadow cast by an upright rod or a boulder or tree reveals the Sun's movement across the sky. If observation is kept up throughout the day, the Sun is seen to rise above the eastern horizon, climb up the sky in a circle inclined at some angle to the plane defined by the horizon and **culminate**, i.e. reach a maximum altitude above the line joining the north to the south points, then descend in a mirror image of its forenoon path to set on the western horizon. If the Moon is seen during that day, it will appear to imitate the Sun's behaviour in rising and setting.

1.3.2 Night

As darkness falls, the first stars become visible above the eastern horizon. With the ending of twilight the fainter stars can be seen and, as the hours pass, the stellar groups rise from the eastern horizon, reach their maximum altitude like the Sun, then set or become dim and invisible as daylight returns. The impression of being on a flat plane surmounted by a dark revolving bowl to which the stars are attached is strong, especially when it is seen that there are many stars in a particular region of the sky that revolve, never rising, never setting, about a hub or pivot. These stars are said to be **circumpolar**. It is then clear that those other stars that rise and set do so simply because their circular paths about this pole are so big that they intersect the horizon.

The Moon also revolves across this upturned bowl. Although the Moon appears to have an angular motion across the sky similar to that of the stars, careful observation over a few hours reveals that it moves slightly eastwards relative to the star background.

Occasionally a bright object, called a **meteor**, shoots across the sky in a second, looking like a fast-moving or 'falling star'. It may be too that faintly luminous sheets are seen, hanging down the bowl of the heavens like great curtains. These are the **aurorae**[W 1.1].

If our observer is watching at any time after October 4, 1957, it is quite likely that one or more faint specks of light will be seen to cross the sky, taking a few minutes to do so, their presence giving reminder that man-made satellites are now in orbit about the Earth. Indeed, one of the latest satellites—the International Space Station[W 1.2]—is exceedingly bright—as bright as the brightest planet Venus—and bears testament to the continual development of manned orbiting laboratories.

1.4 A month

The month is the next period of any significance to our watcher. During this time, the ideas about the heavens and their movements change. It will be noted that after a few nights the first group of stars seen above the eastern horizon just after sunset is markedly higher at first sight, with other groups under it becoming the first stars to appear. Indeed, after a month, the first group is about thirty degrees above the eastern horizon when the first stars are seen after sunset. It is then apparent that the Sun must shift its position against the stellar background as time passes. The rate is slow (about one degree per day—or about two apparent solar diameters) compared with its daily, or diurnal, movement about the Earth.

The Sun is not the only object to move independently of the stellar patterns. A few nights' observations of the Moon's position against the stars (its **sidereal** position) show that it too moves but at a much faster rate, about thirteen degrees per day, so that it is seen to make one complete revolution of the stellar background in twenty-seven and one-third days, returning to the same constellation it occupied at the beginning of the month. In addition, its shape changes. From a thin crescent, like a reversed 'C', seen in the west just after sunset, it progresses to the phase we call first quarter about seven days later. At this phase, the Moon's terminator is seen to be almost a straight line. Fourteen days after new moon, it is full and at its brightest, appearing at its highest in the sky about midnight. Seven days later it has dwindled to third quarter and rises before the Sun, a pale thin crescent once more, a mirror image of its phase just after new moon. Twenty-nine and one-half days after new moon, it is new once more.

It was a fairly easy matter for the ancients to ascertain that the Moon was nearer the Earth than the stars. Frequently the Moon was seen to blot out a star, occulting it until it reappeared at the other edge

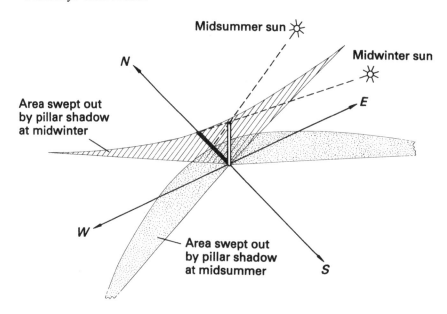

Figure 1.1. The change in length of a shadow according to the time of day and the time of year.

of the Moon's disc. And occasionally the Moon was eclipsed, the Earth progressively blocking off the sunlight until the satellite's brightness had diminished to a dull, coppery hue. An even more alarming, but rarer, occurrence took place at times during daylight: the Moon revealed its unseen presence near the Sun by eclipsing the solar disc, turning day into night, causing birds to seek their nests and creating superstitious fear in the mind of primitive man.

The observer who studies the night sky for a month or so also discovers something new about the one or two star-like objects noted that do not twinkle. Careful marking of their positions with respect to neighbouring stars shows that they too are moving against the stellar background. There does not seem to be much system, however, about these movements. In the course of a month, one may move in the direction the Moon travels in, while a second object, in another part of the sky, may move in the opposite direction. Indeed, towards the end of this month's observing sessions, either object may cease to move, seem almost to change its mind and begin to retrace its steps on the celestial sphere. These wanderers, or **planets** ('planet' is a Greek word meaning 'wanderer'), are obviously of a different nature from that of the fixed, twinkling stars.

1.5   A year

A year's patient observing, by day and night, provides the watcher with new concepts. For example, the Sun's daily behaviour, moving easterly bit by bit, is linked to the seasonal changes.

Each day, for most observers, the Sun rises, increases altitude until it culminates on the meridian at apparent noon, then falls down the sky until it sets on the western horizon. We have seen that this progress can be studied by noting the changes in direction and length of the shadow cast by a vertical rod stuck in the ground (see figure 1.1).

As the days pass, the minimum daily length of shadow (at apparent noon) is seen to change, becoming longest during winter and shortest during summer. This behaviour is also linked with changes in the rising and setting directions of the Sun. Six months after the Sun has risen between north and east and setting between north and west, it is rising between south and east and setting between south and west. Another six months has to pass before the solar cycle is completed, with the Sun once more rising between north and east and setting between north and west.

All this could be explained by supposing that the Sun not only revolved with the stars on the celestial sphere about the Earth in one day (its diurnal movement) but that it also moved much more slowly along the path among the stars on the celestial sphere, making one revolution in one year, returning to its original position with respect to the stars in that period of time. We have already seen that the observer who notes over a month what group of stars is first visible above the eastern horizon after sunset will have already come to the conclusion that the Sun moves relative to the stars. Now it is seen that there is a regular secular progression right round the stellar background and that when the Sun has returned to its original stellar position, the seasonal cycle is also completed.

The Sun's stellar route was called the **ecliptic** by the ancients. The groups of stars intersected by this path were called the houses of the Zodiac. The ecliptic is found to be a great circle inclined at about $23\frac{1}{2}$ degrees to the equator, the great circle on the sky corresponding to the projection of the Earth's equator, intersecting it at two points, the vernal and autumnal equinoxes, 180 degrees apart.

It was quite natural, then, for the ancients to worship the Sun. Not only did it provide light and warmth by day against the evils of the night but, in addition, its yearly progression was intimately linked to the seasons and so also to seed time and harvest. It was, therefore, necessary to keep track of progress to use it as a clock and a calendar. To this end, the science of sundial-making began, ramifying from simple obelisks that throw shadows on a fan of lines radiating from their bases, to extremely ingenious and complicated erections in stone and metal. Up to the 19th century, these constructions rivalled most pocket-watches in accuracy as timekeepers.

For calendrical purposes, lines of standing stones could be set up, pointing to the midsummer, midwinter and equinoctial rising and setting points of the Sun. In the British Isles, there still remain hundreds of such solar observatories, witnesses to our forefathers' preoccupation with the Sun-god.

The observer who watches the night sky throughout a year counts about thirteen revolutions of the stellar background by the Moon in that time. Over that period of time, it is not apparent that any simple relationship exists between the sidereal period of revolution of the Moon, the period of its phases and the year (the time it takes the Sun to perform one complete circuit of the ecliptic). That knowledge comes after much more extended observation, certainly measured in decades.

It would be noticed, however, that the Moon's sidereal path is very little inclined to the ecliptic (about five degrees) and if records were kept of the points of the ecliptic crossed by the Moon, it might be realized that these points were slipping westwards at a rate of about twenty degrees per year (see figure 1.2).

More information, too, would be acquired about the star-like objects that do not twinkle and which have been found in the course of a month to have a slow movement with respect to the stellar background. These planets, like the Moon, would never be seen more than a few degrees from the plane of the ecliptic, yet month after month they would journey through constellation after constellation. In the case of one or two, their paths would include narrow loops, though only one loop would be observed for each of these planets in the course of the year.

The year's observations would not add much to the observer's knowledge of the stars, except to confirm that their positions and brightnesses relative to each other did not alter and that each star, unlike the Sun, had its own fixed rising and setting direction, unless it was circumpolar. It is possible, however, that in a year, the extra-careful watcher might have cause to wonder if the conclusions about stars were without exception for, by regular comparison of the brightness of one star with respect to that of neighbouring ones, it might be discovered that a few stars were variable in brightness. This was certainly known to the Arabian astronomers of the Middle Ages. The appearance of a **nova** might even be observed, i.e. a star appearing in a position where one had not been previously noted. This occurrence might well lead to doubt about the knowledge of the now familiar constellations—in any event it could bring about the decision to make a star map for future use if the phenomenon happened again. It is also possible that in the course of a year the observer might see a **comet**, a star-like object

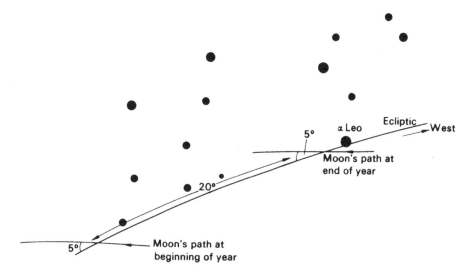

Figure 1.2. The Moon's sidereal path crosses the ecliptic twice each month at an angle of about 5°. For successive lunations the crossing points move westward, covering about 20° over a year. The Constellation of Leo is shown to give an indication of the scale of the movement.

with a long luminous tail. The development of the tail and the movement of the comet head could be detected from night to night.

Our observer by now must have come to tentative conclusions concerning the heavenly phenomena studied and noted. The interpretations, and the use made of the world-picture, will be constrained by the culture of the time. A man of Neolithic times and a Greek of Athens' golden era would develop entirely different cosmologies from identical observations. And a hunter or farmer has different needs, astronomically speaking, from a sailor.

Chapter 2

Ancient world models

First theories were necessarily simple. The Earth was a flat plane with rivers, hills, seas and land, fixed, eternal. The heavenly bodies revolved, passing from east to west. But if the land continued indefinitely, how could the Sun that set in the west be the same Sun that rose in the east the next morning? Perhaps, the Babylonians reasoned, the Earth was flat but finite with a circle of ocean beyond which a ring of mountains supported the heavens, the firmament. Then, if doors were provided in the base of this great solid half-sphere on the eastern and western sides, the celestial bodies would be able to slip through the western doors on setting and be transported in some miraculous way to the east to reappear as ordained.

The Babylonians were skilled astronomers though their world-picture was naïve. They observed the positions of the Sun, Moon, planets and stars for many centuries with great accuracy. They found that they could predict eclipses. Their observations were motivated by their belief that the future of human beings could be predicted from celestial configurations and events such as eclipses or the appearance of comets. Because of this, kings kept court astrologers and the wealthy paid for horoscopes. This belief in astrology, found in all nations, should have withered away with alchemy and the search for the philosopher's stone but even today there are many who set great faith in this pseudo-science. It is perhaps needless to say that modern astronomy demonstrates how ludicrous such beliefs are.

The Egyptians, astronomers almost as skilled as the Babylonians, had equally simple world-pictures. They noticed that the yearly inundation of the Nile valley coincided with the days when the star *Sirius* could be seen best in the morning twilight. This linking of celestial and earthly events spurred on their development of astrology and brought religion into the picture. The Sun-god descended at night, passing beneath the Earth to visit the dead.

Farming people were more interested in the solar cycle since it was linked with seed time and harvest. Seafaring peoples like the Phoenicians and the Minoans used the rising and setting directions of the stars as navigational aids. It may well have been as an aid to memory that the stars were grouped in constellations, embodying myths current at that time.

As is to be expected, the ancient Chinese civilizations produced schools of astronomy and cosmological theories. Serious Chinese astronomy probably began prior to 2000 BC although details of events in that era are largely legendary. The story of the two Chinese astronomers, Ho and Hi, executed for failing to predict an eclipse of the Sun in 2137 BC is possibly apocryphal and may refer to two astronomical colleges of a much later date destroyed in civil strife. Reliable historical details begin about 1000 BC. A farming people required a calendar and so the lengths of month and year were quickly ascertained. A year of $365\frac{1}{4}$ days was certainly used by 350 BC.

By that date, the Chinese constellation figures, 122 in number and quite different from those handed down to us by the Greeks, had been mapped out, the Sun's path—the ecliptic—being divided into 12 regions. The size of a region was not only connected with the heavenly arc inhabited by the Sun each month but also with the yearly journey of the planet Jupiter. The other planetary motions were also studied. As in the west, a pseudo-science of astrology developed from such studies. China was the centre or hub of the flat Earth with heavenly and human events in close harmony: not only did celestial events guide and control men, in particular the Emperor and his court but the decisions and actions of such powerful rulers influenced the state of Heaven.

As mathematical knowledge grew and more accurate astronomical instruments for measuring altitudes and angles were developed in succeeding centuries, the movements of the Sun, Moon and planets were systematized in remarkably accurate tables for prediction purposes. Cometary appearances were noted, among them several apparitions of Halley's comet, and by the 14th century AD the state of Chinese astronomy compared favourably with that of the Arabs in the West.

In various other places where a civilization had developed, astronomical schools flourished. The ravages of time and barbarism have sadly destroyed most of the works of such schools, though happily some traces remain to tell us of the heights of thought their practitioners achieved. For example, we shall see later how ingenious were the steps megalithic man took to keep track of the Sun and Moon. This remarkable civilization flourished in Western Europe in the third and second millennia BC.

Observations of eclipses were also recorded by early American Indians as, for example, by Mayans. A sundial remaining in the 'lost city', Macchu Piccu, provides us with evidence that the Incas of Peru used solar observations to some purpose. The 'Puerta del Sol' at Tiahuanaco, Bolivia, tells us of solar observations prior to the Incas.

However, very few of the ideas and notions of astronomy and cosmology from any of these civilizations have had an influence on the development of our understanding of the astronomical Universe. Our starting points find their origins mainly in ancient Greece.

A completely new departure in mankind's contemplation and interpretation of the heavens came with the flowering of Greek civilization. Many of their thinkers had extraordinarily original minds, were mentally courageous and devoted to rational thought. They were not afraid of questioning cherished beliefs and of following unsettling, disturbing trains of thought.

Many of them dismissed the 'common-sense' picture of solid, flat Earth and god-controlled Heaven. They saw that a spherical Earth poised in space solved a lot of problems. Those stars and planets not seen during the night were simply on the other side of the Earth. Stars were not seen during the day because the dazzling bright Sun blotted out their feeble light. The Moon caused solar eclipses. Pythagoras, in the 6th century BC, taught that the movements of all the heavenly bodies were compounded of one or more circular movements.

In the next century, Philolaus, a follower of Pythagoras, suggested the bold idea that the Earth was not the centre of the Universe and, indeed, that it *moved*. At the centre of the Universe there was a gigantic fire. Around this fire revolved the Earth, Moon, Sun and planets in that order, in circles of various sizes. He also postulated a body called the Anti-Earth to bring the total of moving bodies up to the sacred number of ten. This Anti-Earth revolved about the central fire within the Earth's orbit and was never seen from the Earth because the Earth faced outwards towards the home of the gods— Olympus—situated beyond the sphere of the fixed stars. Philolaus also believed that the Sun was not self-luminous but shone by the light it absorbed from Olympus and the central fire.

In contrast to this, Anaxagoras taught that the Sun was a mass of glowing metal comparable in size with Greece itself. Aristarchus, in the 3rd century BC, agreed with Philolaus that the Earth moved and taught that it rotated on its axis, thus explaining the diurnal motion of the heavens. Moreover, he said, the Sun is a star and the Earth revolves round it, all other stars being very much farther away.

Aristarchus, like Anaxagoras, had ideas about the relative sizes of Sun, Moon and Earth. The Sun's diameter had to be about seven times the diameter of the Earth, a figure far removed from the modern one but embodying the right idea, namely that the Earth is much smaller than the Sun.

Eratosthenes of Alexandria, living about 230 BC, used solar observations and a knowledge of geometry and geography to calculate the circumference of the Earth, obtaining a value within a few per cent of today's accepted figure.

He knew that at the summer solstice the Sun passed through the zenith at Syene in Upper Egypt,

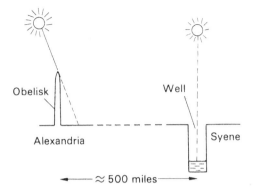

Figure 2.1. The observations of Eratosthenes.

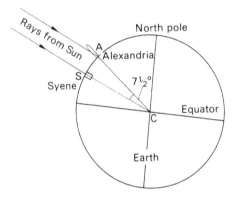

Figure 2.2. The interpretation of the measurements of Eratosthenes.

being reflected at the bottom of a well. At Alexandria, at the same longitude as Syene, the obelisk at the same solar solstice, cast a shadow at noon, showing by its length that the Sun's altitude was $82\frac{1}{2}$ degrees (figure 2.1). He also knew the distance between Syene and Alexandria. Eratosthenes then made the assumptions that the Sun was very far away and that the Earth was spherical. The Sun's rays arriving at Syene and Alexandria could then be taken to be parallel and the angle the Sun's direction made with the vertical at Alexandria ($7\frac{1}{2}°$) would, therefore, be the angle subtended at the Earth's centre C by the arc from Syene to Alexandria (figure 2.2). It was then a simple calculation to find the length of the Earth's circumference by asking what distance would subtend an angle of 360° if the distance from Alexandria to Syene subtended an angle of $7\frac{1}{2}°$ at the Earth's centre.

Other outstanding Greek astronomers and mathematicians such as Hipparchus, Thales, Apollonius, Aristotle and Ptolemy also put forward world-pictures, or cosmologies, that arouse admiration for the way their minds managed to successfully break free from their environment and catch glimpses of the truth. For example, Hipparchus discovered the precession of the equinoxes, noted by the secular change in position of the solar crossing point of its ecliptic path over the celestial equator at the times of the spring and autumnal equinox. He measured the Sun's distance and went a considerable way towards providing theories to account for the motions of Sun and Moon.

Finally, as Greek civilization decayed, the last and perhaps the most influential thinker of them all embodied the work of many of the predecessors in the *Almagest*. Ptolemy, who lived during the second century AD, not only collected and discussed the work of Greek astronomers but carried out original researches himself in astronomy, geography, mathematics, music, optics and other fields of study. His great astronomical work, the **Almagest**, survived the Dark Ages of Western civilization, influencing astronomical thought right up to and beyond the invention of the telescope in the early years of the seventeenth century. The Ptolemaic System describing the apparent motions of the Sun, Moon and planets is discussed in section 12.2.

During the Dark Ages astronomy flourished within the Islamic Empire, once the latter had been stabilized. Ptolemy's **Almagest** was translated into Arabic in 820 AD and thereafter guided the researches of Muslim scientists. They measured astronomical phenomena more precisely than ever before, amassing a wealth of information that proved of inestimable value to Western astronomers when Europe emerged from the Dark Ages. Many of the terms used in modern astronomy come from the Arabic, for example 'zenith', 'nadir', 'almanac', while the names of well-known stars such as *Algol*, *Aldebaran*, *Altair* and *Betelgeuse* are also of Arabic origin. In addition, the Muslim mathematicians introduced spherical trigonometry and Arabic numerals, including a sign for zero—'algebra' is another Arabic word.

They do not seem, however, to have left us new cosmologies. They were content to accept the world-pictures of the Greeks into their custody until the Western world awoke intellectually once again and began anew the study of natural science, including astronomy.

Chapter 3

Observations made by instruments

3.1 The subjectivity of simple measurements

One of the drawbacks of making astronomical observations by eye, with or without the advantage of supplementary equipment, is that they are very subjective. When results taken by several observers are compared, inconsistencies become apparent immediately. For example, if several observers time a **lunar occultation** (i.e. the disappearance of a star behind the lunar disc) at a given site by using stop-watches which are then compared with the observatory master clock, the timed event will have a small range of values. If several occultation timings are taken by the same group of observers, an analysis of the spread of values of each timing will show that certain observers are consistently later than others in operating the stop-watch. Each observer can be considered to have a **personal equation** which must be applied to any observation before comparing it with measurements taken by other observers.

The problem is complicated further as the personal equation of any observer can be time-dependent. This might be a short-term variation depending on the well-being of the observer or it may be a long-term drift which only becomes apparent over a period of years as the observer ages. The first recorded example of such effects appears to have been noted by the fifth Astronomer Royal, Maskelyne, when he wrote in 1796 [*Greenwich Observations* **3**]:

> My assistant, Mr David Kinnebrook, who had observed transits of stars and planets very well in agreement with me all the year 1794 and for a great part of [1795], began from the beginning of August last to set them down half a second later than he should do according to my observations; and in January [1796] he increased his error to eight-tenths of a second. As he had unfortunately continued a considerable time in this error before I noticed it, and did not seem to me likely ever to get over it and return to the right method of observing, therefore, although with reluctance, as he was a diligent and useful assistant to me in other respects, I parted with him.

It is mainly due to this episode that the concept of the personal equation was explored some years later. In Maskelyne's account of the reasons for Kinnebrook's dismissal, he uses the term 'right method of observing', meaning by this that there were discrepancies between Kinnebrook's results and his own and that Kinnebrook's method of observation had deteriorated. It could well have been, of course, that the drift in Maskelyne's own personal equation had occurred contributing to the discrepancies, or even accounting for them completely.

An example of short-term variations in the personal equation occurs in the determination of colour differences made directly by eye. It is well known that the sensitivity of colour depends appreciably on the individual observer; some people have poor ability to differentiate colours and may even be 'colour-blind'. The colour sensitivity of each observer also depends on his or her condition. Under normal conditions the average eye is most sensitive to the green region of the spectrum. However, if the observer is removed to a darkened room, the eyes become accustomed to the dark and maximum sensitivity shifts towards the blue. After a period of about half an hour, the effect is very noticeable. If an observer is made to do violent exercise, the slight rise in the bodily temperature causes the sensitivity peak to move away from the normal position towards the red end of the spectrum. Thus, any determination of colour, being dependent on how the observer's eye responds to colour, depends to a great extent on the condition of the observer and the particular circumstances of the observation.

3.2   Instrumentation in astronomy

As in the case of all of the sciences, instrumentation has been developed in astronomy so that the data provided by the observations are no longer subjective. Again, as in other sciences, the application of instrumentation immediately revealed that the scope for measurement is also extended. For example, when Galileo employed the telescope for astronomical observation, a new range of planetary phenomena was discovered and the number of observable stars was greatly increased. Since Galileo's time, the whole range of observable phenomena has continued to grow with the application of each new type of observing equipment.

The instrumentation which was first applied to astronomy was designed so that the actual measurement of record was made by eye. When photographic material became available, the range of possible observation was immediately increased. This has now been further extended by the introduction of solid state devices in the form of CCDs (charge coupled devices). Whereas the eye is capable of being able to concentrate on only a few stars at a time in a star field, the photographic plate or CCD chip is able to record the light from every star in the field simultaneously. For a star to be seen by eye, the brightness must be above a certain threshold: the eye is not able to accumulate the energy it receives over a period of time to form an impression. The photographic plate and CCD, however, are able to do this and, if a time exposure is made, the resultant images depend on the total energy which falls on to the detector. Thus, besides being able to record many images simultaneously, these devices allow faint stars to be recorded which would not normally be seen by eye (see figure 3.1).

The variation of the sensitivity with wavelength of these detectors is also different to the eye. For example, photographic plates of different types are available with a range of spectral sensitivities. Some plates have their peak of sensitivity in the blue while others have their peak in the red. Blue-sensitive plates will obviously give strong images for blue stars and not for the red, while red-sensitive plates give weak images for blue stars and strong images for red. By using two plates of different spectral sensitivity to photograph a star field, the fact that stars are coloured is easily demonstrated. Because of the physical process involved in the detection of radiation by a silicon-based solid state detector, the natural peak sensitivity tends to be in the red end of the spectrum but, again, the colour response of an applied detector can be modified at its manufacture.

Some special photographic materials are sensitive to colours which cannot be seen by the normal eye. The colour range of astronomical observations can be extended into the ultraviolet or the infrared by the choice of a particular photographic emulsion.

Thus, by recording the astronomical observation on a detector other than the eye, it is possible to extend the scope of the observation by looking at many objects simultaneously, by looking at a range of objects which are too faint for the eye to see and by looking at a much broader range of colour.

The range of available detectors has increased greatly since the photographic process was first applied to astronomy. Detectors based on the photoelectric effect have a common application. Detectors specially designed for infrared work can also be attached to optical telescopes. After the discovery that energy in the form of radio waves was arriving from outer space, special telescopes were designed with sensitive radio detectors at their foci and the era of radio astronomy was born. It is also apparent that our own atmosphere absorbs a large part of the energy arriving from outer space but, with the advent of high flying balloons and artificial satellites, these radiations are now available

Galactic longitude = 49°, Galactic latitude = + 74°

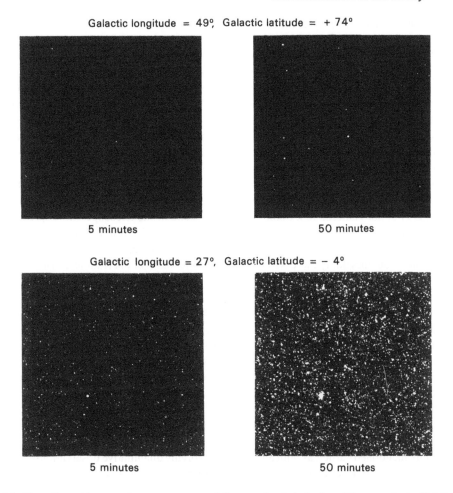

5 minutes   50 minutes

Galactic longitude = 27°, Galactic latitude = − 4°

5 minutes   50 minutes

Figure 3.1. The effect of increased exposure at two different galactic latitudes. (Photography by B J Bok using the 90″ reflector of the Steward Observatory, University of Arizona at Kitt Peak.)

for measurement. New branches of γ-ray, x-ray and infrared astronomy are currently increasing the information that we have concerning the extra-terrestrial bodies.

Although the large range of detectors removes to a great extent the subjectivity of any measurement, special care is needed to avoid the introduction of systematic errors. Each detector acts as a transducer, in that energy with given qualities falls on to the detector and is converted to another form; this new form is then measured. For example, when radiation falls on the sensitive area of a photocathode, the energy is converted in the release of electrons which can be measured as a flow of electric current. The strength of the incident energy can be read as the needle deflection on a meter or converted to a digital form for direct processing by a computer.

The process of converting the incident radiation to a form of energy which is more acceptable for measurement is never one hundred per cent efficient and it is essential that the observer knows exactly how the recording system responds to a given quality and quantity of radiation. In other words, the whole of the equipment which is used to make an observation must be calibrated. The calibration can be calculated either by considering and combining the effects of each of the component parts of the equipment or it can be determined by making observations of assumed known, well-behaved objects. Because of the impossibility of having perfect calibration, systematic errors (hopefully very small) are likely to be introduced in astronomical measurements. It is one of the observer's jobs to ensure that systematic errors are kept below specified limits, hopefully well below the random errors and noise associated with the particular experimental method.

Although every piece of observing equipment improves the process of measurement in some way, the very fact that the equipment and the radiation have interacted means that some of the information contained in the parameters describing the incident radiation does not show up in the final record and is lost. All the qualities present in the incident energy are not presented exactly in the record. Each piece of equipment may be thought of as having an **instrumental profile**. The instrumental profile of any equipment corresponds to the form of its output when it is presented with information which is considered to be perfect.

For example, when a telescope is directed to a point source (perfect information), the shape of the image which is produced (instrumental profile) does not correspond exactly with the source. The collected energy is not gathered to a point in the focal plane of the telescope but is spread out over a small area. The functional behaviour of the 'blurring' is normally referred to as the **point spread function** or **PSF**. For the best possible case, the PSF of the image of a point source is that of a diffraction pattern but inevitably there will be some small addition of aberrations caused by the defects of the optical system or blurring by atmospheric effects. If the recorded image is no larger than that of the instrumental profile, measurement of it gives only an upper limit to the size of the object. Detail within an extended object cannot be recorded with better resolution than the instrumental profile.

For any instrument, there is a limit to the 'sharpness' of the recorded information which can be gleaned from the incoming radiation. This limit set by the instrument, is frequently termed the **resolving power** of the instrument. In all cases there is an absolute limit to the resolving power of any given equipment and this can be predicted from theoretical considerations. Certain information may be present in the incoming radiation but unless an instrument is used with sufficient resolving power, this information will not be recorded and will be lost. When any given piece of equipment is used, it is usually the observer's aim to keep the instrument in perfect adjustment so that its resolving power is as close as possible to the theoretical value.

As briefly mentioned earlier, as with all sciences involving quantitative observations, the measured signal carries **noise** with the consequence that the recording values are assigned uncertainties or errors. One of the ways of describing the quality of measurements is to estimate or to observe the noise on the signal and compare it with the strength of the signal. This comparison effectively determines the **signal-to-noise ratio** of the measurement. Values of this ratio may be close to unity when a signal is just about detectable but may be as high as 1000:1 when precision photometry is being undertaken.

3.3   The role of the observer

Observational astronomy holds a special place in science in that, except for a very few instances, all the knowledge and information has been collected simply by measuring the radiation which arrives from space. It is not like the other laboratory sciences where the experimentalist is able to vary and control the environment or the conditions of the material under investigation. The 'experiment' is going on out in space and the astronomer collects the information by pointing the telescope in a particular direction and then analysing the radiation which is collected.

In interpreting the accumulated data, the reasonable assumption is made that the same physical laws discovered in the laboratory can be applied to matter wherever it is assembled in space. Many of the astronomical measurements, in fact, provide us with means of observing material under a range of conditions which are unattainable in the laboratory. In order to understand these conditions, it is sometimes necessary to provide an extension to the laboratory laws or even consider invoking new laws to describe the observed phenomena.

Laboratory analysis is practised on meteorite samples which are picked up from the surface of the

Earth and on micrometeoritic material which is scooped up by rocket probes in the upper atmosphere. Some thirty years ago the Apollo and Lunakhod missions brought back our first samples of lunar material for laboratory study. Interplanetary space probes have sent and still are sending back new data from the experiments which they carry. They are able to transmit information about the planets that could not have been gained in any other way. Astronomers have also gleaned information about the planets by using radar beams. However, all these active experiments and observations are limited to the inner parts of the Solar System, to distances from the Earth which are extremely small in relation to distances between the stars.

When it comes to stellar work, the experiments, whether on board space vehicles or Earth satellites, or at the bottom of the Earth's atmosphere, are more passive. They involve the measurement and analysis of radiation which happens to come from a particular direction at a particular time. It is very true to say that practically the whole of the information and knowledge which has been built up of the outside Universe has been obtained in this way, by the patient analysis of the energy which arrives constantly from space.

As yet, the greater part of this knowledge has been built up by the observer using ground-based telescopes though in recent years a wide variety of artificial satellite-based telescopes such as the Hubble Space Telescope and Hipparcos have added greatly to our knowledge. The incoming radiation is measured in terms of its direction of arrival, its intensity, its polarization and their changes with time by appending analysing equipment to the radiation collector and recording the information by using suitable devices. The eye no longer plays a primary role here. If the radiation has passed through the Earth's atmosphere, the measurements are likely to have reduced quality, in that they are subject to distortions and may be more uncertain or exhibit an increase of noise. In most cases, however, these effects can be allowed for, or compensated for, at least to some degree.

The task of the observer might be summarized as being one where the aim is to collect data with maximum efficiency, over the widest spectral range, so that the greatest amount of information is collected accurately in the shortest possible time, all performed with the highest possible signal-to-noise ratio. Before the data can be assessed, allowances must be made for the effects of the radiation's passage through the Earth's atmosphere and corrections must be applied because of the particular position of the observer's site and the individual properties of the observing equipment.

It may be noted here also that with the advent of computers, more and more observational work is automated, taking the astronomer away from the 'hands-on' control of the telescope and the interface of the data collection. This certainly takes away some of the physical demands made of the observer who formally operated in the open air environment of the telescope dome sometimes in sub-zero temperatures. Accruing data can also be assessed in real time so providing instant estimates as to its quality and allowing informed decisions to be made as to how the measurements should proceed. In several regards, the application of computers to the overall observational schemes have made the data more objective—but some subtleties associated with operational subjectivity do remain, as every computer technologist knows.

We cannot end this chapter without mentioning the role of the theoretical astronomers. Part of their tasks is to take the data gathered by the observers and use them to enlarge and clarify our picture of the Universe. Their deductions may lead to new observational programmes which will then support their theories or cast doubt upon their validity.

Several comments may be made here.

It goes without saying that an astronomer may be both theoretician and observer, though many workers tend to specialize in one field or the other. Again, it has been estimated that for each hour of data collecting, many hours are spent reducing the observations, gleaning the last iota of information from them and pondering their relevance in our efforts to understand the Universe. The development of astronomical theories often involves long and complicated mathematics, in areas such as celestial mechanics (the theory of orbits), stellar atmospheres and interiors and cosmology. Happily in recent years, the use of the ubiquitous computer has aided tremendously the theoretician working in these fields.

Chapter 4

The nature of the observables

4.1 Introduction

Energy is arriving from space in the form of microscopic bodies, atomic particles and electromagnetic radiation. A great part of this energy is, however, absorbed by the Earth's atmosphere and cannot be observed directly by ground-based observers. In some cases, the absorption processes give rise to re-emission of the energy in a different form. Macroscopic bodies have kinetic energy which is converted into heat; atomic particles interact with the gases in the higher atmosphere and liberate their energy in the form of light, giving rise to such phenomena as the aurorae. Electromagnetic energy of particular frequencies, say in the ultraviolet or x-ray region, is absorbed and re-emitted at other frequencies in the visible region. Thus, besides gaining knowledge about the sources which give rise to the original energy, observation of the re-emitted energy leads to a better understanding of the nature of our atmosphere. However, by far the greater part of our knowledge of astronomical objects is based on the observation of electromagnetic energy which is collected by satellite instrumentation or transmitted directly through the atmosphere and collected by telescope.

4.2 Macroscopic bodies

As the macroscopic bodies penetrate the Earth's atmosphere, the air resists their motion and part of their energy is lost in the form of heat. The heat generated causes the ablated material and the atmospheric path to become ionized and, when the atoms recombine, light is emitted and the rapid progress of the body through the upper atmosphere is seen as a flash of light along a line in the sky. The flash might last for a few seconds. The event is known as a **meteor** (popularly known as a shooting star). The rate of burning of the meteor is not constant and fluctuations in brightness may be seen on its trail, usually with a brightening towards the end of the path. Positional measurements can be made of the meteor and the event can be timed. Simultaneous observations of a meteor at different sites allow determination of its trajectory within the Earth's atmosphere.

On occasions, many meteors can be observed during a relatively short period of time and, by observing their apparent paths across the sky, it is noted that there is a point from which the shower of meteors seems to originate. This point in the sky is known as the **radiant** of the shower. Meteor showers are often annual events and can be seen in the same part of the sky at the same time of the year, although the numbers counted vary widely from year to year. The regular appearances of showers result from the crossing of the Earth's orbit of a fairly tight band of orbits followed by a swarm of meteoritic material.

Meteors can also be detected during the day by radar. As a meteor passes through the upper atmosphere, as has already been mentioned, some of the gases there are ionized. The ionized trail which persists for a short time acts as a good reflector for a radar beam and the effect of any daytime meteor can be displayed on a cathode ray tube. Several daytime showers have been discovered by the use of this technique.

Some of the larger meteors have such large masses that they are incompletely ablated or destroyed in the atmosphere. In this case, the meteor suffers an impact on the Earth's surface. The solid body, or **meteorite**, is frequently available either in the form of a large piece or as scattered fragments. The material can be exposed to the usual analyses in the laboratory.

The smaller meteors or micrometeorites can now also be collected above the Earth's atmosphere by rocket and analysed on return to Earth. It also appears probable that some micrometeorites are continuously percolating through the atmosphere. Because of their size, they attain a low terminal velocity such that any local generated heat by air friction is radiated away at a rate which prevents melting of the particle. Previous micrometeorite sedimentation can be explored by obtaining cores from ancient ice-fields. It is now a difficult problem to separate any fresh contribution from the general dust which is constantly being stirred in the lower atmosphere of the Earth.

4.3 Atomic particles

The atomic particles which arrive in the vicinity of the Earth range from nuclei of atoms of high atomic weight down to individual nuclear particles such as protons and neutrons. The study of these particles is known as **cosmic ray physics**. The analysis of the arrival of such particles tells us about some of the energetic processes occurring in the Universe but so far little has come from these observations in us being able to pinpoint the exact sources which generate the energetic particles. Because of the Earth's magnetic field, any charged particle is deflected greatly from its original direction of travel by the time it arrives at the detector, making it exceedingly difficult to say from which direction in space it originated. At the present time, the Sun is the only body which is definitely known to be a source of particle energy.

It turns out that the basic processes of nuclear (hydrogen) burning within stellar interiors such as the Sun produces the enigmatic neutrino particle. The neutrino has very little interaction with other material and can penetrate great distances through matter. For this reason, the neutrinos generated in the depth of the Sun at a rate $\sim 10^{38}$ s^{-1} pass from the centre to the surface, escaping very readily outwards. At the distance of the Earth, their flux is $\approx 10^{14}$ s^{-1} m^{-2}, this same number (i.e. $\sim 10^{14}$) passing through each person's body per second. Their very low cross section for interaction with other material makes them difficult to detect but some large-scale experiments have been established for this purpose. It must be mentioned that through 'neutrino' observatories, astronomy has helped greatly in our understanding of this particle, particularly in relation to the issue of its mass. Although the general flux from other stars is too low for detection, some 10 neutrinos were detected in 1987 from a supernova in the Large Magellanic Cloud. It is estimated that about 10^9 neutrinos passed through each human being as a result of the event.

4.4 Electromagnetic radiation

4.4.1 The wave nature of radiation

The greatest quantity of information, by far, comes from the analysis of **electromagnetic radiation**. The word describing the quality of the radiation indicates that it has both electric and magnetic properties. As the radiation travels, it sets up electric and magnetic disturbances, which may be revealed by an interaction with materials on which the radiation impinges. In fact, some of the interactions are utilized in detector systems to record and measure the strength of the radiation. For these particular interactions, the energy present in the radiation is transformed into another form which is then suitable for a quantitative assessment.

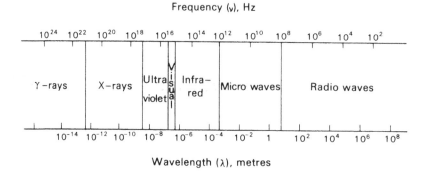

Figure 4.1. The spectrum of electromagnetic radiation.

Thus, any radiation has a strength which can be measured. Quantitative observations of this property can give us information about the source or about the medium through which the radiation has travelled after leaving the source.

Experiments in the laboratory have shown that all electromagnetic radiations have the same type of wave nature. When any radiation passes through a medium, its velocity is reduced by a certain fraction and the wavelength as measured within the medium also reduces by the same fraction. If v is the measured velocity and λ the measured wavelength, their relationship may be written as

$$v = \nu\lambda \qquad (4.1)$$

where ν is a constant of the particular radiation and is known as its **frequency**.

Thus, the electromagnetic spectrum covers an extremely wide range of frequencies. According to the value of the frequency of the radiation, it is convenient to classify it under broad spectral zones, these covering γ-rays, x-rays, ultraviolet light, visible light, infrared radiation, microwaves and radio waves. The spectrum of electromagnetic radiation is illustrated in figure 4.1.

The velocity of any electromagnetic disturbance in free space (vacuum) is the same for radiations of all frequencies. In free space, the fundamental parameter frequency, ν, is related to the wavelength, λ_c, of the radiation and its velocity, c, by the expression:

$$c = \nu\lambda_c. \qquad (4.2)$$

The velocity of electromagnetic radiation in free space has been measured in the laboratory over a wide range of frequencies and, in all cases, the result is close to $c = 3 \times 10^8$ m s^{-1}.

Wavelengths of electromagnetic radiation range from 10^{-14} m for γ-rays to thousands of metres in the radio region. At the centre of the visual spectrum, the wavelength is close to 5×10^{-4} mm or 500 nm. In the optical region, the wavelength is frequently expressed in **Ångström units** (Å) where 1 Å $= 10^{-7}$ mm. Thus, the centre of the visual spectrum is close to 5000 Å.

If the strength of any radiation can be measured in different zones of the spectrum, much information may be gleaned about the nature of the source. In fact, it may not be necessary for measurements to be made over very wide spectrum ranges for the observations to be extremely informative. For example, as we shall see later, measurements of stellar radiation across the visual part of the spectrum can provide accurate values for the temperatures of stars.

Partly for historic reasons, experimenters working in different spectral zones tend to use different terms to specify the exact positions within the spectrum. In the optical region the spectral features are invariably described in terms of wavelength; for radio astronomers, selected parts of the spectrum are normally identified by using frequency, usually of the order of several hundred MHz. By using equation (4.1), it is a simple procedure to convert from wavelength to frequency and *vice versa*.

If, for example, the wavelength of 1 m is involved, then its associated frequency is given by

$$\nu = \frac{c}{\lambda} = \frac{3 \times 10^8}{1} = 100 \times 10^6 = 100 \, \text{MHz}.$$

4.4.2 The photon nature of radiation

There is another aspect to the description of electromagnetic radiation that is important in terms of the atomic processes occurring in astronomical sources and in the process of detection by observational equipment.

At the turn of the twentieth century, it was demonstrated that light also had a particulate nature. Experiments at that time showed that radiation could be considered as being made up of wave packets or photons. The energy associated with each photon can be expressed in the form

$$E = h\nu \tag{4.3}$$

where h is Planck's constant and equal to 6.625×10^{-34} J s. Thus, it can be seen that the photons carrying the most energy are associated with the high frequency end of the spectrum, i.e. the γ-rays— photons associated with the radio spectrum have very low energy.

For many observational circumstances, the flux of energy arriving from faint sources is such that it is the statistical random nature in the arrival of the photons that limits the quality of the measurement. In observations where the source of experimental noise errors is very small, it is perhaps the random arrival of photons that constitute the noise on the measurements. The accuracy of data recorded under such a circumstance is said to be limited by photon counting statistics or by **photon shot noise**. In order to be able to estimate the accuracy to which measurements of brightness or details within the spectrum can be obtained, it is necessary to know the photon arrival rate associated with the generated signal. For this reason, the strengths of observed sources are sometimes referred to in terms of photons s^{-1} rather than in watts. Equation (4.3) is all that is needed to relate the two ways of expressing the amount of energy which is received by the observing equipment. More detail of this topic will be presented in Part 3.

It may also be noted that in the zones covering the high energy end of the spectrum, neither wavelength nor frequency is used to describe the radiation. The more usual units used are those of the energy of the recorded photons. Thus, for example, features occurring in x-ray radiation are normally described in terms of photon energies of order 10 keV.

Equation (4.3) describes the energy of a photon and this can be re-written as

$$E = \frac{hc}{\lambda} \quad \text{J.} \tag{4.4}$$

By remembering the conversion of units such that 1 eV $= 1.6 \times 10^{-19}$ J, the photon energy expressed in eV units is

$$E[\text{eV}] = \frac{6.625 \times 10^{-34} \times \nu}{1.6 \times 10^{-19}} \quad \text{or} \quad \frac{6.625 \times 10^{-34} \times 3 \times 10^8}{1.6 \times 10^{-19} \times \lambda}. \tag{4.5}$$

In order to determine the wavelength associated with a photon of some given energy, consolidation of the numerical parts leads to

$$\lambda[\text{m}] = \frac{1.24 \times 10^{-6}}{E[\text{eV}]} \quad \text{or} \quad \lambda[\text{Å}] = \frac{12\,400}{E[\text{eV}]}. \tag{4.6}$$

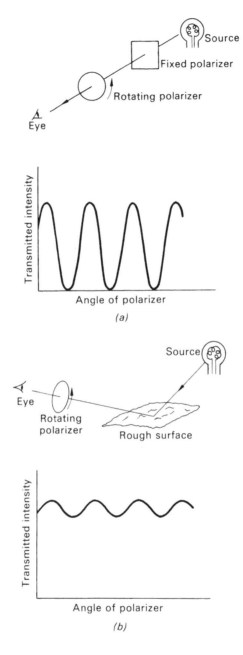

Figure 4.2. (*a*) The artificial production and analysis of totally polarized light. (*b*) The production of partially polarized light as might occur in nature and its analysis.

4.4.3   Polarization

In addition to the strength of any radiation and the variation with frequency, the radiation may have another property. Two apparently identical beams of radiation having the same frequency spread and intensity distribution within that spectral range may interact differently with certain materials or devices. From this we may conclude that radiation has another characteristic. This quality is known as **polarization**. It manifests itself as an orientational quality within the radiation.

The usefulness of polarization as a means of carrying information about a radiating source is sometimes ignored, perhaps as a result of the eye not being directly sensitive to it. However, the simple use of a pair of Polaroid sunglasses reveals that much of the light in nature is polarized to some degree. Rotation of the lenses in front of the eye will demonstrate that the light of the blue sky, light reflected by the sea and light scattered by rough surfaces are all polarized. Measurement of the polarization of the radiation coming from astronomical sources holds much information about the natures of those sources. Its generation may result from scattering processes in a source or by the radiating atoms being in the presence of a magnetic field. Because polarization is essentially an orientational property, its measurement sometimes provides 'geometric' information which could not be ascertained by other observational analyses. In stellar measurements, for example, knowledge of the orientations of magnetic fields may be gleaned.

In the optical region, the simplest polarimetric measurements can be made by placing a plastic sheet polarizer (similar to that comprising the lenses of Polaroid sunglasses) in the beam and measuring the transmitted intensity as the polarizer is rotated. The larger the relative changes in intensity are, the greater the degree of polarization is. If a wholly polarized beam is generated artificially by using a polarizer (see figure 4.2) and this beam is then analysed by a rotating polarizer in the usual way, then the measured intensity will fall to zero at a particular orientation of the analysing polarizer. Although the polarization of the radiation coming from astronomical sources is usually very small, its measurement holds much information about the nature of those sources.

All the parameters which are used to describe radiation, i.e. its strength and its variation across the spectrum, together with any polarization properties, carry information about the condition of the source or about the material which scatters the radiation in the direction of the observer or about the matter which is in a direct line between the original source and the observer. If the observer wishes to gain as much knowledge as possible of the outside universe, measurements must be made of all of the properties associated with the electromagnetic radiation.

Chapter 5

The astronomer's measurements

5.1 Introduction

We have now seen that one of the chief aims of the observational astronomer is to measure the electromagnetic radiation which is arriving from space. The measurements involve:

1. the determination of the direction of arrival of the radiation (see section 5.2);
2. the determination of the strength of the radiation, i.e. the brightness of the source (see section 5.3);
3. the determination of the radiation's polarization qualities (see section 5.4).

All three types of measurement must be investigated over the frequency range where the energy can be detected by the suitably available detectors. They must also be investigated for their dependence on time.

Let us now consider the three types of measurement in a little more detail.

5.2 Direction of arrival of the radiation

Measurements of the direction of arrival of radiation are equivalent to determining the positions of objects on the celestial sphere. In the case of the optical region of the spectrum, the apparent size of each star is smaller than the instrumental profile of even the best recording instrument. To all intents and purposes, stars may, therefore, be treated as point sources and their positions may be marked as points on the celestial sphere. For extended objects such as nebulae and for radiation in the radio region, the energy from small parts of the source can be recorded with a spatial resolution limited only by the instrumental profile of the measuring instrument. Again the strength of the radiation can be plotted on the celestial sphere for the positions where a recording has been made.

In order to plot the positions on the celestial sphere of the sources of radiation, it is obvious that some coordinate system with reference points is needed. For the system to be of real use, it must be independent of the observer's position on the Earth. The coordinate system used has axes known as right ascension (**RA** or α) and declination (**Dec** or δ). RA and Dec can be compared to the coordinate system of longitude and latitude for expressing a particular position on the Earth's surface.

The central part of figure 5.1 depicts the Earth and illustrates the reference circles of the **equator** and the **Greenwich meridian**. The position of a point on the Earth's surface has been marked together with the longitude (λ W) and latitude (ϕ N), angles which pinpoint this position.

The outer sphere on figure 5.1 represents the celestial sphere on which the energy source positions are recorded. The reference circle of the **celestial equator** corresponds to the projection of the **Earth's equator** on to the celestial sphere and the declination ($\delta_\star$) of a star's position is analogous to the latitude angle of a point on the Earth's surface. As the Earth is rotating under the celestial sphere, the projection of the Greenwich meridian would sweep round the sphere, passing through all the stars' positions in

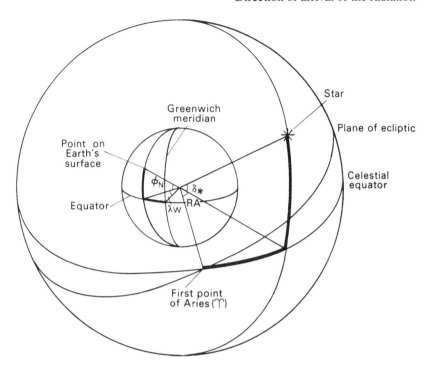

Figure 5.1. Coordinate systems for the Earth (longitude and latitude) and for the celestial sphere (right ascension and declination).

turn. In order to label the stars' positions, therefore, some other meridian must be chosen which is connected directly to the celestial sphere.

During the course of the year, the Sun progresses eastwards round the celestial sphere along an apparent path known as the **ecliptic**. Because the Earth's axis of rotation is set at an angle to the perpendicular to its orbital plane around the Sun, the ecliptic circle is set at the same angle to the celestial equator. The points of intersection of the circles may be used as reference points on the celestial equator; however, it is the intersection where the Sun crosses the equator from south to north which is chosen as the reference point. This position which is fixed with respect to the stellar background is known as the **first point of Aries**, ♈, or the **vernal equinox** (see figure 5.1). The meridian through this point corresponds to RA = 0 hr.

Any stars which happen to be on the observer's meridian (north–south line projected on the celestial sphere) are related in position to the reference meridian on the celestial sphere by **time**. Consequently, a star's position in RA is normally expressed in terms of hours, minutes and seconds of time rather than in degrees, minutes and seconds of arc. By convention, values of RA *increase* in an *easterly* direction round the celestial sphere. As a result, the sky acts as a clock in that the passage of stars across the meridian occurs at later times according to their RA values. A star's position in declination is expressed in degrees, minutes and seconds of arc and is positive or negative depending on whether it is in the northern or southern celestial hemisphere.

For an observer at the bottom of the Earth's atmosphere, the problem of recording the energy source positions in terms of RA and Dec is made difficult by the very existence of the atmosphere. The direction of propagation of any radiation is affected, in general, when it meets a medium where there is a change in the refractive index. In particular, for astronomical observations made in the optical region of the spectrum, the change of direction increases progressively as the radiation penetrates deeper into the denser layers of the atmosphere. The curvature of a beam of light from a star is depicted in

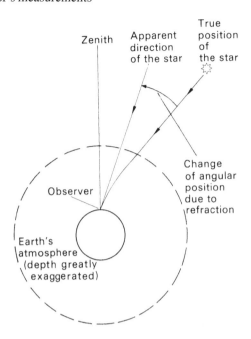

Figure 5.2. Displacement of the position of a star caused by refraction in the Earth's atmosphere. (The effect is exaggerated for clarity in the diagram.)

figure 5.2. Lines illustrating the true direction and the apparent direction of a given star are drawn in the figure. The amount of refraction increases rapidly as the star's position becomes closer to the observer's horizon. At a true altitude of 1 degree, the amount of refraction is approximately a quarter of a degree. There is, however, a simple method for estimating how much a star's position is disturbed by refraction and this can be applied to all observations.

Because of turbulence in the Earth's atmosphere, the apparent direction of propagation moves about by small amounts in a random fashion. Normal positional measurements are, therefore, difficult to make as any image produced for positional determination will be blurred. For the optical region of the spectrum, it is in the first few hundred metres above the telescope aperture which give the greatest contribution to the blurring. It is, therefore, impossible to record a star as a point image but only as a blurred-out patch. For field imaging work this may not be too serious; it should not be difficult to find the centre of the blurred image as this would still retain circular symmetry. In the case of positional measurements by eye (no longer made by professional astronomers), the problem is more serious as the eye is trying to make an assessment of an instantaneous image which is in constant motion.

In the radio region, positional measurements can also be affected by refraction in the ionosphere and in the lower atmosphere. The amount of refraction varies considerably with the wavelength of the radiation which is observed. For the refraction caused by the ionosphere, a typical value of the deviation for radiation at a frequency of 60 MHz is 20 minutes of arc at 5° true altitude. The refraction in the lower atmosphere is mainly due to water droplets and is hence dependent on the weather. The measured effect is approximately twice the amount which is apparent in the optical region. A typical value of the deviation is 0°.5 at a true altitude of 1° and the effect increases rapidly with increasing altitude in the same way as in the case of the optical radiation.

It is obvious that all positional measurements can be improved by removing the effects of refraction and of turbulence and this can now be done by setting up equipment above the Earth's atmosphere, in an orbiting satellite or on the Moon's surface.

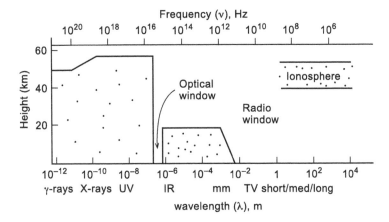

Figure 5.3. The windows of the Earth's atmosphere are depicted with the source heights of the absorptions which prevent the main parts of the electromagnetic spectrum reaching the ground.

5.3 Brightness

5.3.1 Factors affecting brightness

Not all the radiation which is incident on the outer atmosphere of the Earth is able to penetrate to a ground-based observer. The radiations of a large part of the frequency spectrum are either absorbed or reflected back into space and are consequently unavailable for measurement from the ground. The atmosphere is said to possess a **window** in any region of the spectrum which allows astronomical measurements.

Frequencies higher than those of ultraviolet light are all absorbed by a layer of ozone in the stratosphere which exists some 24 km above the Earth. Until the advent of space research, x-rays and γ-rays had not been detected from astronomical objects.

On the other side of the frequency band corresponding to visible radiation, a cut-off appears in the infrared. The absorption in this part of the spectrum is caused by molecules, chiefly water vapour. This cut-off is not very sharp and there are occasional windows in the infrared which are utilized for making observations. The absorption remains practically complete until the millimetre-wave region, where again a window appears. Over a broad part of the radio spectral range, the ionosphere lets through the radiation and the measurement of this form of energy belongs to the realm of the radio astronomer.

The two main windows for observation are depicted in figure 5.3. It may be pointed out that the boundaries are not as sharp as shown in this diagram. Comparison of the spectral widths of the main windows with the whole of the electromagnetic spectrum reveals the large range of frequencies which are unavailable to the ground-based observer and the potential information that is lost.

Above the Earth's atmosphere, however, the full range of the electromagnetic spectrum is available. It has been one of the first tasks of the orbiting observatories and will eventually be that of lunar-based telescopes to make surveys and measurements of the sky in the spectral regions which had previously been unavailable.

For ground-based observations made through the transparent windows, corrections still need to be applied to measurements of the strength of any incoming radiation. This is particularly important for measurements made of optical radiation. By the time a beam of light has penetrated the Earth's atmosphere, a large fraction of the energy has been lost, and stars appear to be less bright than they would be if they were to be viewed above the Earth's atmosphere.

If an observing site is in an area where the air is pure and has little dust or smog content, most of the lost energy in the optical region is scattered out of the beam by the atoms and molecules in the

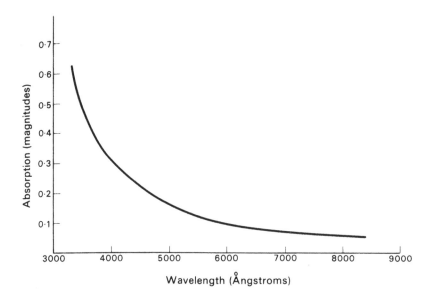

Figure 5.4. Absorption in the visible window caused by scattering from air molecules.

air. This type of scattering is known as **Rayleigh scattering**. According to Lord Rayleigh's theory, atoms and molecules of any gas should scatter light with an efficiency which is inversely proportional to the fourth power of the wavelength (i.e. the scattering efficiency $\propto 1/\lambda^4$). Thus blue light, with a short wavelength, is scattered more easily than red light, which has a longer wavelength. Rayleigh's law immediately gives the reason for the blueness of the daytime sky. During the day, some light, with its broad spectral range, is incident on the atmosphere. As it penetrates towards the ground, part of the energy is scattered in all directions by the molecules in the air. It is this scattered light which the observer sees as the sky. As the scattering process is extremely efficient for the shorter wavelengths, the sky consequently appears blue. When the Moon is observed in the daytime, a blue haze can be seen between it and the observer. This haze is a result of the scattering of some light by the molecules in the atmosphere along the path to the Moon's direction.

Starlight, in its path through the Earth's atmosphere, is weakened by this same process of scattering. As the scattering is wavelength dependent, so must be the weakening. The apparent absorption of starlight, or its **extinction**, is very much stronger in the blue part of the spectrum than in the red (see figure 5.4). Thus the colours of the stars are distorted because of the passage of their radiation through the Earth's atmosphere. If colour measurements are to be attempted, then allowances must be made for the wavelength-selective extinction effects.

The amount of extinction obviously depends on the total number of molecules that the light beam encounters on its passage through the atmosphere, i.e. it depends on the light path. Thus the amount of extinction depends on the altitude of any given star. The light loss is at a minimum for any particular observing site when the star is positioned at the zenith. Even at this optimum position, the total transmission of visible light may only be typically 75%.

As a consequence of the extinction being dependent on a star's altitude, any star will show changes in its apparent brightness during the course of a night as it rises, comes to culmination and then sets. Great care must be exercised when comparisons are made of the brightnesses of stars, especially when the stars cover a wide range of altitudes.

Radio energy is also absorbed on its passage through the layer of electrified particles known as the ionosphere. Further absorption occurs in the lower atmosphere. The amount of absorption is dependent on the particular frequency which is observed. Ionospheric absorption depends on the physical conditions within the layers and, as these are controlled to some extent by the activity on the Sun, the amount of absorption at some frequencies can vary greatly. Refraction caused by the ionosphere can also give rise to the diminution of radio signals. At low altitudes, the ionosphere can spread the radiation in the same way as a divergent lens, thus reducing the amount of energy which is collected by a given radio telescope.

In the lower atmosphere, the radio energy losses are caused by attenuation in rain clouds and absorption by water vapour and oxygen and they are thus dependent on the weather. However, lower atmosphere absorption effects are usually unimportant at wavelengths greater than 100 mm.

Returning again to the optical region, simple naked eye observations reveal that the light from a star suffers rapid variations of apparent brightness. This twinkling effect is known as **intensity scintillation** and the departures of intensity from a mean level are known as **scintillation noise**. Scintillation is caused by turbulence in the Earth's atmosphere rather than by fluctuations in the scattering and absorption; it is certainly not an inherent property of the stars. Minute temperature differences in the turbulent eddies in the atmosphere give rise to pockets of air with small differences in refractive index. Different parts of the light beam from a star suffer random disturbances in their direction of travel. For brief instances, some parts of the energy are refracted beyond the edge of the telescope collecting area, causing a drop in the total energy which is collected. There are other instances when extra energy is refracted into the telescope aperture. Thus, the energy which is collected shows rapid fluctuations in its strength. The magnitude of the effect decreases as the telescope aperture increases, there being a better averaging out of the effect over a larger aperture. Scintillation is very noticeable to the naked eye as its collecting aperture is only a few millimetres in diameter.

Brightness measurements are obtained by taking mean values through the scintillation noise and they are, therefore, subject to uncertainties of a random nature. The magnitude of these uncertainties depends on the strength of the scintillation noise, which in turn depends on the quality of the observing site, on the telescope system used and on the altitude of the star.

Scintillation is also encountered in the radio region of the spectrum and is mainly caused by inhomogeneities in the ionosphere. The fluctuations of signal strength which are observed are less rapid than those recorded in the optical region.

Although scintillation effects are detrimental to obtaining accurate measurements of a source's brightness, studies of the form of scintillation noise itself, in both the optical and radio regions, are useful in gaining information about the Earth's atmosphere, upper winds and the ionosphere.

Although allowances can be made successfully for the effects of atmospheric extinction, it is obvious that absolute brightness measurements, brightness comparisons and colour measurements could be made more easily, with less risk of large random and systematic errors, above the Earth's atmosphere. Again, such measurements can be made by equipment on an orbiting platform or at a lunar observing station.

5.3.2   The magnitude system

The energy arriving from any astronomical body can, in principle, be measured absolutely. The brightness of any point source can be determined in terms of the number of watts which are collected by a telescope of a given size. For extended objects similar measurements can be made of the surface brightness. These types of measurement can be applied to any part of the electromagnetic spectrum.

However, in the optical part of the spectrum, absolutely brightness measurements are rarely made directly; they are usually obtained by comparison with a set of stars which are chosen to act as standards. The first brightness comparisons were, of course, made directly by eye. In the classification introduced by Hipparchus, the visible stars were divided into six groups. The brightest stars were labelled as being of *first* magnitude and the faintest which could just be detected by eye were labelled as being of *sixth* magnitude. Stars with the brightnesses between these limits were labelled as *second*, *third*, *fourth* or *fifth* magnitude, depending on how bright the star appeared.

The advent of the telescope allowed stars to be recorded with magnitudes *greater* than sixth; catalogues of the eighteenth century record stars of seventh, eighth and ninth magnitude. At the other end of the scale, photometers attached to telescopes revealed that some stars were brighter than the first magnitude classification and so the scale was extended to include *zero* and even *negative* magnitude stars. The range of brightnesses amongst the stars revealed that it is necessary to subdivide the unit of magnitude. The stars visible to the naked eye could have magnitudes of -0.14, $+2.83$ or $+5.86$, say, while stars which can only be detected with the use of a telescope could have magnitudes are of $+6.76$, $+8.54$ or even as faint as $+23$.

A more complete description of magnitude systems together with the underlying mathematical relationships is reserved for Part 3.

5.4   Polarization

The presence of any polarization in radiation can be detected by using special devices which are sensitive to its orientation properties. They are placed in the train of instrumentation between the telescope and the detector and rotated. If the recorded signal varies as a device rotates, then the radiation is polarized to some extent. The greater the variation of the signal, the stronger is the amount of polarization in the beam.

In the case of measurements in the optical region, the polarization-sensitive devices would be the modern equivalents of the Nicol prism and retardation plates made of some birefringent material. Brightness measurements are made after placing these devices in the beam of light collected by the telescope. Although the Earth's atmosphere affects the brightness of any light beam, the polarization properties are unaltered by the beam's passage through the atmosphere; at least, any disturbance is smaller than would be detected by current polarimetric techniques. However, since polarization determinations result from brightness measurements, the atmosphere will reduce their quality because of scintillation and transparency fluctuations.

For the radio region, the receiving antenna such as a dipole is itself inherently sensitive to polarization. If the radio radiation is polarized, the strength of the recorded signal depends on the orientation of the antenna in the beam. Simple observation of TV receiving antenna on house roof tops demonstrates the dipole's orientational sensitivity. In some locations the dipole and the guiding rods are in a horizontal plane. In other areas they are seen to be in a vertical plane. This difference reflects the fact that the transmitting antenna may radiate a TV signal with a horizontal polarization while another at a different location may radiate with a vertical polarization, this helping to prevent interference between the two signals. Obviously, to obtain the best reception, the receiving antenna needs to be pointed in the direction of the transmitter and its orientation set (horizontal or vertical) to match the polarizational form of the signal.

Unlike the optical region, the polarization characteristics of radio radiation are altered significantly by passage through the Earth's atmosphere or ionosphere. The magnitude of the effects are very dependent on the frequency of the incoming radiation. Perhaps the most important effect is that due to Faraday rotation: as polarized radiation passes through the ionospheric layers, all the component vibrational planes of the waves are rotated resulting in a rotation of the angle describing the polarization. The total rotation depends on the physical properties within the layers, the path length of the beam and the frequency of the radiation. According to the frequency, the total rotation may be a few degrees or a few full rotations. It can thus be a difficult task to determine the original direction of vibration of any polarized radio radiation as it arrives from above the Earth's atmosphere.

As was the case for positional and brightness measurements, the quality of polarization measurements can be greatly improved if they are made above the Earth's atmosphere, either from an orbiting space laboratory or from the Moon's surface. In addition, orbiting satellites have been used to house radio transmitters and direct measurements made on the Faraday rotation effect have been used to explore the properties of the ionosphere.

5.5   Time

If measurements of the positions, brightnesses and polarization of astronomical sources are repeated, the passage of time reveals that, in some cases, the position of the source or some property of its radiation changes. These time variations of the measured values are of great importance in determining many of the physical properties of the radiating sources. It is, therefore, very necessary that the times of all observations and measurements must be recorded. The accuracy to which time must be recorded obviously depends on the type of observation which is being attempted.

It would be out of place here to enter into a philosophical discussion on the nature of time. However, it might be said that some concept which is called **time** is necessary to enable the physical and mechanical descriptions of any body in the Universe and its interactions with other bodies to be related. One of the properties of any time scale which would be appealing from certain philosophical standpoints is that time should flow evenly. It is, therefore, the aim of any timekeeping system that it should not show fluctuations in the rate at which the flow of time is recorded. If fluctuations are present in any system, they can only be revealed by comparison with clocks which are superior in accuracy and stability. Timekeeping systems have changed their form as clocks of increased accuracy have been developed; early clocks depended on the flow of sand or water through an orifice, while the most modern clocks depend on processes which are generated inside atoms.

About a century ago, the rotation of the Earth was taken as a standard interval of time which could be divided first into 24 parts to obtain the unit of an hour. Each hour was then subdivided into a further 60 parts to obtain the minute, each minute itself being subdivided into a further 60 parts to obtain the second. This system of timekeeping is obtained directly from astronomical observation, and is related to the interval between successive appearances of stars at particular positions in the sky. For practical convenience, the north–south line, or meridian, passing through the observatory is taken as a reference line and appearances of stars on this meridian are noted against some laboratory timekeeping device. As laboratory pendulum clocks improved in timekeeping precision, it became apparent from the meridian transit observations that the Earth suffered irregularities in the rate of its rotation. These irregularities are more easily shown up nowadays by laboratory clocks which are superior in precision to the now old-fashioned pendulum clock.

At best, a pendulum clock is capable of accuracy of a few hundredths of a second per day. A quartz crystal clock, which relies on a basic frequency provided by the vibrations of the crystal in an electronic circuit, can give an accuracy better than a millisecond per day, or of the order of one part in 10^8; and this is usually more than sufficient for the majority of astronomical observations. Even more accurate sources of frequency can be obtained from atomic transitions. In particular, the clock which relies on the frequency which can be generated by caesium atoms provides a source of time reference which is accurate to one part in 10^{11}. The caesium clock also provides the link between an extremely accurate determination of time intervals and the constants of nature which are used to describe the properties of atoms.

Armed with such high-precision clocks, the irregularities in the rotational period of the Earth can be studied. Some of the short-term variations are shown to be a result of the movement of the observer's meridian due to motion of the rotational pole over the Earth's surface. Other variations are seasonally dependent and probably result in part from the constantly changing distribution of ice over the Earth's surface. Over the period of one year, a typical seasonal variation of the rotational period may be of the order of two parts in 10^8.

Over and above the minute changes, it is apparent that the Earth's rotational speed is slowing down progressively. The retardation, to a great extent, is produced by the friction which is generated by the tidal movement of the oceans and seas and is thus connected to the motion of the Moon. The effects of the retardation show up well in the apparent motions of the bodies of the Solar System.

After the orbit of a planet has been determined, its positions at future times may be predicted. The methods employed make use of laws which assume that time is flowing evenly. The predictions, or **ephemeris positions**, can later be checked by observation as time goes by. If an observer uses the rotation of the Earth to measure the passage of time between the time when the predictions are made and the time of the observation, and unknowingly assumes the Earth's rotational period to be constant, it is found that the planets creep ahead of their ephemeris positions at rates which are proportional to their mean motions. The phenomenon is most pronounced in the case of the Moon.

Suppose that a time interval elapses between the time the calculations are performed and the time that the ephemeris positions are checked by observation. The time interval measured by the rotation of the Earth might be counted as a certain number of units. However, as the Earth's rotation is continuously slowing down and the length of the time unit is progressively increasing in comparison with the unit of an evenly-flowing scale, the time interval corresponds to a larger number of units on an evenly-flowing scale. Unknown to the observer who takes the unit of time from the Earth's rotation, the real time interval is actually longer than he/she has measured it to be and the planets, therefore, progress further along their orbits than is anticipated. Thus, the once unexplained 'additional motions' of the planets and the Moon are now known to be caused by the fact that the Earth's rotational period slows down during the interval between the times of prediction and of observation.

It is now practice to relate astronomical predictions to a time scale which is flowing evenly, at least to the accuracy of the best clocks available. This scale is known as **Dynamical Time** (DT).

Chapter 6

The night sky

6.1 Star maps and catalogues

Already in the previous chapters, mention has been made of the names of some of the constellations and stars. Inspection of star maps shows that areas of the sky are divided into zones marked by constellations. Within a constellation it is usual to find an asterism (a pattern of stars) which is readily recognizable. Also some of the stars may carry a name, usually with an Arabic origin, a Greek letter or a number to provide individual identification. At first sight, the nomenclature may seem to be haphazard but this is simply the legacy of history.

In the early days, the stars were listed according to their places within a constellation and designated by a letter or number. The constellation zones are quite arbitrary in relation to the stellar distribution but their designations persist for the ease of identification of the area of sky under observation. It is, however, important to have some appreciation of the background to the system and its current use.

To describe the philosophy behind the development in star catalogues we can do no better than to quote the relevant section of *Norton's Star Atlas*.

> The origin of most of the constellation names is lost in antiquity. Coma Berenices was added to the old list (though not definitely fixed till the time of Tycho Brahé), about 200 BC; but no further addition was made till the 17th century, when Bayer, Hevelius, and other astronomers, formed many constellations in the hitherto uncharted regions of the southern heavens, and marked off portions of some of the large ill-defined ancient constellations into new constellations. Many of these latter, however, were never generally recognized, and are now either obsolete or have had their rather clumsy names abbreviated into more convenient forms. Since the middle of the 18th century when La Caille added thirteen names in the southern hemisphere, and sub-divided the unwieldy Argo into Carina, Malus (now Pyxis), Puppis, and Vela, no new constellations have been recognized. Originally, constellations had no boundaries, the position of the star in the 'head', 'foot', etc, of the figure answering the needs of the time; the first boundaries were drawn by Bode in 1801.

> The star names have, for the most part, being handed down from classical or early medieval times, but only a few of them are now in use, a system devised by Bayer in 1603 having been found more convenient, viz., the designation of the bright stars of each constellation by the small letters of the Greek alphabet, α, β, γ, etc, the brightest star being usually made α, the second brightest β—though sometimes, as in Ursa Major, sequence, or position in the constellation figure, was preferred. When the Greek letters were exhausted, the small Roman letters, a, b, c, etc, were employed, and after these the capitals, A, B, etc—mostly in the Southern constellations. The capitals after Q were not required, so Argelander utilized R, S, T, etc, to denote variable stars in each constellation, a convenient index to their peculiarity.

The fainter stars are most conveniently designated by their numbers in some star catalogues. By universal consent, the numbers of Flamsteed's *British Catalogue* (published 1725) are adopted for stars to which no Greek letter has been assigned, while for stars not appearing in that catalogue, the numbers of some other catalogue are utilized. The usual method of denoting any lettered or numbered star in a constellation is to give the letter, or Flamsteed number, followed by the genitive case of the Latin name of the constellation: thus α of Canes Venatici is described as α Canum Venaticorum.

Flamsteed catalogued his stars by constellations, numbering them in the order of their right ascension. Most modern catalogues are on this convenient basis (ignoring constellations), as the stars follow a regular sequence. But when right ascensions are nearly the same, especially if the declinations differ much, in time 'precession' may change the order: Flamsteed's 20, 21, 22, 23 Herculis, numbered 200 years ago, now stand in the order 22, 20, 23, 21.

For convenience of reference, the more important star catalogues are designated by recognized contractions.

With the application of detectors on telescopes replacing the unaided eye, the resulting catalogues are forced to provide extensive listings of stars containing hundreds and thousands of entries. A famous catalogue based on photographic records of the sky made by Harvard University is known as the *Henry Draper Catalogue*. Each successive star is given the number according to increasing right ascension with the prefix HD. For example, HD 172167 is at once known by astronomers to denote the star numbered 172167 in that catalogue. This particular star is bright and well-known, being *Vega* or α Lyrae. It also appears in all other catalogues and may, for example, be known as 3 Lyrae (Flamsteed's number), Groombridge 2616 and $AGK_2 + 38\ 1711$ (from *Zweiter Katalog der Astronomischen Gesellschaft für das Äquinoktum* 1950).

Returning to the quotation from *Norton's Star Atlas*:

Bode's constellation boundaries were not treated as standard, and charts and catalogues issued before 1930 may differ as to which of two adjacent constellations a star belongs. Thus, Flamsteed numbered in Camelopardus several stars now allocated to Auriga, and by error he sometimes numbered the star in two constellations. Bayer, also, sometimes assigned to the same star a Greek letter in two constellations, ancient astronomers having stated that it belonged to both constellation figures: thus β Tauri $= \gamma$ Auriga and α Andromedae $= \delta$ Pegasi.

To remedy this inconvenience, in 1930 the International Astronomical Union standardized the boundaries along the Jan 1st, 1875, arcs of right ascension and declination, having regard, as far as possible, to the boundaries of the best star atlases. The work had already been done by Gould on that basis for most of the S. Hemisphere constellations.

The IAU boundaries do not change in their positions among the stars and so objects can always be correctly located, though, owing to precession, the arcs of right ascension and declination of today no longer follow the boundaries, and are steadily departing from them. After some 12 900 years, however, these arcs will begin to return towards the boundaries, and 12 900 years after this, on completing the 25 800-year precessional period will approximate to them, but not exactly coincide.

Nowadays, as well as recording the stars' positions for a particular epoch, a general catalogue will also list various observed parameters of each star. For example, the annual changes in right ascension and declination may be given. Other headings might include proper motion, annual parallax, radial velocity, apparent magnitude, colour index and spectral type. Special peculiarities may also be supplied—for example, if the star is a binary system. It may be noted, too, that according to

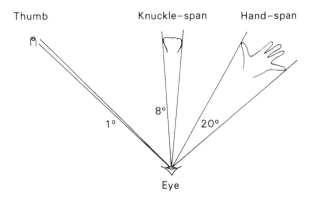

Figure 6.1. The hand as a means of estimating angles.

the IAU convention, the names of constellations are usually referred to by a standardized three-letter abbreviation.

6.2 Simple observations

There are many aids to help provide information about what is available for view in the night sky at any particular time. Many PC software packages[W 6.1,W 6.2] offer active demonstrations of the behaviour of the night sky according to the observer's location and the local time. These can be very informative as motions which, in reality, may take some months to execute can be simulated on the screen and speeded up to take just a few seconds. Thus, for example, the apparent motions of the Moon and the complex planetary paths may be readily appreciated. It is a relatively easy matter to learn which stars are in the sky at a particular time and where the planets are relative to the stellar background. If a planetarium is available, constellation identification can be learned very quickly, especially if a pattern projector is attached for highlighting each constellation.

Familiarity with the night sky, however, is best gained by spending a few hours on different nights observing the 'real' panorama. The true feeling of being under a hemispherical rotating dome with stars attached can only be obtained by outdoor activity. Appreciation of the angular scale associated with the well-known asterisms and identification of the constellation patterns is best gained by direct experience. On the early occasions, it is useful to be armed with a star atlas (such as *Norton's Star Atlas*) or a simple planisphere. This latter device is a hand-held rotatable star map with a masking visor, allowing the correct part of the sky to be seen according to the season and the local time.

In the previous chapters, reference has already been made to angular measure and, in the first place, it is useful to have an appreciation of angular scales as projected on the night sky. For example, the angular sizes of the Sun and Moon are approximately the same being $\sim \frac{1}{2}°$. If either the Sun or full Moon is seen close to the horizon it is readily appreciated that their apparent diameters would need to be extended 720 times to 'fill' the 360° of the full sweep around the horizon. Rough estimates of larger angles between stars, so providing the impression of just how large an area a particular constellation covers, can be made by using the 'rule of thumb' technique as practised by artists.

If the arm is fully extended, different parts of the hand can be used to provide some simple angular values. Typical values are indicated in figure 6.1 but a system and scale should be developed individually by comparing observations with a star map. In the first place it will be noted that the angular extent of the thumb at arm's length is about 1°. This means that if the Moon is in the sky it should be easily blocked from view by the use of the thumb. Figure 6.1 indicates that the knuckle-span

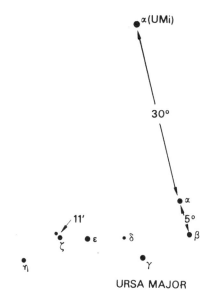

Figure 6.2. The bright starts of Ursa Major.

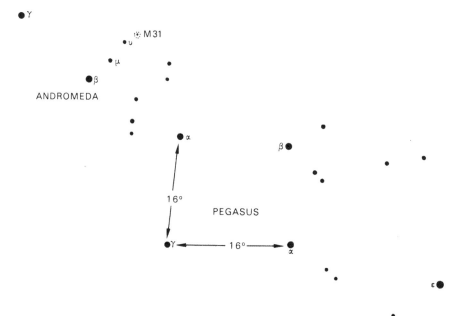

Figure 6.3. The bright stars of Andromeda and Pegasus. The position of the Andromeda Galaxy (NGC 224, M31) is indicated.

is ~8° and that the full hand-span is ~15°. Obviously these values depend on the individual and they must be checked out against some more easily recognizable asterisms.

To start with, this can be done by examining the well-known constellations. For example, in Ursa Major, the seven-star asterism of the Plough can be examined (see figure 6.2). For most northern hemisphere observers the stars are circumpolar and can, therefore, be seen at any time in the year.

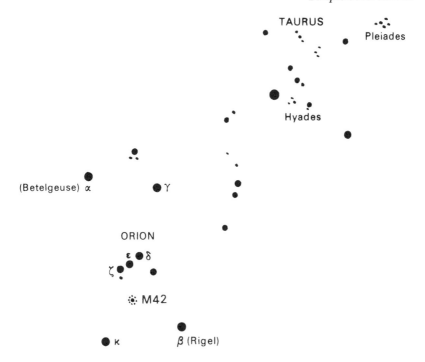

Figure 6.4. The major features of Orion and Taurus.

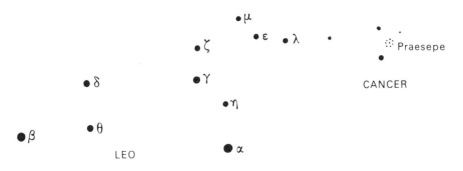

Figure 6.5. The bright stars of Leo.

Close inspection of the pattern reveals that it is actually made up of eight stars, as there are two *Mizar* and *Alcor* (ζ and 80 UMa) separated by 11 minutes of arc. The separation of the stars *Dubhe* and *Mirak* (α and β UMa) is about 5°: the distance between *Polaris* (α UMi) and *Dubhe* (α UMa) is close to 30°. It may be noted that the northern hemisphere sky appears to rotate or pivot about a point very close to *Polaris*. There is no equivalent 'pole star' in the southern hemisphere. The sides of the Square of Pegasus (see figure 6.3) are approximately 16° across the sky.

Over the course of a few weeks, take note of the changes in rising and setting times of the constellations. This can be done by noting the times when a particular group of stars is at the same position in the sky relative to a particular position of a land-mark as seen from some regular observing point. Better still, fairly accurate transit time records can be made over a few nights by using a couple of vertical poles fixed in the ground a few metres apart.

For northern hemisphere observers, it may be noted that the star cluster known as the *Pleiades* (see figure 6.4) may be seen rising in the east in the autumn. As winter progresses the rising time becomes earlier and earlier. The constellation *Orion* (also depicted in figure 6.4) transits in the north–south meridian round about midnight in February. A few months later, it will be noted that *Leo* (see figure 6.5) claims this position.

In the southern hemisphere skies, more bright stars are found than in the northern hemisphere. Also there is the beautiful spectacle of two extensive hazy patches known as the Magellanic Clouds. Two immediate differences are apparent that are initially disturbing to any traveller who changes hemispheres. It may be noted that objects appear to rise on the left-hand side in the N hemisphere with the observer's back to the pole star and on the right-hand side in the S hemisphere with the observer's back to the S pole. Startling too is the fact that a N hemisphere visitor to the southern skies sees the markings on the Moon's face upside down!

A useful starting point is for the student to observe a clear night sky with unaided eye or with binoculars. The scope of the observations and the features that might be noted are as follows:

1. The stars do not have the same brightness. By using a star chart with the magnitude scale, or by getting stellar magnitude from a catalogue, estimate the faintest star that can be seen. Does this vary from night to night? Can faint extended sources, such as the Andromeda Galaxy, be seen? Try the effect of averted vision, i.e. do not look exactly at the source but slightly away from the direct line of sight.
2. Note that there are few stars that can be seen close to the horizon due to the extinction by the Earth's atmosphere. Compare the apparent brightness to stars which are catalogued with the same magnitudes, choosing the stars so that one is close to the zenith and the other close to the horizon.
3. The stars are not randomly distributed in the sky. Note the way that the stars are grouped together. Particular clusters to pay attention to are the Pleiades and Praesepe.
4. The stars twinkle or scintillate. Note that the effect diminishes according to the altitude of a star above the horizon. There is usually very little noticeable effect for stars close to the zenith.
5. Check out that any bright 'star' that does not twinkle is a planet by consultation of an astronomical almanac for the particular time of the year.

If a 35 mm camera is available, it is instructive to take photographs of stars. Colours of stars show up well if colour film is used, although the record of the colour values may not be exact. It is not necessary for the camera to be made to follow the diurnal motion of the stars. A time exposure (a few minutes) should be made by placing the camera on the rigid support, opening the shutter and allowing the stars to drift by. A pattern of star trails will be recorded. They should be easily identifiable from the star map and the recorded colours should be compared with some database. Note that the brighter the star is, the thicker the trail will be.

There is no better way to gain confidence in understanding the basic celestial coordinate systems and, at the same time, experiencing the excitement of finding the famous stars, star clusters, galaxies, etc than by using a small equatorially mounted telescope, if one is available.

For completeness, basic star maps are provided for both the northern and southern hemispheres for the four seasons (see figures 6.6, 6.7, 6.8 and 6.9).

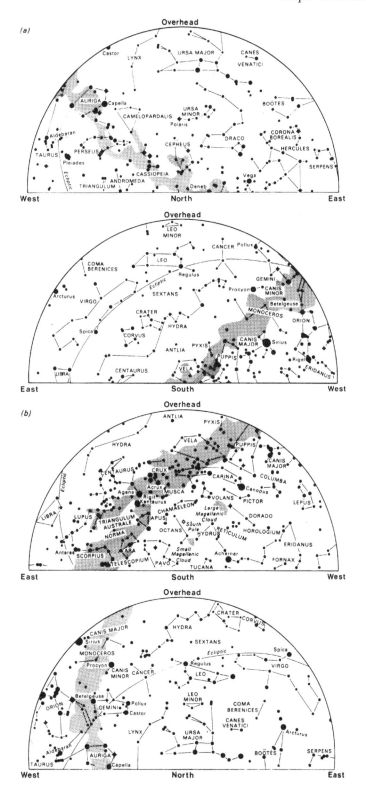

Figure 6.6. The evening constellations for February–April: (*a*) northern hemisphere and (*b*) southern hemisphere.

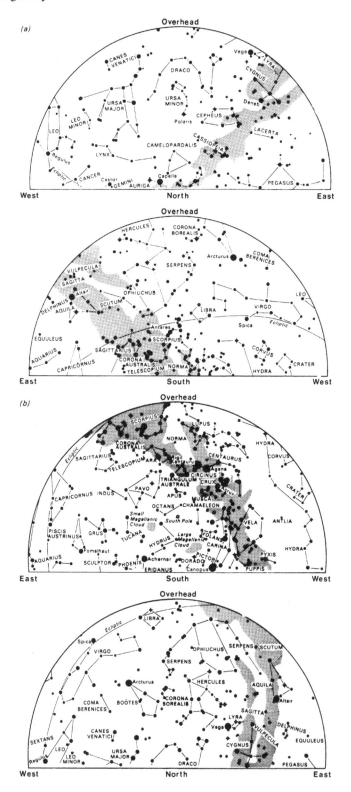

Figure 6.7. The evening constellations for May–July: (*a*) northern hemisphere and (*b*) southern hemisphere.

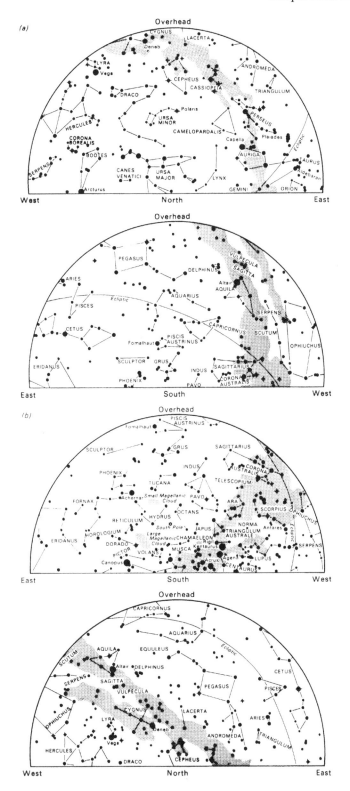

Figure 6.8. The evening constellations for August–October: (*a*) northern hemisphere and (*b*) southern hemisphere.

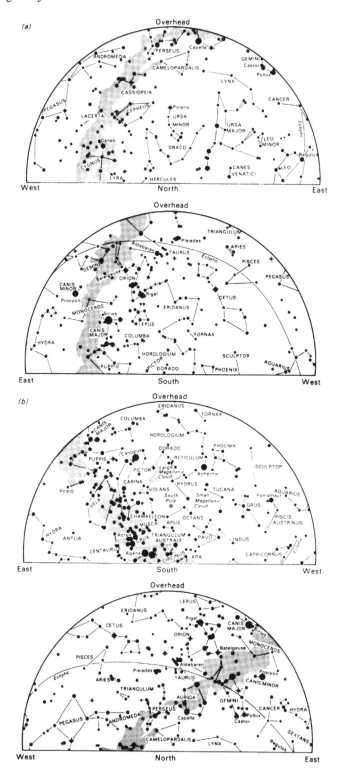

Figure 6.9. The evening constellations for November–January: (*a*) northern hemisphere and (*b*) southern hemisphere.

PART 2

THE CELESTIAL SPHERE AND ELEMENTARY CELESTIAL MECHANICS

Chapters 7–14

PROGRAMME: This part deals with the concepts used in the discussion and interpretation of positional measurements and sets out the basic principles of celestial mechanics.

Positional measurements require reference systems: these reference systems require unambiguous definition while the measurements themselves have to be 'distilled' to remove known instrumental errors and other systematic effects before they can be used to provide data about the celestial body under observation. Celestial mechanics, in its turn, is a branch of astronomy devoted to the study of the orbits of planets and satellites. Its predictions of these bodies' positions can be compared with observations that have undergone the reduction processes just mentioned.

The history of the development of astronomy is closely bound up with the development of these classical concepts. Perhaps no individual who contributed to the concepts should be singled out but it must be noted that the formulation of Newton's laws of motion and his law of gravitation did more than anything to give us our understanding of the causes underlying the orbital movements of the bodies of the Solar System. Indeed, from Newton's time onwards, a new approach to science took place that was concerned as much with causes as with observing the phenomena.

Chapter 7

The geometry of the sphere

7.1 Introduction

We have seen that the observer who views the heavens at night gets the impression that they are at the centre of a great hemisphere onto which the heavenly bodies are projected. The moon, planets and stars seem to lie on this celestial hemisphere, their directions defined by the positions they have on its surface. For many astronomical purposes the distances are irrelevant so that the radius of the sphere can be chosen at will. The description of the positions of bodies on it, taking into account positional changes with time, necessarily involves the use of special coordinate and timekeeping systems. The relationship between positions of bodies requires a knowledge of the geometry of the sphere. This branch of astronomy, called **spherical astronomy**, is in one sense the oldest branch of the subject, its foundations dating back at least 4000 years. Its subject matter is still essential and never more so than today, when the problem arises of observing or calculating the position of an artificial satellite or interplanetary probe. We, therefore, begin by considering the geometry of the sphere.

7.2 Spherical geometry

The geometry of the sphere is made up of great circles, small circles and arcs of these figures. Distances along great circles are often measured as angles since, for convenience, the radius of the sphere is made unity.

A **great circle** is defined to be the intersection with the sphere of a plane containing the centre of the sphere. Since the centre is equidistant from all points on the sphere, the figure of intersection must be a circle by definition.

If the plane does not contain the centre of the sphere, its intersection with the sphere is a **small circle**.

In figure 7.1, $ABCDA$ is a great circle. If two points P and Q are chosen to be 90° away from all points on the great circle (by drawing the diameter POQ perpendicular to the plane $ABCDA$), they are said to be the **poles** of the great circle $ABCDA$.

The small circle $EFGHE$ was obtained by choosing a point K on the diameter PQ and letting the plane through K and perpendicular to PQ cut the sphere. It is easily shown that the figure produced by this procedure is a circle.

Let $PFBQP$ be any great circle through the poles P and Q. Then, by the construction of $EFGH$, KF is perpendicular to OK, so that the triangle KFO is right-angled at K.

By Pythagoras,

$$OF^2 = OK^2 + KF^2. \tag{7.1}$$

But OF and OK are both constant (OF is a radius of the sphere and OK is a constant) and, therefore, by equation (7.1), KF is constant.

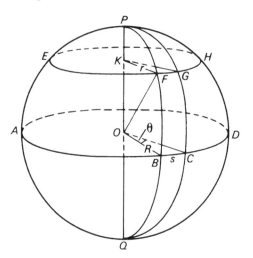

Figure 7.1. The basis of spherical geometry.

By construction, F is any point on figure $EFGHE$ so that all points on $EFGHE$ must be equidistant from K. Since they also lie on a plane, $EFGHE$ must be a circle, centre K.

P and Q are also the poles of circle $EFGHE$.

Draw another great circle $PGCQP$ through the poles P and Q to intersect the small circle $EFGHE$ and great circle $ABCDA$ in G and C respectively.

Then the angle between the tangents at P to the great circles $PFBQP$ and $PGCQP$ is said to be the **spherical angle** at P or angle GPF or angle CPB. A spherical angle is defined *only* with reference to two intersecting great circles.

If three great circles intersect one another so that a closed figure is formed by three arcs of the great circles, it is called a **spherical triangle** provided that it possesses the following properties:

1. Any two sides are together greater than the third side.
2. The sum of the three angles is greater than 180°.
3. Each spherical angle is less than 180°.

In figure 7.1, therefore, figure PBC is an example of a spherical triangle but figure PFG is not, the latter being excluded because one of its sides (FG) is the arc of a small circle. It may be noted in passing that $\triangle PBC$ is a special case where two of the angles, namely $\angle PBC$ and $\angle PCB$, are right angles.

The sides of a spherical triangle are expressed in angular measure.

The length s of the arc BC is given in terms of the angle θ it subtends at the centre of the sphere and the sphere's radius R by the relation

$$s = R \times \theta$$

where θ is expressed in radians.

It should be remembered that

$$2\pi \text{ radians} = 360 \text{ degrees.}$$

Other useful relationships are:

$$1 \text{ radian} \approx 57 \cdot 3 \text{ degrees } (57°3)$$
$$\approx 3438 \text{ arc minutes } (3438')$$
$$\approx 206\,265 \text{ arc seconds } (206\,265'').$$

If the radius of the sphere is taken as unity,

$$s = \theta$$

showing that the length of a great circle arc on a sphere of unit radius is equal to the angle (in circular measure) subtended by this arc at the centre of the sphere.

The length of a small circle arc, such as FG, is related simply to the length of an arc of the great circle whose plane is parallel to that of the small circle.

In figure 7.1, let r be the radius of the small circle $EFGHE$. Then

$$FG = r \times \angle FKG.$$

Also

$$BC = R \times \angle BOC.$$

Both OB and KF lie on plane $PFBQ$; KF also lies on plane $EFGH$ while OB lies on plane $ABCD$. Therefore, KF must be parallel to OB, since plane $EFGH$ is parallel to plane $ABCD$. Similarly, KG is parallel to OC. Then

$$\angle FKG = \angle BOC.$$

Hence,

$$FG = BC \times \frac{r}{R}.$$

In the plane triangle KOF, right-angled at K, $KF = r$; $OF = R$. Hence,

$$FG = BC \times \sin KOF.$$

But $\angle POB = 90°$ so that we may write alternatively

$$FG = BC \cos FB.$$

If the radius of the sphere is unity,

$$PF = \angle POF = \angle KOF$$

and

$$FB = \angle FOB,$$

so that we have

$$FG = BC \sin PF$$

and

$$FG = BC \cos FB. \tag{7.2}$$

7.3   Position on the Earth's surface

To illustrate these concepts we consider the Earth. A point on the surface of the Earth is defined by two coordinates, **longitude** and **latitude**, based on the **equator** and a particular meridian (half of a great circle) passing through the North and South Poles and Greenwich, England. The equator is the great circle whose poles are the North and South Poles. The longitude, λ, of the point is measured east or west along the equator. Its value is the angular distance between the meridian passing through the point and the Greenwich meridian. The longitude may be expressed in angular measure or in time units related to each other by table 7.1.

Table 7.1. Conversion of angular measure to time.

$360° = 24^{\mathrm{h}}$	
$1° = 4^{\mathrm{m}}$	$1^{\mathrm{h}} = 15°$
$1' = 4^{\mathrm{s}}$	$1^{\mathrm{m}} = 15'$
$1'' = (1/15)^{\mathrm{s}}$	$1^{\mathrm{s}} = 15''$

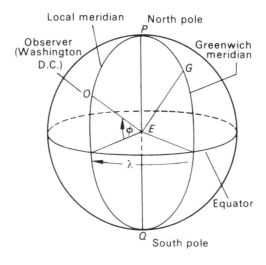

Figure 7.2. The definition of longitude and latitude.

For example, the longitude of Washington, DC, is $5^{\mathrm{h}} 08^{\mathrm{m}} 15^{\mathrm{s}}78$ west of Greenwich ($77° 03' 56''7$ W of Greenwich). Longitude is measured up to 12^{h} ($180°$) east or west of Greenwich.

The latitude, ϕ, of a point is the angular distance north or south of the equator, this angle being measured along the local meridian. Washington, DC, has a latitude $38° 55' 14''0$ N (see figure 7.2). An alternative quantity is **co-latitude**, given by

$$\text{co-latitude} = 90° - \text{latitude}.$$

Because the Earth is not a true sphere, the situation is more complicated than the simple one outlined above, though the latter is accurate enough for most purposes.

When a plumb-line is suspended by an observer at a point on the Earth's surface, its direction makes an angle with the plane of the Earth's equator. This angle is called the **astronomical latitude**, ϕ. The point where the plumb-line's direction meets the equatorial plane is not, in general, the centre of the Earth. The angle between the line joining the observer to the Earth's centre and the equatorial plane is the **geocentric latitude**, ϕ' (see figure 7.3).

There is yet a third definition of latitude. Geodetic measurements on the Earth's surface show local irregularities in the direction of gravity due to variations in the density and shape of the Earth's crust. The direction in which a plumb-line hangs is affected by such anomalies and these are referred to as **station error**. The **geodetic** or **geographic latitude**, ϕ'', of the observer is the astronomical latitude corrected for station error.

The geodetic latitude is, therefore, related to a reference spheroid whose surface is defined by the mean ocean level of the Earth. If a and b are the semi-major and semi-minor axes of the ellipse of

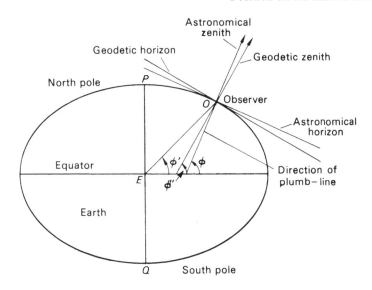

Figure 7.3. A geoid illustrating the difference between astronomical zenith and geodetic zenith.

revolution forming the 'geoid', the **flattening** or **ellipticity**, ϵ, is given by

$$\epsilon = \frac{a - b}{a} = 1 - (1 - e^2)^{1/2}$$

where e is the eccentricity of the ellipse.

Various such reference spheroids exist. The dimensions of the well-known Hayford geoid, for example, are:

$$a = 6378{\cdot}388 \text{ km} = 3963{\cdot}35 \text{ miles}$$
$$b = 6356{\cdot}912 \text{ km} = 3950{\cdot}01 \text{ miles}$$
$$\epsilon = 1/297$$

hence

$$e = 0{\cdot}081\,99.$$

The **geocentric longitude**, λ, is the same as the **geodetic longitude** which is the angular distance east or west measured along the equator from the Greenwich meridian to the meridian of the observer.

If two places on the Earth's surface have the same latitude, they are said to lie on the same **parallel of latitude**. Thus, in figure 7.4, two places A and B, both of latitude ϕ N, lie on the parallel of latitude AB.

Their distance apart, measured along the small circle arc AB, is called the **departure**. In the example, we assume the Earth to be spherical. Angle DOB is the latitude ϕ, so that

$$AC = BD = \phi.$$

If the longitudes of A and B are λ_A W and λ_B W respectively, then their difference in longitude is $\lambda_A - \lambda_B$ and

$$CD = \angle COD = \lambda_A - \lambda_B.$$

Then, by equation (7.2),

$$AB = CD \cos BD$$

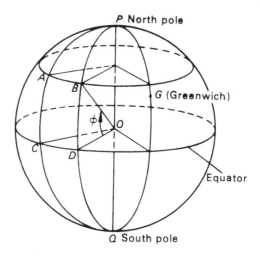

Figure 7.4. An illustration of a parallel of latitude.

i.e.

$$\textbf{departure} = \textbf{difference in longitude} \times \textbf{cos(latitude)}. \qquad (7.3)$$

Distance on the Earth's surface is usually measured in nautical miles, a **nautical mile** being the great circle distance subtending an angle of one minute of arc at the Earth's centre. Although the unit may sound 'old fashioned', it is still used in air transport and in 'down range' distances at rocket launches. Because the Earth's surface is not absolutely spherical, the length of the nautical mile varies, being 6046 feet at the equator and 6108 feet at the poles. A mean value of 6080 feet is used in the UK but the International Nautical Mile used by the Admiralty and most other nations, measures 1·8520 km being defined to be equal to one minute of arc at a latitude of 45°.

A rate of 1 nautical mile per hour is termed 1 **knot**, the unit in which a ship's speed is usually measured.

The difference in longitude may be expressed in minutes of arc, this number being equal to the number of nautical miles. The departure can then be calculated from equation (7.3).

It is to be noted that the difference in longitude is formed algebraically taking east longitudes to be of opposite sign to west longitudes.

7.3.1 GPS satellites

In past centuries and up to the 1990s, optical instruments such as the sextant (see section 20.7.2) were used to measure the angular positions of stars from which an observer's geodetic position could be calculated. With the success of achieving accurate locational fixes by the Global Positioning System (GPS), optical instruments, being very dependent on the skills of the observer and with subsequent data reduction procedures, are now superceded. Small hand-held GPS detector systems are readily available which are capable of determining an operator's position to within a few metres.

The GPS system involves some 32 artificial satellites, each carrying a code and an accurate clock. Their transmissions provide the code with the embedded time accurate to a microsecond. The onboard clocks are regularly updated and corrected from a ground control station, to an accuracy that includes effects involving the principles of general relativity.

The receiver picks up the transmitted signals at the ground and decodes them. After comparing the apparent times provided by each clock, the time differences allow the various satellite–observer distances to be determined. With the detection of four satellites, the information is sufficient to determine the four unknowns of latitude, longitude, height and the time. The observer's coordinates are referred to maps based on the World Geodetic Survey (WGS84), the new terrestrial frame adopted in 1987 as reference for the broadcast orbits. The WGS84 frame takes the Earth's equatorial radius as 6378·137 km with a value of ϵ of $1/298\cdot257\,223\,563$.

Geodetic positions based on older standard maps are related to the regional geodetic datum represented by the spheroid that approximates to the geoid of the locality. Data for the reduction of GPS geodetic coordinates to the local conventional systems are continually undergoing revision.

Care, therefore, must be taken in comparing the latitudes and longitudes provide by GPS relative to printed maps as a result of the differences between the reference systems. In the UK, the systematic difference between the Ordinance Survey positions and a GPS value can have an apparent location shift of about 140 m.

Example 7.1. If the longitude of A is $48°$ W and the longitude of B is $28°$ W, what is the difference in longitude?

$$\text{Difference in longitude} = 48° - 28° = 20°.$$

Example 7.2. If the longitude of A is $60°$ W and the longitude of B is $80°$ E, what is the difference in longitude?

$$\text{Difference in longitude} = 60° - (-80°) = 140°.$$

Example 7.3. Calculate the length of the nautical mile in feet, given that the Earth is a sphere of radius 3960 miles.

$$360° \equiv 2 \times \pi \times 3960 \text{ miles}$$

or

$$360 \times 60' \equiv 2 \times \pi \times 3960 \times 5280 \text{ feet}.$$

Hence,

$$1 \text{ minute of arc} \equiv \frac{2 \times \pi \times 3960 \times 5280}{360 \times 60} \text{ feet} \equiv 6082 \text{ feet}.$$

Or,

$$1 \text{ nautical mile} = 6082 \text{ feet}.$$

Example 7.4. Two places A and B have latitude and longitude $(35° \, 28'$ N, $44° \, 32'$ W) and $(35° \, 28'$ N, $13° \, 30'$ W) respectively. Calculate the departure between them.

$$\text{Difference in longitude} = 44° \, 32' - 13° \, 30' = 31° \, 02' = 1862'$$
$$\text{Departure} = 1862 \cos(35° \, 28')$$
$$= 1517 \text{ nautical miles}.$$

7.4 Spherical trigonometry

7.4.1 The formulas

Just as the formulas of plane trigonometry can be used to perform calculations in plane geometry, special trigonometrical formulas for use in spherical geometry can be established. There are many such formulas but four in particular are more often used than any of the others. They are the relations between the sides and angles of a spherical triangle and are invaluable in solving the problems that arise in spherical astronomy.

In figure 7.5, ABC is a spherical triangle with sides AB, BC and CA of lengths c, a and b, respectively and with angles $\angle CAB$, $\angle ABC$ and $\angle BCA$, hereafter referred to as angles A, B and C respectively. The four formulas are:

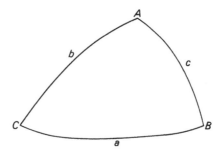

Figure 7.5. Symbols describing a spherical triangle.

1. *The cosine formula*

$$\cos a = \cos b \cos c + \sin b \sin c \cos A. \tag{7.4}$$

There are obviously two variations of this formula, namely

$$\cos b = \cos c \cos a + \sin c \sin a \cos B \tag{7.5}$$

and

$$\cos c = \cos a \cos b + \sin a \sin b \cos C. \tag{7.6}$$

2. *The sine formula*

$$\frac{\sin A}{\sin a} = \frac{\sin B}{\sin b} = \frac{\sin C}{\sin c}. \tag{7.7}$$

This formula must be used with care since, in being given a, b and B, for example, it is not possible to say whether A is acute or obtuse, unless other information is available, i.e. the formula gives A or $(180 - A)$.

3. *The analogue of the cosine formula*

$$\sin a \cos B = \cos b \sin c - \sin b \cos c \cos A. \tag{7.8}$$

Again there are obvious variations of this formula, five in number.

4. *The four-parts formula*

$$\cos a \cos C = \sin a \cot b - \sin C \cot B \tag{7.9}$$

with five other variations. This formula utilizes four consecutive parts of the spherical triangle and is often stated as follows:

$$\cos(\text{inner side}) \cos(\text{inner angle})$$
$$= \sin(\text{inner side}) \cot(\text{other side}) - \sin(\text{inner angle}) \cot(\text{other angle}).$$

It is important to have an appreciation of all these formulas and to know which are the most appropriate for solving any particular problem. Although it is not fashionable for students to be able to reproduce the proofs, derivations of the four formulas are now presented in turn for completeness.

7.4.2 Proof of cosine formula

In figure 7.6, let ABC be a spherical triangle with sides AB, BC and CA of lengths c, a and b respectively. We wish to prove that

$$\cos a = \cos b \cos c + \sin b \sin c \cos A.$$

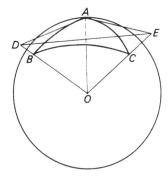

Figure 7.6. Proof of the cosine formula.

Draw tangents at A to the great circle arcs AB and AC to meet the radii OB and OC produced at D and E respectively. Join D and E. Join O and A, O being the centre of the sphere.

Then triangles ADE, ODE are plane triangles, though not in the same plane. By definition, $\angle DAE$ is the spherical angle A. Also $\angle DOE$ is the angle subtended by the arc BC. Hence, $\angle DOE = a$.

In $\triangle DAE$ we have

$$DE^2 = AD^2 + AE^2 - 2AD.AE \cos A. \tag{7.10}$$

In $\triangle DOE$ we have

$$DE^2 = OD^2 + OE^2 - 2OD.OE \cos a. \tag{7.11}$$

Hence, by subtraction

$$2OD.OE \cos a = (OD^2 - AD^2) + (OE^2 - AE^2) + 2AD.AE \cos A. \tag{7.12}$$

Now $\triangle DAO$ is a plane triangle right-angled at A, since DA is a tangent to the circle AB at A and OA is a radius of that circle. Hence, by Pythagoras,

$$OD^2 - AD^2 = AO^2.$$

Similarly, by considering $\triangle OAE$, it is seen that

$$OE^2 - AE^2 = AO^2.$$

Hence, equation (7.12) becomes

$$OD.OE \cos a = AO^2 + AD.AE \cos A$$

or

$$\cos a = \frac{OA}{OD}.\frac{OA}{OE} + \frac{AD}{OD}.\frac{AE}{OE} \cos A.$$

In $\triangle DAO$, OA/OD is the cosine of the angle $\angle DOA$ while AD/OD is the sine of that angle. But $\angle DOA = c$. Hence, $OA/OD = \cos c$; $AD/OD = \sin c$.

Similarly $OA/OE = \cos b$; $AE/OE = \sin b$.

Hence,

$$\cos a = \cos b \cos c + \sin b \sin c \cos A. \tag{7.4}$$

7.4.3   Proof of sine formula

Rewrite the cosine formula (7.4) as

$$\sin b \sin c \cos A = \cos a - \cos b \cos c.$$

Squaring, we obtain

$$\sin^2 b \sin^2 c \cos^2 A = \cos^2 a - 2 \cos a \cos b \cos c + \cos^2 b \cos^2 c.$$

Now

$$\begin{aligned}
\sin^2 b \sin^2 c(1 - \sin^2 A) &= \sin^2 b \sin^2 c - \sin^2 A \sin^2 b \sin^2 c \\
&= (1 - \cos^2 b)(1 - \cos^2 c) - \sin^2 b \sin^2 c \sin^2 A.
\end{aligned}$$

Hence,

$$\sin^2 b \sin^2 c \sin^2 A = 1 - \cos^2 a - \cos^2 b - \cos^2 c + 2 \cos a \cos b \cos c. \tag{7.13}$$

But the right-hand side of equation (7.13) is a symmetrical function of a, b and c so that we can write:

$$\sin^2 a \sin^2 b \sin^2 C = \sin^2 b \sin^2 c \sin^2 A = \sin^2 c \sin^2 a \sin^2 B.$$

Dividing throughout by $\sin^2 a \sin^2 b \sin^2 c$ we obtain

$$\frac{\sin^2 A}{\sin^2 a} = \frac{\sin^2 B}{\sin^2 b} = \frac{\sin^2 C}{\sin^2 c}.$$

Taking the square root and remembering that in a spherical triangle all sides and angles are less than $180°$ in value, we obtain

$$\frac{\sin A}{\sin a} = \frac{\sin B}{\sin b} = \frac{\sin C}{\sin c}. \tag{7.7}$$

7.4.4   Proof of the analogue to the cosine formula

Starting as before with equation (7.4), we have

$$\sin b \sin c \cos A = \cos a - \cos b \cos c.$$

Using equation (7.6), we obtain

$$\sin b \sin c \cos A = \cos a - \cos b(\cos a \cos b + \sin a \sin b \cos C)$$

so that

$$\sin b \sin c \cos A = \cos a \sin^2 b - \sin a \sin b \cos b \cos C.$$

Dividing throughout by $\sin b$, we get the required formula, namely

$$\sin c \cos A = \cos a \sin b - \sin a \cos b \cos C.$$

7.4.5   Proof of the four-parts formula

Let the four consecutive parts be B, a, C and b. Then side a, between the angles, is the 'inner side', b being the 'other'; angle C, between sides a and b, is the 'inner angle' with B the 'other'.

We want to prove that

$$\cos a \cos C = \sin a \cot b - \sin C \cot B.$$

From the cosine formulas (7.5) and (7.6), we have

$$\cos b = \cos c \cos a + \sin c \sin a \cos B \tag{7.5}$$

$$\cos c = \cos a \cos b + \sin a \sin b \cos C. \tag{7.6}$$

Substituting the right-hand side of equation (7.6) for $\cos c$ in equation (7.5), we obtain

$$\cos b(1 - \cos^2 a) = \cos a \sin a \sin b \cos C + \sin c \sin a \cos B.$$

Substituting $\sin^2 a$ for $(1 - \cos^2 a)$ and dividing throughout by $\sin a \sin b$, we get

$$\cot b \sin a = \cos a \cos C + \frac{\sin c}{\sin b} \cos B. \tag{7.14}$$

Using the sine formula (7.7) it is seen that

$$\frac{\sin c}{\sin b} = \frac{\sin C}{\sin B}.$$

Hence, equation (7.14) becomes

$$\cos a \cos C = \sin a \cot b - \sin C \cot B. \tag{7.9}$$

7.5   Other formulas of spherical trigonometry

1. If $s = (a + b + c)/2$,
$$\sin \frac{A}{2} = \left(\frac{\sin(s - b) \sin(s - c)}{\sin b \sin c} \right)^{1/2}.$$

    Obviously there are two variations of this, for $\sin B/2$ and $\sin C/2$.
2. By writing $180 - a$ for A, $180 - b$ for B, $180 - c$ for C, $180 - A$ for a, etc, in the four main formulas, other useful formulas can be obtained. For example, using the cosine formula (7.4), we obtain
$$\cos A = -\cos B \cos C + \sin B \sin C \cos a.$$

Proofs of these formulas will not be given here.

7.6   The small spherical triangle

When a spherical triangle becomes smaller and smaller on a sphere of fixed radius, its angles remain finite but its sides tend to zero length. The triangle, in fact, approximates more and more to a plane triangle and we would expect that the formulas of spherical trigonometry would degenerate into the well-known formulas of plane trigonometry. This indeed is the case.

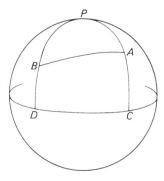

Figure 7.7. Example 7.5—flight route from Prestwick to Gander.

As an example take the cosine formula, namely

$$\cos a = \cos b \cos c + \sin b \sin c \cos A.$$

If θ is any angle, in circular measure,

$$\cos \theta = 1 - \frac{\theta^2}{2!} + \frac{\theta^4}{4!} - \cdots$$

$$\sin \theta = \theta - \frac{\theta^3}{3!} + \frac{\theta^5}{5!} - \cdots. \qquad (7.15)$$

If θ is sufficiently small, we may write

$$\cos \theta = 1 - \theta^2/2 \qquad \sin \theta = \theta.$$

Now let the spherical triangle ABC shrink so that the cosine formula may be written as

$$1 - \frac{a^2}{2} = \left(1 - \frac{b^2}{2}\right)\left(1 - \frac{c^2}{2}\right) + b.c.\cos A$$

or

$$1 - \frac{a^2}{2} = 1 - \frac{b^2}{2} - \frac{c^2}{2} + \frac{b^2 c^2}{4} + b.c.\cos A.$$

We can neglect $b^2 c^2/4$ since it is of smaller order than $a^2/2$, etc.

Hence,

$$a^2 = b^2 + c^2 - 2bc \cos A$$

the well-known formula of plane trigonometry.

Example 7.5. A plane, moving with a speed of 500 knots, flies the great circle route from Prestwick (4° 36′ W, 55° 30′ N) to Gander (54° 24′ W, 48° 34′ N). What distance does it cover and how long does the flight take?

Let A and B represent Prestwick and Gander respectively in figure 7.7 so that the great circle arc AB is the flight route. If the meridians PBD and PAC are drawn from the north pole P through B and A to the equator DC, triangle PBA is a spherical triangle. Applying the cosine formula, we may write

$$\cos AB = \cos AP \cos BP + \sin AP \sin BP \cos APB$$

$$BP = 90° - 48° 34′ = 41° 26′ = 41°\!.4333$$

$$AP = 90° - 55° 30′ = 34° 30′ = 34°\!.5000$$

$$\angle APB = 54° 24′ - 4° 36′ = 49° 48′ = 49°\!.8000.$$

Although the data are given to an accuracy of 1 arc minute, it is advisable to carry one or two extra figures to avoid rounding-off error vitiating the results.

The calculation then proceeds:

$$\cos AB = \cos 34°5000 \cos 41°4333 + \sin 34°5000 \sin 41°4333 \cos 49°8000$$

giving $AB = 30°7061 = 30° 42'$ to the nearest minute.

Hence,

$$AB = 30° 42' = 1842'.$$

The distance AB is, therefore, 1842 nautical miles.

$$\text{Time taken} = \frac{\text{distance}}{\text{speed}} = \frac{1842}{500} \text{ hr}$$

$$= 3^h68 = 3^h 41^m0.$$

Problems—Chapter 7

Note: In all these problems, assume the Earth is spherical.

1. Complete the following table, where A and B are two places on the Earth's surface.

Longitude of A	Longitude of B	Difference in Longitude
56° W	24° W	
49° W	29° E	
172° W	87° E	
36° 42′ W	57° 37′ W	
22° 34′ W	39° 43′ E	
68° 15′ E	39° 57′ E	
54° 28′ E	179° 41′ W	

2. Complete the following table where X and Y are two places on the same meridian of longitude.

Latitude of X	Latitude of Y	Difference in latitude
26° N	36° N	
38° N	29° S	
21° 33′ S	38° 47′ S	
89° 19′ N	75° 22′ S	
19° 43′ S	57° 27′ N	

3. Taking the pairs of places X and Y in question 2, calculate their distance apart in nautical miles.
4. Calculate the departure in nautical miles between the pairs of places U and V in the following table:

Longitude of U	Longitude of V	Latitude of both U and V	Departure
24° 47′ W	13° 34′ W	45° 27′ N	
39° 18′ W	21° 27′ E	20° 54′ S	
18° 57′ E	174° 41′ E	87° 27′ N	
179° 48′ W	134° 31′ E	53° 34′ S	

5. Solve completely the spherical triangle ABC, given

(i) $a = 37° 48'$, $b = 29° 51'$, $C = 74° 37'$.

(ii) $b = 98° 23'$, $c = 76° 39'$, $A = 52° 23'$.

   (iii) $c = 30° 57'$, $a = 127° 08'$, $B = 138° 19'$.
   (iv) $a = 90°$, $b = 74° 39'$, $C = 165° 29'$.

6.   Two places A and B on the same parallel of latitude 38° 33′ N are 123° 19′ apart in longitude. Calculate in nautical miles (i) their distance apart along the parallel, (ii) the great circle distance AB.

7.   Two seaports are on the same parallel of latitude 42° 27′ N. Their difference in longitude is 137° 35′. Ship A and ship B sail at 20 knots from one port to the other. Ship A sails along the parallel of latitude; ship B sails the great circle route connecting the two ports. Calculate the time difference in their arrival times if they leave port together.

8.   Assuming the Earth to be a sphere of radius 6378 km calculate the great circle distance in kilometres between London (51° 30′ N, 0°) and New York (40° 45′ N, 74° W).

   Find also the most northerly latitude reached on the great circle arc.

9.   Two radio astronomy observatories wish to set up a radio link. Their longitudes and latitudes are respectively +2° 18′.4, +53° 14′.2 (Jodrell Bank, England) and +79° 50′.2, +38° 26′.3 (Green Bank, West Virginia, USA). In what angular direction to the north must the antenna at the first station be set to achieve the best transmission (i.e. the great circle direction)?

Chapter 8

The celestial sphere: coordinate systems

8.1 Introduction

In the 4000 years during which astronomy has developed, various coordinate and timekeeping systems have been introduced because of the wide variety of problems to be solved. The coordinate systems have particular reference to great circles by which the direction of any celestial body can be defined uniquely at a given time. The choice of origin of the system also depends on the particular problem in hand. It may be the observer's position on the surface of the Earth (a **topocentric** system) or the Earth's centre (a **geocentric** system) or the Sun's centre (a **heliocentric** system) or, in the case of certain satellite problems, the centre of a planet (a **planetocentric** system). Indeed, in these modern days of manned space-flight, the origin can be a spacecraft (again a **topocentric** system) or centre of the Moon (a **selenocentric** system).

The time system used may be based on the movement of the Sun, on the Earth's rotation or on Dynamical Time (previously known as Ephemeris Time) which is related to the movements of the planets round the Sun and of the Moon about the Earth.

We now consider in some detail a number of coordinate and timekeeping systems.

8.2 The horizontal (alt-azimuth) system

This is the most primitive system, most immediately related to the observer's impression of being on a flat plane and at the centre of a vast hemisphere across which the heavenly bodies move.

In figure 8.1, the observer at O, northern latitude ϕ, can define the point directly opposite to the direction in which a plumb-line hangs as the **zenith**, Z. The plumb-line direction is known as the **nadir**, leading to the Earth's centre, if we assume the Earth to be spherical. On all sides, the plane stretches out to meet the base of the celestial hemisphere at the horizon.

Because of the Earth's rotation about its north–south axis PQ, the heavens appear to revolve in the opposite direction about a point P_1 which is the intersection of QP with the celestial sphere. Because the radius of the sphere is infinite compared with the radius of the Earth, this point is indistinguishable from a point P_2, where OP_2 is parallel to QPP_1. P_2 then said to be the **north celestial pole** and all stars trace out circles of various sizes centred on P_2. Even *Polaris*, the Pole Star, is about one degree from the pole—a large angular distance astronomically speaking when we remember that four full moons (the angular diameter of the Moon is $\sim 30'$) could be laid side-by-side within *Polaris'* circular path about the north celestial pole.

The points N and S are the points where the great circle from the zenith through the north celestial pole meets the horizon, the north point, N, being the nearer of the two to the pole.

Since OP_2 is parallel to CP_1,

$$\angle ZOP_2 = \angle OCP_1 = 90 - \phi.$$

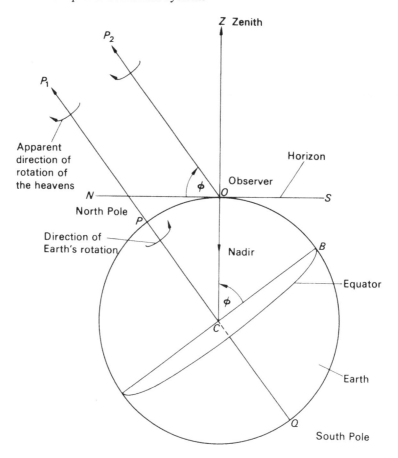

Figure 8.1. Definitions related to the observer's position on the Earth.

Hence, $\angle NOP_2 = \phi$, since $\angle ZON$ is a right angle. That is,

The altitude of the pole is the latitude of the observer.

The horizontal system of coordinates has the observer at its origin so that it is a strictly local or topocentric system.

The observer's celestial sphere is shown in figure 8.2 where Z is the zenith, O the observer, P is the north celestial pole and OX the instantaneous direction of a heavenly body. The great circle through Z and P cuts the horizon $NESAW$ at the north (N) and south (S) points. Another great circle WZE at right angles to the great circle $NPZS$ cuts the horizon in the west (W) and east (E) points. The arcs ZN, ZW, ZA, etc, are called **verticals**. The points N, E, S and W are the **cardinal points**. It is to be noted that west is always on the left hand of an observer facing north. The verticals through east and west are called **prime verticals**; ZE is the **prime vertical east**, ZW is the **prime vertical west**.

The two numbers that specify the position of X in this system are the **azimuth**, A, and the **altitude**, a. Azimuth is defined in a number of ways and care must be taken to find out which convention is followed in any particular use of this system.

For example, the azimuth may be defined as the angle between the vertical through the south point and the vertical through the object X, measured westwards along the horizon from $0°$ to $360°$, or the angle between the vertical through the north point and the vertical through the object X, measured

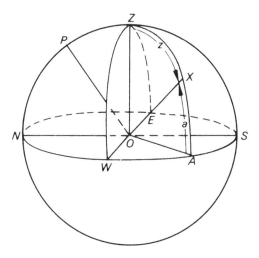

Figure 8.2. The observer's celestial sphere.

eastwards or westwards from 0° to 180° along the horizon. **A third definition commonly used is to measure the azimuth from the north point eastwards from 0° to 360°.** This definition will be kept in this text and is, in fact, similar to the definition of **true bearing**. For an observer in the southern hemisphere, the **azimuth is measured from the south point eastwards from 0° to 360°**.

The **altitude**, a, of X is the angle measured along the vertical circle through X from the horizon at A to X. It is measured in degrees. An alternative coordinate to altitude is the **zenith distance**, z, of X, indicated by ZX in figure 8.2. Obviously,

$$a = 90 - z.$$

The main disadvantage of the horizontal system of coordinates is that it is purely local. Two observers at different points on the Earth's surface will measure different altitudes and azimuths for the same star at the same time. In addition, an observer will find the star's coordinates changing with time as the celestial sphere appears to rotate. Even today, however, many observations are made in the **alt-azimuth** system as it is often called, ranging from those carried out using kinetheodolites to those made by the 250 feet (76 metre) radio telescope at Jodrell Bank, England. In the latter case, a special computer is employed to transform coordinates in this system to equatorial coordinates and *vice versa*. A similar solution is also employed with the large 6 metre (236 in) optical telescope at the Zelenchukskaya Astrophysical Observatory in the Caucasus Mountains. The William Herschel Telescope (4.2 m) on La Palma (Canary Islands) is also an alt-azimuth arrangement.

8.3   The equatorial system

If we extend the plane of the Earth's equator, it will cut the celestial sphere in a great circle called the **celestial equator**. This circle intersects the horizon circle in two points W and E (figure 8.3). It is easy to show that W and E are the west and east points. Points P and Z are the poles of the celestial equator and the horizon respectively. But W lies on both these great circles so that W is 90° from the points P and Z. Hence, W is a pole on the great circle ZPN and must, therefore, be 90° from all points on it—in particular from N and S. Hence, it is the west point. By a similar argument E is the east point.

Any great semicircle through P and Q is called a **meridian**. The meridian through the celestial object X is the great semicircle $PXBQ$ cutting the celestial equator in B (see figure 8.3).

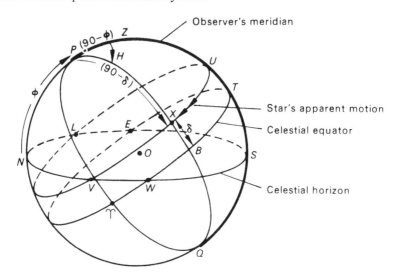

Figure 8.3. The equatorial system.

In particular, the meridian $PZTSQ$, indicated because of its importance by a heavier line, is the **observer's meridian**.

An observer viewing the sky will note that all natural objects rise in the east, climbing in altitude until they **transit** across the observer's meridian then decrease in altitude until they set in the west. A star, in fact, will follow a small circle parallel to the celestial equator in the arrow's direction. Such a circle (UXV in the diagram) is called a **parallel of declination** and provides us with one of the two coordinates in the equatorial system.

The **declination**, δ, of the star is the angular distance in degrees of the star from the equator along the meridian through the star. It is measured north and south of the equator from $0°$ to $90°$, being taken to be positive when north. The declination of the celestial object is thus analogous to the latitude of a place on the Earth's surface, and indeed the latitude of any point on the surface of the Earth when a star is in its zenith is equal to the star's declination.

A quantity called the **north polar distance** of the object (X in figure 8.3) is often used. It is the arc PX. Obviously,

$$\text{north polar distance} = 90° - \text{declination}.$$

It is to be noted that the north polar distance can exceed $90°$.

The star, then, transits at U, sets at V, rises at L and transits again after one rotation of the Earth. The second coordinate recognizes this. The angle ZPX is called the **hour angle**, H, of the star and is measured from the observer's meridian westwards (for both north and south hemisphere observers) to the meridian through the star from 0^h to 24^h or from $0°$ to $360°$. Consequently, the hour angle increases by 24^h each sidereal day for a star (see section 9.2).

8.4 Southern hemisphere celestial spheres

To clarify the ideas introduced in the previous sections, we consider the celestial sphere for an observer in the southern hemisphere. Let the latitude be ϕ S. Then the celestial pole above the horizon is the south celestial pole Q. We proceed as follows:

1. Draw a sphere.
2. Insert the zenith, Z, and the horizon, $NWSE$, a great circle with Z as one of its poles.

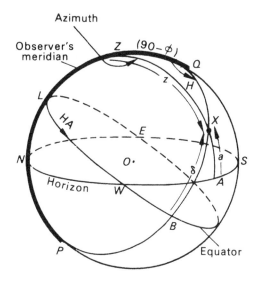

Figure 8.4. A southern hemisphere celestial sphere.

3. In figure 8.4 we have placed W on the front of the diagram. The convention that when facing north, west is on your left hand dictates the placing of N, S and E.
4. Insert Q, the *south* celestial pole, between the *south* point S and the zenith Z and such that the altitude SQ of the pole is the latitude ϕ of the observer. We can then insert the north celestial pole P directly opposite.
5. Put the celestial equator in the diagram, remembering that P and Q are its poles.
6. Insert the observer's meridian $QZNP$, according to the rule that it runs from the pole in the sky *through the zenith* and horizon to the pole below the Earth.
7. Put an arrowhead on the equator with HA beside it to show that the hour angle is measured *westwards* from the observer's meridian.

   Let us suppose we are interested in the position of a star X.
8. Draw the vertical ZXA and the meridian $QXBP$ through its position. Then,

$$\text{the azimuth of } X = \text{arc } SEA = A° \text{ E of S}$$
$$\text{the altitude of } X = \text{arc } AX \quad = a°$$
$$\text{the zenith distance of } X = \text{arc } ZX \quad = 90 - a$$
$$\text{the declination of } X = \text{arc } BX \quad = \delta° \text{ S}$$
$$\text{the hour angle of } X = \text{arc } LB \quad = H^{\text{h}}$$
$$\text{the south polar distance of } X = \text{arc } QX \quad = 90 - \delta.$$

Note: (a) that the azimuth is specified E of S to avoid ambiguity; (b) that the declination is labelled S again to avoid ambiguity.

8.5 Circumpolar stars

Consider the celestial sphere for an observer in latitude ϕ N (figure 8.5). The parallels of declination of a number of stars have been inserted with arrowheads to show that as time passes their hour angles increase steadily as the celestial sphere rotates.

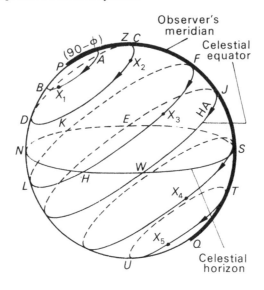

Figure 8.5. Circumpolar stars.

The stars can be put into three classes:

(a) stars that are above the horizon for all values of their hour angle,
(b) stars that are below the horizon for all values of their hour angle,
(c) stars that are seen to rise and set.

Stars in class (a) are **circumpolar** stars. Examples of these in figure 8.5 are stars X_1 and X_2. Star X_1 transits at A north of the zenith in contrast to star X_2's transit which is south of the zenith. These transits are referred to as **upper transit** or **upper culmination**. Both stars also transit below the pole; such transits are described as **below pole** or at **lower culmination**.

Now CD is the parallel of declination of star X_2. In order that the star is circumpolar, then, we must have

$$PD < PN$$

that is

$$90 - \delta < \phi.$$

In order that the upper transit should be south of the zenith we must have

$$PC > PZ$$

that is

$$90 - \delta > 90 - \phi$$

or

$$\phi > \delta.$$

Stars in class (b) are never seen by the observer. The ancients who introduced the constellations were unaware of such stars, thus explaining why a roughly circular area of the celestial sphere in the vicinity of the south celestial pole is not represented in the ancient constellations.

In the diagram, star X_4, of declination δ S, is the limiting case. Now

$$JS = 90 - \phi \qquad SQ = 90 - \delta.$$

Also

$$JQ = 90°.$$

Hence, we have

$$180 - \phi - \delta = 90$$

or

$$\phi + \delta = 90. \tag{8.1}$$

Hence, $\delta = 90 - \phi$ is the limiting declination of a star if it is to remain below the horizon.

If $\delta < 90 - \phi$, the star comes above the horizon. By putting a value in equation (8.1) for the declination of the stars at the edge of the roughly circular area not represented in the ancient constellations, it is found that the constellation-makers must have lived in a latitude somewhere between 34° and 36° N.

Most stars are found in class (c), that is they rise and transit, then set. For example, star X_3 moves along its parallel of declination (the small circle $FHLK$), setting at H, rising at K and transiting at F.

8.6 The measurement of latitude and declination

Let us suppose an observer in latitude ϕ N observes a circumpolar star of declination δ, for example star X_2 in figure 8.5. Using a *meridian circle* or *transit instrument* (section 20.5), the zenith distances ZC and ZD of the star at upper and lower culmination are measured. Now

$$PC = PZ + ZC$$

i.e.

$$90 - \delta = 90 - \phi + ZC.$$

Hence,

$$\phi - \delta = ZC. \tag{8.2}$$

Also,

$$PD = ZD - ZP$$

or

$$90 - \delta = ZD - 90 + \phi.$$

Hence,

$$\phi + \delta = 180 - ZD. \tag{8.3}$$

In equations (8.2) and (8.3) we have two equations in the two unknown quantities ϕ and δ. In principle, therefore, we can solve to obtain values of ϕ and δ.

In practice, a number of circumpolar stars are observed. Each star gives a pair of equations but only one extra unknown, namely the star's declination. Thus, with six stars, twelve equations in seven unknowns given by the latitude and the six declinations have to be solved. This is done by means of a mathematical procedure such as the method of least squares.

Example 8.1. In a place of latitude 48° N, a star of declination 60° N is observed. What is its zenith distance at upper and lower culmination?

In figure 8.6,

$$PZ = 90 - \phi = 42°$$
$$PD = 90 - \delta = 30°.$$

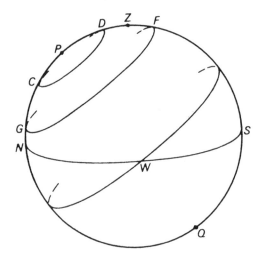

Figure 8.6. Examples 8.1 and 8.2—zenith distances at upper and lower culmination.

Hence,

$$DZ = 42° - 30° = 12°.$$

The zenith distance at upper culmination is, therefore, 12° and is north of the zenith.

$$ZC = ZP + PC = 90 - \phi + 90 - \delta$$
$$= 180 - 48 - 60$$
$$= 72°.$$

The zenith distance at lower culmination is, therefore, 72°.

Example 8.2. The zenith distances of a star at upper culmination (south of the zenith) and lower culmination are 24° and 74° respectively. Calculate the latitude of the observer and the declination of the star.

Let the latitude and declination be ϕ and δ degrees respectively. Then

$$PF = PZ + ZF$$

that is

$$90 - \delta = 90 - \phi + ZF$$

or

$$\phi - \delta = 24°. \tag{8.4}$$

Also,

$$PG = ZG - PZ$$

giving

$$90 - \delta = ZG - 90 + \phi$$

or

$$\phi + \delta = 180 - 74° = 106°. \tag{8.5}$$

Adding equations (8.4) and (8.5) we obtain

$$2\phi = 130° \quad \text{or} \quad \phi = 65° \text{ N}.$$

Substituting 65° for ϕ in equation (8.4) gives $\delta = 41°$ N.

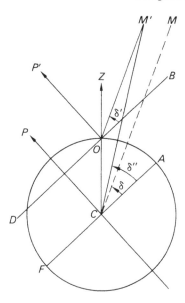

Figure 8.7. The geocentric celestial sphere showing the effect of parallax by moving the centre of a celestial sphere from an observer to the centre of the Earth.

8.7   The geocentric celestial sphere

So far we have assumed that the celestial sphere is centred at the observer. In the case of measurements made in the alt-azimuth system of coordinates, we have seen that as altitude and azimuth are linked to an observer's latitude and longitude, there are as many pairs of coordinates for a star's position at a given time as there are observers, even though the size of the Earth is vanishingly small compared with stellar distances.

Even in the equatorial system, the multitude of observing positions scattered over the Earth raises problems. For example, let us consider the declination of a star.

To the observer at O (see figure 8.7), the direction of the north celestial pole OP' is parallel to the direction CP in which it would be seen from the Earth's centre. Likewise the plane of the celestial equator DOB is parallel to the celestial equator obtained by extending in the Earth's equatorial plane FCA to cut the celestial sphere. A star's direction, as observed from O, would also be parallel to its direction as observed from C. Thus, OM' is parallel to CM where OM' is the direction to a particular star. So far, no problem is raised: the star's declination, δ', as measured at O, is $\angle BOM'$, equal to the **geocentric** declination, δ, or $\angle ACM$.

There are a number of celestial objects, however, where a problem does arise. These objects are within the Solar System—for example, the Sun, the planets and their satellites, comets and meteors and, of course, artificial satellites and other spacecraft. None of these can be considered to be at an infinite distance. A shift in the observer's position from O to C, therefore, causes an apparent shift in their positions on the celestial sphere. We call such a movement (due, in fact, to a shift in the observer's position) a **parallactic** shift. Obviously, this apparent angular shift will be the greater, the closer the object is to the Earth. In particular, if M' is such an object, its declination, δ', given by $\angle BOM'$ will no longer be equal to its geocentric declination, δ'', given by $\angle ACM'$. In passing, it may be mentioned that $\angle OM'C$ is called the **parallactic angle**.

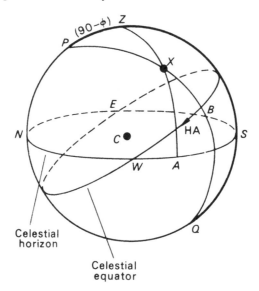

Figure 8.8. The geocentric celestial sphere and the position of a star.

In the almanacs such as *The Astronomical Almanac*[1], information about the positions of celestial objects, including the planets, is given for various dates throughout the year. It would be impossible to tabulate the declinations of a planet at a particular date for all possible observers. Hence, such information is tabulated for a hypothetical observer stationed at the Earth's centre. Various correction procedures are available so that any observer can convert the tabulated geocentric data to topocentric (or local) data.

From now on, unless otherwise stated, we will consider any celestial sphere to be a standard geocentric one. Thus, in figure 8.8 we have a geocentric celestial sphere for an observer in latitude ϕ N. The zenith is obtained by drawing a straight line from the Earth's centre through the observer on the Earth's surface to intersect the celestial sphere at Z. The celestial horizon $NWSE$ is the great circle with Z as one of its poles, while the celestial equator is the intersection of the celestial sphere by the plane defined by the terrestrial equator. The observer's meridian is the great semicircle $PZSQ$.

For a star, X, then

(i)   its azimuth is the arc $NESA$,
(ii) its altitude the arc AX,
(iii) its zenith distance the arc ZX;
(iv) its hour angle is $\angle ZPX$,
(v)   its declination the arc BX
(vi) and its north polar distance the arc PX.

8.8 Transformation of one coordinate system into another

A common problem in spherical astronomy is a wish to obtain a star's coordinates in one system, given the coordinates in another system. The observer's latitude is usually known.

For example, we may wish to calculate the hour angle of H and declination δ of a body when its azimuth (east of north) and altitude are A and a. Assume the observer has a latitude ϕ N.

---

[1] Formerly the *Astronomical Ephemeris*, also published in the United States under the same title. Many similar almanacs are available in other countries.

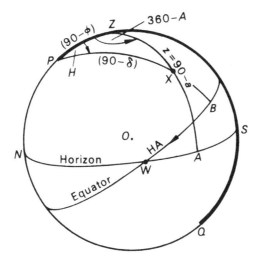

Figure 8.9. The conversion of azimuth and altitude to hour angle and declination.

The required celestial sphere is shown in figure 8.9 where X is the body's position.

In spherical triangle PZX, we see that we require to find arc PX and angle ZPX. We calculate PX first of all, using the cosine formula because we know two sides PZ, ZX and the included angle PZX.

Hence, we may write,

$$\cos PX = \cos PZ \cos ZX + \sin PZ \sin ZX \cos PZX$$

or

$$\sin \delta = \sin \phi \sin a + \cos \phi \cos a \cos A.$$

This equation enables δ to be calculated.

A second application of the cosine formula gives

$$\cos ZX = \cos PZ \cos PX + \sin PZ \sin PX \cos ZPX$$

or

$$\sin a = \sin \phi \sin \delta + \cos \phi \cos \delta \cos H.$$

Re-arranging, we obtain

$$\cos H = \frac{\sin a - \sin \phi \sin \delta}{\cos \phi \cos \delta}$$

giving H, since δ is now known.

Alternatively, using the four-parts formula with ZX, $\angle PZX$, PZ and $\angle ZPX$, we obtain

$$\cos PZ \cos PZX = \sin PZ \cot ZX - \sin PZX \cot ZPX$$

or

$$\sin \phi \cos A = \cos \phi \tan a + \sin A \cot H$$

giving

$$\tan H = \frac{\sin A}{\sin \phi \cos A - \cos \phi \tan a}.$$

We consider the problem in reverse by means of a numerical example.

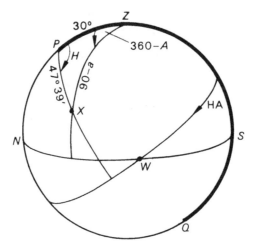

Figure 8.10. Example 8.3—conversion of hour angle and declination to azimuth and altitude.

Example 8.3. A star of declination $42° 21'$ N is observed when its hour angle is $8^h 16^m 42^s$. If the observer's latitude is $60°$ N, calculate the star's azimuth and altitude at the time of observation.

It is often a great help to sketch as accurately as possible a celestial sphere diagram of the problem. This provides a visual check on deductions about quadrants in which an angle lies.

Since $PX = 90 - \delta$, we see that its value is $47° 39'$.

We convert the hour angle value of $8^h 16^m 42^s$ to angular measure by means of table 7.1 (see page 48).

$$8^h 16^m 42^s = 8^h + 16^m + 42^s$$
$$= (8 \times 15)° + (16/4)° + (40/4)' + (2 \times 15)''$$
$$= 120° + 4° + 10' + 30''$$
$$= 124° 10'.5.$$

Hence, $H = \angle ZPX = 124° 10'.5$, in figure 8.10 where X is the star.

$$PZ = 90° - \phi = 90° - 60° = 30°.$$

Applying the cosine formula to $\triangle PZX$, we may write

$$\cos ZX = \cos 30° \cos 47° 39' + \sin 30° \sin 47° 39' \cos 124° 10'.5$$

or

$$\sin a = \cos 30° \cos 47° 39' - \sin 30° \sin 47° 39' \cos 55° 49'.5.$$

The calculation then proceeds as follows:

$$\sin a = \cos 30° \cos 47°.6500 - \sin 30° \sin 47°.6500 \cos 55°.8253$$

giving, on reduction,

$$a = 22° 04'.6.$$

Applying the cosine formula once more to $\triangle PZX$, we have

$$\cos 47° 39' = \cos 30° \cos(90 - a) + \sin 30° \sin(90 - a) \cos(360 - A)$$

or

$$\cos 47° 39' = \cos 30° \sin 22° 04.6 + \sin 30° \cos 22° 04.6 \cos A$$

giving

$$\cos A = \frac{\cos 47°65 - \cos 30° \sin 22°0760}{\sin 30° \cos 22°0760}.$$

On reduction we find that $A = 41°2847$ or $360° - 41°2847$, that is $A = 41° 17.1$ or $318° 42.9$.

It is obvious from the diagram and the value of the hour angle that the correct value is $318° 43'$ east of north to the nearest minute.

Check: Using the sine formula, we may write

$$\frac{\sin H}{\sin(90 - a)} = \frac{\sin(360 - A)}{\sin 47° 39'}$$

that is

$$\sin H = \frac{-\cos a \sin A}{\sin 47° 39'} = \frac{-\cos 22°0760 \sin 41°2847}{\sin 47°65}$$

giving, on reduction, $H = 180° - 55°8248$ since $\sin H$ is negative. Hence, $H = 124°1752 = 124° 10.5$.

We see that this agrees with the original value of H.

8.9   Right ascension

We have seen that in the equatorial system, one of the coordinates of the star, namely the declination, is constant with time. The other, the hour angle, changes steadily with the passage of time and so is unsuitable for use in a catalogue of stellar positions.

The problem is solved in a manner analogous to the way in which places on the Earth's surface are defined uniquely in position, although the Earth is rotating on its axis. Latitude is defined with respect to the terrestrial equator. In spherical astronomy, declination, referred to the celestial equator, carries out the same task in fixing the place of a celestial object. The longitude of a place on the Earth's surface is defined with respect to a meridian through a particular geographical position, namely the Airy Transit Instrument at Greenwich, England, and the meridian through the place in question.

The Greenwich meridian cannot be used for celestial position-fixing. Because of the Earth's rotation under the celestial sphere, the projection of the Greenwich meridian sweeps round the sphere, passing through each star's position in turn. Some other meridian must be chosen which is connected directly to the celestial sphere.

If a point, ♈, fixed with respect to the stellar background, is chosen on the celestial equator, its angular distance from the intersection of a star's meridian and the equator will not change, in contrast to the changing hour angle of that star. In general, then, all celestial objects may have their positions on the celestial sphere specified by their declinations and by the angles between their meridians and the meridian through ♈. The point chosen is the **vernal equinox**, also referred to as the **First Point of Aries**, and the angle between it and the intersection of the meridian through a celestial object and the equator is called the **right ascension** (RA) of the object. Right ascension is measured from 0^h to 24^h or from $0°$ to $360°$ along the equator from ♈ eastwards, i.e. in the direction opposite to that in which the hour angle is measured. Like the definition of hour angle, this convention holds for observers in both northern and southern hemispheres. In drawing a celestial sphere it is advisable not only to mark the observer's meridian heavily, inserting on the equator a westwards arrow with HA (hour angle) beside it but also to mark on the equator an eastwards arrow with RA (right ascension) beside it.

We now show that the choice of ♈ as a reference point is closely connected with the Sun's yearly journey round the stellar background of the celestial sphere.

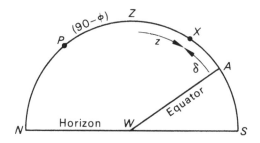

Figure 8.11. The Sun on the meridian.

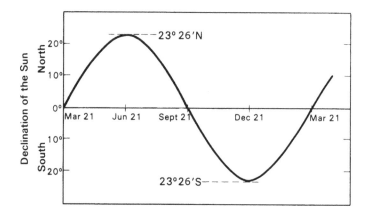

Figure 8.12. The variation of the Sun's declination through the year.

8.10   The Sun's geocentric behaviour

We have already seen that an observer studying the Sun and stars during a year comes to certain conclusions about the Sun's behaviour. The Sun appears to revolve about the Earth once per day in company with the stellar background but also has a slower motion with respect to the stellar background, tracing out a yearly path—the **ecliptic**—among the stars.

More precisely, let us suppose that an observer in a northern latitude (i) measures the meridian zenith distance of the Sun every day, i.e. at apparent noon, and (ii) notes at apparent midnight the constellations of stars that are on the meridian; and that he/she carries out this series of observations for one year. It should be noted that at **apparent noon**, the Sun's hour angle has a value zero; at **apparent midnight** its value is 12^h.

Then, from the midday measurement of zenith distance, the observer can keep track of the Sun's declination changes throughout the year. Assuming the latitude ϕ to be known, it can be seen (see figure 8.11), that if X is the Sun's position at transit on a particular day,

$$PA = 90° = 90 - \phi + z + \delta$$

or

$$\delta = \phi - z$$

where δ and z are the Sun's declination and zenith distance respectively.

A graph of declination against date is obtained, illustrated in figure 8.12.

The record of those stars appearing on the observer's meridian at apparent midnight throughout the year shows that the Sun must make one complete revolution of the stellar background in that time.

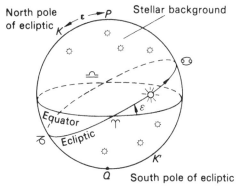

♈   First point of Aries (Ram)   March 21st Vernal equinox

♋   Cancer (Crab)   June 21st Summer solstice

♎   Libra (Scales)   September 21st Autumnal equinox

♑   Capricornus (Goat)   December 21st Winter solstice

Figure 8.13. The solar path along the ecliptic.

Together with the declination record, it enables the Sun's yearly path on the stellar background to be traced out. This path, the ecliptic, is shown in figure 8.13.

The yearly journey takes the Sun through the twelve houses of the Zodiac, ancient constellations embodying the stars within a few degrees (8°–9°) of the ecliptic. In order, they are given here, with symbols and meanings:

Aries	Taurus	Gemini	Cancer	Leo	Virgo
♈	♉	♊	♋	♌	♍
Ram	Bull	Twins	Crab	Lion	Virgin

Libra	Scorpius	Sagittarius	Capricornus	Aquarius	Pisces
♎	♏	♐	♑	♒	♓
Scales	Scorpion	Archer	Goat	Water-bearer	Fishes

The word 'Zodiac' means 'circle of the animals' and is the region of the celestial sphere in which the Sun, the Moon and those planets known to the ancients (Mercury, Venus, Mars, Jupiter and Saturn) are found.

Because of the effects of precession and as a result of the International Astronomical Union's (IAU) adoption of constellation boundaries (see chapter 6) which are not exactly 30° long, the Sun's passage through the constellations does not correspond to its passage through the signs of the Zodiac. According to the IAU boundary definitions, the Sun travels 44° through *Virgo*, only 7° through *Scorpius* and 18° through *Ophiuchus*, a constellation for which there is no Zodiacal sign.

It is, therefore, seen that for half the year the Sun is below the plane of the terrestrial equator and, consequently, that of the celestial equator. Its declination is south during this time, achieving its maximum southerly value about December 21st. This is the time of the **winter solstice** in the northern hemisphere (and the **summer solstice** for those people living in the southern hemisphere). The word 'solstice' means 'the standing still of the Sun' and refers to the pause in the Sun's progress in declination. Around March 21st the Sun crosses the equator, its declination changing from south to north. This is the **spring equinox** (**autumnal equinox** for southern hemisphere inhabitants). The word 'equinox' means 'equal day and night' because at the time of occurrence of the vernal and autumnal equinoxes, day and night are equal in length, everywhere on the Earth.

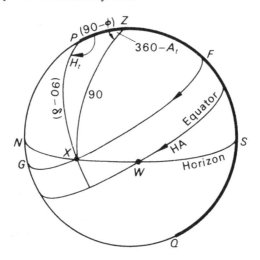

Figure 8.14. The diurnal motion of the Sun.

For six months after March 21st the Sun's declination is positive, that is northerly, becoming a maximum around June 21st, the date of the summer solstice in the northern hemisphere, thereafter decreasing to zero about September 21st, the autumnal equinox for the northern hemisphere.

The ecliptic, therefore, is a great circle defining a plane that intersects the plane of the celestial equator at an angle of about $23°\,26'$. This angle, usually denoted by ε, is called the **obliquity of the ecliptic**. The two points of intersection of the ecliptic with the equator are the First Point of Aries, ♈, and Libra, ♎.

The zero point, from which right ascension is measured, is ♈, the point at which the Sun crosses the equator from south to north on its yearly journey round the ecliptic. Right ascension is, therefore, measured along the equator in the same direction in which the Sun travels round the ecliptic. In one year, consequently, the Sun's right ascension increases from 0^h (March 21st) through 6^h (June 21st), 12^h (September 21st), 18^h (December 21st) to 24^h (the succeeding March 21st).

8.11   Sunset and sunrise

Because of the changing declination of the Sun throughout the year, the hour angle and azimuth of sunset and sunrise will also change.

Strictly speaking, the Sun's declination is an ever-changing quantity, except at the solstices when it has stationary values but the changes are so slow that for many problems it is sufficiently accurate to consider the declination to be constant throughout the day. The Sun's diurnal path, to this degree of accuracy, is along a parallel of declination such as illustrated by the small circle FG in figure 8.14.

Let the Sun's declination be δ N on a particular day and let the Sun set on that day at X, so that $PX = 90° - \delta$ and $ZX = 90°$.

The Sun's hour angle at setting is $H_t = \angle ZPX$; its azimuth at setting is A_t given by $A_t = 360° - \angle PZX$. Its zenith distance ZX, at setting is $90°$.

Using the cosine formula in $\triangle PZX$, we have

$$\cos 90 = \cos(90 - \phi)\cos(90 - \delta) + \sin(90 - \phi)\sin(90 - \delta)\cos H_t$$

or

$$\cos H_t = -\tan\delta\tan\phi \tag{8.6}$$

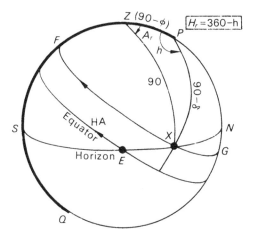

Figure 8.15. The celestial sphere illustrating sunrise.

showing that in a given latitude, the value of the Sun's hour angle at setting depends on the Sun's declination.

Thus, if the latitude is north and the declination is positive,

$$6^{\text{h}} \leq H_t \leq 12^{\text{h}}.$$

If the latitude is north and the declination is negative,

$$0^{\text{h}} \leq H_t \leq 6^{\text{h}}.$$

To obtain the azimuth at sunset, we apply the cosine formula to $\triangle PZX$ again. Thus,

$$\cos(90 - \delta) = \cos(90 - \phi)\cos 90 + \sin(90 - \phi)\sin 90 \cos(360 - A_t)$$

or

$$\cos A_t = \frac{\sin \delta}{\cos \phi}.$$

Note that for sunset, the azimuth, measured eastwards from north, is necessarily greater than 180°. For northern latitudes, if the declination is positive,

$$270° \leq A_t \leq 360°$$

but if the declination is negative,

$$180° \leq A_t \leq 270°.$$

The total number of hours of daylight in a given 24^{h} period is obtained approximately by doubling H_t, the hour angle of sunset (see equation (8.6)), for it is easily seen from the diagram for sunrise (figure 8.15) that the hours of daylight for sunrise to apparent noon equal H_t, being given by h or $\angle ZPX$.

The hour angle at sunrise, H_r, is given by

$$H_r = 360 - h$$

where

$$\cos h = -\tan \phi \tan \delta.$$

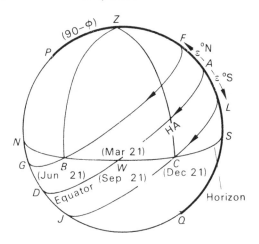

Figure 8.16. The azimuth limits of sunset.

The azimuth at sunrise, A_r, is given by

$$\cos A_r = \frac{\sin \delta}{\cos \phi}.$$

It is obvious that for northern latitudes, with positive declinations,

$$0 \leq A_r \leq 90°$$
$$12^h \leq H_r \leq 18^h$$

while for negative declinations,

$$90° \leq A_r \leq 180°$$
$$18^h \leq H_r \leq 24^h.$$

Thus, in figure 8.16 for a place in northern latitude ϕ, the sunset point on the western horizon will move back and forth between B and C where FG and LJ are the most northerly and southerly parallels of declination of the Sun respectively, and $AF = AL = \varepsilon$, the obliquity of the ecliptic.

A corresponding cycle in the azimuth at the sunrise point takes place along the eastern horizon.

8.12   Megalithic man and the Sun

The Sun, as god and giver of light, warmth and harvest, was all-important. He was a Being to be worshipped and placated by ritual and sacrifice. On a more practical note, his movements provided a calendar by which ritual and seed time and harvest could be regulated. Throughout the United Kingdom, for example, between the middles of the third and second millennia BC, vast numbers of solar observatories were built. Some were simple, consisting of a single alignment of stones; others were much more complicated, involving multiple stone circles with outlying stones making ingenious use of natural foresites along the horizon to increase their observational accuracy. A surprising number of such megalithic sites still exist, the most famous of which are Stonehenge, in England, and Callanish, in the Outer Hebrides, Scotland.

One example will be sufficient to illustrate megalithic man's ingenuity. It exists at Ballochroy on the west coast of the Mull of Kintyre, Scotland.

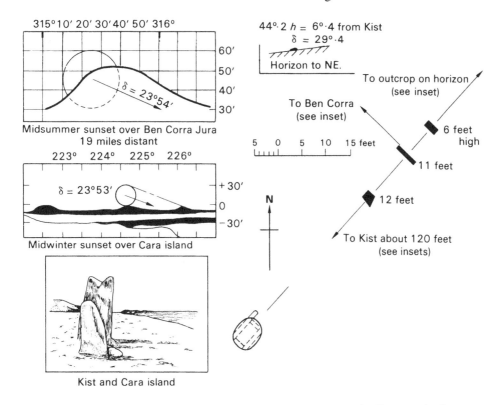

Figure 8.17. Megalithic man's sighting stones at Ballochroy (55° 42′ 44″ N, 5° 36′ 45″ W).

Three large stones (see figure 8.17) are set up in line close together; a stone kist (a grave built of stone slabs) is found on this line to the south-west, about 40 m from the stones. The slabs are set parallel to each other. Looking along the flat face of the central stone we see the outline of Ben Corra in Jura, some 30 km away. Looking along the direction indicated by the line of stones and the kist, we see off-shore the island of Cara.

About 1800 BC, the obliquity of the ecliptic was slightly different in value from the value it has now. It had, in fact, a value near to 23° 54′. This would, therefore, be the maximum northerly and southerly declination of the Sun at that era on midsummer's and midwinter's day respectively (for the northern hemisphere). The corresponding values of the midsummer and midwinter sunset azimuths indicate that on midsummer's day, from a pre-determined position near the stones, the Sun would be seen to set behind Ben Corra in the manner indicated in figure 8.17 and such that momentarily a small part of its upper edge would reappear further down the slope. Midwinter's day would be known to have arrived when the megalithic observers saw the Sun set behind Cara Island in the manner shown.

We do not know the exact procedure adopted by the megalithic astronomers but it is likely to have been based on the following method. As the Sun sets on the western horizon, the observer, by moving along the line at right angles to the Sun's direction, could insert a stake at the position he/she had to occupy in order to see the upper edge or limb of the Sun reappear momentarily behind the mountain slope indicated by the stone alignment (figure 8.17). This procedure would be repeated for several evenings until midsummer's day. On each occasion the stake would have to be moved further left or a fresh one put in, but by midsummer's eve, a limit would be reached because the Sun would be setting at its maximum declination north. On the evenings thereafter, the positioning of the pegs would be retraced. In this way it would be known when midsummer's day occurred. It may be noted in passing that with a distant foresight such as the mountain 30 km or so away, even a shift of 12 arc sec in the

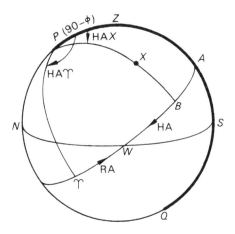

Figure 8.18. The celestial sphere illustrating local sidereal time.

Sun's setting the position will mean a stake-shift of the order of 2 m. This demonstrates the sensitivity of the method and the ingenuity of this ancient culture.

8.13   Sidereal time

Right ascension, together with declination, forms a coordinate system for stellar positions useful in constructing star catalogues, in contrast to the alt-azimuth and equatorial systems where one or both coordinates change rapidly with time.

The First Point of Aries, Υ, being a point on the stellar background, will rotate with the heavens like a star, transiting and rising and setting. We can, therefore, give a precise meaning to the phrase, 'the hour angle of Υ (HAΥ)'. It is the angle which the meridian through Υ makes with the observer's meridian, $\angle ZP\Upsilon$ in figure 8.18. It is also called the **local sidereal time** (LST). Hence,

$$\text{HA}\Upsilon = \text{LST}.$$

If X is the position of a star, its meridian PX meeting the equator at the point B, then we have:

$$\text{right ascension of } X = \text{arc } \Upsilon B$$
$$\text{hour angle of } X = \angle ZPX = \text{arc } AB.$$

But

$$A\Upsilon = \Upsilon B + BA$$

hence,

$$\text{hour angle of } X + \text{right ascension of } X = \text{local sidereal time}$$

or

$$\mathbf{HAX + RAX = LST}. \tag{8.7}$$

This is an important relationship for X can be any celestial object—star, Sun, Moon, planet, even an artificial satellite or spacecraft. If any two of the three quantities in equation (8.7) are known, the third can be calculated. We will consider the implications of this later when we look at sidereal time in more detail.

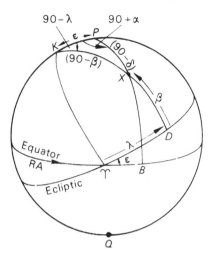

Figure 8.19. Ecliptic coordinates.

8.14 The ecliptic system of coordinates

This system is specially convenient in studying the movements of the planets and in describing the Solar System. The two quantities specifying the position of an object on the celestial sphere in this system are ecliptic longitude and ecliptic latitude. In figure 8.19 a great circle arc through the pole of the ecliptic K and the celestial object X meets the ecliptic in the point D. Then the **ecliptic longitude**, λ, is the angle between Υ and D, measured from $0°$ to $360°$ along the ecliptic in the eastwards direction, that is in the direction in which right ascension increases. The **ecliptic latitude**, β, is measured from D to X along the great circle arc DX, being measured from $0°$ to $90°$ north or south of the ecliptic. It should be noted that the north pole of the ecliptic, K, lies in the hemisphere containing the north celestial pole. It should also be noted that ecliptic latitude and longitude are often referred to as **celestial latitude** and **longitude**.

The point of intersection Aries (Υ) of the celestial equator and the ecliptic is often referred to as the **ascending node**, since an object travelling in the plane of the ecliptic with the direction of increasing right ascension (eastwards) passes through Aries from southern to northern declinations. By similar reasoning, Libra ($\simeq$) is called the **descending node**.

The origins most often used with this system of coordinates are the Earth's centre and the Sun's centre since most of the planets move in planes inclined only a few degrees to the ecliptic.

It is often required to convert from the ecliptic system to equatorial coordinates, i.e. the system of right ascension and declination or *vice versa*. This may be achieved by considering the spherical triangle KPX in figure 8.19, where $\angle KPX = 90° + \alpha$, α being the right ascension of X, or ΥB, while BX is the object's declination, δ.

Let us suppose α, δ are known, also the obliquity of the ecliptic, and it is required to calculate, λ, β. Then, using the cosine formula,

$$\cos(90 - \beta) = \cos\varepsilon\cos(90 - \delta) + \sin\varepsilon\sin(90 - \delta)\cos(90 + \alpha)$$

or

$$\sin\beta = \cos\varepsilon\sin\delta - \sin\varepsilon\cos\delta\sin\alpha. \tag{8.8}$$

Applying the cosine formula once more, we have

$$\cos(90 - \delta) = \cos\varepsilon\cos(90 - \beta) + \sin\varepsilon\sin(90 - \beta)\cos(90 - \lambda)$$

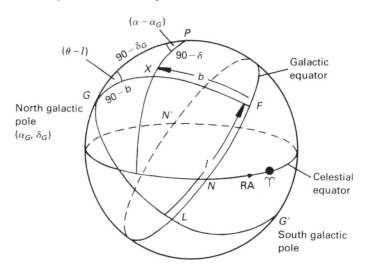

Figure 8.20. Galactic coordinates.

that is

$$\sin \delta = \cos \varepsilon \sin \beta + \sin \varepsilon \cos \beta \sin \lambda$$

or

$$\sin \lambda = \frac{\sin \delta - \cos \varepsilon \sin \beta}{\sin \varepsilon \cos \beta}. \tag{8.9}$$

Values for λ may be obtained directly from α, δ by substituting for β in equation (8.9), so providing a formula for the calculation of λ given by

$$\tan \lambda = \frac{\sin \alpha \cos \varepsilon + \tan \delta \sin \varepsilon}{\cos \alpha}. \tag{8.10}$$

The quadrant associated with λ can be elucidated by noting the signs of the numerator and denominator in either of the equations (8.9) or (8.10).

Alternatively, these identities could have been derived using the the four-parts formula. The solution of the problem in reverse (given λ, β, ε; find α, δ) is left to the reader.

8.15 Galactic coordinates

In the same way that it is convenient to use ecliptic coordinates in problems dealing with motions of planets about the Sun, it is often convenient in studies of the distribution and movements of the bodies in the stellar system to which the Sun belongs, to use a coordinate system based on the observational fact that the Galaxy is lens-shaped with the Sun in or near to the median plane of the lens.

The fact that the Milky Way is a band of light occupying a great circle supports this view.

We can take the Galaxy as being symmetrically distributed on either side of the galactic equator LNF (figure 8.20) which intersects the celestial equator in the two points N and N'. These points are referred to as the **ascending** and **descending nodes** respectively, since an object travelling in the galactic equator and passing through N in the direction of increasing RA, ascends from the southern to the northern hemisphere. In passing through N', it descends from the northern to southern hemisphere. Similarly, the north galactic pole G is the galactic pole lying in the northern hemisphere.

Any object X (RA $= \alpha$, Dec $= \delta$) has galactic coordinates in latitude and longitude.

Table 8.1. Position of the Galactic Pole.

IAU galactic pole (N) ($b^{II} = +90°$)	
$\alpha = 12^h 51^m 4$	$\delta = +27° 07'\!.7$ (Epoch 2000)
$\alpha = 12^h 49^m 0$	$\delta = +27° 24'\!.0$ (Epoch 1950)
$\alpha = 12^h 46^m 6$	$\delta = +27° 40'\!.0$ (Epoch 1900)
Ohlsson galactic pole (N) ($b^I = +90°$)	
$\alpha = 12^h 40^m$	$\delta = +28°\!.0$ (Epoch 1900)

Galactic longitude, l, is measured along the galactic equator to the foot of the meridian from G through the object from 0° to 360° in the direction of increasing right ascension. Prior to 1959, the zero was the ascending node N (the *Ohlsson System*); since 1959, it is L, the point where the semi-great circle from G at the position angle $\theta = PGL = 123°$ ($= 90° + 33°$) meets the galactic equator. This seemingly arbitrary angle is taken so that L lies in the direction of the galactic centre. Thus, the galactic longitude l of X (figure 8.20) is $\angle LNF$ and $\angle PGX$ is ($\theta - l$).

Galactic latitude, b, is measured north and south of the galactic equator from 0° to 90° along with the semi-great circle from the north galactic pole through the object to the equator. Thus, the galactic latitude b of X is arc FX and is north.

To differentiate between the old (Ohlsson) and newer (IAU) systems of galactic coordinates, it is usual to label l and b with superscripts I and II respectively. Table 8.1 summarizes the position of the north galactic pole in the two systems, the changes with time of the IAU system being caused by precession of the equinoxes (see section 11.9).

A typical conversion problem is to find the galactic longitude l and galactic latitude b of an object X of known right ascension α and declination δ, given the coordinates of the north galactic pole $G(\alpha_G, \delta_G)$ and the position angle θ.

Spherical triangle PGX shows that as before the cosine formula, applied twice, can obtain the desired quantities. Thus,

$$\cos(90 - b) = \cos(90 - \delta_G)\cos(90 - \delta) + \sin(90 - \delta_G)\sin(90 - \delta)\cos(\alpha - \alpha_G)$$

or

$$\sin b = \sin \delta_G \sin \delta + \cos \delta_G \cos \delta \cos(\alpha - \alpha_G) \tag{8.11}$$

giving b. Also,

$$\cos(90 - \delta) = \cos(90 - \delta_G)\cos(90 - b) + \sin(90 - \delta_G)\sin(90 - b)\cos(\theta - l)$$

that is

$$\sin \delta = \sin \delta_G \sin b + \cos \delta_G \cos b \cos(\theta - l)$$

or, rearranging,

$$\cos(\theta - l) = \frac{\sin \delta - \sin \delta_G \sin b}{\cos \delta_G \cos b}. \tag{8.12}$$

Again, as in the case of the conversion of α, δ to ecliptic coordinates, values for $\sin b$ and $\cos b$ taken from equation (8.11) may be substituted into equation (8.12) leading to the identity

$$\tan(\theta - l) = \frac{\tan \delta \cos \delta_G - \cos(\alpha - \alpha_G)\sin \delta_G}{\sin(\alpha - \alpha_G)}. \tag{8.13}$$

Knowing the value of θ, we can calculate l.

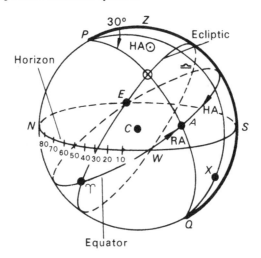

Figure 8.21. Example 8.4—the celestial sphere for an observer in latitude 60° N at LST 9^h on June 21st.

We now consider some celestial sphere problems embodying the concepts of the last few sections in this chapter. Since the problems require estimations only, no calculation by spherical trigonometrical formulas is required. But by drawing a fair-sized celestial sphere (at least 100 mm in diameter), by remembering that foreshortening occurs in such drawings (see the 10° marks on arc NW in figure 8.21), and allowing for it in estimating angles and arcs, by remembering that all great circles are foreshortened into ellipses containing the centre of the sphere, values within a few degrees of the correct answers may easily be obtained. The very useful convention of using dotted lines for arcs on the back of the sphere should also be adhered to and the use of coloured pencils to distinguish one great circle from another is advised.

Example 8.4. Draw the celestial sphere for an observer in latitude 60° N at local sidereal time 9^h on June 21st. Put in the horizon, the equator, the four cardinal points, the observer's meridian, the ecliptic, the Sun and the position of a star X (right ascension 8^h, declination 50° S). What is the estimated hour angle of the Sun?

Draw the sphere. Insert the zenith, Z, and the horizon. The latitude $\phi = 60°$ N, hence $90 - \phi = 30°$ and P can be placed so that $PZ = 30°$. Hence, Q and the equator may be put in. Thicken PZQ to indicate the observer's meridian. N is the point of intersection with the horizon of the vertical from Z through P. Now S, W and E can be put in the diagram, remembering that when facing north, W is on the observer's left-hand side. Put arrowheads in with RA and HA beside them on the equator to indicate the directions in which right ascension and hour angle are measured.

Insert ♈, using the fact that

$$HA♈ = LST = 9^h.$$

Insert ♎, the other point of intersection of the ecliptic with the equator. It must be on the other side of the sphere through the centre C.

Remember

(i) that the obliquity of the ecliptic is ∼23$\frac{1}{2}$° and
(ii) that the Sun passes through ♈ from south declination to north declination in the direction of increasing right ascension.

Hence, insert the ecliptic.

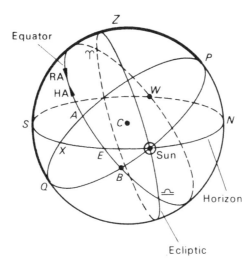

Figure 8.22. Example 8.5—the celestial sphere for latitude 30° N at the time of rising of *Sirius*.

Insert the Sun, ☉, knowing that on June 21st the Sun's right ascension is about 6^h, that is $\Upsilon A \equiv 90°$.

Insert the star X according to the information given.

From figure 8.21, the hour angle of the Sun (HA☉) is estimated to be about 3^h.

In fact, since RA☉ $= 6^h$ and LST $= 9^h$, we could have obtained the HA☉ from the relation

$$\text{LST} = \text{HA}☉ + \text{RA}☉$$

or

$$\text{HA}☉ = \text{LST} - \text{RA}☉ = 9^h - 6^h = 3^h.$$

Example 8.5. Draw the celestial sphere for latitude 30° N, showing the star *Sirius* (right ascension $6^h\,40^m$, declination 17° S) at rising and draw the ecliptic. Estimate from your diagram the approximate date when *Sirius* rises with the Sun.

Again a sphere is drawn (see figure 8.22), the zenith and horizon are inserted and we note that ZP this time is 60°. Because we are considering the rising of a star and of the Sun, it is more convenient to have the east point E on the front of the diagram. This dictates the position of the north point N so that the west point is on the observer's left-hand side when facing north. W and S are now inserted. The north celestial pole can now be put in between N and Z and such that $NP = 30°$. Q and the equator are drawn in, the observer's meridian PZQ is indicated and arrowheads with RA and HA are added to the equator, their directions being fixed by remembering that hour angle is always measured from the observer's meridian westwards.

Sirius is rising, therefore it must be on the horizon: it has declination 17° S so it must lie on a parallel of declination set 17° from the equator in the southern hemisphere. Hence, *Sirius* is at X, where $AX = 17°$.

We now insert the ecliptic. Υ is found by noting that it must be on the equator at a point such that the right ascension of X, ΥA, is $6\frac{2}{3}^h$. Having fixed Υ to accord with this, ♎ is then put in. Using the convention illustrated by figure 8.23 and remembering that the value of the obliquity is about $23\frac{1}{2}°$, we draw in the ecliptic.

The Sun must be at ☉ since it is rising and always lies on the ecliptic.

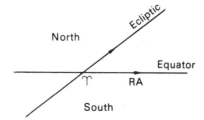

Figure 8.23. Example 8.5—the convention of the ascending node of the ecliptic (♈) as viewed from the outside of the celestial sphere.

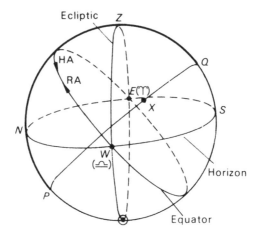

Figure 8.24. Example 8.6—the celestial sphere for latitude 30° S showing the ecliptic.

Its right ascension, ♈B, is, therefore, about $8\frac{2}{3}^{\text{h}}$. But its right ascension increases by 24^{h} in 12 months or 2^{h} per month.

The Sun's right ascension being 6^{h} about June 21st, it will have increased a further $2\frac{2}{3}$ hours in 1 month 10 days after June 21st, that is about August 1st.

Hence, the approximate date when *Sirius* rises with the Sun for the given latitude is August 1st.

Example 8.6. Draw the celestial sphere for an observer in latitude 30° S, showing the Sun, the ecliptic and the First Point of Aries at apparent midnight on June 21st. Estimate the local sidereal time. Show also the position of a star of right ascension 13^{h} and declination 30° S.

First we draw the sphere and insert zenith Z and the horizon. Inserting the four cardinal points N, E, S, W, we place the south celestial pole Q at an altitude of 30° above the south point since, by definition, the great circle from Z through Q cuts the horizon in the south point.

We place the north celestial pole P in the diagram, insert the equator and the observer's meridian QZP. Arrowheads are put in with HA and RA beside them to indicate the directions in which hour angle and right ascension are measured, remembering that hour angle is always measured from the observer's meridian westwards.

It is June 21st so that the Sun's declination is 23° 26′ N; it is apparent midnight, i.e. HA⊙ = 12^{h}. Hence, in figure 8.24, the Sun's position is given by ⊙.

The Sun's right ascension on June 21st is $6^{\text{h}} \equiv 90°$. Hence, the ecliptic must pass through ⊙ and intersect the equator 90° before and behind ⊙ in right ascension. The only possible points are W and

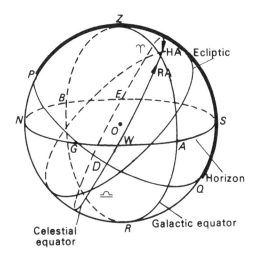

Figure 8.25. Example 8.7—the celestial for latitude 28° N displaying galactic coordinates.

E. Draw in the ecliptic. By making use of the definition embodied in figure 8.23, we can see that E is ♈ and W is ♎.

$$\text{The local sidereal time} = \text{HA}♈ = 18^{\text{h}}.$$

The star's position X is then inserted according to the information given.

Example 8.7. Draw the celestial sphere for latitude 28° N, and insert the north galactic pole G (declination 28° N) when it is setting. Insert the galactic equator at this instant. Estimate the hour angle of G at setting and state the local sidereal time, the right ascension of G being $12^{\text{h}} 47^{\text{m}}$. Insert the First Point of Aries in the diagram and draw the ecliptic.

Show that, once in each sidereal day, the galactic equator and the horizon coincide in this latitude. At what sidereal time does this occur? At what approximate date would this occur at midnight?

As usual we put the zenith Z and the horizon into the sphere. The problem deals with setting of a celestial object, therefore we wish the west point in W on the front of the diagram (figure 8.25). This dictates the positioning of the other cardinal points N, E and S. The altitude of the pole being the latitude of the observer, P is placed 28° above N and then the horizon, the south celestial pole Q, the observer's meridian and the arrowheads for HA and RA can be inserted.

The galactic pole G is setting, so is placed on the horizon such that GD is 28° N. In drawing the galactic equator at this moment we remember that G is its pole, i.e. all points on the galactic equator are 90° from G. We choose A and B to be two such points; Z and the nadir R must also be 90° from G for they are the poles of the great circle on which G lies, namely the horizon. Hence, the galactic equator can now be drawn in, passing through the zenith, Z.

$$\text{The hour angle of } G = \angle ZPD \approx 7^{\text{h}}.$$
$$\text{LST} = \text{RA}G + \text{HA}G = 12^{\text{h}} 47^{\text{m}} + 7^{\text{h}} = 19^{\text{h}} 47^{\text{m}} \text{ approximately.}$$

♈ is placed so that the arc ♈$ED = 12^{\text{h}} 47^{\text{m}}$; Libra (♎) is then inserted and, using figure 8.23 and the obliquity value of $23\frac{1}{2}°$, the ecliptic can be put in.

Since G in the course of a **sidereal day** (the time between two successive passages of ♈ across the observer's meridian) follows the parallel of declination 28° N, and $PZ = 90 - \phi = 62°$, Z lies on his parallel. Hence, once per sidereal day, G coincides with Z, i.e. the pole of the galactic equator coincides with the pole of the horizon. By definition, then, the galactic equator and horizon must coincide at this time.

The hour angle of G at which this occurs is zero. But the RAG is $12^h 47^m$. Hence,

$$\text{LST} = \text{RA}G + \text{HA}G = 12^h 47^m + 0^h = 12^h 47^m.$$

To occur at midnight, the hour angle of the Sun would be 12^h. But

$$\text{RA}\odot = \text{LST} - \text{HA}\odot$$
$$= 12^h 47^m - 12^h = 47^m.$$

The Sun's right ascension increases by 24^h in 12 months or 2^h per month. It, therefore, increases from 0^h (March 21st) to $47^m \sim \frac{3}{4}^h$ in approximately 11^d. The date is, therefore, around April 1st.

Problems—Chapter 8

Note: In the following problems, assume (i) a spherical Earth, (ii) the obliquity of the ecliptic to be $23° 26'$. Although problems 5 to 16 require the use of spherical trigonometry, the student without trigonometry will find it instructive to try to solve them approximately by drawing the relevant celestial spheres. The student's estimates can be compared with the answers given in the Answer Appendix.

1. Draw the celestial sphere for an observer in latitude 45° S, putting in the observer's meridian, the four cardinal points and a star of azimuth 300° E of S and altitude 30°. *Estimate* the star's right ascension and declination if the local sidereal time at that instant is 9^h. Insert the ecliptic. If it is also apparent midnight, *estimate* the date. On what date (approximately) will the star set when the Sun sets?

2. Draw the celestial sphere for an observer in latitude 55° S, showing the positions of two stars X (altitude 40°, azimuth 130° E of S) and Y (HA = 19^h, Dec = 40° S). *Estimate* from your diagram the hour angle and declination of X and the altitude and azimuth of Y.

    If the local sidereal time is 10^h, sketch the ecliptic in your diagram; *estimate* the celestial longitudes and latitudes of the two stars and *estimate* the approximate date on the assumption that the Sun is rising at this moment.

3. Draw the celestial sphere for an observer in latitude 30° N, putting in the horizon, equator, zenith, north and south celestial poles and the observer's meridian.

    Show the positions of two stars X and Y as follows:

    $$X : \text{hour angle } 3^h; \text{ declination } 64° \text{ N}$$
    $$Y : \text{azimuth } 120° \text{ W}; \text{ altitude } 20°.$$

    From the diagram *estimate* (i) the azimuth and altitude of X, (ii) the hour angle and declination of Y. If the right ascension of X is 6^h, insert the ecliptic.

    A traveller states that when he/she was in that latitude the Sun passed through his/her zenith. Give a reason for believing or disbelieving him/her. Show that star X is a circumpolar star.

4. Draw the celestial sphere for an observer in latitude $23° 26'$ N, inserting the horizon, equator, zenith, north and south celestial poles and the observer's meridian. Insert the First Point of Aries and the ecliptic when the Sun is rising on June 21st. Put in the Sun's position when it transits on that day (i.e. at apparent noon). *Estimate* the local sidereal time at apparent noon, also the altitude and azimuth at that time of a star X whose right ascension and declination are 10^h and 65° N respectively. Is the star circumpolar?

5. Show that the celestial longitude λ and latitude β of a star can be expressed in terms of its right ascension α and declination δ by the formulas

    $$\sin \beta = \sin \delta \cos \varepsilon - \cos \delta \sin \varepsilon \sin \alpha$$
    $$\cos \beta \cos \lambda = \cos \delta \cos \alpha$$
    $$\cos \beta \sin \lambda = \sin \lambda \sin \varepsilon + \cos \delta \cos \varepsilon \sin \alpha$$

where ε is the obliquity of the ecliptic.

If $\alpha = 6^{\mathrm{h}}$, $\delta = 45°$ N, calculate λ and β.

6. When the vernal equinox rises in azimuth 90° E of N, find the angle the ecliptic makes with the horizon for an observer in latitude 60° N.

7. Show that the point on the horizon at which a star rises is

$$\sin^{-1}(\sec \phi \sin \delta)$$

north of east where ϕ is the observer's latitude and δ is the declination of the star.

8. *Calculate* the azimuths of the star *Procyon* (declination $= 5°$ N) when its zenith distance is 80° as seen by an observer in latitude 56° N.

9. Show that the ecliptic longitude λ of a star of right ascension α and declination δ is given by

$$\tan \lambda = \sin \varepsilon \tan \delta \sec \alpha + \cos \varepsilon \tan \alpha$$

where ε is the obliquity of the ecliptic.

The pole of the galactic equator has coordinates RA $= 12^{\mathrm{h}} 51^{\mathrm{m}}$, Dec $= 27° 08'$ N. *Calculate* its ecliptic longitude and latitude, also the dates when the Sun crosses the galactic equator assuming the Sun to move in a circular orbit about the Earth.

10. In north latitude 45° the greatest azimuth (east or west) of a circumpolar star is 45°. Prove that the star's declination is 60° N.

11. *Calculate* the hour angle of *Vega* (declination 38° 44' N) when on the prime vertical west in latitude 50° N. For what latitudes is *Vega* circumpolar?

12. Geomagnetic meridians are defined similarly to geographic meridians but with respect to the geomagnetic pole (79° N, $4^{\mathrm{h}} 40^{\mathrm{m}}$ W). *Calculate* the angle between the geomagnetic and geographical meridians at Glasgow (56° N, $0^{\mathrm{h}} 17^{\mathrm{m}}$ W).

13. Given that H_1, H_2 are the hour angles of a star of declination δ on the prime vertical west and at setting respectively for an observer in north latitude, show that

$$\cos H_1 \cos H_2 + \tan^2 \delta = 0.$$

14. Given that the ecliptic longitude of the Sun is 49° 49', *calculate* the hour angle of setting of the Sun and the local sidereal time at the instant of setting for an observer in latitude 54° 55' N.

15. *Calculate* the azimuth of the Sun at rising on midsummer's day at Stonehenge (latitude 51° 10' N) at a time when the obliquity of the ecliptic was 23° 48'.

16. The zenith distances of a circumpolar star at upper and lower culmination are found to be 13° 7' 16'' and 47° 18' 26''.

Calculate the latitude of the observer and the declination of the star.

Chapter 9

The celestial sphere: timekeeping systems

9.1 Introduction

Primitive people based their sense of the passage of time on the growth of hunger or thirst and on impersonal phenomena such as the changing altitude of the Sun during a day, the successive phases of the Moon and the changing seasons. By about 2000 BC civilizations kept records and systematized the impersonal phenomena into the day, the month and the year. We have seen that emphasis was given to the year as a unit of time by their observation that the Sun made one revolution of the stellar background in that period of time.

Since everyday life is geared to daylight, the Sun became the body to which the system of timekeeping used by day was bound. The apparent solar day was then the time between successive passages of the Sun over the observer's meridian or the time during which the Sun's hour angle increased by 24^h ($360°$). In a practical way the Sun was noted to be on the meridian when the shadow cast by a vertical pillar was shortest.

However, the apparent diurnal rotation of the heavens provided another system of timekeeping called sidereal time, based on the rotation of the Earth on its axis. The interval between two successive passages of a star across the observer's meridian was then called a **sidereal day**. Early on in the history of astronomy it was realized that the difference between the two systems of timekeeping—solar and stellar—was caused by the orbital motion of the Sun relative to Earth. Thus, in figure 9.1, if two successive passages of the star over the observer's meridian define a sidereal day (the star being taken to be at an infinite distance effectively from the Earth), the Earth will have rotated the observer O through $360°$ from O_1 to O_2. In order that one apparent solar day will have elapsed, however, the Earth will have to rotate until the observer is at O_3 when the Sun will again be on the meridian. Since the Earth's radius vector SE sweeps out about $1°$ per day and the Earth rotates at an angular velocity of about $1°$ per 4 minutes, the sidereal day is consequently about 4 minutes shorter than the average solar day.

We will now consider these systems in greater detail.

9.2 Sidereal time

We have seen that the First Point of Aries (Vernal Equinox ♈) is the reference point chosen on the rotating celestial sphere to define the sidereal day. The time between successive passages of the vernal equinox across the observer's meridian is **one sidereal day**. So the hour angle of the vernal equinox increases from 0^h to 24^h and the **local sidereal time** (LST) is defined as the hour angle of the vernal equinox (HA♈). The LST, as its name implies, depends upon the observer's longitude on the Earth's surface.

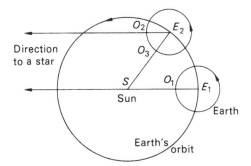

Figure 9.1. The time interval between successive transits of a star.

We had by equation (8.7)

$$\text{LST} = \text{HA}X + \text{RA}X \tag{9.1}$$

where X was any object.

If the LST is known and the right ascension α and declination δ of the object have been computed for that time, then the hour angle H and δ are known, giving the object's direction on the celestial sphere.

In an observatory there is usually a master clock keeping the local sidereal time of that longitude. Since the hour angle of a star is zero when it transits on the observer's meridian, the star's right ascension at that time is the LST. In the past, a careful check on the clock error—no clock is perfect—and the rate of change of the error was then made by observing frequently the sidereal times of transit of well-known stars and comparing them with their right ascensions. Such stars were called 'clock stars' and these observations were once part of the routine work at any observatory. Nowadays, the LST is usually generated by running a small computer which receives a time-embedded radio signal, the program set according to the longitude of the observatory.

It should be noted that the position of the vernal equinox, Υ, amid the stars on the celestial sphere changes because of precession and nutation (see chapter 11). The gravitational effects of the Sun and Moon cause continuous small changes to the obliquity of the ecliptic. The positions of the equinoxes, Υ and $\simeq$, the intersections of the ecliptic with the equator, also alter, with the equinoxes moving backwards with time, the steady mean backwards rate being termed **general precession**. In addition, nutation adds variations to this rate. If the nutational variations are neglected, the position of Υ at any time defines the Greenwich Mean Sidereal Time (GMST). If the nutational variations are taken into account, the position of Υ at that time is the actual position and defines Greenwich Apparent Sidereal Time (GAST). The value of the difference between the two times is referred to as the equation of the equinoxes, $\mathcal{E}_\Upsilon$, and is expressed as

$$\mathcal{E}_\Upsilon = \text{GAST} - \text{GMST}.$$

Daily values for both GAST and GMST are tabulated in *The Astronomical Almanac*. In 2000, for example, the value of $\mathcal{E}_\Upsilon$ fluctuated about a value of -1^s. In calculating the hour angle of any object at a given location and time, or the LST, the tabulated values of GAST should be used.

The time between transits of a celestial object over the Greenwich meridian and the local observer's meridian is equal to the longitude of the local observer. This is easily seen from figure 9.2 where the geocentric celestial sphere, north celestial pole P, is shown with the Earth North Pole, p. Greenwich, g, is shown and its zenith G, the meridian through G, namely PGB, being the Greenwich observer's meridian. An observer in longitude λW is indicated by O with the zenith and observer's meridian given by Z and PZA. The vernal equinox is shown as Υ as usual and a celestial object is seen transiting at X.

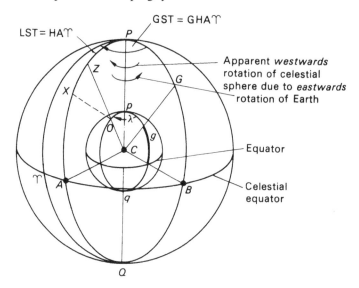

Figure 9.2. The relation between local sidereal time and Greenwich sidereal time.

Then the Greenwich hour angle of X is $\angle GPX$ which is the longitude λW of the observer.

The Greenwich hour angle of the vernal equinox is $\angle GP\Upsilon$ which is equal to the hour angle $\angle XP\Upsilon$ plus the longitude λW of the observer.

Or

$$\text{GHA}\Upsilon = \text{HA}\Upsilon + \lambda\text{W}.$$

But the hour angle of Υ is the LST, therefore we may write

$$\text{GAST} = \text{LST} + \lambda\text{W}.$$

If Υ were any celestial object $\star$, we would have

$$\text{GHA}\star = \text{HA}\star + \lambda\text{W}.$$

It is easily seen that if the longitude of the local observer had been east we would have written

$$\text{GHA}\star = \text{HA}\star - \lambda\text{E}.$$

If we agree that a western longitude is positive and eastern is negative, we may combine these relations to give

$$\text{GHA}\star = \text{HA}\star + \lambda \tag{9.2}$$

a very important expression often remembered by the mnemonic:

Longitude east, Greenwich least,
Longitude west, Greenwich 'best'.

The sidereal day at a given place begins when the vernal equinox Υ transits across the observer's meridian and ends 24 sidereal hours later when Υ is once more on the meridian. Thus, one sidereal day is the time taken for one rotation of the Earth about its axis with respect to the stellar background. The sidereal clock in an observatory should, therefore, read 0^h when Υ is on the meridian. We have the following table:

1 sidereal day $=$ 24 sidereal hours

1 sidereal hour $=$ 60 sidereal minutes

1 sidereal minute $=$ 60 sidereal seconds.

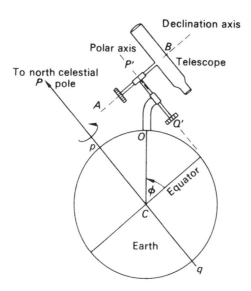

Figure 9.3. An equatorial telescope mounting.

The right ascensions and declinations of celestial objects are catalogued. This information, together with the knowledge of the LST, enables an equatorially-mounted telescope to be pointed at an object of interest, even if it is invisible to the unaided eye. Anticipating section 22.2, it may be said that such a telescope is free to rotate about an axis parallel to the Earth's rotational axis ($P'Q'$ in figure 9.3) and also about an axis perpendicular to it (AB in figure 9.3).

A graduated circle at A on the declination axis enables the telescope to be swung about this axis and then clamped so that it sweeps along the correct parallel of declination for the object of interest when the telescope is rotated about the other axis $P'Q'$, known as the **polar axis**. The object will be in the telescope's field of view when the other graduated circle at Q' is set to the value of the object's hour angle at that time.

The correct angle is obtained from a knowledge of the local sidereal time and the object's right ascension. Thus, using equation (9.1),

$$HA\star = LST - RA\star.$$

The telescope drive (a motor that turns the telescope about the polar axis) is then locked in to compensate for the apparent rotation of the heavens due to the Earth's rotation about its axis pq. In this way, the object, once caught in the field of view of the telescope, is kept there, for hours if necessary.

Example 9.1. A star has right ascension $9^h\,46^m\,12\overset{s}{.}7$. When it transits on two consecutive nights the times on the clock are $9^h\,46^m\,21\overset{s}{.}0$ and $9^h\,46^m\,18\overset{s}{.}6$. Calculate the clock error and the rate.

The problem is best attempted by setting out the data in tabular form.

Let RA$\star$ denote the star's right ascension. Now by equation (9.1),

$$LST = HA\star + RA\star.$$

But at transit, HA$\star = 0$, hence LST $=$ RA$\star$. Hence, by defining the error as 'clock time $-$ LST', we have:

LST			Clock time			Error	Rate
h	m	s	h	m	s		
9	46	12·7	9	46	21·0	8^s3 fast	-2^s4 per
9	46	12·7	9	46	18·6	5^s9 fast	sidereal day

The clock loses 2^s4 each sidereal day.

Example 9.2. Calculate the hour angle of a star with RA $18^h 24^m 42^s$, given that the LST is $4^h 13^m 22^s$. Now by equation (9.1),

$$HA\star = LST - RA\star.$$

Hence, we may write:

	h	m	s
LST	4	13	22
Add 24^h to give	28	13	22
RA$\star$	18	24	42
HA$\star$	9	48	40

Note: since the RA$\star$ is greater than the LST, we have added 24^h to the LST before subtracting the RA$\star$.

9.3 Mean solar time

One of the methods by which the ancients measured the passage of time was by means of the Sun's diurnal movement across the sky. Sundials were used to do this and the timekeeping system involved is referred to as **apparent solar time**. One **apparent solar day** was the time between successive passages of the Sun across the observer's meridian (the shadow of the sundial gnomon was shortest then and lay on a north–south line). An hour of apparent solar time was the interval of time it took the Sun's hour angle to increase by one hour; at apparent noon the Sun reached the observer's meridian while apparent midnight corresponded to a time when the Sun's hour angle had a value of 12 hours. Ordinary daily activities are so linked with the Sun's position in the sky that it was natural that civil timekeeping should be based on the solar movement.

Unfortunately there is one major problem in using apparent solar time.

If the length of the apparent solar day is measured by an accurate sidereal clock, it is found to vary throughout the year. There are two main reasons for this:

(i) The Sun's apparent orbit about the Earth is an ellipse in which equal angles are not swept out by the line joining the Sun to Earth (the radius vector) in equal times. (In fact, Hipparchus knew that the two halves of the year from equinox to equinox differed by almost 8 days in length—see table 9.3, page 104.)

(ii) The path of the Sun is in the ecliptic which is inclined at an angle of $23\frac{1}{2}°$ approximately to the equator along which the Sun's hour angle is measured.

Astronomers overcome this problem to obtain **mean solar time** by introducing a fictitious body called the **mean sun**. The mean sun moves along the *equator* in the direction of increasing right ascension at a constant angular velocity which is equal to the mean angular velocity of the real Sun in the ecliptic (about 1 degree per day) during its yearly journey round this great circle. Since the mean sun increases its right ascension at a constant rate of about 1 degree per day and the vernal equinox (from which right ascension is measured) increases its hour angle at a constant rate of 24^h per sidereal

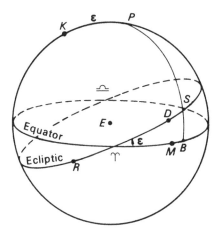

Figure 9.4. Positions of the Sun S, the mean sun M, the dynamical mean sun D and the direction of perigee R.

day, the time between successive passages of the mean sun over the observer's meridian is constant. This time interval is called a **mean solar day**.

To define the relationship between the positions of the Sun and mean sun at any instant, we have to consider the concept in more detail. The problem is to obtain a body related to the Sun but which moves so that its right ascension increases at a constant rate. It is solved in two stages corresponding to the two reasons (i) and (ii) given earlier which produce the irregularities in apparent solar time.

Stage (i). The Sun is said to be at **perigee** when it is nearest to the Earth. This occurs once per year about January 1st. A fictitious body called the **dynamical mean sun** is introduced which starts off from the perigee direction with the Sun, moves along the ecliptic with the mean angular velocity of the Sun and consequently returns to the perigee at the same time as the Sun.

Stage (ii). When this dynamical mean sun, moving in the ecliptic, reaches the vernal equinox Υ, the **mean sun** starts off along the equator and, as we have seen, moves in that great circle with the Sun's mean angular velocity, returning to Υ at the same time as the dynamical mean sun.

Figure 9.4 shows the positions at some epoch of the Sun S, the mean sun M, the dynamical mean sun D and the direction of perigee R on the geocentric celestial sphere.

The meridian from the north celestial pole P through the Sun meets the equator at B so that the arc ΥB is the Sun's right ascension (RA$\odot$). The right ascension of the mean sun (RAMS) is the arc ΥM.

The quantity BM is referred to as the **equation of time** ($\mathcal{E}$) and defined by the relation

$$\mathcal{E} = \text{RAMS} - \text{RA}\odot. \tag{9.3}$$

The equation of time varies from a value of about $-14\frac{1}{4}^{\text{m}}$ to about $+16\frac{1}{4}^{\text{m}}$ throughout the year as the mean sun and the Sun get ahead of each other in their movement round the equator, as shown in figure 9.5.

Many sundials carry a table of values of the equation of time at various dates throughout the year, enabling the observer to deduce civil time from the apparent solar time given by the sundial.

The mean solar day is divided into 24 mean solar hours, each hour into 60 mean solar minutes, each minute into 60 mean solar seconds. When the mean sun is on the meridian (HAMS $= 0^{\text{h}}$) it is said to be **mean noon**. When the HAMS $= 12^{\text{h}}$, it is **mean midnight**.

We note that

$$\text{LST} = \text{HA}\star + \text{RA}\star.$$

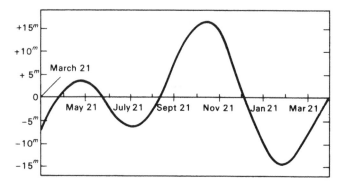

Figure 9.5. Values of the equation of time throughout the year.

If $\star$ denotes the Sun ($\odot$) and the mean sun (MS), in turn we have

$$\text{LST} = \text{HA}\odot + \text{RA}\odot = \text{HAMS} + \text{RAMS}. \tag{9.4}$$

Hence, by equations (9.3) and (9.4), we can write

$$\mathcal{E} = \text{HA}\odot - \text{HAMS}. \tag{9.5}$$

In the special case of the Greenwich meridian, we have Greenwich Mean Time (GMT), more often referred to as Universal Time (UT) such that

$$\text{UT} \equiv \text{GMT} = \text{GHAMS} \pm 12^{\text{h}} = \text{GHA}\odot - \mathcal{E} + 12^{\text{h}}. \tag{9.6}$$

If the UT of transit of the Sun at Greenwich (UT$\odot$G) is introduced, we can rewrite equation (9.6):

$$\text{UT}\odot\text{G} = 12^{\text{h}} - \mathcal{E} \tag{9.7}$$

since at transit, GHA$\odot$ = 0.

With respect to the timing of the Sun's transit, a term called the **ephemeris transit** is used. This quantity is tabulated for each day of the year in *The Astronomical Almanac*. It is the dynamical time of transit of the Sun across the **ephemeris meridian**, a terrestrial meridian $1 \cdot 002\,738 \Delta T$ east of Greenwich (for a discussion of the nature of dynamical time, see section 9.9).

It should be noted that this quantity is not exactly equal to the Universal Time of the Sun's transit at Greenwich because (a) it is a different meridian and (b) Universal Time and Terrestrial Dynamic Time, TDT (formerly Ephemeris Time), have different time scales.

Now ΔT itself is defined by the relation

$$\Delta T = \text{Terrestrial Dynamic Time} - \text{Universal Time}. \tag{9.8}$$

The Universal Time of transit of the Sun across the ephemeris meridian (UT$\odot$E) is then given by

$$\text{UT}\odot\text{E} = \text{TDT} - \Delta T.$$

Hence, the Universal Time of transit of the Sun across the Greenwich meridian (UT$\odot$G) will be given by

$$\text{UT}\odot\text{G} = \text{TDT} - \Delta T + 1 \cdot 002\,738 \Delta T$$

or

$$\text{UT}\odot\text{G} = \text{TDT} - 0 \cdot 002\,738 \Delta T.$$

For navigational purposes based on solar positional measurements, it is, therefore, accurate enough to rewrite equation (9.7) as

$$\text{Ephemeris transit} = 12^{\text{h}} - \mathcal{E} \tag{9.9}$$

since the equation of time is a slowly varying quantity and ΔT is only of the order of 66 seconds at present (2000).

Hence, by equation (9.6),

$$\text{GHA}\odot = \text{GHAMS} + 12^{\text{h}} - \text{Ephemeris transit}. \tag{9.10}$$

9.4 The relationship between mean solar time and sidereal time

We have seen that the mean solar day is about 4 minutes longer than the sidereal day. A more precise relationship will now be developed.

Equation (9.4) gives

$$\text{LST} = \text{HAMS} + \text{RAMS}. \tag{9.11}$$

Let the values of these quantities be T_1, H_1 and R_1 at a particular epoch and T_2, H_2 and R_2 one mean solar day later.

Then by equation (9.11), we have

$$T_1 = H_1 + R_1$$
$$T_2 = H_2 + R_2.$$

Subtracting, we obtain

$$T_2 - T_1 = (H_2 - H_1) + (R_2 - R_1). \tag{9.12}$$

But $H_2 - H_1 = 24^{\text{h}}$, since 1 mean solar day has elapsed.

The Sun's mean angular rate is n, where

$$n = 360°/365\tfrac{1}{4} \text{ days}$$

or

$$n = 24^{\text{h}}/365\tfrac{1}{4} \text{ days}$$

so that the increase in the mean sun's right ascension in 1 mean solar day is $24^{\text{h}}/365\tfrac{1}{4}$. This must be the quantity $(R_2 - R_1)$. Hence, equation (9.12) becomes

$$T_2 - T_1 = 24^{\text{h}} + 24^{\text{h}}/365\tfrac{1}{4}$$
$$= 24^{\text{h}}(1 + 1/365\tfrac{1}{4})$$
$$= 24^{\text{h}}(366\tfrac{1}{4}/365\tfrac{1}{4}) \text{ of sidereal time.}$$

This interval of sidereal time is equal to 24 hours of mean solar time so that we obtain the relation:

$$24^{\text{h}} \text{ mean solar time} = 24^{\text{h}} \times \frac{366\tfrac{1}{4}}{365\tfrac{1}{4}} \text{ sidereal time.}$$

Alternatively,

$$24^{\text{h}} \text{ sidereal time} = 24^{\text{h}} \times \frac{365\tfrac{1}{4}}{366\tfrac{1}{4}} \text{ mean solar time.}$$

Table 9.1. Conversion of mean solar time into sidereal time.

24^h mean solar time	$\equiv$	$(24^h + 3^m 56\overset{s}{.}556)$	sidereal time
1^h mean solar time	$\equiv$	$(1^h + 9\overset{s}{.}8565)$	sidereal time
1^m mean solar time	$\equiv$	$(1^m + 0\overset{s}{.}1643)$	sidereal time
1^s mean solar time	$\equiv$	$(1^s + 0\overset{s}{.}0027)$	sidereal time

Table 9.2. Conversion of sidereal time into mean solar time.

24^h sidereal time	$\equiv$	$(24^h - 3^m 55\overset{s}{.}910)$	mean solar time
1^h sidereal time	$\equiv$	$(1^h - 9\overset{s}{.}8296)$	mean solar time
1^m sidereal time	$\equiv$	$(1^m - 0\overset{s}{.}1638)$	mean solar time
1^s sidereal time	$\equiv$	$(1^s - 0\overset{s}{.}0027)$	mean solar time

More exactly,

$$1 \text{ mean solar day} = 24^h 03^m 56\overset{s}{.}5554 \text{ of sidereal time}$$
$$1 \text{ sidereal day} = 23^h 56^m 04\overset{s}{.}0905 \text{ of mean solar time.}$$

Tables for the conversion of mean solar time to or from sidereal time are printed in several almanacs. Any conversion, however, can be performed using tables 9.1 and 9.2.

Alternatively, we may proceed as in the following example: convert $6^h 42^m 10^s$ MST to sidereal time.

$$10^s = 10/60^m = 0\overset{m}{.}1667 \qquad 42\overset{m}{.}1667 = 42\cdot1667/60^h = 0\overset{h}{.}702\,78.$$

Hence,

$$6^h 42^m 10^s \text{ MST} = 6\overset{h}{.}702\,78 \text{ MST} = 6\overset{h}{.}702\,78 \times 366\tfrac{1}{4}/365\tfrac{1}{4} \text{ ST} = 6\overset{h}{.}721\,13 \text{ ST.}$$

Now

$$0\overset{h}{.}721\,13 = 0\cdot721\,13 \times 60^m = 43\overset{m}{.}2678 \qquad 0\overset{m}{.}2678 = 0\overset{s}{.}2678 \times 60 = 16\overset{s}{.}1.$$

Hence, the required interval of sidereal time is $6^h 43^m 16\overset{s}{.}1$.

9.5 The civil day and timekeeping

The Greenwich meridian is regarded as the standard meridian on the Earth for timekeeping using mean solar time. Just as for any other observer's meridian, the Greenwich hour angle of the mean sun is zero when the mean sun transits across the meridian through the zenith at Greenwich; i.e. at mean noon

$$\text{GHAMS} = 0^h$$

while at mean midnight,

$$\text{GHAMS} = 12^h.$$

For convenience, however, a civil day begins at midnight. The **Greenwich Mean Time** (GMT) is then zero hours. But the GHAMS is then 12 hours in value. By the time the GHAMS is 24 hours or zero hours once more (i.e. by mean noon) the GMT is 12^h.

It is, therefore, seen that the relation between the GHAMS and the GMT is given by

$$\text{GMT} = \text{GHAMS} \pm 12^h \tag{9.13}$$

the plus or minus being used if the GHAMS is less than or greater than 12^h respectively.

In observatories throughout the world, events are often recorded in GMT since it provides an unambiguous universal time service. For this reason it is more often referred to as **Universal Time** (UT).

The UT time scale forms the basis of civil timekeeping and is defined by a formula relating it to Greenwich Mean Sidereal Time. Thus, UT is determined directly from stellar observations and consequently its progress contains irregularities due to variations in the rotational rate of the Earth. The time scale generated by a particular observatory is designated UT0 and this can be affected due to variations in the local meridian resulting from the small motions of the rotational pole relative to the geographic pole. After correction, the scale is designated as UT1. Hence, we may rewrite equation (9.8) as

$$\Delta T = \text{TDT} - \text{UT1}. \tag{9.14}$$

The time scales that are available by most broadcast time services are based on Coordinated Universal Time (UTC) which is maintained to within 0^s90 of UT1. As described in section 9.9, the most precisely used laboratory-generated time scale, more accurate than any time scale obtained from astronomical observations, is International Atomic Time, the acronym written as TAI (Temps Atomique Internationale). UTC is set to differ from TAI by an *integral* number of seconds and, as UT1 and TAI have different rates, the occasional leap second is added to UTC, normally at the end of June or December.

Unless the local longitude is near the Greenwich meridian, civil time systems differ from UT, the surface of the Earth having by international agreement been divided into **standard time zones** for this purpose.

This convention gives a clock-time related approximately to the Sun's position in the sky and also avoids the necessity of a moving observer continually adjusting his/her watch, adjustment only being required when a zone line is crossed.

Within each zone the same civil mean time, called **Zone Time** (ZT) or **Standard Time** (ST) is used and the zones are defined by meridians of longitude, each zone being $15°$ (1^h) wide. The Greenwich Zone (Zone 0) has bounding meridians $0^h 30^m$ W and $0^h 30^m$ E and keeps the time of the Greenwich meridian, namely GMT (UT). Zone $+1$ has boundaries $1^h 30^m$ W and $0^h 30^m$ W, keeping the time of meridian 1^h W. Zone -1 has boundaries $1^h 30^m$ E and $0^h 30^m$ E, keeping the time meridian 1^h E. The division of the Earth's surface in this way is continued east and west until Zones $+12$ and -12. According to the previous definition, both the zones would keep the time of 12^h W, which is also 12^h E. The convention is made at the zone from $11^h 30^m$ W to 12^h W is Zone $+ 12$, while the zone from $11^h 30^m$ E to 12^h E is Zone -12. The meridian separating them is called the **International Date Line** where a given day first begins.

It should be added that the actual Date Line, for geographical reasons, does not follow faithfully the 12^h meridian but makes local detours to include in one hemisphere parts of countries that will be placed in the other if the Line did not deviate in this way.

In the same way, small countries keep the Zone Time of one zone even though part of them lies in the neighbouring zone.

In large countries such as the USA and Russia, more than one zone is involved. For example, in the United States, four time zones are used, the mean times being called Eastern, Central, Mountain and Pacific Times, based on the meridians 5^h, 6^h, 7^h and 8^h west of Greenwich.

It is well known that ships crossing the Date Line from east to west omit one day while others crossing from west to east add one day—a circumstance that puzzled the great navigator Magellan's

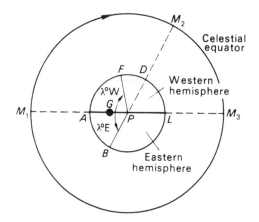

Figure 9.6. The Greenwich date and zone time.

sailors when they arrived home one day too late by their reckoning, and which enabled Jules Verne's character Phileas Fogg to win his wager to journey round the World in 80 days.

In many problems it is necessary to convert from a particular zone time and date to the corresponding GMT and date. The **Greenwich Date** is the name given to a GMT plus the date at Greenwich and we now consider the relationship between a Greenwich Date and a zone time and date.

9.6   The Greenwich date and zone time (with date)

In figure 9.6 the Earth is seen as viewed from above the North Pole P, the meridian of longitude PGA through Greenwich G meeting the equator in A. The International Date Line PL cuts the equator at L, its southern half being, of course, unseen.

The standard meridian PB, of longitude $\lambda°$ E, likewise meets the equator at B, where $\angle APB = \lambda°$ E.

Let a civil day, for example November 3rd, begin at the Date Line. It does so when the HAMS is 12^h for that meridian, i.e. the mean sun is at M_1, above the point A on the equator.

November 3rd will not begin for places in the zone keeping the zone time of the standard meridian PB until the mean sun has moved round the celestial sphere to be at M_2, above D, where D is $180°$ distant from B round the equator.

For the Greenwich meridian and Zone 0, November 3rd will begin even later, when the mean sun has revolved until it is above L, at M_3.

The Greenwich Date is then November 3rd, GMT 00^h 00.

But the zone time with date at B is by then November 3rd, $ZT\lambda^h$, since $\angle DPL = \angle APB = \lambda$ E.

The required relation is thus

$$GMT = ZT - \lambda E$$

where λ is now expressed in hours.

By taking a western standard meridian at F, say, where $\angle APF = \lambda W$, it is easily shown that the relation is given by

$$GMT = ZT + \lambda W$$

λ being expressed in hours.

We may, therefore, combine these rules by adopting as before the convention that a western longitude is positive and an eastern one is negative. Thus,

$$GMT = ZT + \lambda. \tag{9.15}$$

Whereas ordinary clocks and watches keep the zone time of the particular zone in which their owners live, if continuous radio signals with time codes are not available, special chronometers are used by navigators to give GMT (or UT). These GMT chronometers are very accurate and knowledge of their error (fast on GMT or slow on GMT) and the rate is obtained by periodically checking their reading against the time signals broadcast throughout the world.

It was the development of such reliable timekeepers by John Harrison in the 18th century that solved the outstanding problem in navigation. The latitude of a ship at sea can be found by making the necessary astronomical observations (see section 8.6) but to determine the longitude involves a knowledge of the GMT. Until Harrison's marine chronometers were built, no clock taken to sea remained accurate enough.

Applying the correction to a chronometer reading, therefore, gives the GMT. It should be noted, however, that a reading on a chronometer with a 12-hour dial contains an ambiguity not present with a 24-hour clock face.

9.7   The tropical year and the calendar

The year used in civil life is based on the **tropical year**, defined as the interval in time between successive passages of the Sun through the vernal equinox and equalling 365·2422 mean solar days. Because this is not an exact integer, the number of days in a year is made variable so that the seasons do not drift relative to the regular dates as time progresses. For convenience, the calendar year contains an integral number of days, either 365 or 366. Every fourth year, called a **leap year**, has 366 days, excepting those century years (such as 1900 AD) whose numbers of hundreds (in this case 19) are indivisible by four exactly. The year 2000 AD was, therefore, a leap year. These rules give a mean civil year equal in length to 365·2425 mean solar days, a figure very close to the number of mean solar days in a tropical year. On this scheme, the calendar is accurate to 1 day in 3323 years with the result that some particular leap year designated by the rule will need to be abandoned about 3000 years from now.

The presently-used calendar year is the **Gregorian**, just defined, and introduced by Pope Gregory in 1582. Previously, the **Julian** calendar had been used in which simply *every* fourth year was a leap year of 366 days, February 29th being the extra day. This gave an average value for the length of the civil year of 365·25 mean solar days. By 1582, the discrepancy between the number and the length of the tropical year (365·2422 mean solar days) had led to the considerable error of over 12 days. The introduction of the Gregorian calendar removed this error.

Unfortunately, political and religious obstacles caused the introduction to be carried out in different countries at different epochs. For example, the change took place in Great Britain in 1752, an act of Parliament in 1751 stating that the year should begin on January 1st (instead of March 25th as had been the custom in England) and that the day following September 2nd, 1752, should be September 14th, thus dropping 11 days. The change in Russia took place even later; the October Revolution of 1917 took place in November, according to countries using the Gregorian calendar.

9.8   The Julian date

The irregularities in the present calendar (unequal months, days of the week having different dates from year to year) and the changes from the Julian to the Gregorian calendar make it difficult to compare lengths of time between observations made many years apart. Again, in the observations of variable stars, it is useful to be able to say that the moment of observation occurred so many days and fractions of a day after a definite epoch. The system of **Julian Day Numbers** was, therefore, introduced to reduce computational labour in such problems and avoid ambiguity. January 1st of the year 4713 BC was chosen as the starting date, time being measured from that epoch (mean noon on January 1st,

4713 BC) by the number of days that have elapsed since then. The **Julian Date** is given for every day of the year in *The Astronomical Almanac*. Tables also exist for finding the Julian Date for any day in any year.

For example, the Julian Date for June 24th, 2000, is 245 1719·5 when June 24th begins; again the time of an observation made on June 24th, 1962, at 18^h UT is JD 245 1720·25.

Time may also be measured in **Julian Centuries**, each containing exactly 36 525 days.

Orbital data for artificial Earth satellites are often referred to epochs expressed in **Modified Julian Date Numbers** in which the zero point in this system is 17·0 November, 1858. Hence,

$$\text{Modified Julian Date} = \text{Julian Date} - 240\,0000{\cdot}5 \text{ days.}$$

For some astronomical observations, the fact that the Earth is in orbit about the Sun causes a difficulty when accurate timings of events are required. Observed delays and advances relative to predicted times of satellite eclipses by Jupiter were noted by Roemer, the effect turning out to be a milestone in connection with the determination of the velocity of light (see section 11.3). Resulting from the Earth's motion about the Sun, the effect is also a problem with timed measurements related to any celestial body. For example, for a star lying close to the ecliptic, the path length that the light has to travel will change from night to night according to the Earth's orbital motion. Over a six-month period, the difference in timings can be as much as ~ 16 minutes, this being the time it takes for light to cross the diameter of the Earth's orbit. For a star nearer to the ecliptic pole, the path length variation will be much smaller.

In order to compensate for the effects, times of astronomical observations are sometimes expressed in terms of an **Heliocentric Julian Date** or **HJD**, the Julian Date of the record transposed to a timing that would have been obtained at the centre of the Sun. As the effect is related to the Earth's orbit, it will be appreciated that the correction formula will involve the ecliptic coordinates of the observed object and those of the Sun. Quite simply, the correction conversion formula may be written as

$$\text{HJD} = \text{JD} - 0{\overset{d}{\cdot}}0058 \cos \beta \cos(\odot - \lambda) \tag{9.16}$$

where $\odot$ is the solar longitude on the ecliptic and β, λ are the source's ecliptic latitude and longitude respectively.

9.9   Dynamical time

Both mean solar time and sidereal time are based on the rotation of the Earth on its axis. Until comparatively recently, it was thought that, apart from a slow secular increase in the rotation period due to tidal friction, the Earth's period of rotation was constant. Tidal friction, due to the Moon's gravitational effect, acts as a break on the Earth's rotation.

The development and use of very accurate clocks revealed that other variations occur in the period of the Earth's rotation. These small changes in general take place abruptly and are not predictable. Since UT is based on observations of the transits of celestial objects made from the irregularly rotating Earth, it must differ from a theoretical time that flows on uniformly. This time is the Newtonian time of celestial mechanics, that branch of astronomy which deals with and predicts the movements of the Sun, Moon and planets. Hence, their positions as published in **ephemerides** (tables of predicted positions) based on celestial mechanics theories are bound to **Terrestrial Dynamical Time**, this replacing **Ephemeris Time** in 1984.

The value of (Terrestrial) Dynamical Time at a given instant is obtained by very accurate observations of abrupt variations in the longitudes of the Sun, Moon and planets due to corresponding variations in the Earth's rate of rotation. It is estimated that to define Dynamical Time correctly to one part in 10^{10}, observations of the Moon are required over five years. In practice, laboratory atomic clocks may be used to give approximate values of Dynamical Time, their readings being subsequently

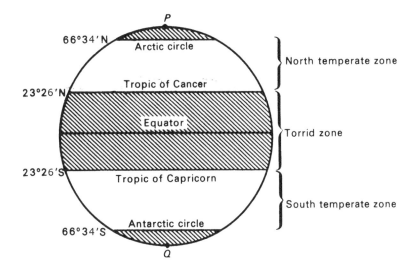

Figure 9.7. The geographical zones.

corrected by long series of astronomical observations. The quantity that is in fact determined is ΔT, where

$$\Delta T = \text{Dynamical Time} - \text{Universal Time}.$$

This quantity is tabulated in *The Astronomical Almanac*. At present (2000) its value is about 66^s, increasing at about 1 second per year. The relationship between Dynamical Time and International Atomic Time (TAI), based on careful analysis of atomic processes, is, for practical purposes, taken to be

$$\text{TDT} = \text{TAI} + 32^s184.$$

International Atomic Time (TAI) is the most precisely determined time scale available for astronomical use. TAI has the SI second as its fundamental unit. It is defined as the duration of 9 192 631 770 cycles of the radiation from the transition between two hyperfine levels in the ground state of caesium (^{133}Cs). There are 86 400 SI seconds in the mean solar day.

9.10   The Earth's geographical zones

We can now understand readily why the Earth's surface is divided into the well-known zones indicated in figure 9.7 by parallels of latitude $23° 26'$ and $66° 34'$ north and south of the equator; also why zones and parallels are named in the way they are.

Within the Torrid Zone, bounded by the Tropics of Cancer and Capricorn, an observer has the Sun in the zenith, or very nearly so, on two days of the year.

If the observer's latitude is numerically less than $23° 26'$ (the obliquity of the ecliptic ε and the maximum declination of the Sun), there will be two occasions each year when

$$90 - \phi = 90 - \delta$$

where δ is the Sun's declination (see figure 9.8).

The limiting cases are when the observer's latitude is $23° 26'$ N (Tropic of Cancer) and $23° 26'$ S (Tropic of Capricorn). In order that the Sun should pass through the zenith of an observer in latitude $23° 26'$ N, it must have its maximum northerly declination—this occurs about June 21st when the Sun

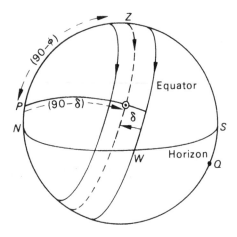

Figure 9.8. The conditions for the Sun being overhead.

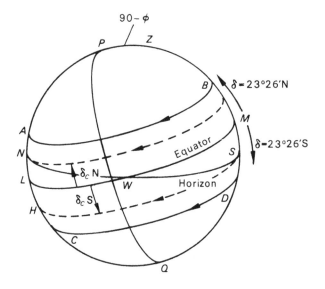

Figure 9.9. The conditions for the midnight Sun and Polar night.

is in Cancer. Similarly, for an observer in latitude 23° 26′ S, the Sun must have its maximum southerly declination to pass through the zenith—this occurs about December 21st when the Sun is in Capricorn.

Within the zone bordered by the Arctic Circle (latitude 66° 34′ N) and including the North Pole, the Sun will remain below the horizon for at least 24^h at some part of the year and at another part six months later, will be above the horizon for at least 24^h.

When the latitude ϕ is greater than 66° 34′ N, the maximum northerly (AB) and maximum southerly (CD) parallels of declination of the Sun are, respectively, entirely above and below the horizon, as seen in figure 9.9.

Consider the Sun's declination to be positive. Now, $\angle NWL = 90 - \phi$, so that the critical declination of the Sun at which it will remain above the horizon and not set is δ_c N, given by

$$\delta_c \, \text{N} = 90 - \phi.$$

Hence, the period of the year during which the Sun's declination δ has values

$$\delta_c \, \text{N} \leq \delta \leq 23° \, 26' \, \text{N}$$

is the period during which the Sun never sets even at midnight. This is the phenomenon known as the **midnight Sun**. It is obvious that June 21st will lie at the middle of this period.

It is easily seen from figure 9.9 that while the Sun's declination has values between δ_c S and $23° \, 26'$ S, the Sun will remain below the horizon. Again, the numerical value of δ_c S is given by

$$\delta_c = 90 - \phi.$$

This period of time when the Sun does not rise, known as the **polar night**, is around December 21st, the winter solstice in the northern hemisphere.

Obviously for a person in latitudes further south than the Antarctic Circle ($66° \, 34'$ S), corresponding phenomena are observed, though the date halfway through the period of the midnight Sun will be December 21st; the date June 21st will lie at the middle of the time interval during which the Sun is never seen.

At the North and South Poles, the observer experiences six-month periods with the Sun continuously above the horizon alternating with six-month periods when the Sun is below, neither rising nor setting.

9.11   The seasons

Time is also measured according to the passage of the seasons. Spring, summer, autumn and winter have always loomed large in mankind's life and philosophy, linked as they are to seed time and harvest, good weather and bad and the length of the day. They are, of course, also directly connected with the Sun's passage round the ecliptic.

In the northern hemisphere spring, summer, autumn and winter are defined to begin when the Sun respectively reaches the **vernal equinox**, the **summer solstice**, the **autumnal equinox** and the **winter solstice**.

Thus, in **spring**, the Sun's right ascension increases from 0^h to 6^h while its declination increases from $0°$ to $23° \, 26'$ N.

In **summer**, the RA$\odot$ increases from 6^h to 12^h, its declination decreasing from $23° \, 26'$ N to $0°$.

In **autumn**, the RA$\odot$ increases from 12^h to 18^h, its declination changing from $0°$ to $23° \, 26'$ S.

In **winter**, the RA$\odot$ increases from 18^h to 24^h, its declination changing from $23° \, 26'$ S to $0°$, at which time spring begins again.

In the southern hemisphere autumn begins when the Sun's declination changes from south to north, i.e. about March 21st. The seasons, autumn, winter, spring and summer succeed each other in the same order in which they follow each other in the northern hemisphere.

As the Sun's rate of increase in right ascension is not a constant, the seasons are of unequal length. For any year, their lengths may be calculated from data tabulated in *The Astronomical Almanac*. One approach is to determine by interpolation the times when the Sun's declination is exactly zero (the equinoxes) and when it achieves the maximum and minimum values. The latter are, however, difficult to calculate exactly with accuracy using simple interpolation techniques.

The best and simplest method is to determine the time intervals between the apparent Sun passing through the exact right ascension points of 0^h, 6^h, 12^h, 18^h and 24^h. This can be done by noting the value of the right ascension at 00^h (DT) for the day the Sun passes through each of the previously noted exact points. By also noting the right ascension values for the appropriate following day at 00^h, the times for the Sun being at the exact points can be obtained by simple linear interpolation. The time intervals between each of the required exact points then provide values for the lengths of each season.

Table 9.3. The lengths of the northern hemisphere seasons in the year 1999 to 2000.

Season	Days	Hours
Winter	88	23·86
Spring	92	18·21
Summer	93	15·66
Autumn	89	20·17

Comparison of each of the seasons' lengths shows that there are variations amounting to a few minutes from year to year as a result of irregularities in the motion of the apparent Sun relative to the equatorial coordinate frame. For the year covering 1999 to 2000, the lengths of the seasons are given in table 9.3.

A simple and only slightly less accurate way of determining the lengths of the seasons involves calculations of the equation of time, $\mathcal{E}$, on the dates corresponding to the equinoxes and solstices. Consider the development of the general formula for such calculations by taking spring as an example.

From earlier we had

$$\mathcal{E} = \text{RAMS} - \text{RA}\odot \tag{9.3}$$

where $\mathcal{E}$ is the equation of time.

Let the values of the equation of time at the beginning and end of spring be $\mathcal{E}_1$ and $\mathcal{E}_2$ while the values of the RAMS at those times are R_1 and R_2. The corresponding values of the RA$\odot$ are 0^h and 6^h by definition. Then by equation (9.3),

$$\mathcal{E}_1 = R_1 - 0^h,$$
$$\mathcal{E}_2 = R_2 - 6^h$$

giving

$$R_2 - R_1 = \mathcal{E}_2 - \mathcal{E}_1 + 6^h.$$

But the mean sun increases its right ascension by 24^h in close to $365\frac{1}{4}$ days. Therefore the length of spring in days is given by

$$\left(\frac{\mathcal{E}_2 - \mathcal{E}_1 + 6}{24}\right) \times 365\tfrac{1}{4}. \tag{9.17}$$

Values of the equation of time precise to 1 s may be derived from *The Astronomical Almanac*. Using equation (9.9), namely

$$\text{ephemeris transit} = 12^h - \text{equation of time} \tag{9.9}$$

and taking the relevant values of the ephemeris transit from the almanac at the beginning of the day when the Sun achieves 0^h and 6^h, we find for the year 2000 that

$$\mathcal{E}_2 - \mathcal{E}_1 = 5^m\,33^s\!6 = 0^h\!.092\,67.$$

The expression (9.17) can then be evaluated and it is found that the length of spring in the northern hemisphere in 2000 is $92^d\,17^h\!3$, a value not very different to that given in table 9.3. Values for the lengths of the other seasons can be obtained by obtaining the difference of the equations of time, $\mathcal{E}_2 - \mathcal{E}_1$, at the end and beginning of each season.

The amount of heat received from the Sun and the average daily temperature have very little to do with the varying distance of the Earth from the Sun. In the northern hemisphere, in fact, the Earth is

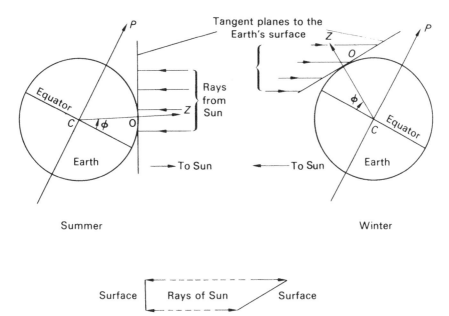

Figure 9.10. The heating of the Earth's surface.

further from the Sun in summer than in winter. The main factors determining the heating effect of the Sun throughout the year are the number of hours of daylight per day and the meridian altitude of the Sun.

We have seen (section 8.11) that when the Sun's declination δ is positive, in spring and summer, the number of hours of daylight is greater than 12, rising to a maximum as δ tends to its maximum northerly value around June 21st. When δ is negative, during autumn and winter, less than 12 hours of daylight are experienced, the minimum number occurring when δ, around December 21st, has its maximum southerly value.

The meridian altitude of the Sun is higher in spring and summer than in autumn and winter. The heat rays from the Sun will, therefore, strike the Earth's surface more directly in spring and summer than in autumn and winter (figure 9.10) with the result that a beam of given cross section from the Sun will cause greater heating of the smaller area it illuminates. It can easily be seen that the amount of energy arriving per unit area is proportional to $\cos(\phi - \delta)$.

Furthermore, it will be seen from section 19.7.2 that the Earth's atmosphere absorbs incoming radiation and that the extent of the absorption increases rapidly as the altitude of the source decreases. Consequently, a larger fraction of the heat rays is lost to absorption in the winter months than in the summer.

9.12 Twilight

There is one phenomenon that lengthens the fraction of the day given over to daylight and which is important where astronomical observations are concerned.

Even after the Sun has set, some sunlight is received by the observer, scattered and reflected by the Earth's atmosphere. As the Sun sinks further below the horizon, the intensity of this light diminishes. The phenomenon is called **twilight** and is classified as **civil**, **nautical** and **astronomical** twilight. Civil twilight is said to end when the Sun's centre is 6° below the horizon, nautical twilight ends when the centre is 12° below the horizon, while astronomical twilight ends when the centre of the Sun is 18°

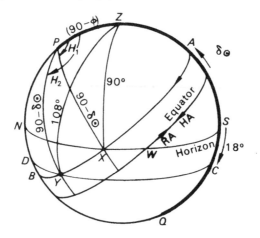

Figure 9.11. The calculation for twilight.

below the horizon. Corresponding definitions hold for morning twilight: thus civil morning twilight, for example, begins when the Sun's centre is 6° below the horizon.

Twilight is a nuisance, astronomically speaking, often preventing the observation of very faint celestial objects. We shall see later that in some latitudes during part of the year, twilight is indeed continuous throughout the night, evening and morning twilight merging because the Sun's centre at all times of the night is less than 18° below the horizon.

An accurate enough approximation to the duration of twilight is to calculate twice the difference between the hour angles of the Sun's centre when it is 18° below the horizon and when it is on the horizon.

In figure 9.11, let H_1, H_2 be the values of the Sun's hour angle when its centre is on the horizon at X and 18° below the horizon at Y respectively, its declination having value $\delta_\odot$ N. Let the observer's latitude be ϕ N. Then

$$PZ = 90 - \phi \qquad PX = PY = 90 - \delta_\odot \qquad ZX = 90 \qquad ZY = 108°.$$

In $\triangle PZX$, using the cosine formula, we have

$$\cos 90 = \cos(90 - \phi)\cos(90 - \delta_\odot) + \sin(90 - \phi)\sin(90 - \delta_\odot)\cos H_1$$

giving

$$\cos H_1 = -\tan\phi\tan\delta_\odot.$$

In $\triangle PZY$, again using the cosine formula, we obtain

$$\cos 108 = \sin\phi\sin\delta_\odot + \cos\phi\cos\delta_\odot\cos H_2$$

or

$$\cos H_2 = \frac{\cos 108 - \sin\phi\sin\delta_\odot}{\cos\phi\cos\delta_\odot}.$$

The duration of astronomical twilight is then given by $H_2 - H_1$. It is seen that the duration depends upon the Sun's declination and the latitude of the observer.

The limiting case when evening twilight ends as morning twilight begins will occur when B in figure 9.11 coincides with D, i.e. the Sun's centre is exactly 18° below the horizon at apparent midnight.

For this to happen, $ZB = 108°$. But

$$ZB = ZP + PB$$

or

$$108 = 90 - \phi + 90 - \delta_\odot$$

or

$$\delta_\odot = 72 - \phi. \tag{9.18}$$

If, for a given northern latitude ϕ N, the Sun's declination is equal to or greater than the value of δ given by equation (9.18), twilight will be continuous throughout the night. Thus, in high northern latitudes for a part of the year and around June 21st, observing conditions are impaired by this phenomenon.

There is, of course, a similar period around December 21st in high southern latitudes.

Example 9.3. An astronomer makes an observation on April 4th, 2000, at UT $10^h 25^m 30^s$. The observer is in longitude $96° 30'$ W. On looking at *The Astronomical Almanac*, it is noted that at 0^h UT, April 4th, the Greenwich apparent sidereal time (GAST) was $12^h 50^m 27^s$. Calculate the LST of the observation.

We break the calculation into two parts:
(i) Find the GAST of the observation.

To do this we know that the observation was made at UT $10^h 25^m 30^s$. We know the GAST at UT 0^h. The interval of mean solar time to be converted into an interval of sidereal time is, therefore, $10^h 25^m 30^s - 0^h$ or $10^h 25^m 30^s$.

Using table 9.1, we proceed as follows:

10^h MST	$=$	$10^h + 98^s565$	$=$	10^h	1^m	38^s565 sidereal time
25^m MST	$=$	$25^m + 4^s107$	$=$		1^m	4^s107 sidereal time
30^s MST	$=$	$30^s + 0^s081$	$=$			30^s081 sidereal time
				10^h	27^m	12^s753 sidereal time

To the nearest second, therefore,

$$10^h 25^m 30^s \text{ MST} = 10^h 27^m 13^s \text{ sidereal time.}$$

Hence, the GAST of the observation is

$$10^h 27^m 13^s + 12^h 50^m 27^s = 23^h 17^m 40^s.$$

(ii) Find the LST from the GAST and the observer's longitude.
Using equation (9.2),

$$\text{GAST} = \text{LST} + \lambda.$$

We convert the longitude $96° 30'$ W to hours, minutes and seconds by means of table 7.1.
Thus,

$$96° 30' = 90° + 6° + 30' = 6^h + 24^m + 2^m = 6^h 26^m.$$

Remembering the mnemonic 'Longitude west, Greenwich best', we write

	h	m	s
GAST	23	17	40
Longitude (W)	6	26	0
LST	16	51	40

Example 9.4. An observation of the Sun was made by a ship's navigator in zone −9 at approximate zone time $3^h 8^m$ on March 4th. By the ship's chronometer, the observation was made at $18^h 12^m 4^s$. The chronometer was $2^m 14^s$ fast on GMT. Calculate the correct Greenwich Date (GD).

The first step is to find the approximate GD. Thus,

	h	m	s	Date
Approximate zone time	3	08	0	March 4th
Zone (E)	−9			
Approximate GD	18	08	0	March 3rd

Note that since the zone is east, the GD is less, so that the 9^h are subtracted. To do this, we call the approximate ZT $27^h 08^m 0^s$ March 3rd.

We now use the chronometer time, applying the error:

	h	m	s	
Chronometer time	18	12	4	
Error (fast)		2	14	
Correct GD	18	9	50	March 3rd

The date 'March 3rd' has been obtained from the approximate GD. If the chronometer had had a 12-dial, it would have read $6^h 12^m 4^s$. The value of the approximate GD would have told the navigator to add 12 hours to the chronometer reading.

Example 9.5. A passenger on a cruise liner celebrates his birthday one evening with a party. It is June 18th. The ship is in zone −12 and at zone time $23^h 36^m$, June 18th, it crosses the date line. Show that the passenger may legitimately have another birthday party for the same birthday.

Just before crossing, the zone time was	$23^h 36^m$, June 18th
The zone time was	−12
Hence, the Greenwich Date was	$11^h 36^m$, June 18th
Immediately after crossing, the zone is	+12
Therefore, the zone time is	$23^h 36^m$, June 17th

Thus, by crossing the date line from east to west, June 18th becomes June 17th and when the passenger wakens next morning it is his birthday again!

Example 9.6. Calculate the Sun's hour angle for an observer in longitude 103° 40′ W, keeping the mean solar time of zone +7, who has made an observation of the Sun at approximate zone time $16^h 30^m$, June 1st. The chronometer time was $23^h 31^m 20^s$, the chronometer error being $1^m 10^s$ slow on GMT. The ephemeris transit of the Sun for June 1st was $11^h 57^m 41^s$ (taken from *The Astronomical Almanac*).

By equation (9.10),

$$GHA\odot = GHAMS + 12^h - \text{Ephemeris transit.}$$

Hence,

$$GHA\odot = GHAMS + 2^m 19^s. \tag{A}$$

We proceed by setting up the following scheme.

	h	m	s	Date
Approximate ZT	16	30	0	June 1st
Zone	+7			
Approximate GD	23	30	0	June 1st, using equation (9.15)
Chronometer time	23	31	20	
Error (slow)		+1	10	
Correct GD	23	32	30	June 1st
Hence, GHAMS is	11	32	30	using equation (9.13)
		+2	19	
GHA⊙	11	34	49	using (A)
Longitude (W)	−6	54	40	
HA⊙	4	40	9	using equation (9.2).

In the second last line, the longitude has been converted, thus

$$103° \, 40' = 6 \times 15° + 13° + 40' = 6^h + 52^m + 160^s = 6^h \, 54^m \, 40^s.$$

Example 9.7. Find the LST and the hour angle of the star *Regulus* (RA $10^h \, 5^m \, 11^s$) from the following data: Zone, +4; approximate zone time, $3^h \, 14^m$ January 4th; chronometer reads $7^h \, 12^m \, 56^s$ at the time of observation; chronometer error, $2^m \, 5^s$ slow on GMT; longitude of observer, $58° \, 20'$ W at UT 0^h, January 4th, the value of the GAST is $6^h \, 53^m \, 34^s$.

We proceed in the usual way:

	h	m	s	Date
Approximate ZT	3	14	0	January 4th
Zone	+4			
Approximate GD	7	14	0	January 4th
Chronometer time	7	12	56	
Error (slow)		+2	5	
Correct GD	+7	15	1	January 4th

The interval of mean solar time to be converted into sidereal time is, therefore, $7^h \, 15^m \, 01^s − 0^h = 7^h \, 15^m \, 01^s$.

Then,

$$7^h \, 15^m \, 01^s = 7^h \, 15^m 01667 = 7^h25028 \text{ MST}.$$

Hence,

$$7^h25028 \text{ MST} = 7^h25028 \times 366\tfrac{1}{4}/365\tfrac{1}{4} \text{ ST} = 7^h27013 \text{ ST}.$$

Converting to hours, minutes and seconds, we get

$$7^h27013 = 7^h \, 16^m2078 = 7^h \, 16^m \, 12^s47 \text{ or } 7^h \, 16^m \, 12^s$$

to the nearest second.

	h	m	s	
The GAST at UT 0^h, January 4th is	6	53	34	
Add the sidereal time interval of	7	16	12	
Hence, the GAST at observation time is	14	09	46	
Longitude (W) is, after conversion,	3	53	20	
Therefore, the LST is	10	16	26	using equation (9.2)
But the RA of the star is	10	05	11	
Hence, the HA of the star is	0	11	15	using equation (9.1)

The longitude $58° \, 20'$ was converted in the usual way to $3^h \, 53^m \, 20^s$.

Problems—Chapter 9

Note: Assume (i) a spherical Earth, (ii) the obliquity of the ecliptic to be 23° 26′.

1. What is the lowest latitude at which it is possible to have a midnight Sun?
2. What is the latitude of a place at which the ecliptic can coincide with the horizon?
3. Suppose the Earth rotated with the same angular velocity as at present but in the opposite direction, what would be the length of a mean solar day? How many mean solar days would there be in a year?
4. Complete the following table:

Hour angle of star	Longitude of observer	Greenwich hour angle of star
$2^h\,46^m\,21^s$	30° W	
$18^h\,24^m\,40^s$	65° E	
$1^h\,19^m\,46^s$	121° E	
$23^h\,04^m\,57^s$	42° 37′ E	

5. Convert the mean time interval of $6^h\,46^m\,21^s$ to the corresponding interval of sidereal time.
6. Convert the sidereal time interval of $23^h\,13^m\,47^s$ to the corresponding interval of mean solar time.
7. Find the Zone Time on February 3rd when *Procyon* (RA $7^h\,36^m\,10^s$) crosses the meridian at Ottawa (longitude 75° 43′ W), given that at GMT 0^h, February 3rd, the Greenwich sidereal time is $8^h\,48^m\,8^s$. The zone is +5.
8. Find the latitude of a place at which astronomical twilight just lasts all night when the Sun's declination is 16° N.
9. Calculate the duration of evening astronomical twilight for a place in latitude 50° N when the Sun's declination is 5° 20′ N.
10. At mean noon on a certain date, the sidereal time was 14 hours. What will the sidereal time be at mean noon 50 days after, in the same place? Take the length of a tropical year to be $365\frac{1}{4}$ days.
11. Find the Sun's hour angle for an observer in longitude 39° 30′ W, given the following data:
    Zone +3; approximate Zone Time of observation $8^h\,20^m$ May 14th;
    chronometer time (corrected) $11^h\,21^m\,47^s$; equation of time $+3^m\,45^s$.
12. Find the hour angle of *Vega* (RA $18^h\,34^m\,52^s$) for an observer in longitude 126° 34′ E from the following data:
    Zone −8; approximate Zone Time of observation $6^h\,30^m$ February 2nd; chronometer time $22^h\,29^m\,58^s$; chronometer error (slow on GMT) $1^m\,35^s$; for GMT 0^h February 2nd, Greenwich sidereal time was $8^h\,45^m\,9^s$.
13. At approximately 2.30pm Zone Time, an observer keeping the time of Zone −8 made an observation of the Sun on December 10th. The observer's position was 55° N, 122° 30′ E. The chronometer reading was $6^h\,32^m\,45^s$, its error (fast on GMT) being $4^m\,22^s$. Given that the ephemeris transit was $11^h\,52^m\,34^s$, calculate the Sun's hour angle.
    Taking the Sun's declination to be −22° 55′, obtain, to the nearest minute, the GMT of sunset for the observer on December 10th.
14. Calculate the interval of sidereal time, for a place in latitude 45° N, between the passage of a star over the prime vertical west and its setting, given the star's zenith distance on the prime vertical west is 45°.
15. Taking the apparent orbit of the Sun to be circular and in the ecliptic, show that the equation of time $\mathcal{E}$ is given by
$$\mathcal{E} = \cot^{-1}(\cot\alpha\cos\varepsilon) - \alpha$$
where α is the Sun's right ascension and ε is the obliquity of the ecliptic. On which dates would $\mathcal{E}$ vanish in this case?
16. Calculate to the nearest minute the interval of mean time elapsing between the setting of the Sun and of Venus on March 31st, at Washington, DC (38° 55′ N, 77° 04′ W) given that the Sun's declination is 4° 20′ N, the equation of time is -4^m, the right ascension and declination of Venus are $3^h\,33^m$ and 21° 56′ N, and the Greenwich sidereal time at GMT 0^h on March 31st is $12^h\,33^m$.
17. *The Astronomical Almanac* for 2000 provides the following entries:

Date (00^h)		Apparent RA$\odot$		Date (00^h)		Apparent RA$\odot$	
June 21	05^h	59^m	41^{s}15	Sep 22	11^h	57^m	23^{s}07
June 22	06^h	03^m	50^{s}68	Sep 23	12^h	00^m	58^{s}59

Calculate the length of summer in 2000.

18. At the summer solstice and at the autumnal equinox the values of the equation of time are $-1^m 27^s$ and $7^m 20^s$ respectively. Calculate the length of summer.

19. The longitude of Columbia University, New York, is $4^h 55^m 50^s$ W. The sidereal time at mean noon at Greenwich on a certain day is $17^h 23^m 08^s$. Show that on the same day when the sidereal time at Columbia University is $20^h 08^m 04^s$, the hour angle of the mean sun at the same place is $2^h 43^m 41^s$.

Chapter 10

The reduction of positional observations: I

10.1 Introduction

In general, astronomical observations of an object's position undergo a process of reduction. This procedure removes known instrumental errors and other systematic effects in order to provide data about the celestial body that is as objective as possible.

Such reduced observations, independent of the observer's position, are then suitable for catalogue purposes or for comparison so that changes in the body's position with time can be derived. The raw observations may be the altitude (or zenith distance) and azimuth of the object, its hour angle and declination, or its position (on a photographic plate or CCD frame) with respect to a stellar background. In addition, a time is noted at which the observation was made; this time may be a Universal Time (UT) or a Local Sidereal Time (LST).

If the altitude and azimuth of the object are measured, the first corrections to be applied are known instrumental errors. This entails a frequent calibration of the observing instrument since such errors are not, in general, static.

The philosophy behind the correction procedures described here is that they are related to small effects. They are, therefore, to the first order in small quantities, independent of each other and can be applied in any order. The end product will usually be a geocentric equatorial position for the object or a heliocentric ecliptic position or even, in the case of a star cluster or galaxy, a galactocentric equatorial position.

We now describe the corrections in turn.

10.2 Atmospheric refraction

10.2.1 The laws of refraction

When a ray of light passes from a transparent substance of one density into another transparent substance of a different density, the ray changes direction. It is said to be refracted and the amount of the deviation from its original direction depends upon the relative densities of the substances.

Let a ray of light, AB, passing through a vacuum, meet the upper boundary PQ of a plane parallel slab of glass at an angle i to the normal BN to the slab (figure 10.1). The angle NBA is called the **angle of incidence**. The ray is refracted upon entering the slab so that it leaves the lower boundary RS of the slab at C. Angle MBC, or r, is called the **angle of refraction** and is less than i. The ray's path on leaving at the point C is along the line CD, at an angle ZCD to the normal ZC.

Then the first law of refraction states that the incident ray AB, the normal BN and the refracted ray BC are coplanar.

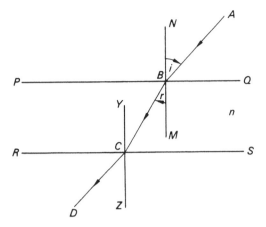

Figure 10.1. Refraction of a light beam by a slab.

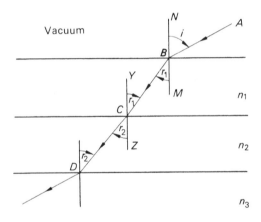

Figure 10.2. Refraction by a series of slabs.

The second law (Snell's law) states that

$$\frac{\sin i}{\sin r} = n \qquad (10.1)$$

where n is called the **index of refraction** of the substance making up the slab. Since $r < i$, $n > 1$.

If YCZ is the normal at C, $\angle YCB = \angle MBC = r$ (PQ being parallel to RS) and hence, since ray paths are reversible, $\angle ZCD = i$. In other words, the emergent ray CD will be parallel to AB but not collinear with it.

For a number of plane, parallel slabs of indices of refraction $n_1, n_2, n_3, \ldots, n_j$, we can extend relation (10.1). In figure 10.2, the ray of light passes from slab 1, of refractive index n_1, into slab 2, of refractive index n_2. Then by Snell's law,

$$\sin i = n_1 \sin r_1.$$

Also, for the second slab,

$$\sin i = n_2 \sin r_2.$$

Hence, for j slabs, we have

$$\sin i = n_1 \sin r_1 = n_2 \sin r_2 = n_3 \sin r_3 = \cdots = n_j \sin r_j. \qquad (10.2)$$

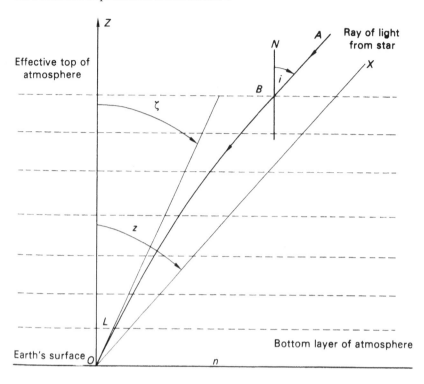

Figure 10.3. Refraction by the Earth's atmosphere of light from a star.

10.2.2 Astronomical refraction

Aircraft, balloons, high altitude rockets and artificial satellites have been used to measure the way in which the density of the Earth's atmosphere diminishes with the increase of height above the Earth's surface. The density is still appreciable enough at 150 km height to produce changes in the orbit of an artificial Earth satellite due to air-drag. In addition, studies of aurorae show that there is still some atmosphere up to a height of 800 km. A ray of light entering the Earth's atmosphere from outer space will be bent or refracted so that the observed direction of the source of light must be different from its true direction. It is found, however, that the air-density falls off so swiftly with height that above the 100 km level, no appreciable refraction takes place. Since the radius of the Earth is 6372 km, evaluation of atmospheric refraction can neglect the curvature of the atmosphere so long as we deal with zenith distances less than about 45 degrees.

In figure 10.3, the atmosphere is, therefore, taken to be made up of a large number of thin parallel layers of different densities, the density being greatest at the Earth's surface and constant within each layer. Above the atmosphere is the vacuum of space. A ray of light from a star meets the topmost layer of the atmosphere at B, the angle of incidence being $\angle NBA$, or i, and is thereafter refracted in successive layers until it reaches the observer at O. Its direction in the last layer being LO, the star appears to lie in the direction OL, of apparent zenith distance $\angle ZOL$, or ζ. In fact, if OX is drawn parallel to BA, angle ZOX, or z, is the true zenith distance of the star at the time of observation. Since the direction of increasing density is downwards, the index of refraction also increases in that direction, so that the star is displaced towards the zenith along the great circle through Z and X, where X is the true position of the star.

If n is the index of refraction of the bottom layer we have, from equation (10.2),

$$\sin i = n \sin r$$

where $\angle ZOL = r$.

But $\angle ZOL = \zeta$. Also AB is parallel to OX, hence $i = z$ and

$$\sin z = n \sin \zeta. \tag{10.3}$$

Let R, defined by

$$R = z - \zeta \tag{10.4}$$

be the **angle of refraction**, the correction that has to be applied to the apparent zenith distance ζ to obtain the true zenith distance z.

Eliminating z from equations (10.4) and (10.3), we obtain, on expanding,

$$\sin R \cos \zeta + \cos R \sin \zeta = n \sin \zeta. \tag{10.5}$$

Now R is a small angle, so that we may write

$$\sin R = R \qquad \cos R = 1. \tag{10.6}$$

If R is expressed in seconds of arc, equation (10.5) becomes, using (10.6),

$$R \cos \zeta = 206\,265(n-1) \sin \zeta$$

or

$$R = 206\,265(n-1) \tan \zeta.$$

If we let $k = 206\,265(n-1)$, it is found that k is $60''3$, at the standard temperature and pressure, $0\,°C$ and 1000 mbar (760 mm Hg); the value of k for other pressures and temperatures can be found from formulas or tables. Then z, the true zenith distance, is given by

$$z = \zeta + R. \tag{10.7}$$

The formula

$$R = k \tan \zeta \tag{10.8}$$

is valid for zenith distances less than 45° and is a fairly good approximation up to 70°. Beyond that, a more accurate formula taking into account the curvature of the Earth's surface is required, while for zenith distances near 90° special empirical tables are used.

A further problem related to accurate corrections to positional measurements is the fact that the refractive power, $(n-1)$, of the atmospheric air exhibits dispersion, i.e. its value is wavelength dependent. At wavelengths of 400, 500, 600 and 700 nm the corresponding values of k are approximately $60''4$, $57''8$, $57''4$ and $57''2$. By substituting these values in equation (10.8) it is immediately apparent that, at a zenith distance of 45°, a stellar image which should be essentially point-like appears as a tiny spectrum with a length ~ 3 arc sec, with the blue end towards the zenith and the red end towards the horizon. Under good seeing conditions (see section 19.7.3), this effect is clearly visible when telescopic stellar images are inspected by eye.

For radio measurements, refraction depends strongly upon the frequency employed. The lower atmosphere produces refraction effects approximately twice as large as the optical effect, decreasing rapidly with increasing angle of elevation. The ionosphere also refracts radio waves due to induced motion of charged particles in amounts dependent on the air-density gradient. If N is the electron density per cubic metre and ν is the frequency in cycles per second (Hz), then the local effective dielectric constant n (which varies throughout the ionosphere) may be expressed by

$$n = \left(1 - \frac{81N}{\nu^2}\right)^{1/2}.$$

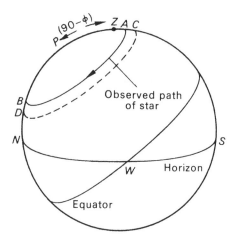

Figure 10.4. Measurement of an angle of refraction.

As height increases above the Earth's surface, electron density increases then falls off again. It may become so large that $81N/\nu^2 \geq 1$, making n an imaginary number or zero. In such cases a radio signal is reflected and cannot penetrate the ionosphere from the inside or from the outside. In other cases when the frequency is high enough, penetration takes place without bending of the signal. If we assume that the ionosphere consists of concentric shells about the Earth, Snell's law enables the path of the radio signal to be calculated from the relation

$$n\rho \sin i = \text{constant}$$

where ρ is the radius of curvature of the shell of dielectric constant n and i is the angle of incidence of the signal.

Study of ionospheric refraction by comparison of optical and radio tracking of artificial satellites has yielded valuable data concerning the ionosphere.

Having applied the correction for refraction, the topocentric altitude and azimuth may be converted into topocentric equatorial coordinates (hour angle and declination), as in section 8.8. The application of the LST, using equation (8.7), enables the topocentric right ascension to be found, if required.

10.2.3   Measurement of the constant of refraction

The constant of refraction, k, may be measured by using the transits of a circumpolar star. It has been seen that the effect of refraction is to displace the star towards the zenith along the vertical through star and zenith. Thus, in figure 10.4, an observer in latitude ϕ notes that a star has upper and lower culminations at A and B respectively. Let the observed zenith distances of A and B be ζ_A and ζ_B respectively. In the absence of refraction, the upper and lower transits would have been seen to be at positions C and D on the celestial sphere, where $PC = PD = 90 - \delta$, δ being the star's declination. *Note*: if the declination of the star was such that the upper transit had been north of the zenith, the observed path of the star would have intersected the parallel of declination it would have followed in the absence of refraction.

Now by equation (10.8),

$$CA = k \tan \zeta_A \qquad DB = k \tan \zeta_B$$

so that we have

$$ZC = \zeta_A + k \tan \zeta_A \qquad ZD = \zeta_B + k \tan \zeta_B.$$

But

$$ZC = PC - PZ = 90 - \delta - (90 - \phi) = \phi - \delta.$$

Hence,

$$\phi - \delta = \zeta_A + k \tan \zeta_A. \tag{10.9}$$

Similarly,

$$ZD = ZP + PD = 90 - \phi + 90 - \delta = 180 - \phi - \delta$$

so that

$$180 - \phi - \delta = \zeta_B + k \tan \zeta_B. \tag{10.10}$$

If the observer's latitude were accurately known, the two equations (10.9) and (10.10) in the two unknowns δ and k could be solved to yield values of δ and k. But because of small changes in the Earth's crust, small variations take place in the latitude of the telescope used. In practice, therefore, at least two circumpolar stars are observed, within a short period of time, so that two more equations are obtained, namely

$$\phi - \delta' = \zeta'_A + k \tan \zeta'_A. \tag{10.11}$$
$$180 - \phi - \delta' = \zeta'_B + k \tan \zeta'_B. \tag{10.12}$$

The four equations (10.9)–(10.12) are now solved to give values of δ, δ', ϕ and k.

10.2.4 Horizontal refraction

When the Sun or Moon is observed rising or setting, the observed zenith distance of its centre is $90°$. For such a zenith distance, the refraction amounts to $35'$ and is called the **horizontal refraction**. Since the angular diameter of these bodies is about $30'$, they are, in fact, below the horizon when their centres are seen to be on the horizon. In the case of the Sun, therefore, horizontal refraction lengthens the time interval during which it is daylight, sunrise and sunset taking place earlier and later respectively than they would occur if refraction were absent. Tables which provide accurate sunrise and sunset times for any location take account of horizontal refraction—a point not mentioned in section 8.11.

Table 10.1 shows how quickly refraction diminishes with altitude. Because of this, an extended body like the Sun is decidedly oval in shape near rising and setting, the refraction of the upper and lower limbs being different.

Table 10.1. Variation of apparent altitude with angle of refraction.

Apparent altitude	Angle of refraction
$0°$	$35' \, 21''$
$1°$	$24' \, 45''$
$2°$	$18' \, 24''$
$3°$	$14' \, 24''$
$4°$	$11' \, 43''$
$10°$	$5' \, 18''$
$30°$	$1' \, 41''$
$60°$	$0' \, 34''$
$90°$	$0' \, 0''$

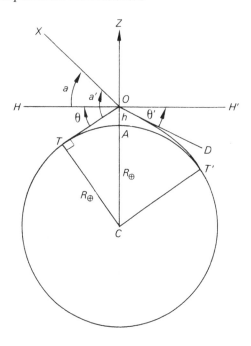

Figure 10.5. The angle of dip.

10.3   Correction for the observer's altitude

So far, the horizon plane has been taken to be the plane tangential to the Earth's surface at the observer's position and such that the line from observer to zenith is perpendicular to it. The altitude of a heavenly body has been defined as the angle from the foot of that vertical line in the horizontal plane to the body's position on the celestial sphere. Most professional observatories are at some height above sea-level and, as a consequence, the apparent horizon is greater than 90° from the zenith. This means that rising and setting times of any object are earlier and later respectively than if the observatory had the same latitude and longitude but was at sea-level. Any altitude measurement made relative to the apparent horizon needs correction to allow for the observer's height above sea-level. In the days of sextant navigation (see section 20.7.2) or when using a theodolite or alt-azimuth-mounted telescope for tracking artificial satellites, such corrections were very important.

If the observer is at some point O, h metres above point A at sea-level, the horizon, instead of being in a plane HOH' (figure 10.5), is at some angle θ below it where $\angle HOT$ is θ and OT is the tangent to the Earth's surface at T. The angle θ is called the **angle of dip**. Corrections to basic altitude measurements can be corrected by applying the angle of dip.

The observed altitude a' of a star X, say, is given by $\angle XOT$ and is related to the true altitude a by

$$a = a' - \theta. \tag{10.13}$$

Let the Earth's radius be $R_\oplus$ metres. Then

$$CT = CA = R_\oplus$$

and

$$CO = R_\oplus + h.$$

Triangle OTC is right-angled at T; $\angle HOC = 90°$; therefore, $\angle TOC = 90 - \theta$.

Now

$$\sin TOC = \cos\theta = \frac{R_\oplus}{r_\oplus + h}.$$

But θ is a small angle so we may write

$$\cos\theta = 1 - \frac{\theta^2}{2}.$$

Hence,

$$1 - \frac{\theta^2}{2} = \frac{R_\oplus}{R_\oplus + h}.$$

Then

$$\frac{\theta^2}{2} = \frac{R_\oplus + h}{R_\oplus + h} - \frac{R_\oplus}{R_\oplus + h} = \frac{h}{R_\oplus + h}$$

so that

$$\theta = \sqrt{\frac{2h}{R_\oplus + h}}. \qquad (10.14)$$

Now h is much less than $R_\oplus$ so we may write

$$\theta = \sqrt{\frac{2h}{R_\oplus}} \text{ rad.}$$

If θ is now expressed in minutes of arc, and we may take 1 radian to be 3438 minutes of arc,

$$\theta = 3438\sqrt{\frac{2h}{R_\oplus}} \text{ minutes of arc.}$$

Taking the unit distance to be the metre, we find that $R_\oplus = 6{\cdot}372 \times 10^6$ m, so that

$$\theta = 1{\cdot}93\sqrt{h} \text{ minutes of arc}$$

h being given in metres.

When refraction is taken into account, the path of the ray from the horizon at T' is curved as shown and, therefore, appears to come from a direction OD, so that the distance to the horizon is greater and the angle of dip is less.

The angle of dip θ' is then found to be given by the expression

$$\theta' = 1{\cdot}78\sqrt{h} \text{ minutes of arc}$$

where h is in metres.

It is of interest to consider at this stage a quantity related to the angle of dip, namely the distance to the apparent horizon.

In figure 10.5, this is the distance OT, neglecting refraction. Now

$$\cos TOC = \frac{OT}{OC}$$

or

$$OT = (R_\oplus + h)\sin\theta = (R_\oplus + h) \times \theta$$

since θ is small.

By equation (10.14), then,

$$OT = (R_\oplus + h)\sqrt{\frac{2h}{R_\oplus + h}} = \sqrt{2h(R_\oplus + h)}.$$

But h is much less than $R_\oplus$, so that the distance d to the apparent horizon is given by

$$d = \sqrt{2R_\oplus h}.$$

If h is expressed in metres and d in kilometres, we have

$$d = 3{\cdot}57\sqrt{h} \text{ km.} \tag{10.15}$$

Taking refraction into account, the distance d' of the apparent horizon is now given by OT' in figure 10.5. The expression (10.15) is modified to

$$d = 3{\cdot}87\sqrt{h} \text{ km} \qquad (d \text{ in metres}). \tag{10.16}$$

10.4 Geocentric parallax

In section 8.7, it was seen that the only practical way to give predictions of the positions of celestial objects within the Solar System was by using a geocentric celestial sphere. By doing this, the multitude of different directions of a particular object, say the Moon, as seen at one time by all observers scattered over the finite-sized Earth could be reduced to one direction only, that seen by a hypothetical observer at the Earth's centre. This is particularly necessary where artificial Earth satellites are concerned. Any satellite is so near the Earth that its apparent zenith distances, as measured by two observers in different geographical positions not too far apart, can differ by tens of degrees. For the Moon itself, this difference can be of order one degree, i.e. approximately twice its apparent angular diameter. For the nearby planets the differences can amount to a considerable fraction of 1 minute of arc. For the Sun the variation in apparent zenith distance at any one time due to differing geographical position of the observer is a few seconds of arc. Even the smallest of these quantities is much larger than the accuracy to which astronomical observations are taken.

The problem, therefore, arises of reducing topocentric observations of a body's position to the corresponding geocentric positions or of abstracting geocentric positions from an almanac and transforming them to the particular observer's local coordinate system. This problem is bound up historically with the attempts made in past centuries to measure accurately the Moon's distance and the value of the astronomical unit (section 13.8), i.e. the mean Earth–Sun distance in kilometres.

The principle employed was essentially the same as the age-old one used by the surveyor in obtaining the distances of specific points without actually measuring to these points with ruler or chain.

For example, in figure 10.6 we have a broad river whose width we wish to measure. Opposite a tree or post C on the far bank, we measure out a baseline AB. With a protractor we measure angles CAB and CBA. The information we now have (two angles and the included side) enables us to draw the triangle ABC to a suitable scale or, by the sine-formula, to calculate the distance BC across the river. The angle ABC, or angle p, is called the **parallactic angle**. It is, of course, equal to 180° minus the sum of the angles at A and B and is essentially the apparent change in the direction of the tree or post C as one moves from A to B. That this is so is easily seen if we draw BD parallel to AC. Then $\angle CBD = p$.

It is important that the base-line is not too small in comparison to the distance to be measured. Essentially, the quantity measured is the parallactic angle p, the angle subtended at the object by the base-line. If our accuracy of measurement is 1 second of arc, the order of accuracy in measuring the required distance is indicated in table 10.2. It is seen that the accuracy of the result falls off rapidly with

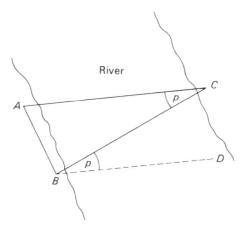

Figure 10.6. The use of parallax on the terrestrial scale.

Table 10.2. Measurement of distance by parallax.

For a base-line one unit long		
Distance to be measured (in units)	Order of size of parallactic angle	Order of accuracy result 1 part in
1	60°	200 000
10	6°	20 000
100	0·6° ~ 30′	2000
1000	0·06° ~ 3′	200

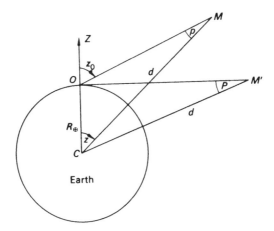

Figure 10.7. Geocentric and horizontal parallax.

increasing distance, the length of the base-line remaining constant. Much of the trouble in the use of this surveying method to measure distances within the Solar System lies in the smallness of available base-lines with respect to the distances involved.

Consider an observer at O (figure 10.7) on the surface of the Earth, centre C, the zenith being in the direction OZ. The measurement of the **apparent zenith distance** ZOM of the Moon, or z_0, is made. For a hypothetical observer at C, with the same zenith, the Moon's zenith distance is z, given by $\angle ZCM$, and called the **true** or **geocentric zenith distance**.

Let the Earth's radius be $R_\oplus$ with d the geocentric distance of the Moon. Angle OMC is the angle of parallax, p, namely the angle subtended at the object by the base-line OC. Then, from $\triangle OCM$,

$$\frac{\sin p}{R_\oplus} = \frac{\sin(180 - z_0)}{d}$$

or

$$\sin p = \frac{R_\oplus}{d} \sin z_0. \tag{10.17}$$

Suppose that the Moon happens to be on the observer's horizon, i.e. it is rising or setting. Its apparent zenith distance is $90°$ and it is given in figure 10.7 as the point M', where $CM' = d$ and $\triangle OCM'$ is right-angled at O. Let the angle of parallax $OM'C$ in this special case be denoted by P. Then

$$\sin P = \frac{R_\oplus}{d}. \tag{10.18}$$

We may, therefore, rewrite equation (10.17) in the form

$$\sin p = \sin P \sin z_0. \tag{10.19}$$

Now even for the Moon, the nearest of all natural celestial objects, P is only of order one degree, so that for the Moon, Sun and planets, both p and P are small angles at all times. Hence, equation (10.19) becomes

$$p = P \sin z_0. \tag{10.20}$$

The angle P, defined by equation (10.18), is called the **horizontal parallax**. If it is known and the apparent zenith distance z_0 is measured by an observer, the angle of parallax p can be calculated from equation (10.20) and used to obtain the geocentric or true zenith distance z, using the relation

$$z = z_0 - p. \tag{10.21}$$

Because z is less than z_0, it is obvious that the effect of parallax is to displace the object away from the zenith, in contrast to the effect of refraction.

Now the distances of the Sun, Moon and planets vary throughout the year so that their horizontal parallaxes also vary. Values of P are therefore given in *The Astronomical Almanac* for the Sun and Moon for every day of the year; formulas are also given enabling the horizontal parallax of the planets to be found.

Because the Moon revolves about the Earth in an elliptic orbit, a quantity called the **mean horizontal parallax** can be defined, given by equation (10.18) when the Moon's mean geocentric distance is inserted for d. Its value is $3423''$.

The corresponding quantity for the Sun is the **solar parallax**. Its accepted value is $8''794\,148$ and it is important since a knowledge of it and the size of the Earth enables equation (10.18) to be used to give the Earth's mean distance from the Sun. One of the major tasks in 18th, 19th and also 20th century astronomy has been the improvement of our knowledge of the value of the solar parallax. Methods of measuring the distances of the Sun, Moon and planets will be discussed later.

10.5   The semi-diameter of a celestial object

To the naked eye, both Sun and Moon are objects of finite size and, therefore, subtend an angle at the observer. The same is true of the planets but the angle is too small to be appreciated by the unaided eye.

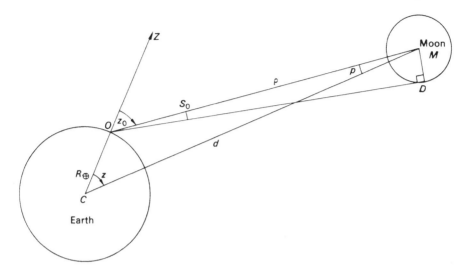

Figure 10.8. The semi-diameter of the Moon.

A quantity called the **angular semi-diameter** (or **semi-diameter**) of a celestial object can be defined to be the angle subtended at the Earth's centre by the radius of the object. Values of the semi-diameter or diameter for Sun, Moon and planets (except Pluto) are tabulated in *The Astronomical Almanac*. Obviously a knowledge of the distance of a celestial object and the measurement of its angular semi-diameter enables its linear semi-diameter, or radius, to be found.

To obtain the semi-diameter S of the Moon, say, the observer at O in figure 10.8 measures the Moon's apparent zenith distance $z_0(\angle ZOM)$ and the apparent semi-diameter $S_0(\angle MOD)$. Let $R_\oplus$ and $R_{\leftmoon}$ denote the radii of Earth and Moon respectively and let p be the value of the angle of parallax OMC at the time of observation, C and M being the centres of Earth and Moon respectively.

If ρ and d are the distances of the Moon's centre from the observer and the Earth's centre, we have, by definition,

$$\sin S = \frac{R_{\leftmoon}}{d}.$$

But

$$\frac{R_{\leftmoon}}{d} = \frac{R_{\leftmoon}}{R_\oplus} \cdot \frac{R_\oplus}{d} = \frac{R_{\leftmoon}}{R_\oplus} \sin P$$

so that

$$\sin S = \frac{R_{\leftmoon}}{R_\oplus} \sin P. \tag{10.22}$$

But both S and P are small angles, hence equation (10.22) may be written as

$$S = \frac{R_{\leftmoon}}{R_\oplus} \times P. \tag{10.23}$$

Now

$$\sin S_0 = \frac{R_{\leftmoon}}{\rho} = \frac{R_{\leftmoon}}{R_\oplus} \cdot \frac{R_\oplus}{d} \cdot \frac{d}{\rho} = \frac{d}{\rho} \cdot \frac{R_{\leftmoon}}{R_\oplus} \sin P$$

or

$$S_0 = \frac{d}{\rho} S. \tag{10.24}$$

In $\triangle OMC$, using the sine formula, we have

$$\frac{\sin COM}{d} = \frac{\sin OCM}{\rho}.$$

Hence,

$$S = \frac{\rho}{d} S_0 = S_0 \frac{\sin OCM}{\sin COM}.$$

Now $\angle COM = 180° - z_0$, and $\angle OCM = z_0 - p$, where $p = P \sin z_0$, using equations (10.20) and (10.21). Hence,

$$S = S_0 \frac{\sin(z_0 - P \sin z_0)}{\sin z_0}. \tag{10.25}$$

Measurement of the semi-diameter of a celestial object, therefore, gives a measure of its distance from the Earth in some arbitrary unit. In the cases of the Sun and the Moon, their semi-diameters are so large that careful measurements give quite accurate information about their changing distances from the Earth and so enable values of the eccentricities of their elliptical orbits about the Earth to be obtained. For the planets, with semi-diameters only a few minutes of arc, or less, measurement of their varying distances by means of a check on their changing semi-diameters is too rough to provide useful data.

When measuring the apparent zenith distance of a celestial object of finite angular size (e.g. the Moon) it is usually more accurate to measure to the upper or lower edge (or limb) of the object and then add or subtract the semi-diameter of the object.

10.6 Measuring distance in the Solar System

10.6.1 The Moon

The Moon is the only natural object in the Solar System whose distance can be found accurately using the classical parallax method.

In principle, it involves the simultaneous measurement of the meridian zenith distance of the Moon at two observatories with the same longitude but widely separated in latitude. The wide latitude separation provides a reasonably long base-line

Let observatories O_1 and O_2 have latitudes ϕ_1 N and ϕ_2 S. Let the meridian zenith distances of the Moon be z_{10} and z_{20} at these two observatories, that is $Z_{10} = \angle Z_1 O_1 M$ and $Z_{20} = \angle Z_2 O_2 M$ (see figure 10.9).

Then if z_1 and z_2 are the true (i.e. geocentric) zenith distances for observatories O_1 and O_2 respectively, with p_1 and p_2 the parallactic angles $O_1 MC$ and $O_2 MC$, we may write

$$\phi_1 + \phi_2 = z_1 + z_2 \tag{10.26}$$

$$z_1 = z_{10} - p_1 \qquad z_2 = z_{20} - p_2 \tag{10.27}$$

$$p_1 = P \sin z_{10} \qquad p_2 = P \sin z_{20} \tag{10.28}$$

where P is the value of the Moon's horizontal parallax when the observations are made.

By equations (10.28),

$$P = (p_1 + p_2)/(\sin z_{10} + \sin z_{20}).$$

By equations (10.27),

$$p_1 + p_2 = z_{10} + z_{20} - (z_1 + z_2).$$

By equation (10.26), then, we have

$$P = [z_{10} + z_{20} - (\phi_1 + \phi_2)]/(\sin z_{10} + \sin z_{20}) \tag{10.29}$$

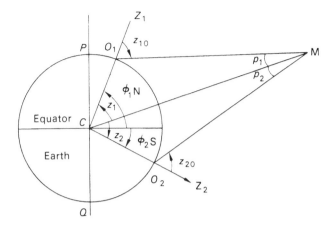

Figure 10.9. Measurement of the Moon's distance by observations from two sites.

so that P may be calculated from known and measured quantities. Knowing the radius of the Earth, the geocentric distance CM of the Moon can be found from equation (10.18).

In practice, the procedure is complicated by a number of factors. The observatories are never quite on the same meridian of longitude so that a correction has to be made for the change in the Moon's declination in the time interval between its transits across the two observatories' respective meridians.

In addition, the Moon is not a point source of light. Agreement, therefore, has to be made between the observatories as to which crater to observe, a correction thereafter giving the distance between the centres of Earth and Moon.

Due allowance must be made for the individual instrumental errors and the local values of refraction. In the case of the Moon, its parallax is so large that any residual error is not great.

Since the advent of radar, laser methods and lunar probes, including lunar artificial satellites, the Moon's distance can be measured very accurately. The limiting accuracy is partly due to the accuracy with which the velocity of light is known but also involves the accuracy with which the Earth observing stations' geodetic positions are known. Uncertainties are of the order of 0·1 of a *metre*, the average distance being 384 400 *kilometres*. See also section 15.9.

10.6.2 The planets

It will be shown later (section 12.7) that it is relatively easy to obtain accurately the distances of the planets from the Sun in units of the Earth's distance from the Sun. If, therefore, the distance of any planet from the Earth can be accurately measured in kilometres at any time, all the planetary heliocentric distances can be found in kilometres. In particular, the **astronomical unit** (the mean distance of the Earth from the Sun) can be obtained in kilometres.

All the classical parallax methods are hopelessly inaccurate where the distance of a planetary body is concerned. The Moon's distance is about thirty times the Earth's diameter, which is essentially the length of the available base-line. The distance of the asteroid Eros, at its closest approach, however, is some 20 000 000 km, or about 1500 times the length of the base-line. Table 10.2 (p 121) shows that it is not possible to obtain the geocentric parallax of Eros to high accuracy by this method.

Some improvement in accuracy is obtained by using one observatory, instead of two. In this way, the factors of different instrument errors and weather conditions, which control refraction, can be avoided. The base-line is provided by taking observations of the asteroid at different times so that in the time interval involved, the observatory moves due to the Earth's diurnal movement and orbital motion. The observing programme is carried out around the time of opposition when the planet is near the observer's meridian at midnight. Observations are made a few hours before midnight and a few hours after midnight. Before the development of radar, many observatories carried out such observing programmes, notably between 1900 and 1901 and between 1930 and 1931, when Eros was in opposition. The value of the solar parallax (see section 10.4) derived from these programmes was probably not any more accurate than one part in a thousand.

Nowadays, the use of powerful radio telescopes as radar telescopes has enabled the accuracy to be increased by a factor of at least one hundred. The geocentric distance of the planet Venus has been measured repeatedly by this method. It consists essentially of timing the interval between transmission of a radar pulse and the reception of its echo from Venus. This interval, with a knowledge of the velocity of electromagnetic radiation (the speed of light) enables the Earth–Venus distance to be found. Thus, if EV, c and t are the Earth–Venus distance, velocity of radio waves and time interval respectively,

$$EV = \tfrac{1}{2}ct.$$

Various corrections have to be made to derive the distance Venus-centre to Earth-centre. For example, the distance actually measured is the distance from the telescope to the surface of Venus. The effect of the change of speed of the radar pulse when passing through the ionosphere also has to be taken into account.

An even more accurate value of the solar parallax has been obtained by tracking Martian artificial satellites such as Mariner 9 over extended periods of time. Range and range rate measurements (i.e. line of sight distance and speed by radio tracking) allow the distance Earth-centre to Mars-centre to be found.

10.7   Stellar parallax

10.7.1   Stellar parallactic movements

The direction of a star as seen from the Earth is not the same as the direction when viewed by a hypothetical observer at the Sun's centre. As the Earth moves in its yearly orbit round the Sun, the geocentric direction (the star's position on a geocentric celestial sphere) changes and traces out what is termed the **parallactic ellipse**. Thus, in figure 10.10, the star X at a distance d kilometres is seen from the Earth at E_1 to lie in the direction E_1X_1 relative to the heliocentric direction SX'. Six months later, the Earth is now at the point E_2 in its orbit and the geocentric direction of the star is E_2X_2.

The concepts used in stellar parallax are analogous to those used in geocentric parallax. Thus, if we looked upon the Earth's orbital radius a as the base-line, we can define the star's **parallax** as P, given by

$$\sin P = \frac{a}{d}. \tag{10.30}$$

Because stellar distances are so great compared with the Earth's orbital radius, we can assume that the Earth's orbit to be circular. Angle P is small so that we can write equation (10.30) as

$$P = 206\,265\frac{a}{d}$$

where P is in seconds of arc.

Again, if we draw E_1Y parallel to SX in figure 10.10 and let $\angle YE_1X = \angle E_1XS = p$, we have, from $\triangle E_1XS$,

$$\frac{\sin p}{a} = \frac{\sin XE_1S}{d}$$

or

$$\sin p = \frac{a}{d}\sin XE_1S.$$

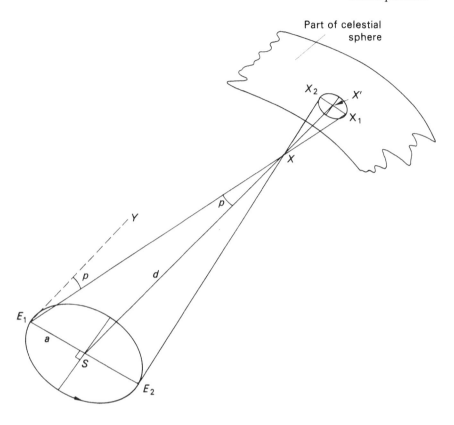

Figure 10.10. The parallactic ellipse (the size is exaggerated for clarity).

Using equation (10.30) and letting $\angle X E_1 S = \theta$, we obtain

$$\sin p = \sin P \sin \theta$$

or

$$p = P \sin \theta \tag{10.31}$$

since both p and P are small angles.

Thus, the star is displaced towards the Sun along the great circle through the star's heliocentric position and the Sun by an amount p, given by equation (10.31), where P is the star's parallax and θ is the angle between the Sun's direction and the star's. Whereas P remains constant, p and θ vary because of the Earth's movement in its orbit about the Sun.

10.7.2 The parallactic ellipse

We consider in more detail the path traced out on a geocentric celestial sphere by a star throughout the year. In figure 10.11, the star X has *heliocentric* right ascension and declination α and δ but because of stellar parallax it is displaced towards the Sun S_1 by an amount XX_1, given by equation (10.31), namely

$$XX_1 = P \sin X S_1. \tag{10.32}$$

Since the Sun moves round the ecliptic in one year and the star is always displaced by parallax towards the Sun, the star's path $X_1 X_2 X_3 \ldots$ during the year will be a closed curve on the celestial sphere, enclosing the star's heliocentric position X.

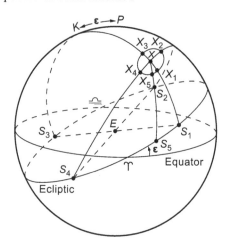

Figure 10.11. The projection of the parallactic ellipse on the celestial sphere.

Let S_1 be the point defined by the intersection of the great circle through the north pole of the ecliptic K and X with the ecliptic. Then S_1 is the position of the Sun when its angular distance from X is least. Let S_2 be the Sun's position three months later. By equation (10.31), we have

$$XX_2 = P \sin XS_2. \tag{10.33}$$

Three months later, the Sun is at the point S_3 and the star's displacement to X_3 is given by

$$XX_3 = P \sin XS_3. \tag{10.34}$$

But S_3 is the Sun's position six months after it occupied the position S_1. Hence, their longitudes are 180° different and

$$XS_3 = 180° = XS_1.$$

By equations (10.32) and (10.34), it is, therefore, seen that $XX_1 = XX_3$ in size.

In a similar manner, it can be seen that if S_4 is the position reached by the Sun six months after it left S_2, the parallactic displacement XX_4 of the star is equal to XX_2 in size but in an opposite direction.

Now S_2 and S_4 are the positions of the Sun that are farthest from the star's heliocentric position X, that is $S_4S_1 = S_2S_1 = 90°$, so that the star's parallactic shifts at such times are greatest. Any other shift, such as XX_5, will lie between XX_1 and XX_2 in size.

Hence, the apparent path of the star due to stellar parallax is an ellipse whose major axis is parallel to the ecliptic and has a value P arc sec. The value of the minor axis depends on the star's position. For example, if the star happened to be at K, the north pole of the ecliptic, the minor axis would be of length P arc sec, since all solar positions would be 90° from the star's heliocentric direction K. In other words the ellipse would be a circle. If, however, the star lay on the ecliptic, the ellipse would degenerate into an arc of the ecliptic, the minor axis vanishing.

10.7.3 The measurement of stellar parallax

After the publication (in 1543) of the Copernican theory of the Universe, which stated that the observed behaviour of planets could be as easily explained if it was assumed that the Earth revolved about the Sun, repeated attempts were made to measure the parallaxes of the brighter, and presumably nearby, stars. Almost 300 years were to elapse, however, before success was achieved, first by Bessel at Konigsberg in 1838, then by Henderson at the Cape of Good Hope and by F Struve at Dorpat soon after. The values they obtained showed why it took three centuries to detect the parallactic movements of stars.

For the star 61*Cygni*, Bessel found a parallax of $0''\!.314$; Henderson measured the parallax of α *Centauri* to be almost three-quarters of one second of arc while Struve showed that of *Vega* to be about one-tenth of 1 second of arc. These are very small angles. In fact, only 23 stars are known with parallaxes of $0''\!.24$ or greater, *Proxima Centauri* having a parallax close to $0''\!.75$.

The modern method of measuring a star's parallax involves the use of photographic plates or CCD detectors. In principle, records are taken six months apart of the area of the sky surrounding the star. If the star is near enough, the shift of the Earth from one side of its orbit to the other should produce a corresponding apparent shift of the star against the very faint stellar background (see figure 10.10). This shift will change the right ascension and declination of the star and it is essentially these changes in coordinates that are measured. Because the shifts are very small, they are measured using faint reference stars, so faint that they are presumably far enough away for their own parallactic displacements to be negligible. To fix our ideas, let us consider one such reference star only, with right ascension α_R and declination δ_R.

Let the heliocentric right ascension and declination of the parallax star be α and δ and let its apparent coordinates be α_1, δ_1 and α_2, δ_2 at the times the first and second records are taken.

Now the change in a star's right ascension due to parallax will be given by an expression of the form

$$\alpha' - \alpha = P \times F$$

where α', α are the star's apparent and heliocentric right ascensions, P is its parallax and F is a function of the star's equatorial coordinates, the Sun's longitude and the obliquity of the ecliptic. This function will have a particular value at any given date and this value, from a knowledge of the form of the function, can be calculated.

Let its values be F_1 and F_2 when the two records were made. Then

$$\alpha_1 - \alpha = P \times F_1 \qquad \alpha_2 - \alpha = P \times F_2.$$

Subtracting, we obtain

$$\alpha_1 - \alpha_2 = P(F_1 - F_2)$$

or, introducing the reference star's right ascension,

$$(\alpha_1 - \alpha_R) - (\alpha_2 - \alpha_R) = P(F_1 - F_2).$$

The quantities $(\alpha_1 - \alpha_R)$ and $(\alpha_2 - \alpha_R)$ are the differences between the right ascensions of the parallax star and the reference star and can be measured on a suitable measuring engine or with reference to the pixel grid of the detector. Hence,

$$P = \frac{(\alpha_1 - \alpha_R) - (\alpha_2 - \alpha_R)}{F_1 - F_2}.$$

In practice, several plates or frames are taken at each epoch and more than one reference star is used, the two epochs (separated by six months) being chosen so that the most advantageous value of $F_1 - F_2$ is obtained. The practical limit to this method from Earth-based telescopes is quickly reached. Only parallaxes greater than $0''\!.01$ can be measured at all reliably and only a few thousand stars have had their parallaxes measured in this way.

A major step forward in accuracy was the launching of the artificial Earth satellite *Hipparcos* by the European Space Agency in August 1989. Its 0.30 m telescope measured the positions, proper motions[1] and parallaxes of about 120 000 stars to an accuracy of better than $0''\!.002$. It also measured the brightnesses and colours of more than one million stars.

---

[1] A star's proper motion is its heliocentric angular shift in one year due to its space velocity relative to the Sun.

10.7.4   The parsec

We have seen that the distances of the stars are so great that any measured parallax is less than 1 second of arc. There is, therefore, a need to introduce a unit of length for use in describing such distances that will lead to convenient numerical values.

The **parsec** is the more usual unit used by astronomers: it is the distance of a celestial body whose *par*allax is 1 *sec*ond of arc. Now 1 second of arc is approximately $1/206\,265$ of a radian; therefore, a distance of 206 265 astronomical units will be the distance at which one astronomical unit subtends 1 second of arc. Hence, we may write

$$1 \text{ parsec} = 206\,265 \text{ AU}$$

or, using the accepted value of the AU, namely 149.6×10^6 km,

$$1 \text{ parsec} = 30.86 \times 10^{12} \text{ km approximately.}$$

A larger unit, the **kiloparsec** ($= 10^3$ parsec) is often used in expressing the distances of stars or the size of galaxies.

In popular books on astronomy, the **light-year** is often employed as a unit of length. As its name implies, it is the distance travelled by light in 1 year. The velocity of light is 299 792 km s^{-1} and there are 31.56×10^6 s in 1 year. Hence,

$$1 \text{ light-year} = 9.46 \times 10^{12} \text{ km, approximately}$$

so that

$$1 \text{ parsec is equal to about } 3.26 \text{ light-years.}$$

The use of the light-year as the unit of length does emphasize one important aspect of astronomical observations, namely that distant objects are seen not as they are now but as they were at a time when the light entering the observer's telescope left them. In the case of galaxies, this 'timescope' property of a large telescope is particularly important, enabling information to be obtained about the Universe in its remote past.

10.7.5   Extrasolar planets

Of accelerating interest is the possible detection of minute cyclical positional shifts in a star's position caused by the presence of planets in orbit about it. The detection of extrasolar planetary systems is of great importance but it offers extreme technical challenges requiring regular measurements of very small changes of a star's coordinates relative to other local field stars.

A feel for the problem can be appreciated by considering our own solar system as providing an example. Jupiter is the most massive planet orbiting the Sun. The centre of mass based on this two-body system is approximately at 1 solar radius (696 000 km) from the centre of the Sun. As seen from a large distance, say at some star, the Sun would appear to move by its diameter over an interval of time of one-half the orbital period of Jupiter. If the observer were at 10 parsecs (relatively close for a star), the apparent displacement over this time interval would be

$$= \frac{2 \times 6.96 \times 10^5}{10 \times 30.86 \times 10^{12}} \text{ radian}$$
$$= \frac{4.51 \times 10^{-9}}{206\,265} \text{ arc sec}$$
$$= 0.''00022$$

an extremely small angle.

Recent observations using other principles suggest that some stellar systems do indeed have planets of the order of the same size as Jupiter but orbiting much closer to the parent body with periods of revolution of a few days. With such shorter timescales, the necessary measurements would accrue more easily but unfortunately the amount of motion of the star about the centre of gravity of the system also reduces. As techniques improve in sensitivity, no doubt extrasolar planets will eventually be detected and monitored by refined positional measurements.

Problems—Chapter 10

Note: Take the value of the constant of refraction to be $60''3$, the value of horizontal refraction to be $35'$.

1. The observed zenith distance of a star was $25°$. Calculate its true zenith distance.
2. Neglecting the effect of refraction, find in kilometres the distance of the sea-horizon from the top of a cliff 200 m above sea-level.
3. The top of a mountain 1000 m in height can just be seen from a ship approaching the land on which the mountain is situated. If the observer's eye is 30 m above sea-level, how far is the ship from the mountain?
4. Show that at the spring equinox, for a place in latitude ϕ, the Sun will become visible at the top of the tower of height h feet about $13.94\sqrt{h}\sec\phi$ seconds earlier than it will at the foot of the tower. (Neglect refraction.)
5. The Moon's apparent zenith distance was measured to be $43°\,28'$. Calculate its true zenith distance if the value of the Moon's horizontal parallax was $60'$. (Neglect refraction.)
6. The maximum and minimum values of the Moon's semi-diameter are $16'5$ and $14'8$. Calculate the eccentricity of the Moon's orbit.
7. A geostationary satellite, in orbit about the Earth's equator, is at a distance from the Earth's centre of 4.2×10^4 km. Calculate its geocentric parallax (i.e. the positional angular shift between observing it at the equator and from one of the Earth's poles). Assume the Earth's radius $= 6.38 \times 10^3$ km.
8. What is the parallax of a star at the distance of (i) 25 parsecs, (ii) 94 light-years?
9. The parallaxes of two stars are $0''074$ and $0''047$. The stars have the same right ascensions, their declinations being $62°$ N and $56°$ N respectively. Calculate the distances of the stars from the Sun and the distance between them, in parsecs.
10. On a given day the Moon's horizontal parallax is $58'7$ and its semi-diameter is $16'0$. Find (i) the radius of the Moon in kilometres, (ii) the horizontal parallax when the semi-diameter is $15'0$, given that the Earth's radius is 6372 km.
11. Calculate (taking refraction into account) the dip of the sea-horizon and its distance when the observer's eye is 26.5 m above sea-level.
12. The Sun's semi-diameter and horizontal parallax are $16'$ and $8''79$ respectively. Calculate the Sun's diameter and distance given that the Earth's radius is 6372 km.
13. Calculate the displacements due to parallax in right ascension and declination of a star of parallax $0''38$ in right ascension 18^h and declination $72°\,30'$ N on (i) March 21st, (ii) June 21st, (iii) September 21st, (iv) December 21st.
14. A circumpolar star transits north of the zenith at upper culmination. If its apparent zenith distances are $17°\,14'\,32''$ and $67°\,29'\,51''$ respectively at upper and lower transit, calculate its declination and the latitude of the observer.
15. A star of parallax P has ecliptic longitude and latitude λ and β respectively. Show that the displacement in latitude due to annual parallax vanishes when the Sun's longitude $\lambda_\odot$ is given by

$$\lambda_\odot = \lambda \pm \frac{\pi}{2}.$$

   Prove that the maximum displacement in latitude is

$$P \sin \beta.$$

16. Find for approximately how many days the midnight Sun (Sun's upper limb above the horizon at midnight) is visible at a point on the Arctic Circle (a) neglecting refraction, (b) taking horizontal refraction into account. (Sun's angular semi-diameter $= 16'$; obliquity of the ecliptic $= 23°\,26'$; horizontal refraction $= 35'$.)

17. A planet (declination 30° N) is observed from a station in latitude 60° N six hours before and after meridian passage. The shift in position relative to the stellar background is found to be 5 seconds of arc. If the geocentric motion of the planet can be neglected, calculate its geocentric distance as a fraction of the Earth's distance from the Sun. (Solar parallax 8″.79.)

18. Find the true zenith distance of the Moon's centre from an observation that gave the apparent altitude of the Moon's lower limb above the sea-horizon to be 27° 45′.2. (Height of eye = 4·88 m; angular semi-diameter of Moon = 15′.4; horizontal parallax of Moon = 56′.5; constant of refraction = 60″.3.)

Chapter 11

The reduction of positional observations: II

11.1 Introduction

Throughout the 17th and 18th centuries, many attempts were made to measure stellar parallaxes. All failed. The success of Newtonian science in explaining the movements of the planets (including the Earth) about the Sun made the failure to detect the apparent shift of the brighter, and presumably nearer, stars due to the Earth's annual journey in its heliocentric orbit all the more exasperating. The improvement in accuracy in measuring stellar positions in those centuries was remarkable and yet no stellar parallax was observed. One answer, of course, was that the stars were so far away that in comparison with the distance to even the nearest one, the diameter of the Earth's orbit about the Sun was minute, so minute that the apparent shift of the star was too small to be detected.

Some hope that this view was false occurred in 1718 when Halley compared modern observations of stellar positions with those made by Hipparchus and Ptolemy. Hipparchus had observed about 140 BC while Ptolemy had been active during the 2nd century AD

Halley noticed that, even allowing for observational errors, the positions of the bright stars *Arcturus, Procyon* and *Sirius* were now different from those they had occupied one and a half millennia before. Halley suggested that these so-called fixed stars were not only moving through space with their own velocities but that in all likelihood every star, if observed long enough, would be seen to be in motion. This discovery of Halley's encouraged other astronomers to persevere with their attempts to measure stellar distances.

Among those astronomers was James Bradley. Quite often research with a particular aim in science leads to the discovery of something quite unexpected but as important as, or more important than, the original investigated effects. Bradley's attempt to measure stellar parallax was without success but he did make two important discoveries through it; one was **aberration**; the other was **nutation**.

11.2 Stellar aberration

By 1725 it had become obvious that any parallactic shift would be very small. Bradley, working at first with Molyneux, set up his telescope with great care. He used a meridian circle strapped vertically to a pillar so that it would remain rigidly fixed in position. The star he chose was γ *Draconis* because it transited almost exactly in the zenith. It was also bright and, therefore, would, with any luck, be among the nearest stars. Hence, the star entered the field of his vertically-mounted telescope each night and crossed it to disappear off the other side. Any change in the star's declination would be seen as a change in the star's path across the field, the path being 'higher' or 'lower' than usual. Moreover, because γ *Draconis* transited so near to the zenith, the correction for refraction would be very small and this was advantageous since the value of k used in the refraction formula might be in error.

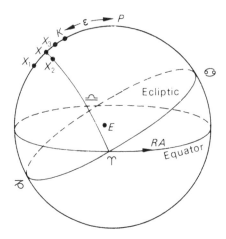

Figure 11.1. Transit observations of γ *Draconis*—the predicted effects of parallactic shift.

The coordinates for γ *Draconis* are RA $\sim 18^{\rm h}$, Dec $\sim 51°$. If we assume its right ascension to be exactly $18^{\rm h}$, its *heliocentric* position is at X in figure 11.1.

We have seen (section 10.7) that, because of stellar parallax, a star should be displaced towards the Sun by an amount $\Delta\theta$, where

$$\Delta\theta = P \sin\theta$$

P being the star's annual parallax and θ being the angle between the geocentric directions of star and Sun.

On December 21st, therefore, when the Sun is at ♑, the star will be displaced to X_1, where

$$XX_1 = P \sin X♑.$$

Now this will result in an apparent decrease of the star's declination by an amount XX_1, measured as an increase in the star's zenith distance by that amount, since Bradley observed the star when it transited.

Three months later, when the Sun is at ♈, the parallactic shift will be to a point X_2, in a direction that will be effectively along the parallel of declination through X. The star's declination will then be unaffected. About June 21st, when the sun has reached ♋, the displacement of the star will be to X_3, where

$$XX_3 = P \sin X♋ = P \sin(180 - X♑) = XX_1.$$

In this case the star's declination will be increased by the amount XX_1, resulting in a decrease in the star's zenith distance when Bradley measured it on the meridian.

Again, on or about September 21st, when the Sun was at ♎, the star's declination would be unaffected by the parallactic shift.

If we let q denote the angular quantity XX_1, Bradley expected that he would obtain a series of readings for the star's declination of the form shown in the first line of table 11.1.

Bradley did indeed find a small shift in the star's declination, the total change being of the order of 40 seconds of arc. It was oscillatory with a period of one year, the measured declinations being shown in the second line of table 11.1, with $q_1 \sim 20''$. As can be seen, however, the expected and observed declinations were three months out of phase with each other. Over the course of the next two years, Bradley observed a number of other stars in different parts of the sky and found that they too showed small yearly motions about their mean positions. In all cases, the maximum displacement was about

Table 11.1. The discovery of stellar aberration.

Date	December 21st	March 21st	June 21st	September 21st
Expected declination	$\delta - q$	δ	$\delta + q$	δ
Measured declination	δ	$\delta - q_1$	δ	$\delta + q_1$

40 seconds of arc and the time-tables of these stellar shifts were always three months out of step with the parallactic motion Bradley sought to detect.

It is said that Bradley hit upon the correct interpretation of these results during a pleasure trip on the river Thames. The sailing boat Bradley and his friends were in tacked to and fro. Bradley's attention was caught by the pennant at the masthead and by the way its direction changed every time the boat tacked. The crew told him that the pennant's direction was due to a combination of wind velocity and the boat's velocity at that moment.

Bradley realized the significance of the sailors' remarks.

Replace the wind by light coming from a star; replace the boat by the Earth moving around the Sun, the shifting course of the boat being now the continually changing direction of motion of the Earth in its circular orbit. Then the pennant indicates the apparent direction in which the star is seen, a direction dependent upon the velocity of light from the star and the velocity of the Earth in its orbit.

Bradley's own words are instructive:

> At last I conjectured that all the phenomena hitherto mentioned proceeded from the progressive motion of light and the Earth's annual motion in its orbit. For I perceived that, if light was propagated in time, the apparent place of a fixed object would not be the same when the eye is at rest, as when it is moving in any other direction than that of the line passing through the eye and object; and that when the eye is moving in different directions, the apparent place of the object would be different.

The fact that light travelled with a finite speed was known to Bradley and he also knew its approximate speed.

11.3   The velocity of light

In 1675, the Danish astronomer Roemer had measured the velocity of light by noting variations in the times of eclipses of Jupiter's satellites. Galileo himself had observed on many occasions that the satellites he had discovered disappear into the shadow cast by Jupiter and had suggested that, by constructing tables of the eclipse times, they could be used as astronomical clocks for finding ships' longitudes at sea. The disappearance or reappearance of a satellite could be looked upon as a light signal. If the velocity of light was infinitely large, these light signals would occur at regular intervals since the orbits of the satellites about Jupiter were very nearly circular.

Roemer found that the eclipses did not take place at absolutely regular time intervals. All four satellites' eclipses were sometimes early and sometimes late. He also noticed that the amounts by which they were early or late depended upon the positions of Jupiter and the Earth: eclipses were earliest when Jupiter was nearest the Earth and latest when Jupiter and the Earth were on opposite sides of the Sun. Roemer concluded correctly that the discrepancies were due to the velocity of light being finite, so that the light signal indicating the beginning or end of an eclipse took time to cross space from the vicinity of Jupiter to the Earth. This time varied because of the changing Earth–Jupiter distance. The maximum discrepancy between the earliest and latest eclipse times was then the time it

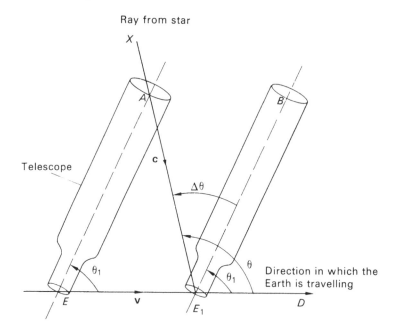

Figure 11.2. The effect of aberration.

took light to cross the diameter of the Earth's orbit. Roemer, from a rough knowledge of this distance and the time interval, calculated the velocity of light, obtaining a value not far removed from the modern value of 299 792·5 km s^{-1}.

11.4   The angle of aberration

Bradley remembered this work of Roemer, neglected for half a century. He also knew that the velocity of the Earth in its heliocentric orbit was about 30 km s^{-1} in a direction always at right angles to its radius vector (assuming the Earth's orbit to be circular). He now had all the information required to explain the phenomena he had observed and produce a formula to predict them.

In everyday life we are familiar with a number of examples embodying the principle involved. For instance, in a stationary car on a rainy day we see the raindrops stream downwards. But when the car moves, the raindrops' paths slant so that they appear to be coming from a direction between directly overhead and the direction in which the car is travelling.

In figure 11.2, let light from a star X enter a telescope at A so that the observer sees the star in the middle of the field of view. The telescope, because of the Earth's orbital velocity **v** km s^{-1}, is moving in the direction EE_1. By the time the light travelling with velocity **c** km s^{-1} ($c = 299\,792\cdot5$ km s^{-1}) has reached the foot of the telescope, the telescope has moved into the position E_1B. To the observer, the star appears to lie in the direction E_1B because the telescope has had to be tilted slightly away from the true direction of the star towards the instantaneous direction in which the observer is moving.

The **angle of aberration**, $\Delta\theta$, is obtained by considering $\triangle AEE_1$ in which the distances EE_1 and AE_1 are proportional to v and c respectively. If we take $\angle AE_1D = \theta$ and $\angle AED = \theta_1$, then

$$\frac{\sin AEE_1}{AE_1} = \frac{\sin EAE_1}{EE_1}$$

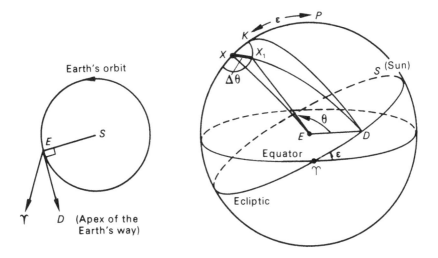

Figure 11.3. The aberrational ellipse.

or

$$\frac{\sin\theta_1}{c} = \frac{\sin(\theta - \theta_1)}{v}.$$

Hence,

$$\sin(\theta - \theta_1) = \frac{v}{c}\sin\theta_1.$$

Remembering that $\theta - \theta_1 = \Delta\theta$ is a small angle (the quantity $v/c \sim 1/10\,000$), and expressing $\Delta\theta$ in seconds of arc, we may write

$$\Delta\theta = 206\,265\frac{v}{c}\sin\theta$$

or

$$\Delta\theta = \kappa\sin\theta. \tag{11.1}$$

The constant κ is called the **constant of aberration**; its value is $20''496$. The angle of aberration $\Delta\theta$ obviously depends upon the value of θ, the angle between the star's direction and the instantaneous direction in which the observing is travelling. This is the direction D in which the Earth is moving, a direction known as the **apex of the Earth's way** (figure 11.3) and which has a longitude always 90° less than that of the Sun (ΥDS). The Sun possessed that longitude three months previously, which Bradley realized was the explanation why his observations were always three months out of step with the ones he expected. Just as in the case of stellar parallax, the star appears to move in an ellipse about its heliocentric position X and just as in the case of parallax, it may easily be shown that the major axis of this ellipse, of value $40''992$, is parallel to the ecliptic. The shape of this **aberrational ellipse** obviously depends upon the star's position on the celestial sphere. For a star on the ecliptic, the ellipse degenerates into an arc of length 2κ and for a star at the pole of the ecliptic, the ellipse is a circle.

11.5   The constant of aberration

The constant of aberration is officially defined as being equal to the ratio of the speed of a planet of negligible mass, moving in a circular orbit of unit radius, to the velocity of light. It is expressed in seconds of arc by multiplying this ratio by the number of seconds of arc in 1 radian. A value of $20''496$ is now adopted.

If we consider the Earth's orbit as circular, the Earth's distance from the Sun, a, to be $149\,600 \times 10^6$ m, the number of seconds in 1 sidereal year, T, to be $31 \cdot 56 \times 10^6$ and the velocity of light, c, to be $299\,792 \cdot 5 \times 10^3$ m s^{-1}, we have

$$\kappa = 206\,265 \frac{v}{c} = 206\,265 \frac{2\pi a}{cT}$$

giving

$$\kappa = 20'' \cdot 492.$$

In fact, the Earth's orbit has eccentricity $e = 0 \cdot 016\,74$ and the more correct expression for κ is

$$\kappa = 206\,265 \frac{2\pi a}{cT(1 - e^2)^{1/2}}. \tag{11.2}$$

From section 10.4, we may write

$$P = 206\,265 \frac{R_\oplus}{a}. \tag{11.3}$$

where P is the solar parallax (measured in seconds of arc) and $R_\oplus$ is the Earth's radius.

By equations (11.2) and (11.3), we have

$$\kappa P = \frac{2\pi R_\oplus (206\,265)^2}{cT(1 - e^2)^{1/2}}. \tag{11.4}$$

The quantities on the right-hand side of this expression being known accurately, it was formerly thought that the accurate measurement of κ would enable an accurate value to be obtained for the solar parallax and hence of the scale of the Solar System. Unfortunately the measurement of κ, by means of meridian circle observations in which the observed declinations of stars were found, is tied in with a phenomenon called the **variation of latitude**. The Earth's crust shifts slightly with respect to the Earth's axis of rotation so that the latitude of an observatory is not quite fixed. Modern methods of measuring the solar parallax and the value of the astronomical unit by radar are much more accurate and so the expression is best used to determine a value of κ from a knowledge of P.

In *The Astronomical Almanac* and other almanacs, tables are provided enabling the observer to allow for the effects of aberration on the right ascension and declination of the stars. Quantities C and D, depending upon the Sun's longitude, are given for every day of the year. Then if (α, δ) and (α_1, δ_1) are the equatorial coordinates of the positions of the star unaffected and affected by aberration,

$$\alpha_1 - \alpha = Cc + Dd$$
$$\delta_1 - \delta = Cc' + Dd'$$

where c, d, c', d' are functions of the star's coordinates.

11.6   Diurnal and planetary aberration

The rotation of the Earth on its axis produces aberration in an amount large enough to be taken into account in precise observational work.

Let an observer be at a place of latitude ϕ and let $R_\oplus$, c and T be the Earth's radius, the velocity of light and the number of seconds in one sidereal day. Then the observer O has velocity v due to the Earth's rotation, given by

$$v = \frac{2\pi R_\oplus \cos \phi}{T}$$

as easily seen by figure 11.4.

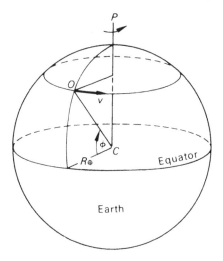

Figure 11.4. The velocity of an observer due to the rotation of the Earth.

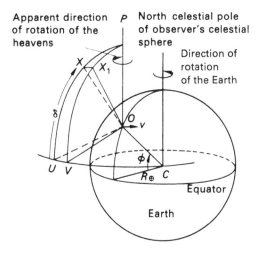

Figure 11.5. The effect of diurnal aberration.

The constant of diurnal aberration κ_1 is then given in seconds of arc by

$$\kappa_1 = 206\,265 \frac{v}{c}.$$

Evaluating, it is found that

$$\kappa_1 = 0''32 \cos\phi. \tag{11.5}$$

Let us consider an observation made of a star of declination δ when it crosses the meridian. As the observer moves in an easterly direction due to the Earth's rotation, diurnal aberration will cause the star to be seen east of its geocentric position. The star's observed time of transit is, therefore, late by an amount Δt, obtained in the following way.

The star's position at apparent transit is X_1 (figure 11.5) but its true position is X, where

$$XX_1 \approx 0''32 \cos\phi.$$

Now if XX_1 is an arc of the star's parallel of declination,

$$XX_1 = UV \cos \delta.$$

But UV is the delay in transit time since it is the difference in the hour angles of X and X_1. Hence, the delay time Δt is given by

$$\Delta t = UV = XX_1 \sec \delta = 0''32 \cos \phi \sec \delta$$

that is

$$\Delta t = 0^s021 \cos \phi \sec \delta.$$

The declinations at transit, and, therefore, the meridian zenith distances, of the stars are not affected by diurnal aberration.

In the case of planets, the light entering the terrestrial observer's eye comes from the direction the planet occupied at the epoch t_0 at which the light left the planet. In addition to the normal correction for stellar and diurnal aberration, therefore, a further correction is required to obtain the planet's direction at the epoch t_1 at which it is observed. The time interval $(t_1 - t_0)$ is obtained from the planet's geocentric distance divided by the velocity of light. If $d\alpha/dt$ and $d\delta/dt$ are the rates of change in geocentric right ascension and declination of the planet then the required corrections due to **planetary aberration** are given by $(t_1 - t_0) \, d\alpha/dt$ and $(t_1 - t_0) \, d\delta/dt$.

It may also be mentioned in passing that in the case of an artificial Earth satellite with an orbital velocity of order 7 km s^{-1}, the aberration in its observed position is of order 5 arc sec and must be allowed for in precise tracking.

11.7   Precession of the equinoxes

It seems certain that the constellation figures such as *Ursa Major, Orion, Andromeda,* etc, are at least four and a half millennia old and were created by people living between latitudes 34° and 36° north.

The arguments leading to such definite conclusions are remarkably straight-forward. It has been seen in section 8.5 that for an observer living in latitude $\phi°$ N, a region of the celestial sphere of angular radius ϕ degrees about the south celestial pole remains unseen below the observer's horizon. When we examine the distribution of these ancient constellations over the celestial sphere, we see (section 8.5) that they avoid a roughly circular area in the vicinity of the south celestial pole. This area must have remained below the horizon of the constellation-makers. The centre of this roughly oval-shaped area is difficult to determine but corresponds with the approximate position of the south celestial pole some thousands of years ago. Its radius is approximately 35°. Again, when we consider the distribution of the constellations, we find them arranged in rings symmetrical with the north celestial pole's position of about 4500 years ago. A more accurate date, in fact, may be found by drawing the rectangle that best fits each ancient constellation, then drawing a great circle that bisects the sides of the rectangle, as shown for two hypothetical constellations in figure 11.6. If there was no system in the origin of the constellations, no system should be found in the result. It is found, however, that most of the bisecting great circles run through a position the north celestial pole occupied in 2100 BC ±300 years.

The conclusion that a group of people living in latitude near 34° N, around 2800 BC, formed for some special purpose (probably navigational) the constellation figures receives quite independent confirmation from another source.

The Greek astronomer Hipparchus about 125 BC examined and criticized a poem by Aratus (of date about 250 BC) dealing with the celestial sphere called the Sphere of Eudoxus, which itself dates from about 400 BC. The ancient constellations appeared on this sphere and Aratus' poem not only described in detail these constellations but made a large number of astronomical statements. Hipparchus commented on the errors in these descriptions and statements and found it necessary to

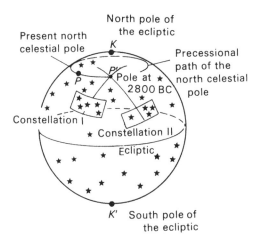

Figure 11.6. The north celestial pole when the constellations were set out.

rearrange a number of the constellation figures, in some cases adding bright stars clearly visible to him but for some reason omitted by the constellation-makers, in other cases pointing out that stars included in other constellations were no longer there. M W Ovenden has examined the large number of astronomical statements in Aratus's poem and shown that they are correct for a north and south celestial pole position of the time 2600 BC ±800 years and a latitude of 36° N ±1½°. Using a different set of statements in Aratus's poem, A E Roy derived a date when they were all correct of 2300 BC ±300 years. All estimates as to the date of establishment of the constellations, therefore, agree within the uncertainties.

It is quite likely, as Ovenden has argued, that Hipparchus' study of the poem of Aratus inspired him to make his great discovery of the **precession of the equinoxes**. Comparison of the celestial longitudes and latitudes of stars measured by him with similar measurements made by Timocharis and Aristillus in Alexandria 150 years before led him to conclude that, whereas the latitudes of the stars were constant, the longitudes had increased by about 50″ per annum. Hipparchus was forced to the conclusion that the relative positions of the ecliptic and the celestial equator were changing. The obliquity ε, as far as he could see, was constant. Because the celestial latitudes of the stars did not change, he deduced that the position of the ecliptic against the stellar background was also fixed. But because the longitude is increased secularly, he had to conclude that the vernal equinox ♈ was slipping backwards along the ecliptic at a rate of 50″ per year. The celestial equator, one of whose points of intersection with the ecliptic is ♈, therefore, had to be precessing. In figure 11.7 the process is illustrated with respect to a star X, of celestial longitude λ and latitude β.

One year later, its latitude is still of value β but its longitude has increased to value $\lambda+50''$ because the celestial equator has precessed, producing a new vernal equinox ♈$_1$ and a new north celestial pole P_1.

The motion of the pole is thus along a small circle, of radius ε, whose pole is the north pole of the ecliptic K. The backward movement of ♈ and ♎ is called the **precession of the equinoxes**, a full revolution occurring in some 26 000 years.

It is obvious from figure 11.7 that the right ascensions and declinations of the stars must change because of the phenomenon of precession and it also shows why the descriptions of the constellations in the poem of Aratus were in error. These constellations were probably formed 2500 years before Hipparchus' time—the centre of the constellation-less gap in the southern hemisphere gives the position occupied by the south celestial pole about 2600 BC, roughly one-tenth of a precessional period before. Some bright stars just above the horizon at the earlier epoch (i.e. able to rise and set diurnally)

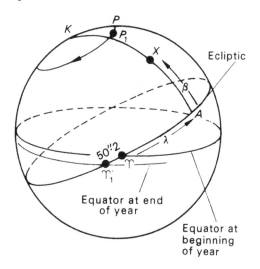

Figure 11.7. The movement of ♈ by precession.

and included in the constellations would, by Hipparchus' era, be below the horizon; other bright stars, unseen by the constellation-makers, would, however, have had their declinations changed sufficiently by the secular phenomenon of precession to appear above Hipparchus' horizon at nights.

It is seen then that the celestial equator and the vernal equinox change their positions. The problem arises of knowing at any time where these important references are. A method of achieving this is outlined in the next section.

11.8 Measuring the positions of ♈ and the celestial equator

Use is made of the fact that the Sun's annual path against the stellar background is the ecliptic. We know that the value of the Sun's maximum northerly declination is the obliquity ε. Measurements of the Sun's meridian zenith distance z at transit on a number of days around the summer solstice and a knowledge of the observer's latitude ϕ give a set of values for δ by means of the equation

$$\delta = \phi - z. \tag{11.6}$$

The maximum in the graph of δ against time is the value of the obliquity.

If a second set of observations of the Sun's meridian zenith distance were carried out near an equinox, its declination will again be calculated from equation (11.6).

From figure 11.8, using the four-parts formula, we obtain

$$\sin \alpha = \tan \delta \cot \varepsilon$$

from which the Sun's right ascension α at the time of the transit can be found.

At transit, the Sun's hour angle is zero so that the value of α is the local sidereal time of the hour angle of ♈. This enables the observatory sidereal clock error to be determined. The sidereal times of transits of stars may then be noted, giving their right ascensions. Their declinations can also be deduced from their meridian zenith distances by equation (11.6). In this way the equatorial coordinates of the stars (i.e. their positions with respect to the celestial equator and the First Point of Aries) can be found or, using the opposite viewpoint, the positions of the equator and ♈ against the stellar background can be determined at any time.

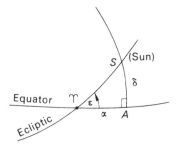

Figure 11.8. Measurement of the position of ♈.

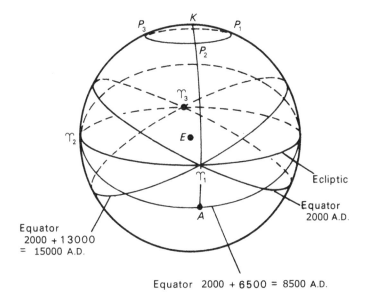

Figure 11.9. The celestial sphere illustrating the precessional movement of the celestial equator.

11.9 Effect of precession on a star's equatorial coordinates

Let us consider now the sort of changes made in the star's coordinates by precession, taking the precessional period to be 26 000 years. In figure 11.9, the north pole K of the ecliptic is set at the top of the diagram, with the ecliptic as shown. This is often a convenient orientation for a diagram in problems on precession where the north celestial pole and the celestial equator shift.

Let P_1, P_2, P_3, be the positions of the north celestial pole now (2000 AD say), one-quarter and one-half of the precessional period later. The corresponding positions of ♈ are ♈$_1$, ♈$_2$, ♈$_3$. The various celestial equators are also inserted.

Consider a star whose present right ascension and declination are 0^h and $0°$ respectively. The star is, therefore, at ♈$_1$ on the celestial sphere.

By 8500 AD, the First Point of Aries and the north celestial pole have moved to positions ♈$_2$ and P_2 because of precession, and the meridian from the north celestial pole of that date through the star will cut the equator of that date at A. The star's equatorial coordinates are then RA $= 6^h$, Dec $= \varepsilon°$ N, since ♈$_2A$ is of length $90°$. That this is so is seen from the fact that ♈$_2$ is the pole of the great circle KP_2♈$_1$.

By 15 000 AD, the positions of the north celestial pole and First Point of Aries are P_3 and Υ_3. The coordinates of the star are now $12^h, 0°$.

Obviously, after three-quarters of the precessional period has passed, the star's coordinates will be $18^h, \varepsilon°$ S, its coordinates returning to their 2000 AD values after one complete precessional period has elapsed.

The present pole star, *Polaris*, is about one degree from the north celestial pole and so is a fair indicator of the pole's position at the present day. If a star-map is obtained and a circle of radius ε ($\sim 23\frac{1}{2}°$) is traced about the north pole of the ecliptic, this circle will pass through the positions that have been or will be occupied by the north celestial pole. For example, some 4000 years ago, the bright star γ *Draconis* was about $4°$ from the pole; and in 12 000 years time, the pole will be within a few degrees of the bright star α *Lyrae (Vega)*.

Star catalogues give the coordinates of stars with respect to a particular equator and vernal equinox, i.e. the positions the equator and equinox occupied at a particular epoch, say, the beginning of the years 1950, 1975 or 2000. The stellar positions in *The Astronomical Almanac* and other almanacs are also with reference to the position of the equator and equinox of a certain date—in this case the beginning of the year for which the almanac has been produced, for example, 2000·0. This equator and equinox are referred to as the **mean equator** and **mean equinox** for 2000·0. Formulas exist, for a time interval of ten years or so, that can provide the **mean coordinates** (α_1, δ_1) of a star at the beginning of a year either before or after the reference epoch 2000·0. Thus, if (α, δ) are the mean coordinates of the star at 2000·0, the mean coordinates at 2001·0 are given by

$$\alpha_1 - \alpha = 3^s075 + 1^s336 \sin\alpha \tan\delta$$
$$\delta_1 - \delta = 20''043 \cos\alpha. \tag{11.7}$$

The problem of computing the changes in the mean coordinates due to precession when the time interval is more than a few years involves a more complicated procedure.

11.10   The cause of precession

Although Hipparchus discovered the phenomena caused by precession in the 2nd century BC, almost two millennia had to elapse before Newton gave an explanation. Newton showed that it was due to the gravitational attractions of Sun and Moon on the rotating, non-spherical Earth.

Newton argued on the following lines. If the Earth was perfectly spherical or if the Sun and Moon always moved in the plane of the celestial equator, their net gravitational attractions would act along lines joining these bodies to the Earth's centre and so would produce no tendency to tilt the Earth's rotational axis.

The situation is quite different, however. Not only is the Earth an oblate spheroid with an equatorial bulge but both Sun and Moon move in apparent orbits inclined to the plane of the Earth's equator. Let us consider the Sun. Its orbital plane is the plane of the ecliptic. Because of the asymmetry of the gravitational forces acting upon the Earth's particles of matter, the resultant solar attraction does not pass through the Earth's centre C but acts along the line DF (see figure 11.10). If the Earth were non-rotating, this force would, in time, tilt the planet until the equator and the ecliptic planes coincided. The Earth behaves, however, like a spinning-top. The spinning-top's weight W acts vertically downwards through its centre of mass G. The axis of spin AB precesses about the vertical AV, always moving at right angles to the plane instantaneously defined by AB and AV so that the top precesses, its spin-axis sweeping out a cone (see figure 11.11).

In the case of the Earth, the resultant gravitational pull of the Sun acting along DF causes the Earth's axis of rotation QP to sweep out a precessional cone of axis CK, CK being the direction to the north pole of the ecliptic, K. The north celestial pole, therefore, traces out a small circle about K, of radius ε, in the precessional period of 26 000 years.

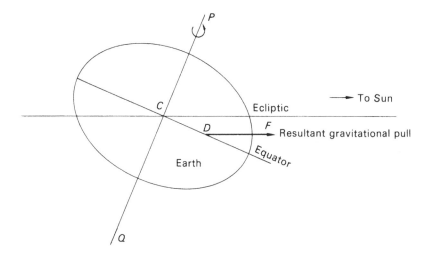

Figure 11.10. The gravitational pull of the Sun on the non-spherical Earth.

The effect of the Sun is complicated by the fact that not only is the Sun's angular distance above the plane of the equator changing throughout the year but its orbit is elliptical so that its gravitational pull varies. The complete analytical description of the Sun's precessional effect is, therefore, very complicated.

The Moon's effect is more complex still but, because its orbit is inclined at only a few degrees to the plane of the ecliptic, its main contribution to precession is similar to that of the Sun. The *uniform* motion of the north celestial pole along the small circle about the north pole of the ecliptic due to the combined effects of Sun and Moon is called **luni-solar precession**. As we have seen, this produces a continuous movement of ♈ along the ecliptic at a rate of 50″.2 per annum (the modern value) and a corresponding movement of the celestial equator.

11.11 Nutation

In attempting to measure stellar parallax, Bradley had been attempting to remove the last main objection to the heliocentric theory of the Solar System. He had failed to detect parallax but in discovering and interpreting correctly the phenomena of stellar aberration he had provided striking evidence that the Earth was in motion about the Sun.

He now knew that any parallactic displacement of the star had to be less than two seconds of arc, so closely did observed shifts and calculated shifts due to aberration agree. But we find him writing in 1729:

> I have likewise met with some small varieties in the declinations of other stars... which do not seem to proceed from the same cause (of aberration).

The small remaining differences between Bradley's observations and calculations did lead him after 19 years of careful work to the discovery of **nutation**. The displacements were found to have a period of $18\frac{2}{3}$ years and were interpreted by Bradley as being due to a minute wobble of the Earth's axis of rotation. He suggested also that the cause of this wobble was the Moon's action on the Earth's equatorial bulge. The plane containing the Moon's orbit about the Earth is itself precessing, so that in a period of $18\frac{2}{3}$, the line of intersection of the Moon's orbital plane with the plane of the ecliptic makes one complete revolution of the ecliptic.

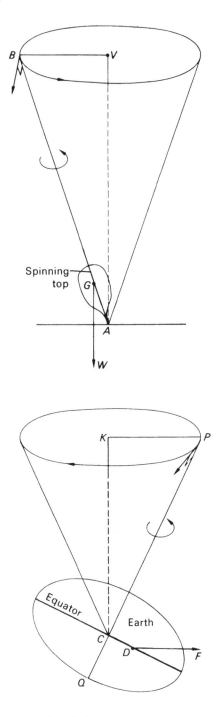

Figure 11.11. The conical movement of a spinning top compared with the precession of the Earth's pole.

In figure 11.12 we exhibit the result of nutation. Due to **luni-solar precession**, the north celestial pole arrives at some time at the point P_1, while the vernal equinox arrives at Υ_1. But because of **nutation**, there is an additional effect so that the true positions of the pole and the vernal equinox are

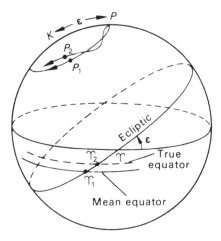

Figure 11.12. The effect of nutation on the position of ♈.

actually P_2 and ♈$_2$. The **true** equator therefore is, in general, different from the *mean* equator and the *true* coordinates of a star are defined with respect to the true equator and vernal equinox.

Formulas exist by which the mean coordinates (α, δ) at 2000·0, say, can be converted into true coordinates (α_1, δ_1) at 2000·0, June 24th, UT $18^h\,00^m$, for example. These formulas are in the form

$$\alpha_1 - \alpha = Aa + Bb + E$$
$$\delta_1 - \delta = Aa' + Bb'. \tag{11.8}$$

where the quantities A, B and E are tabulated in *The Astronomical Almanac* and other almanacs for every day of the year. The other quantities, a, a', b, b', are functions of the star's coordinates.

Because of the nutational wobble in the Earth's axis of rotation, the obliquity of the ecliptic ($K P_2$ in figure 11.12) varies about its mean value $K P_1$. The amplitude on either side of the mean value is about $9''\!.2$.

11.12   The tropical and sidereal years

In section 9.7, it was stated that the year used in civil life is based on the **tropical year**, defined as the interval in time between successive passages of the Sun through the vernal equinox. We now see that because of precession, the vernal equinox moves along the ecliptic at an average rate of $50''\!.2$ per annum in the direction opposite to that in which the Sun moves. The tropical year (of length 365·2422 mean solar days) is, therefore, shorter than the **sidereal year**, defined to be the time it takes the Sun to make one complete circuit of the ecliptic. Its length is found from the relation that

$$\frac{1\ \text{tropical year}}{360° - 50''\!.2} = \frac{1\ \text{sidereal year}}{360°}.$$

giving the length of the sidereal year as 365·2564 mean solar days.

Problems—Chapter 11

Note: Take the precessional period to be 26 000 years, the obliquity of the ecliptic to be $23\frac{1}{2}°$ and the constant of aberration to be $20''\!.49$.

1. A star in the ecliptic has celestial longitude 42°. If the Sun's longitude is 102°, calculate the change in the star's longitude due to aberration.

2. A star has a celestial longitude of 0°. What is the star's apparent longitude on (i) March 21st, (ii) June 21st, (iii) September 21st, (iv) December 21st?

3. Given that the constant of aberration for an observer on Earth is $20''49$ and that the distance of Mars from the Sun is 1.5 astronomical units, calculate the value of the constant of aberration for an observer on Mars assuming circular orbits.

4. Find the present longitude of the point which will be the First Point of Aries 5200 years hence. Show that the north celestial pole will then be about 27° from the present pole.

5. The star 36 *Draconis* is near the north pole of the ecliptic. Neglecting its parallax, at what part of the year will (i) its right ascension, (ii) its north declination, be (a) maximum, (b) minimum, due to aberration?

6. Show that for a star whose true position is on the celestial equator, the change in its declination due to aberration is

$$-\kappa \sin \varepsilon \cos \lambda_{\odot}$$

where κ is the constant of aberration. ε is the obliquity of the ecliptic and $\lambda_{\odot}$ the longitude of the Sun.

7. Calculate the displacements in celestial longitude and latitude, due to aberration, of a star in celestial latitude $+20°$, the Sun's longitude at the instant concerned being the same as that of the star.
   If the stellar parallax is $0''32$, what are the corresponding displacements due to parallax?

8. The celestial latitude and longitude of a star at the present time are $5°\ 14'$ N, $327°\ 47'$. Calculate the time which must elapse before the star becomes an equatorial star.

9. Show that at any place, and at any instant, there is a position of a star such that the effect of annual aberration is equal and opposite to that of refraction. Find the zenith distance of the star at midnight on the shortest day, given that the ratio of the constants of refraction and aberration is 2.85.

10. Assuming the obliquity of the ecliptic to be constant, show that 13 000 years hence a star whose present coordinates are 18^{h} and $47°$ S will then lie on the celestial equator. What will then be its corresponding right ascension?

11. If at a place in north latitude ϕ, the star in problem 10 just comes above the horizon at the present date at upper culmination, find (i) the value of ϕ, (ii) the star's altitude at upper culmination 13 000 years hence at the same place.

12. The present equatorial coordinates of a star are 3^{h} and $0°$. Calculate when next the star's declination will be zero.

13. If the date in the year when *Sirius* rises just after sunset changes by one day in 71 years, calculate (approximately) the value of the constant of precession, and the difference in the lengths of the tropical and sidereal years, given that the length of the tropical year is 365.2422 mean solar days.

14. Calculate (i) the ecliptic coordinates, (ii) the equatorial coordinates, referred to the present pole and equator, of the point on the celestial sphere which will, after one-third of the precessional period, become the north celestial pole.

15. The present ecliptic longitude and latitude of a star are $32°$ and $25°$ N, respectively. Find the declination and right ascension of the star at its minimum distance from the pole and calculate the time when this will occur.

Chapter 12

Geocentric planetary phenomena

12.1 Introduction

On July 21st, 1969, an event unique in the history of mankind took place. For the first time ever, a human being observed the heavens from the surface of a celestial object other than the Earth. When Neil Armstrong stepped onto the Moon's surface, he gave mankind a shift in perspective that until then had had to be imagined, for all previous direct views of the Sun, Moon, planets and stars had been made from one position, namely the Earth.

Observations of celestial objects from the Earth are usually referred to as geocentric observations although, strictly speaking, it should be remembered that any particular observation is made from a point on the Earth's surface and has to be 'corrected' or 'reduced' to the Earth's centre before it can be said to be truly geocentric.

The Earth's changing position round the Sun complicates the apparent behaviour of celestial objects (their geocentric behaviour); they have observed movements that in part are due simply to the velocity of the observer's vehicle—the Earth—just as objects in a landscape viewed from a moving car appear to have movements they do not, in fact, possess.

Paradoxically, the nearest and the farthest celestial objects are the least affected by the Earth's movement, though for different reasons.

The nearest natural object of any size, the Moon, accompanies the Earth round the Sun, moving about the Earth in an orbit that is roughly elliptical. The apparent movement of the Moon against the stellar background is, therefore, almost entirely due to its orbital motion; the component of shift caused by the observer viewing the Moon from the rotating surface of the Earth is small. If we, therefore, observe the Moon's sidereal position at a given time each night, we find that the Moon moves eastwards against the stellar background by approximately 13° per 24 hours. Its sidereal period of revolution about the Earth is found to be about $27\frac{1}{3}$ days.

The farthest celestial objects are the galaxies. They are at distances so great compared to the diameter of the Earth's orbit that the Earth's yearly journey about the Sun cannot affect their observed positions on the celestial sphere. Only 120 000 of the nearest stars in our own galaxy are close enough to have had their annual parallactic shifts measured accurately by the *Hipparcos* satellite (see section 10.7.3).

The Sun's apparent sidereal movement is due principally to the Earth's orbital movement. We have seen that the Sun appears to move in a great circle—the ecliptic—at a rate of about 1 degree per day, returning to any particular stellar position in one year. Because the orbit is an ellipse, equal angles are not swept out in equal times by the line joining the centres of Earth and Sun.

In the case of the planets, their geocentric movements are the most complicated of all, being compounded of their own orbital movements and the Earth's. The ancients observed that these objects not only 'wandered' with respect to the celestial background but occasionally appeared to change their

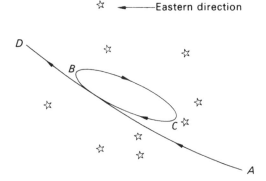

Figure 12.1. A loop in the apparent path of a planet.

minds, retracing their steps for some time before resuming their former progress or curling in a great loop before advancing once again.

In general, the motion was eastwards, or **direct**, that is in the direction of increasing right ascension. In figure 12.1, this direct motion is shown in the sections of the planet's apparent path *AB* and *CD*. Between *B* and *C*, however, the motion is reversed and is said to be **retrograde**. The points *B* and *C*, where the planet is changing its direction of travel, are called **stationary points**.

12.2   The Ptolemaic System

The main theory of the Solar System left by the Greeks to post-Roman Europe was a geocentric one. It was given in Ptolemy's *Almagest* and so bears his name. Its acceptance by astronomers lasted about 15 centuries.

The Ptolemaic theory sought to describe the apparent movement of all the heavenly bodies and indeed predict their future positions. It did so successfully; all the apparent motions of the Sun, Moon, planets and stars were adequately accounted for. In figure 12.2 the main features of the Ptolemaic System are sketched.

The Earth was the fixed centre of the Universe.

The stars were fixed to the surface of a transparent sphere which rotated westwards in a period of one sidereal day.

The Sun and the Moon revolved about the Earth.

The large circles centred on the Earth were called **deferents** and the small circles centred on the large circles or on the line joining Earth and Sun were called **epicycles**. The planets moved in the epicyclic orbits whose centres themselves moved in the directions shown.

Because Mercury and Venus were never seen far from the Sun (they were always evening or morning objects), the centres of their epicycles were fixed on the line joining Sun to Earth.

To agree with observation, the radii joining Mars, Jupiter and Saturn to the centres of their epicycles were always parallel to the Earth–Sun line.

It is worth noting at this point that errors in the descriptions and diagrams of Ptolemy's System by some modern commentators reveal an unjustified contempt for a theory that was remarkably successful in accounting for the known phenomena of the celestial sphere.

This basic theory was indeed a good first approximation to the observed behaviour of the heavenly bodies. Further modifications improved the 'fit'. For example, to account for some of the observed irregularities in the planets' motions, it was supposed that the deferents and epicycles had centres slightly displaced from the Earth's centre and the deferent circles respectively. Slight tilts were given to some of the circles.

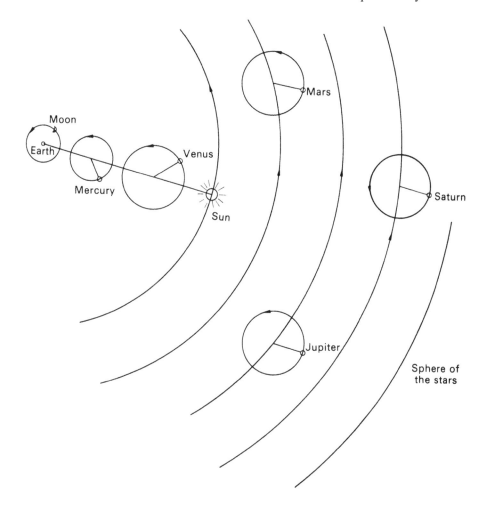

Figure 12.2. The Ptolemaic System.

When the Arabian astronomers of the Middle Ages accumulated more accurate observations of the planets, they found that the Ptolemaic theory had to be modified still further, epicycles being added to the epicycles until the system became cumbersome and, to some at least, unconvincing.

12.3 The Copernican System

Nicholas Copernicus (1473–1543 AD) formulated a theory of the Universe that was heliocentric, i.e. the Sun became the centre of the Universe, the Earth being just another planet. Nearly all the known motions of the planets were accounted for by supposing them to revolve about the Sun, in circular orbits but with their centres slightly displaced from that of the Sun. He also had to keep a few small epicycles.

The Moon revolved about the Earth as before but the Earth rotated on its axis, thus accounting for the diurnal rotation of the heavens.

Copernicus perfected his system over thirty years, his great book *De Revolutionibus Orbium Coelestium* being published only in the year of his death. A preface, not written by Copernicus, suggested that his heliocentric system need not be supposed to be closer to the truth than the Ptolemaic

System but could be looked upon as merely a simpler mathematical device. It is probable, however, that Copernicus himself believed it to be a physical reality. Certainly, it aroused deep psychological and religious opposition, displacing as it did the Earth, the home of mankind, from the centre of the Universe to the status of a satellite.

There were indeed two scientific objections that could be made, illustrating that a little learning is a dangerous thing.

Opponents of the Copernican theory pointed out that if the Earth moved round the Sun (a) the Moon would be left behind, (b) the brighter, and, therefore nearer stars, would show a parallactic yearly shift against the background of the fainter and presumably more distant stars. Neither of these events was observed.

The vindication of the Copernican theory had to await the work of three men, Tycho Brahé (1546–1601 AD), Johannes Kepler (1571–1630 AD) and Galileo Galilei (1564–1642 AD). We will consider Galileo's work first of all.

12.4   The astronomical discoveries of Galileo

In 1609, Galileo built his first telescope and initiated a revolution in astronomy. Among his discoveries with that telescope and improved instruments he constructed in later years were the following:

- He saw the Milky Way resolved into stars.
- He discovered sunspots. From their movements he measured the period of rotation of the Sun to be close to 27 days.
- When he turned his telescopes to the Moon he observed mountains, craters and plains. He also measured the Moon's apparent wobble on its axis, enabling us to see more than half of the lunar surface.
- The planets showed disks of appreciable size.
- Venus showed phases like the Moon.
- He discovered Jupiter had four moons of its own.
- Saturn was seen to have two blurred appendages which Galileo was quite unable to interpret.
- The Pleiades, the star group containing six or seven stars as seen with the unaided eye, had at least 36 stars when viewed with the telescope.
- The three nebulous stars in Praesepe became 40 in number.

Two discoveries, in particular, gave Galileo tremendous satisfaction, enabling him to dismiss the Ptolemaic System in favour of the Copernican one.

To those who said that the Moon would be left behind if the Earth moved, he could point to Jupiter's satellites. Even in the Ptolemaic System, Jupiter moved, yet its four moons were not left behind.

The phases of Venus were even more convincing as an argument in favour of the heliocentric theory.

Undoubtedly, the planets shone by the reflected light of the Sun. Their phases, as observed from the Earth, would depend upon whether or not the Earth was the centre of the Solar System.

In the Ptolemaic System (figure 12.3(a)), Venus could never lie on the other side of the Sun from the Earth (in superior conjunction as we would now put it). In the telescope, it would always remain crescent in phase, and its apparent angular size would not change greatly.

In the Copernican System (figure 12.3(b)), Venus could be in superior conjunction, V_4. Before it became lost in the Sun's glare, it would show an almost full phase, rather like full Moon. When it reappeared some days later on the other side of the Sun, having passed behind the Sun, its phase would again be almost full but would be waning, instead of waxing. As the planet moved round in its orbit from V_4 to V_2 via V_1, not only would the phase wane until a thin crescent was seen just before V_2 but

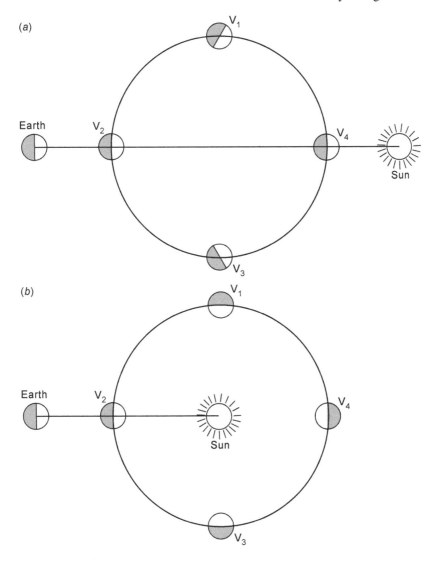

Figure 12.3. The phases of Venus (*a*) the Ptolemaic System (*b*) the Copernican System.

its apparent size would increase markedly as it approached the Earth. After it passed V_2 and entered the other half of its synodic journey, its size would decrease and its phase wax from a thin crescent to almost full Venus once more.

Observations of Venus by Galileo showed that Venus changed phase and angular size exactly in accordance with the Copernican System's predictions.

12.5   Planetary configurations

At this stage it is convenient to define a number of terms frequently used in describing planetary configurations with respect to the Sun and the Earth. In figure 12.4, *E* refers to the Earth and *S* represents the Sun.

Planets are divided into two classes: those whose orbits lie within the Earth's orbit are called

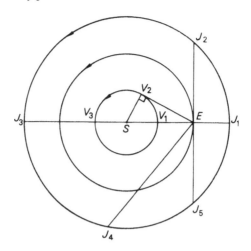

Figure 12.4. Planetary configurations.

inferior planets; those with orbits outside the Earth's orbit are **superior** planets. Thus, in figure 12.4, the letters V and J refer respectively to an **inferior** and a **superior** planet.

We assume for the moment that planets move round the Sun in circular, coplanar orbits. The actual planetary orbits are ellipses of low eccentricity in planes inclined only a few degrees from each other so that the terms defined below are still applicable in the real cases.

A superior planet on the observer's meridian at apparent midnight is said to be in **opposition** (configuration SEJ_1). It is obvious that an inferior planet cannot ever be in opposition.

A planet whose direction is the same as that of the Sun is said to be in **conjunction** (configurations EV_1S, ESV_3, ESJ_3): an inferior planet can be in **superior conjunction** (configuration ESV_3) or in **inferior conjunction** (configuration EV_1S).

The angle the geocentric radius vector of the planet makes with the Sun's geocentric radius vector is called the planet's **elongation** (for example, angle SEV_2 or SEJ_4). It is obvious that an inferior planet has zero elongation when it is in conjunction and **maximum elongation** (less than $90°$) when its geocentric radius vector is tangential to its orbit (configuration SEV_2). The elongation of a superior planet can vary from zero (configuration SEJ_3) to $180°$ (configuration SEJ_1). When its elongation is $90°$, it is said to be in **quadrature** (configurations SEJ_2 and SEJ_5). These quadratures are distinguished by adding **eastern** or **western**. In figure 12.4, the north pole of the ecliptic is directed out of the plane of the paper so that J_5 and J_2 are in eastern and western quadratures respectively.

Another useful term relating the position of a planet with respect to the Sun and Earth is that of **phase angle**. This angle is defined by the Sun, the planet and the Earth. For the inferior planets, it can take a value in the range $0°$ to $180°$. In figure 12.4, the configuration SV_2E depicts a phase angle of $90°$. For superior planets, the phase angle lies in the range $0°$ to a **maximum value**, this being determined according to the distance of the planet from the Sun.

12.6   The synodic period

An important concept in geocentric phenomena is the **synodic period** of a celestial object. This is the time interval between successive similar configurations of the object, the Sun and Earth. To fix our ideas, we consider the synodic period S of a planet.

Let T_1 and T_2 be the **sidereal** periods of revolution of two planets P_1 and P_2 about the Sun. Assume that the planets move in circular, coplanar orbits. In a circular orbit, the radius vector sweeps

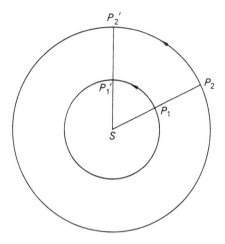

Figure 12.5. Successive similar configurations $(SP_1P_2, SP_1'P_2')$ of two planets.

out equal angles in equal times, that is the planet's angular velocity n is constant. Let the angular velocities of planets P_1 and P_2 be n_1 and n_2 respectively. Then,

$$n_1 = \frac{360°}{T_1} \qquad n_2 = \frac{360°}{T_2} \qquad (12.1)$$

since the sidereal period is the time it takes the planet to make one complete revolution of the stellar background. [*Note*: the stellar background is *fixed*; it only appears to rotate because the Earth rotates on its axis, so that in one sidereal period a planet's heliocentric radius vector (the line joining the centres of planet and Sun) sweeps out 360°.]

Now the planet nearer the Sun has a smaller period of revolution than the one farther away, so that

$$T_1 < T_2.$$

Hence,

$$n_1 > n_2$$

and the radius vector SP_1, therefore, gains on the radius vector SP_2 by $(n_1 - n_2)$ degrees per day.

In figure 12.5 let the configuration SP_1P_2 denote the positions of the Sun and the two planets at a particular epoch. Then, without at the moment specifying which planet is the Earth, it is clear that a synodic period S will have elapsed by the time the next configuration $SP_1'P_2'$ occurs. During the time interval S_1, the radius vector of P_1 will have gained 360° on P_2's radius vector.

But SP_1 gains on SP_2 by $(n_1 - n_2)$ degrees per day, so that it gains 360° in time S, where

$$S \times (n_1 - n_2) = 360°$$

or, using relations (12.1),

$$S \left(\frac{360}{T_1} - \frac{360}{T_2} \right) = 360$$

giving

$$\frac{1}{S} = \frac{1}{T_1} - \frac{1}{T_2}. \qquad (12.2)$$

Case (a). If the planet is inferior, T_1 refers to the planet's sidereal period and T_2 refers to the Earth's.

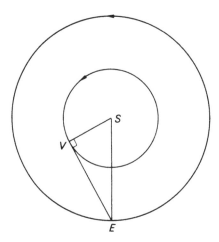

Figure 12.6. Measurement of inferior planet's distance at maximum elongation.

Case (b). If the planet is superior, T_1 refers to the Earth's sidereal period and T_2 refers to the planet's.

Example 12.1. The synodic period of Venus was found to be 583·9 days. If the length of the year is 365·25 days, calculate the sidereal period of Venus.

By relation (12.2),

$$\frac{1}{S} = \frac{1}{T_1} - \frac{1}{T_2}$$

so that we have, since Venus is an inferior planet,

$$\frac{1}{583\cdot9} = \frac{1}{T_1} - \frac{1}{365\cdot25}$$

or

$$\frac{1}{T_1} = \frac{365\cdot25 + 583\cdot9}{583\cdot9 \times 365\cdot25}$$

giving

$$T_1 = 224\cdot7 \text{ days.}$$

12.7 Measurement of planetary distances

It is possible to obtain the relative sizes of the planetary orbits in the following ways:

Case (a). Let the planet be an inferior one. Then we have seen (section 12.5) that the maximum elongation of the planet occurs when the planet's geocentric radius vector EV (figure 12.6) is perpendicular to the planet's heliocentric radius vector SV. By carefully measuring the elongation $\angle SEV$ on a series of nights around maximum elongation, it is possible to obtain a value for the maximum elongation. At this time $\triangle SEV$ is right-angled at V. Hence,

$$\frac{SV}{SE} = \sin(SEV_{\text{max}})$$

where the value of $\angle SEV$ is known.

The quantity SV/SE is, therefore, the distance of the planet from the Sun in units of the Earth's distance from the Sun.

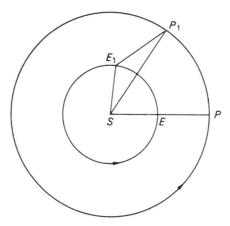

Figure 12.7. Measurement of the distance of a superior planet.

Case (b). The planet is a superior one. In this case the problem is more difficult but can be solved if the planet's synodic period S is known.

Let the planet P be in opposition at a given time with the Sun and the Earth in positions S and E as shown in figure 12.7. The elongation is $180°$.

After t days have elapsed, the Earth's radius vector SE_1 has moved ahead of the planet's radius vector, reducing the elongation from $180°$, at opposition, to the value given by $\angle SE_1P_1$. This value can, of course, be measured.

In t days, $\angle ESP$ will have increased from $0°$, at opposition, to a value θ, given by

$$\theta = (n_\oplus - n_P)t$$

where $n_\oplus$ and n_P are the angular velocities of Earth and planet in their orbits.

Using relations (12.1), we may write

$$\theta = 360\left(\frac{1}{T_\oplus} - \frac{1}{T_P}\right)t$$

where $T_\oplus$ and T_P are the sidereal periods of Earth and planet respectively.

Then, by equation (12.2),

$$\theta = 360\frac{t}{S}.$$

Since t and S are both known, θ can be evaluated, that is $\angle E_1SP_1$ is calculated. Hence, $\angle E_1P_1S$ can be found from the relation,

$$\angle E_1P_1S = 180 - \angle SE_1P_1 - \angle E_1SP_1.$$

Using the sine formula of plane trigonometry, we have

$$\frac{\sin P_1E_1S}{SP_1} = \frac{\sin E_1P_1S}{SE_1}$$

or

$$\frac{SP_1}{SE_1} = \frac{\sin P_1E_1S}{\sin E_1P_1S}$$

giving the distance of the planet from the Sun, again in units of the Earth's distance from the Sun.

Proceeding in this way, it is, therefore, possible to construct an accurate model of the Solar System in terms of the distance of the Earth from the Sun without knowing the scale absolutely. Such a model was obtained in the 16th century. It was a far more difficult problem, only satisfactorily solved with the advent of radar, to measure the scale.

12.8   Geocentric motion of a planet

Once the sidereal period T of a planet is found, also the mean heliocentric distance a, the velocity V of the planet in its orbit (assumed circular) can be calculated. Thus,

$$V = \frac{2\pi a}{T}.$$

For two planets, therefore, the ratio of their orbital velocities is given by

$$\frac{V_2}{V_1} = \left(\frac{a_2}{a_1}\right)\left(\frac{T_1}{T_2}\right) \tag{12.3}$$

where the subscripts 1 and 2 refer to the inner and outer planet respectively. Putting in values, it is found that the velocity of the inner planet is greater than that of the outer.

As we shall see later, Kepler found that for any planet,

$$a^3 \propto T^2$$

or

$$a^3 = kT^2 \tag{12.4}$$

where k is a constant.

Then by equation (12.4),

$$T = \left(\frac{a^3}{k}\right)^{1/2}.$$

According to Kepler, therefore, we may write for the two planets,

$$T_1 = \left(\frac{a_1^3}{k}\right)^{1/2} \qquad T_2 = \left(\frac{a_2^3}{k}\right)^{1/2}.$$

Substituting for T_1 and T_2 in equation (12.3), we obtain,

$$\frac{V_2}{V_1} = \left(\frac{a_1}{a_2}\right)^{1/2}. \tag{12.5}$$

In figure 12.8, the orbits of the Earth and a planet are shown, the radii of the orbits (assumed circular and coplanar) being a and b units respectively. Since the planet is a superior one, b is greater than a.

At opposition, the positions of the planet and Earth are P_1 and E_1 and their velocity vectors are shown as V_P and $V_\oplus$ respectively, tangential to their orbits.

Now by equation (12.5), $V_P < V_\oplus$ and so the angular velocity of the planet as observed from the Earth is

$$\frac{V_P - V_\oplus}{P_1 E_1}$$

and is in a direction opposite to the orbital movement. It is, therefore, *retrograde* at opposition.

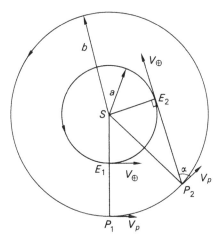

Figure 12.8. The velocities of the Earth and a superior planet produce retrograde motion of the planet (P_1) at opposition and direct motion of the planet (P_2) at quadrature.

At the following quadrature, the positions of planet and Earth are P_2 and E_2, where $\angle SE_2P_2 = 90°$. The Earth's orbital velocity $V_\oplus$ is now along the line P_2E_2 but the planet's velocity V_P has a component $V_P \sin \alpha$, perpendicular to E_2P_2. The other component $V_P \cos \alpha$ lies along the line P_2E_2 and, like the Earth's velocity $V \oplus$, does not contribute to the observed angular velocity of the planet. This geocentric angular velocity at quadrature is, therefore,

$$\frac{V_P \sin \alpha}{E_2 P_2}$$

and is seen to be in the same direction as the orbital movement. It is thus *direct* at quadrature.

12.9   Stationary points

Some time between opposition and quadrature, the planet's geocentric angular velocity must change from being retrograde to being direct. The point where it is neither retrograde nor direct is said to be a **stationary point**. We wish to obtain an expression for the elongation E at a stationary point in terms of the distances of planet and Earth from the Sun. From such an expression and the measured value of the elongation, the planet's heliocentric distance can be calculated.

It is also possible to obtain an expression giving the value of the angle θ between the heliocentric radius vectors of the planet and Earth when the stationary point is reached, in terms of the two planets' distances b and a respectively. This expression, with a knowledge of the synodic period of the planet, enables a prediction to be made of the time of the next stationary point after an opposition of the planet.

In figure 12.9, let the positions of Earth and planet at a stationary point be E and P. The velocity of P relative to E must lie along the geocentric radius vector EP if the planet appears stationary. This velocity is represented by the line PB; the parallelogram $PABC$ is the parallelogram of velocities where $PA = V_P$, $PC = -V_\oplus$ and PB is the resultant of these velocities.

Since the orbits are circular, the angle θ between the heliocentric radius vectors must be the angle PCB between the velocity vectors. Also it is readily seen that $\angle APB = 90 + \phi$.

Extend CP to meet the line SE produced at D. Then $\angle DPA = \theta$ and $\angle EPD = 90 - (\theta + \phi)$. Hence, in $\triangle PCB$, using the sine formula, we have

$$\frac{\sin[90 - (\theta + \phi)]}{V_P} = \frac{\sin(90 + \phi)}{V_\oplus}$$

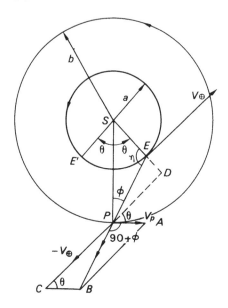

Figure 12.9. Superior planet at a stationary point.

or

$$\cos(\theta + \phi) = \frac{V_P}{V_\oplus} \cos \phi. \tag{12.6}$$

Now in any plane triangle XYZ with sides x, y, z and opposite angles X, Y, Z, we have

$$x = y \cos Z + z \cos Y.$$

Applying this well-known formula to $\triangle SEP$, we may write

$$SP = PE \cos \phi + SE \cos \theta$$

i.e.

$$b = PE \cos \phi + a \cos \theta. \tag{12.7}$$

Applying it again to $\triangle SEP$, we obtain

$$SE = SP \cos \theta + PE \cos SEP$$

or

$$a = b \cos \theta - PE \cos(\theta + \phi) \tag{12.8}$$

since $\angle SEP = 180 - (\theta + \phi)$.
  Equations (12.7) and (12.8) give

$$b - a \cos \theta = PE \cos \phi$$
$$b \cos \theta - a = PE \cos(\theta + \phi)$$

so that

$$\frac{b - a \cos \theta}{b \cos \theta - a} = \frac{\cos \phi}{\cos(\theta + \phi)}. \tag{12.9}$$

Using equations (12.6) and (12.5), we obtain

$$\frac{b - a\cos\theta}{b\cos\theta - a} = \frac{V_\oplus}{V_P} = \frac{b^{1/2}}{a^{1/2}}.$$

Hence, rearranging, we find that

$$\cos\theta = \frac{a^{1/2}b^{1/2}(a^{1/2} + b^{1/2})}{a^{3/2} + b^{3/2}}. \tag{12.10}$$

If b, expressed in units of the Earth's distance from the Sun, has values β, with the value of a being unity, equation (12.10) may be written as

$$\cos\theta = \frac{\beta^{1/2}(1 + \beta^{1/2})}{1 + \beta^{3/2}}. \tag{12.11}$$

By symmetry, it is obvious that when the Earth was at the point E' (figure 12.9), where $\angle ESP = \theta$, there was another stationary point. The total time during which the planet will be seen to move in the retrograde direction is hence the time it takes the Earth's radius vector to advance through an angle 2θ with respect to the planet's radius vector. It will be given by t_R where

$$t_R = \frac{2\theta}{360} \times S = \frac{\theta S}{180} \tag{12.12}$$

S being the planet's synodic period.

The time that elapses between opposition and the next stationary point is obviously $t_R/2$.

The time during a synodic period that the planet's motion is direct is t_D, where

$$t_D = \left(\frac{360 - 2\theta}{360}\right) \times S = \left(1 - \frac{\theta}{180}\right)S. \tag{12.13}$$

Going back to $\triangle SEP$, we may write

$$\frac{a}{b} \cdot \sin\eta = \sin\phi \tag{12.14}$$

where η is the elongation at a stationary point.

Now $\eta = 180 - (\theta + \phi)$, so that from equation (12.6), we have

$$-\cos\eta = \frac{V_P}{V_\oplus}\cos\phi$$

or

$$\frac{V_\oplus}{V_P}\cos\eta = -\cos\phi. \tag{12.15}$$

Squaring and adding equations (12.14) and (12.15), and using relation (12.5), we obtain

$$\frac{a^2}{b^2}\sin^2\eta + \frac{b}{a}\cos^2\eta = 1.$$

Hence, substituting $(1 - \cos^2\eta)$ for $\sin^2\eta$, there results

$$\left(\frac{b}{a} - \frac{a^2}{b^2}\right)\cos^2\eta = 1 - \frac{a^2}{b^2}$$

giving

$$\cos^2 \eta = \frac{a(b^2 - a^2)}{b^3 - a^3}.$$

Now $b^3 - a^3 = (b - a)(b^2 + ab + a^2)$, so that

$$\cos^2 \eta = \frac{a(a + b)}{a^2 + ab + b^2}$$

$$\sin^2 \eta = \frac{b^2}{a^2 + ab + b^2}$$

and

$$\tan^2 \eta = \frac{b^2}{a(a + b)}. \tag{12.16}$$

Now ϕ lies between $0°$ and $90°$, so that a consideration of equations (12.14) and (12.15) shows that, at the stationary point, the elongation η must lie between $90°$ and $180°$. This removes the ambiguity of sign in expression (12.16).

12.10   The phase of a planet

It was already known to the ancients that planets differed from stars by their geocentric motion against the stellar background. In Galileo's telescopes, the planets revealed other differences. Whereas stars remained point-like, planets showed disks whose sizes varied with time. Their appearance was obviously that of a sphere being illuminated from a particular direction and seen from another, and whose distance from the observer also varied. In other words, they exhibited **phases**. We have seen that by studying the sequence of phases and changes in apparent angular size of Venus, Galileo was able to show that the Ptolemaic System was not in accord with the facts.

A general expression for the phase of a planet is now obtained.

In figure 12.10, a planet, centre P, is observed from the Earth, E, when its elongation is $\angle SEP$, S being the Sun's position. We let $\angle SPE = \phi$, the phase angle, as before.

The hemisphere ACB of the planet (where APB is perpendicular to SP) will be illuminated by the Sun. From the Earth, however, only the part CPB of this lit-up hemisphere will be seen, CD being perpendicular to PE.

If we draw BF to meet CD at right angles, then it is obvious that the fraction of the planet's diameter seen illuminated will be CF/CD.

The **phase**, q, is defined to be this fraction. Then

$$q = \frac{CF}{CD} = \frac{CP + PF}{CD}. \tag{12.17}$$

But $PB = CP = PD =$ the radius of the planet. The angle $BPD =$ angle $SPE = \phi$, since angles SPB and EPD are both right angles. Hence, $PF = PB \cos\phi$, and equation (12.17) becomes

$$q = \tfrac{1}{2}(1 + \cos\phi). \tag{12.18}$$

Consider the case of an inferior planet.

At inferior conjunction, $\phi = 180°$, so that $q = 0$; the planet's dark hemisphere is turned towards the Earth, the planet's phase just about to be 'new' (cf the new Moon).

At superior conjunction, $\phi = 0°$, so that $q = 1$; the planet's phase is 'full' (cf the full Moon).

When $\phi = 90°$, $q = 0.5$ and the phase is such that half the illuminated disc is visible (cf the Moon at first and third quarter). Obviously all intermediate values of ϕ from $0°$ to $180°$ are possible, enabling all phases to be seen.

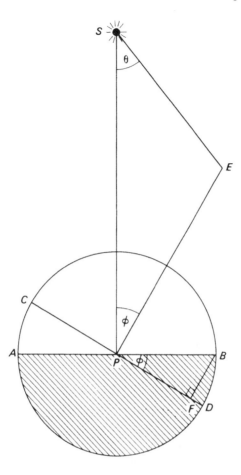

Figure 12.10. The phase of a planet.

In the case of a superior planet, $0 \leq \phi < 90°$, the value of zero occurring when the planet reaches opposition. The phase is, therefore, unity at this time and as ϕ increases to maximum (less than 90°), the phase wanes but never gets as low as one-half. The planet is always seen from the Earth with more than half its disc illuminated and the phase (to borrow another term from the Moon's cycle of phases) is always **gibbous**.

The planet Mars can be seen in a distinctly gibbous appearance but because of the large heliocentric distances of the giant outer planets, values of the phase angle ϕ are so small that these planets appear to be seen as though at the full phase.

The calculation of the phase at any time after the time of inferior conjunction of an inferior planet or the opposition of a superior planet is straightforward.

In figure 12.10, let $\angle ESP = \theta$. Then in $\triangle SEP$, using the sine formula,

$$\frac{SE}{\sin \phi} = \frac{SP}{\sin[180 - (\theta + \phi)]}$$

or, if a and b are lengths of the radius vectors SE and SP respectively,

$$b \sin \phi = a \sin(\theta + \phi).$$

Expanding the right-hand side and rearranging, we obtain

$$\tan \phi = \frac{a \sin \theta}{b - a \cos \theta}. \tag{12.19}$$

Equation (12.19) gives ϕ when a, b and θ are known. The angle θ is readily calculated from the relation

$$\theta = 360 \times \frac{t}{S} \tag{12.20}$$

where S is the planet's synodic period and t is the time that has elapsed since inferior conjunction (inferior planet) or opposition (superior planet).

Having found ϕ, equation (12.18) will then give the phase.

12.11   Improvement of accuracy

The expressions derived in previous sections in this chapter were obtained under the assumption that the planets move about the Sun in circular, coplanar orbits. In fact, the actual planetary orbits are ellipses of low eccentricity in planes inclined a few degrees to each other. The derivation of more accurate expressions taking the eccentricities and inclinations of the orbits into account is beyond the scope of this text and, in any case, gives values that are only a few per cent different from those given by the simple expressions obtained here.

Problems—Chapter 12

Note: Take the length of the sidereal year to be 365·25 days. Assume all orbits are circular and coplanar, unless otherwise stated.

1. A planet's elongation is measured as 125°. Is it an inferior or superior planet?
2. The sidereal period of Mercury is 88 days. What is its synodic period?
3. What is the maximum possible elongation of Venus, given that its distance from the Sun is 0·723 AU?
4. The synodic period of Jupiter is 398·9 days. What is its sidereal period?
5. The heliocentric distance of Venus is 0·723 AU. On a certain day, the planet's phase was $\frac{1}{12}$; calculate its elongation.
6. Calculate the ratio of the Earth's orbital speed to that of the planet Neptune, given that the distance of Neptune from the Sun is 30·06 AU.
7. The synodic period of Mars is 779·9 days and its heliocentric distance is 1·524 AU. Find its phase 85 days after opposition.
8. An asteroid is orbiting the Sun in a circular orbit of radius 4 AU. Calculate the ratio of its angular diameters at opposition and quadrature.
9. The planet Mars reaches a stationary point 36·5 days after opposition, its elongation being measured to be 136° 12′. Given that the planet's orbital period is 687·0 days, calculate the distance of Mars from the Earth in astronomical units at the stationary point, also the planet's phase.
10. During a synodic period of Venus, the elongation η at time t_1 is the same as the elongation at time t_2 ($t_2 > t_1$), the planet being on the same side of the Sun at both times. If the phase at t_2 is three times the phase at t_1, find
    (i)   the value of η,
    (ii) the interval $(t_2 - t_1)$ in days,

    given that the heliocentric distance of Venus is 0·723 AU and its synodic period 583·9 days.
11. Calculate the length of time during which Jupiter has retrograde motion in each synodic period, given that Jupiter's heliocentric distance is 5·2 AU and its sidereal period is 11·86 years.
12. At a point on the Moon's equator, two astronauts notice that their Apollo spacecraft, in a circular equatorial orbit, transits directly overhead with an observed sidereal angular velocity ω_1, and sets a few minutes later

with an observed angular velocity ω_2. If the ratio ω_1/ω_2 is 10·25, calculate the approximate height of the spacecraft above the Moon's surface (in lunar radii) assuming (i) the spacecraft's orbital motion is retrograde to the Moon's direction of rotation, (ii) the spacecraft's orbital period to be very short compared with the Moon's period of rotation.

13. The planet Jupiter, orbital period 11·86 years, was in opposition on 1968, February 10th. When was it next in conjunction? Calculate when it was in opposition in 1969. In what year after 1969 is there no opposition? How frequent are such years?

14. At opposition a superior planet, heliocentric distance b AU, has a geocentric angular velocity against the stars of $-\omega_1$ degrees per day. At the next quadrature, the measured value is $+\omega_2$ degrees per day. Show that

$$\frac{\omega_2}{\omega_1} = \frac{1}{b}\left(\frac{b-1}{b^{1/2}-1}\right).$$

15. Neglecting the sizes of Venus and the Earth, show that a transit of Venus across the Sun's disc lasts about 8 hours. Take the synodic period of Venus to be 584 days, its orbital radius as 0·723 AU and the Sun's angular diameter to be 32′.

Chapter 13

Celestial mechanics: the two-body problem

13.1 Introduction

By his telescope observations, Galileo showed that the Ptolemaic theory failed as an adequate description of planetary geocentric phenomena. The Copernican System, however, was able to embrace the new discoveries regarding the phases of Venus. No-one, however, was able as yet to give any satisfactory explanation of why the movements of the planets were as observed or why the Moon revolved about the Earth. Half a century had to pass before the explanation was given by Sir Isaac Newton (1642–1727 AD). His work was built on the foundations laid by Tycho Brahé (1546–1601 AD), Johannes Kepler (1571–1630 AD) and Galileo Galilei (1564–1642 AD).

13.2 Planetary orbits

13.2.1 Kepler's laws

Johannes Kepler, from the study of the mass of observational data on the planets' positions collected by Tycho Brahé, formulated the three laws of planetary motion forever associated with his name. They are:

(1) **The orbit of each planet is an ellipse with the Sun at one focus.**
(2) **For any planet the radius vector sweeps out equal areas in equal times.**
(3) **The cubes of the semi-major axes of the planetary orbits are proportional to the squares of the planets' periods of revolution.**

Kepler's first law tells us what the shapes of the planetary orbits are and gives the position of the Sun within them.

Kepler's second law states how the angular velocity of the planet in its orbit varies with its distance from the Sun.

Kepler's third law relates the different sizes of the orbits in a system to the periods of revolution of the planets in these orbits.

At the time of their formulation these laws were based upon the most accurate observational material available. The great Danish astronomer, Tycho Brahé, had worked in the pre-telescope era of astronomy but the accuracy of his observations had been of a high standard, certainly high enough in the case of the Martian orbit to convince Kepler that the age-old idea of circular orbits had to be discarded in favour of elliptical ones.

Kepler's laws are still very close approximations to the truth. They hold not only for the system of planets moving about the Sun but also for the various systems of satellites moving about their primaries. Only when the outermost retrograde satellites in the Solar System are considered, or close satellites of

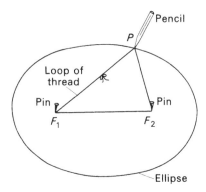

Figure 13.1. Method of drawing an ellipse.

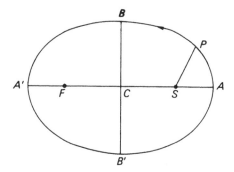

Figure 13.2. An elliptical planetary orbit.

a non-spherical planet, do they fail to describe in their usual highly accurate manner the behaviour of such bodies. Even then, however, they may be used as a first approximation.

We now examine them in more detail.

13.2.2 Kepler's first law

The law states that the orbit of a planet is an ellipse with the Sun at one focus.

A well-known way of drawing an ellipse is to insert two pins some distance apart at F_1 and F_2 in a sheet of paper (see figure 13.1), place a loop of thread over them, hold it taut by means of a pencil and then run the pencil along the path allowed by the tight loop. The figure obtained is an ellipse. The two positions occupied by the pins are called the **foci**.

If the two pins are placed nearer to each other and the operation is repeated with the same loop, it is found that the ellipse is more circular than the previous one. The ellipse, in fact, becomes a circle in the limiting case where the two pins occupy the same position, i.e. only one pin is used.

Figure 13.2 shows an elliptical planetary orbit, with the Sun, S, at one of the foci, as Kepler's first law states. The other focus, F, is often called the empty focus.

Then the line AA' is the **major axis** of the ellipse, C is the centre and, therefore, CA and CA' are the **semi-major axes**. Likewise BB' is the **minor axis**, with CB and CB' the **semi-minor axes**. If a and b denote the lengths of the semi-major and semi-minor axes respectively, then

$$b^2 = a^2(1 - e^2)$$

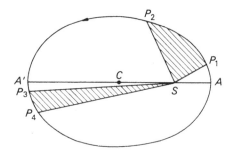

Figure 13.3. An orbital ellipse, illustrating Kepler's second law.

where e is the **eccentricity** of the ellipse, a quantity defined by the relation

$$e = CS/CA.$$

The eccentricity gives an idea of how elongated the ellipse is. If the ellipse is a circle, $e = 0$, since S and F are coincident with C. The other limit for e is unity, obtained when the ellipse is so narrow and elongated that the empty focus is removed to infinity.

In fact, the planetary orbits are almost circular so that e is a small fraction.

Let P be the position of the planet in its orbit. Then SP is the planet's radius vector. The planet is said to be at **perihelion** when it is at A. It is then nearest the Sun, since

$$SA = CA - CS = a - ae = a(1 - e).$$

When the planet is at A' it is said to be at **aphelion**, for it is then farthest from the Sun. We have, then,

$$SA' = CA' + CS = a + ae = a(1 + e).$$

The angle ASP is called the **true anomaly** of the planet.

13.2.3 Kepler's second law

This law states that the rate of description of area by the planet's radius vector is a constant.

Let us suppose that in figure 13.3, the planet's positions at times t_1, t_2, t_3 and t_4 are P_1, P_2, P_3 and P_4. Then between times t_1 and t_2 its radius vector has swept out the area bounded by the radius vectors SP_1, SP_2 and the arc P_1P_2. Similarly, the area swept out by the radius vector in the time interval $(t_4 - t_3)$ is the area SP_3P_4.

Then Kepler's law states that

$$\frac{\text{area } SP_1P_2}{t_2 - t_1} = \frac{\text{area } SP_3P_4}{t_4 - t_3} = \text{constant.} \tag{13.1}$$

If $t_2 - t_1 = t_4 - t_3$, then the area $SP_1P_2 = $ area SP_3P_4.

In particular, if the area is the area of the ellipse itself, the radius vector will be back to its original position and Kepler's second law, therefore, implies that the planet's **period of revolution** is constant.

Let us suppose the time interval $(t_2 - t_1)$ to be very small and equal to interval $(t_4 - t_3)$. Position P_2 will be very close to P_1, just as P_4 will be close to P_3. The area SP_1P_2 is then approximately the area of $\triangle SP_1P_2$ or

$$\tfrac{1}{2}SP_1 \times SP_2 \times \sin P_1SP_2.$$

If $\angle P_1 S P_2$ is expressed in radians, we may write

$$\sin P_1 S P_2 = \angle P_1 S P_2 = \theta_1$$

since $\angle P_1 S P_2$ is very small. Also,

$$S P_1 \approx S P_2 = r_1 \qquad \text{say}$$

so that the area $S P_1 P_2$ is given by

$$\tfrac{1}{2} r_1^2 \theta_1.$$

Similarly, area $S P_3 P_4$ is given by

$$\tfrac{1}{2} r_2^2 \theta_2$$

where $S P_3 = r_2$ and $\angle P_3 S P_4 = \theta_2$. Let $t_4 - t_3 = t_2 - t_1 = t$. Then by equation (13.1),

$$\frac{1}{2} r_1^2 \left(\frac{\theta_1}{t} \right) = \frac{1}{2} r_2^2 \left(\frac{\theta_2}{t} \right) = \text{constant.}$$

But θ/t is the angular velocity, ω, in the limit when t tends to zero.
Hence,

$$\tfrac{1}{2} r_1^2 \omega_1 = \tfrac{1}{2} r_2^2 \omega_2 = \text{constant} \tag{13.2}$$

is the mathematical expression of Kepler's second law.

As shown in figure 13.3, in order that this law is obeyed, the planet has to move fastest when its radius vector is shortest, at perihelion, and slowest when it is at aphelion.

13.2.4 Kepler's third law

In the third law, Kepler obtained a relationship between the sizes of planetary orbits and the periods of revolution. Now it happens that the semi-major axis of a planetary orbit is the average size of the radius vector, or the mean distance, so that an alternative form of the third law is to say that the cube of the mean distance of a planet is proportional to the square of its period of revolution.

Hence, if a_1 and T_1 refer to the semi-major axis and sidereal period of a planet P_1 moving about the Sun,

$$\frac{a_1^3}{T_1^2} = \text{constant} \tag{13.3}$$

the constant being the same for any of the planetary orbits. If a_2, a_3, etc and T_2, T_3, etc refer to the semi-major axes and sidereal periods of the other planets P_2, P_3, etc moving about the Sun, then

$$\frac{a_1^3}{T_1^2} = \frac{a_2^3}{T_2^2} = \frac{a_3^3}{T_4^2} = \cdots = \text{constant.}$$

The most convenient form of the constant is obtained by taking the planet to be the Earth in relation (13.3), expressing the distance in units of the Earth's semi-major axis and the time in years.

Then, for the Earth, $a_1 = 1$, $T_1 = 1$ and so the constant becomes unity. For any other planet, consequently,

$$\boldsymbol{a^3 = T^2} \tag{13.4}$$

showing that if we measure the sidereal period, T, of the planet, we can obtain its mean distance, a, from relation (13.4).

13.3   Newton's laws of motion

Kepler's three laws provide a convenient and highly accurate way of describing the orbits of planets and the way in which the planets pursue such orbits. They do not, however, give any physical reason whatsoever why planetary motions obey these laws. Newton's three laws of motion, coupled with his law of gravitation, provided the reason. We consider the three laws of motion before looking at the law of gravitation.

Newton's three laws of motion laid the foundations of the science of dynamics. Though some, if not all of them, were implicit in the scientific thought of his time, his explicit formulation of these laws and exploration of the consequences in conjunction with his law of universal gravitation did more to bring into being our modern scientific age than any of his contemporaries' work.

They may be stated in the following form

(1) Every body continues in its state of rest or of uniform motion in a straight line except insofar as it is compelled to change that state by an external impressed force.

In other words, in the absence of *any* force (including friction), an object will remain stationary or, if moving, will continue to move with the same speed in the same direction forever.

(2) The rate of change of momentum of the body is proportional to the impressed force and takes place in the direction in which the force acts.

Momentum is mass multiplied by velocity. The second law states that the rate at which the momentum changes will depend upon the size of the force acting on the object and naturally enough also depends upon the direction in which the force acts.

In calculus, d/dt denotes a rate of change of some quantity.

Both velocity v and force F are directed quantities or **vectors**, i.e. they define specific directions and such directed quantities are usually underlined or printed in bold type. Mass m is not a directed quantity. We may, therefore, summarize laws **(1)** and **(2)** by writing

$$\frac{\mathrm{d}(m\boldsymbol{v})}{\mathrm{d}t} = \boldsymbol{F} \tag{13.5}$$

where we have chosen the unit of force so that the constant of proportionality is unity. Since in most dynamical problems the mass is constant, we may rewrite equation (13.5) as

$$m\frac{\mathrm{d}\boldsymbol{v}}{\mathrm{d}t} = \boldsymbol{F} \tag{13.6}$$

or

$$\text{mass} \times \text{acceleration} = \text{impressed force.} \tag{13.7}$$

(3) To every action there is an equal and opposite reaction.

The rocket working in the vacuum of space is an excellent example of this law. The action of ejecting gas at high velocity from the rocket engine in one direction results in the acquiring of velocity by the rocket in the opposite direction (the reaction). It is also found that the momentum given to the gas is equal to the momentum acquired by the rocket in the opposite direction.

13.4   Newton's law of gravitation

One of the most far-reaching scientific laws ever formulated, Newton's law of universal gravitation is the basis of **celestial mechanics**—that branch of astronomy dealing with the orbits of planets and satellites—and **astrodynamics**—the branch of dynamics that deals with the orbits of space probes and artificial satellites. During the two and a half centuries succeeding its formulation, its consequences were investigated by many of the foremost mathematicians and astronomers that have ever lived. Many elegant mathematical methods were created to solve the intricate sets of equations that arose from statements of the problems involving mutually gravitating systems of masses.

The law itself is stated with deceptive simplicity as follows.

Every particle of matter in the universe attracts every other particle of matter with a force directly proportional to the product of the masses and inversely proportional to the square of the distance between them.

Hence, for two particles separated by a distance r,

$$F = G\frac{m_1 m_2}{r^2} \tag{13.8}$$

where F is the force of attraction, m_1 and m_2 are the masses and G is the constant of proportionality, often called the (universal) constant of gravitation.

As a young man in his early twenties, Newton thought of his law and looked for some test of its validity. One such test was to apply it to the Moon's orbit. Could the Moon be kept in its orbit round the Earth by the same force of gravitation that causes an apple to fall to the ground? Indeed, legend has it that Newton's discovery was sparked off when he saw an apple fall to the ground in the orchard of his family home at Woolsthorpe, where he was staying because Cambridge University had been closed for fear of the Great Plague. It is worth while quoting Newton's own words when he was an old man looking back to those Woolsthorpe days.

> And the same year I began to think of gravity extending to the orb of the Moon and having found out how to estimate the force with which a globe revolving within a sphere presses the surface of the sphere, from Kepler's Rule of the periodical times of the Planets being in a sesquialterate proportion of their distances from the centre of their orbs I deduced that the forces which keep the Planets in their orbs must be reciprocally as the squares of their distances from the centres about which they revolve: and thereby compared the force requisite to keep the Moon in her orb with a force of gravity at the surface of the Earth, and found them answer pretty nearly. All this was in the two plague years 1665 and 1666, for in those days I was in the prime of my age for invention, and minded Mathematics and Philosophy more than at any other time since.

Newton, therefore, used Kepler's laws in deducing his law of gravitation. He also knew, from Galileo's experiments, the value of the acceleration of freely falling bodies near the Earth's surface. He knew the size of the Earth, the radius of the Moon's orbit and the sidereal period of revolution of the Moon.

The well-known formula connecting the distance s through which a body falls from rest in time t under the acceleration g due to gravity is

$$s = \tfrac{1}{2}gt^2.$$

In the units Newton used, it is known that the body falls 16 feet in the first second so that $g = 32$ feet per second per second (ft s^{-2}).

At the Moon's distance, objects are 60 times as far away from the Earth's centre as objects at the Earth's surface. Hence, if Newton's law gravitation held, the acceleration due to Earth's gravity at the Moon's distance would be g_M, given by

$$g_M = g(\tfrac{1}{60})^2 = \tfrac{32}{3600} \text{ ft s}^{-2}.$$

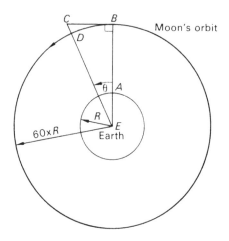

Figure 13.4. The Moon's orbit about the Earth.

In the first second, therefore, a body should fall a distance s_M where

$$s_M = \frac{1}{2} g_M \times 1^2 = \frac{32 \times 12}{2 \times 3600} = 0.0535 \text{ inch.}$$

If gravity did not exist, the Moon would move through space in a straight line (BC in figure 13.4). But because of the Earth's attraction, the Moon is continually falling towards the Earth. Its path, BD, can, therefore, be thought of as being made up of the straight line motion BC and a fall CD. If the time interval between B and D was 1 second, then CD is the distance fallen by the Moon in 1 second. According to Newton's calculation, it should be 0.0535 inch.

Now, $CD = CE - DE = CE - BE$, if we assume the Moon's orbit to be circular.

In the right-angled triangle CBE,

$$\frac{BE}{CE} = \cos CEB = \cos \theta, \qquad \text{say}$$

or

$$CE = BE/\cos \theta.$$

Hence,

$$CD = BE \left(\frac{1}{\cos \theta} - 1 \right).$$

Now θ is a very small angle—it is the angle swept out by the Moon's radius vector in one second—so that we may write

$$\cos \theta = 1 - \frac{\theta^2}{2}.$$

or, by the binomial theorem,

$$\frac{1}{\cos \theta} = 1 + \frac{\theta^2}{2}.$$

Substituting in the expression for CD, we obtain

$$CD = \frac{BE \times \theta^2}{2}.$$

Now

$$\theta = \frac{2\pi}{27 \cdot 322 \times 86\,400}$$

taking the sidereal period of revolution of the Moon to be 27·322 mean solar days and the number of seconds in a day to be 86 400.

The radius BE of the Moon's orbit is $60 \times 3960 \times 5280 \times 12$ inches, taking the Earth's radius to be 3960 miles. Hence, it is found that

$$CD = 0 \cdot 0533 \text{ inch}$$

in satisfactory agreement with the value of 0·0535 inches predicted by the law of gravitation.

In fact, the value obtained by Newton was 0·044 inches. Because of this disappointing disagreement he abandoned the theory as contradicted by the facts. Six years later, in 1671, Picard's measurement of the arc of a meridian in France corrected an error in the previously accepted value of the size of the Earth. Newton heard of this, repeated the calculation and now found agreement. He thereupon resumed the subject, carrying out the brilliant series of research projects that culminated in the publication of the *Principia*.

13.5   The *Principia* of Isaac Newton

In this great work, Newton seized all the astronomical and physical phenomena available, organized them, demonstrated how they followed from his three laws of motion and the law of universal gravitation, predicted new observational data and laid the foundations of mathematical physics so firmly that much of the researches of mathematicians, scientists and astronomers in the next two and a half centuries became efforts to develop logically the consequences of his theory of the World.

Among the contents of the *Principia* were the following:

He proved that a rotating globe such as the Earth would be flattened slightly at the poles due to the centrifugal force in its equatorial regions 'diluting' the force of gravity in these regions. This protuberance of matter at the equator would be acted upon by the Moon and Sun, causing the Earth's axis of rotation to precess slowly like the axis of a spinning top or gyroscope so that the axis sweeps out a cone. The precession of the equinoxes, a consequence of this, had been discovered by Hipparchus about 134 BC but until Newton's time had remained unexplained.

He proved that under the law of gravitation, a planet would move in an ellipse about the Sun under Kepler's laws and that this ellipse would change slightly in size, shape and orientation over the years because of the attractions of the other planets. In this way he accounted for hitherto unexplained changes in the orbits of Jupiter and Saturn. He also showed how to measure the mass of any planet that had one or more satellites.

In the case of the Moon's orbit—an extremely complicated problem that, as Newton once remarked to his friend Edmund Halley, 'made his head ache and kept him awake so often that he would think of it no more'—he was able to prove that certain irregularities in its movement about the Earth were due to the Sun's attraction and he predicted some features that were subsequently found by observation.

He even demonstrated (see figure 13.5) that if a projectile were fired with sufficient velocity from a cannon at the top of a terrestrial mountain so high it was outside the Earth's atmosphere, it would describe a circle or ellipse about the Earth. It would, barring a collision, go on revolving about the Earth indefinitely. In other words, three hundred years before Newtonian science had developed far enough to put the first artificial satellite into orbit, Isaac Newton predicted its possibility.

He also explained the phenomenon of the tides as an additional consequence of the law of gravitation, laying the foundations of all the modern work on that subject. He showed that the orbits of comets were also governed by this law and described how their paths could be calculated.

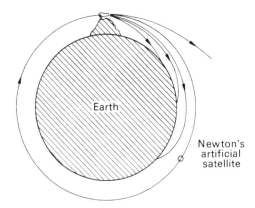

Figure 13.5. Newton's suggestion of an artificial satellite.

In addition to all this, the book contained the solutions of many problems in the motions of fluids, accounts of ingenious experiments, the theory of the calculus and other brilliant researches.

It has been said that when the *Principia* appeared, there were only half a dozen men alive capable of really understanding it. Many probably realized when they bought the book—it sold for some ten or twelve shillings—that here was a milestone in science. They could hardly realize that, in fact, the publication of the *Principia* marked the birth of the modern scientific world, in general, and modern astronomy, in particular.

13.6   The two-body problem

13.6.1   Equations of motion

The two-body problem, first stated and solved by Newton, asks: 'Given at any time the positions and velocities of two massive particles moving under their mutual gravitational force, the masses also being known, provide a means of calculating their positions and velocities for any other time, past or future.'

At any moment, the velocity vector of one of the masses relative to the other, and the line joining them, defines a plane. Now the gravitational force between the bodies acts along the line joining them, so there is no acceleration trying to make them leave this plane. The orbit of each particle, therefore, lies in this plane. We can set up their equations of motion using Newton's laws of motion and his law of gravitation by taking two rectangular axes Ox' and Oy' in the plane, as in figure 13.6.

Let the two particles be P_1 and P_2 of masses m_1 and m_2, and with coordinates (x_1, y_1) and (x_2, y_2) with respect to the axes Ox' and Oy'.

Then the mutual distance apart is r, where

$$r^2 = (x_2 - x_1)^2 + (y_2 - y_1)^2.$$

The magnitude of the force of gravity F is given by equation (13.8), namely

$$F = G\frac{m_1 m_2}{r^2}.$$

Consider particle P_1. It is attracted towards P_2, experiencing an acceleration which can be thought of as resolved into components d^2x_1/dt^2 and d^2y_1/dt^2 along the Ox' and Oy' axes respectively. The force F can similarly be resolved into components along the Ox' and Oy' axes. In the case of particle P_1, the force acts from P_1 to P_2 so that for P_1, the components are:

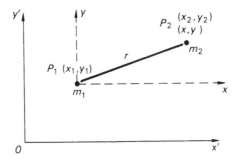

Figure 13.6. The two-body problem presented in a rectangular coordinate frame.

(1) $\dfrac{Gm_1m_2}{r^2}\left(\dfrac{x_2 - x_1}{r}\right)$ along the Ox' axis; and

(2) $\dfrac{Gm_1m_2}{r^2}\left(\dfrac{y_2 - y_1}{r}\right)$ along the Oy' axis.

The quantities $(x_2-x_1)/r$ and $(y_2-y_1)/r$ are called **direction-cosines** and are the relevant factors that F must be multiplied by to give not only the correct size of the components on Ox' and Oy' but also the correct directions. In the case of particle P_2, the correct components are

$$\frac{Gm_1m_2}{r^2}\left(\frac{x_1 - x_2}{r}\right)$$

and

$$\frac{Gm_1m_2}{r^2}\left(\frac{y_1 - y_2}{r}\right)$$

respectively, since for P_2 the force acts from P_2 to P_1.

Now by Newton's laws 1 and 2 (i.e. by equation (13.7)),

$$\text{mass} \times \text{acceleration} = \text{impressed force.}$$

For the first particle P_1, we may then write

$$m_1 \times \frac{\mathrm{d}^2x_1}{\mathrm{d}t^2} = \frac{Gm_1m_2}{r^2}\left(\frac{x_2 - x_1}{r}\right) \tag{13.9}$$

$$m_1 \times \frac{\mathrm{d}^2y_1}{\mathrm{d}t^2} = \frac{Gm_1m_2}{r^2}\left(\frac{y_2 - y_1}{r}\right). \tag{13.10}$$

Equations (13.9) and (13.10) are the so-called differential equations of motion of particle P_1.
For the second particle P_2, we have

$$m_2 \times \frac{\mathrm{d}^2x_2}{\mathrm{d}t^2} = \frac{Gm_1m_2}{r^2}\left(\frac{x_1 - x_2}{r}\right) \tag{13.11}$$

$$m_2 \times \frac{\mathrm{d}^2y_2}{\mathrm{d}t^2} = \frac{Gm_1m_2}{r^2}\left(\frac{y_1 - y_2}{r}\right). \tag{13.12}$$

We may simplify equations (13.9)–(13.12) in the following way. Divide both sides of (13.9) by m_1 and both sides of (13.11) by m_2 to give

$$\frac{\mathrm{d}^2x_1}{\mathrm{d}t^2} = \frac{Gm_2}{r^3}(x_2 - x_1) \tag{13.13}$$

and

$$\frac{d^2x_2}{dt^2} = \frac{Gm_1}{r^3}(x_1 - x_2) \tag{13.14}$$

respectively. Subtract equation (13.13) from equation (13.14), obtaining

$$\frac{d^2(x_2 - x_1)}{dt^2} + \frac{G(m_1 + m_2)}{r^2}(x_2 - x_1) = 0. \tag{13.15}$$

If we take a set of axes P_1x, P_1y through P_1 and the origin, with P_1x and P_1y parallel to Ox' and Oy' respectively, we see that the coordinates x and y of P_2 with respect to these new axes are given by

$$x = x_2 - x_1; \qquad y = y_2 - y_1.$$

Letting

$$\mu = G(m_1 + m_2)$$

equation (13.15) may be written

$$\frac{d^2x}{dt^2} + \mu\frac{x}{r^3} = 0. \tag{13.16}$$

Obviously, by treating equations (13.10) and (13.12) in a similar fashion, we are led to the equation

$$\frac{d^2y}{dt^2} + \mu\frac{y}{r^3} = 0. \tag{13.17}$$

13.6.2   The solution of the two-body problem

It is not within the scope of this text to show how equations (13.16) and (13.17) may be solved. The solution, however, may be written as

$$r = \frac{h^2/\mu}{1 + e\cos\theta} \tag{13.18}$$

where h is a constant which is twice the rate of description of area by the radius vector, e is the eccentricity of the orbit and θ is the true anomaly (see section 13.2.2).

Equation (13.18) is, in fact, the polar equation of a conic section. In obtaining this solution, Newton generalized Kepler's first law, for a conic section can not only be an ellipse but also a parabola or a hyperbola.

We will, however, obtain an important relation, the **energy integral**. This leads to a useful expression giving the velocity of a planet at any point in his orbit.

13.6.3   The energy integral

If we multiply equation (13.16) by dx/dt and (13.17) by dy/dt and add, we obtain the relation

$$\frac{dx}{dt}\frac{d^2x}{dt^2} + \frac{dy}{dt}\frac{d^2y}{dt^2} + \frac{\mu}{r^3}\left(x\frac{dx}{dt} + y\frac{dy}{dt}\right) = 0. \tag{13.19}$$

Now

$$\frac{d}{dt}\left[\frac{1}{2}\left(\frac{dx}{dt}\right)^2 + \frac{1}{2}\left(\frac{dy}{dt}\right)^2\right] = \frac{dx}{dt}\frac{d^2x}{dt^2} + \frac{dy}{dt}\frac{d^2y}{dt^2}.$$

Also, $r^2 = x^2 + y^2$ so that

$$\frac{d}{dt}\left(\frac{1}{r}\right) = -\frac{1}{r^2}\frac{dr}{dt} = -\frac{1}{r^2}\frac{d(x^2 + y^2)^{\frac{1}{2}}}{dt} = -\frac{1}{r^2}\frac{1}{2}\cdot\frac{1}{(x^2 + y^2)^{\frac{1}{2}}}\left(2x\frac{dx}{dt} + 2y\frac{dy}{dt}\right)$$

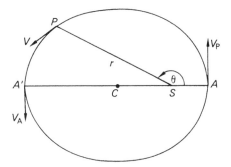

Figure 13.7. The velocity of a planet in an elliptical orbit showing that at perihelion A and aphelion A' the velocity vector is perpendicular to the radius vector.

giving

$$\frac{d}{dt}\left(-\frac{\mu}{r}\right) = \frac{\mu}{r^3}\left(x\frac{dx}{dt} + y\frac{dy}{dt}\right).$$

Hence, equation (13.19) may be written as a perfect differential, namely

$$\frac{d}{dt}\left[\frac{1}{2}\left(\frac{dx}{dt}\right)^2 + \frac{1}{2}\left(\frac{dy}{dt}\right)^2 - \frac{\mu}{r}\right] = 0.$$

Integrating, we obtain

$$\frac{1}{2}V^2 - \frac{\mu}{r} = C \tag{13.20}$$

where C is the so-called energy constant and V is the velocity of one mass with respect to the other since

$$V^2 = \left(\frac{dx}{dt}\right)^2 + \left(\frac{dy}{dt}\right)^2.$$

The first term, $\frac{1}{2}V^2$, is the **kinetic energy**, energy the planet in its orbit about the Sun possesses by virtue of its speed. The second term, $-\mu/r$, is the **potential energy**, energy the planet possesses by virtue of its distance from the Sun.

What equation (13.20) states is that the sum of these two energies is a constant, a reasonable statement since the two-body is an isolated system, no energy being injected into the system or being removed from it. In an elliptic orbit, however, the distance r is changing. Equation (13.20) shows that there is a continual trade-off between the two energies: when one is increasing, the other is decreasing.

If we wish to obtain an expression giving the velocity V of the planet, we must interpret the constant C. This is done in the following section.

13.6.4 The velocity of a planet in its orbit

Let P be the position of a planet in its elliptical orbit about the Sun S at a given time when its velocity is V and its radius vector SP has length r (see figure 13.7).

Let V_P, V_A be the velocities at perihelion A and aphelion A' respectively.

The points A and A' are the only places in the orbit where the velocity is instantaneously at right angles to the radius vector and where, consequently, we may write

$$V = r\omega \tag{13.21}$$

where ω is the angular velocity.

At every point, however, Kepler's second law holds, namely that

$$r^2\omega = h. \tag{13.22}$$

Hence, *at perihelion and aphelion only,* we have

$$V = \frac{h}{r}.$$

For perihelion,

$$V_P = \frac{h}{a(1-e)}.$$

For aphelion,

$$V_A = \frac{h}{a(1+e)}$$

so that

$$\frac{V_P}{V_A} = \frac{1+e}{1-e}. \tag{13.23}$$

Now the energy equation (13.20) is

$$\frac{1}{2}V^2 - \frac{\mu}{r} = C$$

so that at perihelion, we have

$$\frac{1}{2}V_P^2 - \frac{\mu}{a(1-e)} = C \tag{13.24}$$

while at aphelion,

$$\frac{1}{2}V_A^2 - \frac{\mu}{a(1+e)} = C. \tag{13.25}$$

Subtracting equation (13.25) from equation (13.24) and using relation (13.23) to eliminate V_A, we obtain

$$V_P^2 = \frac{\mu}{a}\left(\frac{1+e}{1-e}\right). \tag{13.26}$$

In similar fashion, we obtain

$$V_A^2 = \frac{\mu}{a}\left(\frac{1-e}{1+e}\right). \tag{13.27}$$

Again, equations (13.26) and (13.27) give

$$V_A V_P = \frac{\mu}{a}. \tag{13.28}$$

Subtracting equation (13.24) from equation (13.20) and using (13.26) to eliminate V_P, we obtain, after a little reduction, the required relation

$$V^2 = \mu\left(\frac{2}{r} - \frac{1}{a}\right). \tag{13.29}$$

13.6.5   The period of revolution of a planet in its orbit

Let us suppose the orbit to be circular so that $r = a$. Then expression (13.29) becomes

$$V^2 = \frac{\mu}{a}.$$ (13.30)

But

$$V = \frac{2\pi a}{T}$$

where T is the time it takes the planet to describe its circular orbit. Hence,

$$T = 2\pi \left(\frac{a^3}{\mu}\right)^{\frac{1}{2}}.$$ (13.31)

Although it will not be proved here, the expression (13.31) holds even when the orbit is elliptical, a being the semi-major axis of the orbit.

13.6.6   Newton's form of Kepler's third law

Let two planets revolve about the Sun in orbits of semi-major axes a_1 and a_2, with periods of revolution T_1 and T_2. Let the masses of the Sun and the two planets be M, m_1 and m_2 respectively. Then by equation (13.31),

$$T_1 = 2\pi \left(\frac{a_1^3}{\mu_1}\right)^{\frac{1}{2}}$$

where $\mu_1 = G(M + m_1)$. Also,

$$T_2 = 2\pi \left(\frac{a_2^3}{\mu_2}\right)^{\frac{1}{2}}.$$

Hence,

$$\left(\frac{T_2}{T_1}\right)^2 = \left(\frac{a_2}{a_1}\right)^3 \left(\frac{\mu_1}{\mu_2}\right) = \left(\frac{a_2}{a_1}\right)^3 \left(\frac{M + m_1}{M + m_2}\right).$$ (13.32)

Kepler's third law would have been written as

$$\left(\frac{T_2}{T_1}\right)^2 = \left(\frac{a_2}{a_1}\right)^3.$$

The only difference between this last equation and (13.32) is a factor k, where

$$k = \frac{M + m_1}{M + m_2}.$$

Dividing top and bottom by M, we obtain

$$k = \frac{1 + m_1/M}{1 + m_2/M}.$$

The greatest departure of k from unity arises when we take the two planets to be the most massive and the least massive in the Solar System. The most massive is Jupiter: in this case $m_1/M = 1/1047{\cdot}3$. Of the planets known to Newton, the least massive was Mercury, giving $m_2/M = 1/6\,200\,000$. Hence, to three significant figures, $k = 1$. Kepler's third law is, therefore, only an approximation to the truth, though a very good one. Newton's form of Kepler's third law, namely equation (13.32), is much better.

13.6.7 Measuring the mass of a planet

Let a planet P with orbital semi-major axis a, sidereal period of revolution T and mass m possess a satellite P_1 that moves in an orbit about P with semi-major axis a_1 and sidereal period of revolution T_1. Let the masses of the Sun and satellite be M and m_1 respectively.

Then by equation (13.31), we have for the planet and the Sun,

$$T = 2\pi \left(\frac{a^3}{\mu} \right)^{\frac{1}{2}}. \tag{13.33}$$

where $\mu = G(M + m)$.

For the planet and satellite we have

$$T_1 = 2\pi \left(\frac{a_1^3}{\mu_1} \right)^{\frac{1}{2}} \tag{13.34}$$

where $\mu_1 = G(m + m_1)$.

Dividing equation (13.34) by equation (13.33), we obtain

$$\frac{T_1}{T} = \left[\left(\frac{a_1}{a} \right)^3 \frac{\mu}{\mu_1} \right]^{\frac{1}{2}}$$

or

$$\frac{\mu}{\mu_1} = \left(\frac{T_1}{T} \right)^2 \left(\frac{a}{a_1} \right)^3 = \frac{M + m}{m + m_1}.$$

We may write

$$\frac{M + m}{m + m_1} = \left(\frac{M}{m} \right) \left(\frac{1 + m/M}{1 + m_1/m} \right).$$

But we have seen that the ratio m/M is much less than unity and for the satellites in the Solar System, the ratio of their masses to that of their primary is also much less than unity. Hence,

$$\frac{M}{m} = \left(\frac{T_1}{T} \right)^2 \left(\frac{a}{a_1} \right)^3. \tag{13.35}$$

Only two of the planets, Mercury and Venus, have no known satellites. Earth, Mars, Jupiter, Saturn, Uranus, Neptune and Pluto each have one or more moons and so can have their masses in terms of the Sun's mass calculated by measuring the satellites' orbital semi-major axes and sidereal periods of revolution. Other methods have to be adopted in the cases of Mercury and Venus. The masses of Mercury and Venus were first obtained from their perturbing effects on the orbits of other planets; the mass of Venus is now accurately known from observations of artificial satellites of Venus. The triple flypast of Mercury by the spacecraft Mariner 10 has improved our knowledge of that planet's mass. The mass of Pluto, derived from its moon Charon, is of the order of 0·003 Earth masses.

Example 13.1. The artificial satellite Vanguard 1 was found to have a period of revolution of 134 minutes. It approached the Earth to distance of 660 km and receded to a distance of 4023 km. Calculate the mass of the Earth with respect to the Sun's mass if the semi-major axis of the Earth's orbit is $149 \cdot 5 \times 10^6$ km, the radius of the Earth is 6372 km and the year contains 365·25 mean solar days.

In figure 13.8, the artificial satellite is at **perigee** at A (the point in its orbit nearest the Earth's centre E) and at **apogee** at A' (when it is farthest from the Earth's centre).

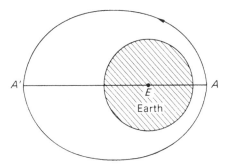

Figure 13.8. Example 13.1—the orbit of an artificial satellite.

The size of the major axis AA' of the ellipse is given by

$$AA' = 4023 + 2 \times 6372 + 660 = 17\,427 \text{ km}$$

so that the semi-major axis a_1 is of length 8713·5 km. The period of revolution is $T_1 = 134$ minutes. For the Earth orbit,

$$a = 149 \cdot 5 \times 10^6 \text{ km} \qquad T = 365 \cdot 25 \times 24 \times 60 = 5 \cdot 2596 \times 10^5 \text{ minutes.}$$

Letting M and m be the masses of Sun and Earth respectively, we have, using equation (13.35),

$$\frac{M}{m} = \left(\frac{134}{5 \cdot 2596 \times 10^5} \right)^2 \left(\frac{149 \cdot 5 \times 10^6}{8713 \cdot 5} \right)^3$$

giving, after a little calculation,

$$\frac{M}{m} = 327\,800.$$

Hence, the Earth's mass is only about three-millionths that of the Sun.

13.6.8 Co-periodic orbits

Rewriting equation (13.31), we have

$$T = 2\pi \left(\frac{a^3}{\mu} \right)^{\frac{1}{2}}. \tag{13.31}$$

Re-arranging equation (13.29), we obtain

$$a = 1 \Big/ \left(\frac{2}{r} - \frac{V^2}{\mu} \right). \tag{13.36}$$

If we eliminate a between equations (13.31) and (13.36), there results

$$T = 2\pi \mu \left(\frac{2\mu}{r} - V^2 \right)^{-3/2}. \tag{13.37}$$

Equations (13.31), (13.36) and (13.37) highlight some interesting properties of elliptical motion.
It is seen that the semi-major axis is a function of the radius vector and the square of the velocity.
If, therefore, a body of mass m_1 is projected at a given distance r from another body of mass m_2 with

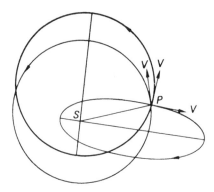

Figure 13.9. Co-periodic orbits.

velocity V, the semi-major axis a of its orbit is independent of the direction of projection and depends only on the magnitude of the velocity. In figure 13.9, all the orbits have the same initial radius vector SP and the same initial velocity magnitude V though the directions of projection are different. All orbits, therefore, have the same semi-major axis a given by equation (13.36).

It is also seen from equation (13.37) that the periods in these orbits must also be the same. If particles were projected from P simultaneously into these orbits, they would all pass through P together on return though the orbits they pursued were quite different in shape and not even necessarily in the same plane, as long as the initial radius vector SP and speed V were the same.

13.7   Solar radiation pressure

In addition to the force of gravity generated by the mass of the Sun, the central force of solar radiation pressure may also affect the orbits of interplanetary material. Its sense is opposite to that of gravity but is generally a very small effect certainly for objects such as planets. However, in the case of small dust particles, say those ejected from comets in forming a tail, for certain sizes of particle, it turns out that there can be an exact balance between solar gravitation and radiation pressure so that the resultant orbit of the ejected particle from the comet is a 'straight line', so fulfilling Newton's first law (see earlier). This circumstance is a beautiful example of astronomy offering a circumstance to verify a law which is very difficult to explore within the laboratory.

Consider the condition of this special case, evaluating the forces on a small spherical grain of radius, a, mass, m_a, and with density, ρ, at a distance, r, from the Sun. The force of gravity may be expressed as

$$F_G = \frac{GM_\odot m_a}{r^2} = \frac{GM_\odot (\frac{4}{3}\pi a^3)\rho}{r^2}.$$

Now the Sun has a luminosity (total energy output) of $L_\odot$ ($= 3 \cdot 8 \times 10^{26}$ W) and the flux intercepted by the particle is given by

$$\left(\frac{L_\odot}{4\pi r^2}\right)(\pi a^2).$$

The photons in the radiation carry momentum and hence give rise to a force acting along the direction of the received light. The force of radiation pressure, F_{RP}, is simply the flux received divided by the speed of light and, therefore, may be written as

$$F_{RP} = \frac{L_\odot}{4\pi r^2}\frac{\pi a^2}{c}.$$

For an exact balance, $F_G = F_{RP}$ and, therefore,

$$\frac{GM_\odot(\frac{4}{3}\pi a^3)\rho}{r^2} = \frac{L_\odot}{4\pi r^2}\frac{\pi a^2}{c}$$

and, hence,

$$a = \frac{L_\odot}{GM_\odot c}\frac{3}{16\pi\rho}.$$

Substituting for the various values and constants and by assuming a typical density of 3000 kg m^{-3}, the balance of forces is achieved for a particle with radius

$$a = \frac{3\cdot8 \times 10^{26}}{6\cdot67 \times 10^{-11} \times 1\cdot99 \times 10^{30} \times 3 \times 10^8}\frac{3}{16\pi \times 3 \times 10^3} \approx 0\cdot2\ \mu\text{m}.$$

Thus, for such tiny dust particles emanating from a comet, no matter the distance from the Sun, they would continue to travel with their velocity at ejection and travel in a straight line without a force acting on them. Measurement of the movement of some parts of the cometary dust tail verifies this. The main body of the comet continues in its orbit about the Sun.

There are proposals and design studies for utilizing radiation pressure as a means of propulsion within the solar system. By making very large 'sails' of light-weight material with high area-to-mass ratio, it is feasible to propel light-weight space vehicles using solar radiation as the 'fuel'. By altering the attack of the sail at any time, the orbit of the vehicle can be readily modified at any time. The principles behind the manoeuvres are referred to as **solar sailing**.

13.8   The astronomical unit

In section 12.7 it was shown how an accurate scale model of the Solar System could be obtained, all the planetary distances being expressed in terms of the Earth's distance from the Sun. Astronomers have found it convenient to use data connected with the Earth's orbit and the Sun as their units of time, distance and mass. Newton's precise statement of Kepler's third law for a planet of mass m_2 revolving about the Sun of mass m_1 may be written in the form (see equation (13.31))

$$k^2(m_1 + m_2)T^2 = 4\pi^2 a^3.$$

Taking the solar mass, the mean solar day and the Earth's mean distance from the Sun as the units of mass, time and distance respectively, the equation becomes

$$k^2(1 + m_2)T^2 = 4\pi^2 a^3$$

where k^2 is written for G, the gravitational constant, and m_2, T and a are all in the units defined earlier. The quantity k is called the **Gaussian constant of gravitation**.

If, as was done by Gauss, the planet is taken to be the Earth and T given the value of 365·256 383 5 mean solar days (the length of the sidereal day adopted by Gauss), while m_2 is taken to be $1/354\,710$ solar masses, k is found to have the value 0·017 202 098 95, the value of a being, of course, unity. This distance, the semi-major axis of the Earth's orbit, was called the **astronomical unit** (AU).

The concept has since been refined. From time to time, various quantities have been determined more accurately but to avoid having to re-compute k and other related quantities every time, astronomers retain the original value of k as absolutely correct. This means that the Earth is treated like any other planet. The unit of time is now the ephemeris day. The Earth's mean distance from the Sun is now taken as 1·000 000 03 astronomical units while the Earth–Moon system's mass is $1/328\,912$ solar masses.

We may note that the precise definition of the **astronomical unit** is given by Kepler's third law,

$$k^2(1 + m)T^2 = 4\pi^2 a^3$$

with the Sun's mass taken to be unity, $k = 0.017\,202\,098\,95$, and the unit of time taken to be one ephemeris day. It is the radius of a circular orbit in which a body of negligible mass, undisturbed by the gravitational attractions of all other bodies, will revolve about the Sun in one Gaussian year of $2\pi/k$ ephemeris days. It has a value of $1.496\,00 \times 10^{11}$ m.

Problems—Chapter 13

Take the length of the year to be 365·25 days.

1. An asteroid's orbit has a semi-major axis of 4 AU. Calculate its sidereal period.
2. Given that the semi-major axis of the orbit of Venus is 0·7233 AU, calculate its sidereal period.
3. The synodic period of Mars is 780 days. Calculate its mean distance from the Sun in astronomical units.
4. The orbital period of Jupiter's fifth satellite about the planet is 0·4982 days and its orbital semi-major axis is 0·001 207 AU. The orbital period and semi-major axis of Jupiter are 11·86 years and 5·203 AU respectively. Find the ratio of the mass of Jupiter to that of the Sun.
5. Suppose a planet existed in a circular orbit about the Sun of radius 0·1 AU. What would be its period of revolution?
6. A communications satellite in a circular equatorial orbit always remains above a point of fixed longitude. Given that the sidereal day is $23^h\,56^m$ long, the year $365\frac{1}{4}$ days in length and the distances of the satellite from Earth's centre and Earth from the Sun are 41 800 km and $149·5 \times 10^6$ km respectively, calculate the ratio of the masses of the Sun and Earth.
7. The minimum and maximum heights of an artificial Venusian satellite above the solid surface of the planet (as measured by radar) are 696 and 2601 km respectively. The satellite's period of revolution is observed to be 104 minutes. If the semi-major axis and sidereal period of the Venusian orbit are 0·723 AU and 0·615 years respectively, calculate the radius of Venus, given that 1 AU is $149·5 \times 10^6$ km and the mass of Venus is 1/403 500 times the Sun's mass.
8. The period of revolution of an artificial satellite of the Moon is $2^h\,20^m$. Its minimum and maximum distances above the lunar surface are, respectively, 80 and 600 km. If the radii of the Moon and of the Moon's orbit are 1738 and 384 400 km respectively and the lunar sidereal period is $27\frac{1}{3}$ days, calculate the semi-major axis and eccentricity of the lunar artificial satellite's orbit and the ratio of the masses of the Earth and Moon.
9. The period of Jupiter is 11·86 years and the masses of the Sun and Jupiter are 330 000 and 318 times that of the Earth respectively. Calculate the change in Jupiter's orbital period if its mass suddenly became the same as that of the Earth.
10. The semi-major axis of the orbit of Mars is 1·524 AU and the orbital eccentricity is 0·093. Assuming the Earth's orbit to be circular and coplanar with that of Mars, calculate (i) the distance of Mars from the Earth at closest approach, (ii) the ratio of the speeds of Mars in its orbit at perihelion and aphelion and (iii) the speed of Mars at perihelion in AU per year.
11. Halley's comet moves in an elliptical orbit of eccentricity 0·9673. Calculate the ratio of (i) its linear velocities, (ii) its angular velocities, at perihelion and aphelion.
12. Two artificial satellites are in elliptical orbits about the Earth and they both have the same value for their semi-major axes. The ratio of the linear velocities at perigee of the two satellites is $\frac{3}{2}$ and the orbital eccentricity of the satellite with the greater perigee velocity is 0·5. Find the orbital eccentricity of the other satellite and the ratio of the apogee velocities of the two satellites.
13. An artificial satellite moves in a circular orbit inclined at 30° to the equator in a period of 2 hr at a height of 1689 km. What is its horizontal parallax? Neglecting the Earth's rotation, find the maximum time for which the satellite can remain above the horizon (a) at a station where the satellite passes directly overhead, (b) at Winnipeg (latitude 50° N). (Earth's radius = 6378 km.)

Chapter 14

Celestial mechanics: the many-body problem

14.1 Introduction

Newton formulated and solved the two-body gravitational problem where two massive particles move in orbits under their mutual gravitational attraction. He applied the formulas he obtained to such problems as the orbit of a planet moving round the Sun or the orbit of the Moon about the Earth. His success was due to two fortunate circumstances. One is the distribution of masses in the Solar System. All planetary masses are small with respect to the Sun's mass just as the masses of satellites are small compared to those of their primaries. Because of this, the mutual gravitational attractions of planets are small in relation to the Sun's force of attraction on each planet. The second is that the diameters of planets are small compared with their distances from each other and from the Sun. To a very good approximation, therefore, the nature of the orbit of a planet about the Sun is a two-body problem, the bodies being a planet and the Sun.

Newton realized, of course, that the other planets' attractions had to be taken into account in describing precisely what orbit a planet would travel in. The problem was, in fact, a many-body problem and Newton was the first to formulate it. In its form where the objects involved are point-masses (that is they have mass but no volume—a physical impossibility but a remarkably useful concept!), the many-body problem may be stated as follows: Given at any time the positions and velocities of three or more massive particles moving under their mutual gravitational forces, the masses also being known, provide a means by which their positions and velocities can be calculated for any time, past or future.

The problem is more complicated when bodies cannot be taken to be point-masses so that their shapes and internal constitutions have to be taken into account as in the Earth–Moon–Sun problem.

In order to discuss the methods that have been invented in celestial mechanics to deal with his problem, we consider first what is meant by the elements of an orbit.

14.2 The elements of an orbit

In figure 14.1, the orbit of a planet P about the Sun S is shown. The orbit is an ellipse, for at the moment we consider the Sun to be the only mass attracting the planet.

Let $S\Upsilon$ be the direction of the First Point of Aries as seen from the Sun, and let the fixed reference plane be the plane of the ecliptic.

Then the planet's orbital plane will intersect the plane of the ecliptic in some line NN_1, called the **line of nodes**. If motion in the orbit is described in the direction shown by the arrowhead, N is called the **ascending node**; N_1 is the **descending node**.

The angle ΥSN is the **longitude of the ascending node**, usually denoted by the symbol Ω.

The angle between the orbital plane and the plane of the ecliptic is called the **inclination**, i.

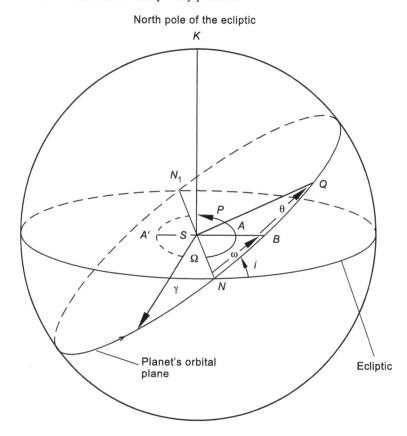

Figure 14.1. The unambiguous orientation in space of a planetary orbit by means of the orbital elements Ω, ω, i (defined in text).

The two angles Ω and i, therefore, fix unambiguously the orientation of the planet's orbital plane in space.

The orbit APA' lies in the orbital plane.

Let the orbit have perihelion A and aphelion A'. The line AA' is the so-called **line of apsides**. If SA is produced it will meet the celestial sphere in a point B. Then the orientation of the orbit in the orbital plane is given by $\angle NSB$, denoted by ω, and called the **argument of perihelion**.

The size and shape of the orbit are given by the **semi-major axis**, a, and the **eccentricity**, e. It remains to fix the position P of the planet in its orbit at any given time t.

This can be done by using the formulas of the two-body problem's solution, if any time τ at which the planet was at perihelion A is known.

The quantity τ is called the **time of perihelion passage**. For example, it may be 2000, April 14th, UT $22^{\mathrm{h}}\,47^{\mathrm{m}}\,12\overset{s}{.}4$.

The six numbers, Ω, ω, i, a, e, τ, are referred to as the six elements of the planetary orbit. It is worth noting that it is the first five that describe the orientation, size and shape of the orbit. The sixth enables the planet's position to be found within the orbit at any time.

14.3   General properties of the many-body problem

If the masses of the planets were vanishingly small compared to the Sun's mass, then the orbit of any planet would be unchanging and the six elements would be constant. Indeed, Kepler's three laws are the solution to the many-body problem in such a case. But the planetary masses are by no means negligible and, in the case of comets, near approaches to planets can occur so that, in general, the problem is much more complicated.

In the past three centuries, it has inspired (and frustrated!) many eminent astronomers and mathematicians. It is perhaps not obvious that even the three-body problem is of a much higher degree of complexity than the two-body problem. But if we consider that each body is subject to a complicated variable gravitational field due to its attraction by the other two, such that close encounters with either may be brought about, the result of each near-collision being an entirely new type of orbit, we see that it would require a general formula of unimaginable complexity to describe all the consequences of all such encounters.

In point of fact, several general and useful statements may be made concerning the many-body problem and these were proved quite early on in its history. They were known to Euler (1707–83) but since then no further overall properties have been discovered or are likely to be.

The statements follow from the only known integrals of the differential equations and refer to the centre of mass of the system, the total energy of the system and its total angular momentum. Without saying anything about the trajectories of the individual particles, the following statements can be made:

(a)   The centre of mass of the system moves through space with constant velocity, i.e. it moves in a straight line at a fixed speed.
(b)   The total energy of the system (the sum of all the kinetic energies and potential energy) is constant. Thus, although there is a continual trade-off among the members in kinetic energy and potential energy, the total energy is unaffected.
(c)   The total angular momentum of the system is constant.

In addition to these properties, particular solutions of the three-body problem that exist when certain relationships hold among the velocities and mutual distances of the particles were found by Lagrange. He showed that if the three bodies occupy the vertices of an equilateral triangle, their speeds being equal in magnitude and inclined at the same angle to each mutual radius vector, they will remain in an equilateral triangle formation, though the triangle will rotate and may change its size. Lagrange also showed that if the three bodies are placed on a straight line at mutual distances depending upon the ratios of their masses, they will remain on that line, though it will rotate. Although these equilateral triangle and collinear solutions of the three-body problem were thought to be of theoretical interest only at the time of their presentation, it was subsequently discovered that they occur in the Solar System.

Two groups of asteroids, called the Trojans, revolve about the Sun in Jupiter's orbit, so that their periods of revolution equal that of Jupiter. In their orbit about the Sun, they oscillate about one or other of the two points 60° ahead or behind Jupiter's heliocentric position. Among Saturn's moons, Telesto and Calypso remain 60° ahead or behind the more massive Tethys while Helene, in Dione's orbit, keeps 60° ahead of Dione.

14.4   General perturbation theories

It has been seen that, because the planets' mutual attractions are so much smaller than the Sun's attraction upon them, the planets' orbits, to a high degree of approximation, are ellipses about the Sun. This two-body approximation has been the starting point in many attempts to obtain theories of the planets' motions. The two-body orbit of a planet about the Sun is supposed to vary in size, shape and orientation as if it were a soft plasticine ring moulded by the spectral fingers of the other planets' gravitational fields.

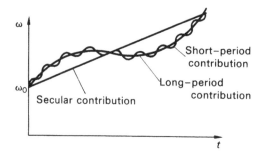

Figure 14.2. Typical behaviour of the argument of perihelion ω.

Now the elements of a planet's orbit describe not only its size, shape and orientation but also give information enabling the planet's position at any time to be calculated. These elements must vary if the orbit is perturbed. By complicated and tedious mathematical operations, formulas for the changes in the elements are obtained. These formulas are long sums of sines and cosines plus secular terms. The **periodic** part of the formula for each element may contain hundreds of such trigonometrical functions of time. The **secular** terms are not periodic terms but cause changes proportional to the time and occur in all elements including the semi-major axis, if the mathematical operations are carried out to a high enough degree of approximation. Indeed, **acceleration** terms also occur.

Such analytical expressions, valid for a period of time, are called **general perturbations**. They enable some deductions to be made regarding the past and future states of the planetary system though it must be emphasized that no results valid for an arbitrarily long time may be obtained in this way. About one thousand years is the estimated interval of time within which such planetary theories give accurate answers.

The method of general perturbations has also been applied to satellite systems, to asteroids disturbed by Jupiter and to the orbits of artificial Earth satellites.

By making certain assumptions, for example that the values of the semi-major axes of the planetary orbits do not vary widely, analytical perturbation theories may be obtained that describe what happens to the planetary elements for much longer times, for the order of millions of years. It is found that the semi-major axes, the eccentricities and inclinations suffer only periodic changes in size whereas the other three elements (the longitude of the ascending node, the argument of perihelion and the time of perihelion passage) have secular changes as well as periodic ones.

To fix our ideas regarding the nature of periodic and secular terms in the formula describing the behaviour of an element, say ω, let us write:

$$\omega = \omega_0 + \lambda t + b \sin t + B \sin \frac{t}{A}$$

where t is the time, ω_0, λ, b, B and A are constants, A being much greater than unity.

The term λt is a secular term, growing from age to age, $b \sin t$ is a short-period term since its argument t will take it swiftly through its cycle; and $B \sin(t/A)$ is a long-period term since its argument t/A will cause the cycle to occur much more slowly than that of the short-period term. In addition, the amplitude B may be of much greater size than b. The remaining quantity ω_0 is simply the value ω had when t was zero.

Then figure 14.2 sketches roughly the behaviour of ω.

It should be remembered that, in fact, there are many short- and long-period terms in the expressions for the long-term changes in the planetary orbital elements.

The long-term behaviour of the semi-major axis a, sketched in figure 14.3, shows how an analytical theory, limited to a time interval of only one thousand years because of its attempt to give

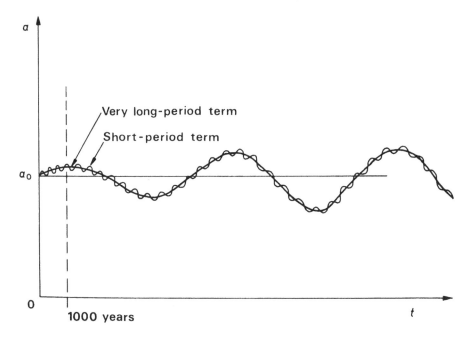

Figure 14.3. Long-term behaviour of the semi-major axis *a*.

highly accurate predictions of planetary positions, might produce secular terms that, in fact, were merely the leading terms in the series standing for sines and cosines of very long period. Thus, for example,

$$\sin \alpha t = \alpha t - \frac{\alpha^3 t^3}{6} + \frac{\alpha^5 t^5}{120} - \cdots .$$

Although the terms on the right-hand side of this expression grow with time *t*, the value of $\sin \alpha t$ is never found outside the range -1 to $+1$.

14.5 Special perturbation theories

14.5.1 General principles

A different approach to the many-body problem is that of **special perturbations**, a method most workers in celestial mechanics before the days of high-speed computers shrank from using, since it involved the step-by-step numerical integration of the differential equations of motion of the bodies from the initial epoch to the epoch at which the bodies' positions were desired. Each step consists essentially of calculating the bodies' positions, velocities and mutual gravitational attractions at time t_1, say, their positions and velocities at time t_2, where $(t_2 - t_1)$ is a small time interval, perhaps a few hours, perhaps a few days, depending upon the problem. The new positions give the new values of the mutual gravitational attractions, so that the calculation can be carried forward to a new time t_3.

The great advantage of this method is that it is applicable to any problem involving any number of bodies. Nowadays, in both celestial mechanics and astrodynamics (the study of the trajectories followed by spacecraft, artificial satellites and interplanetary probes), special perturbations are applied to all sorts of problems, especially since many modern problems fall into regions in which general perturbation theories are absent. In astrodynamics, in particular, the solution is often required in a very short time: a high-speed computer suitably programmed will provide the answer. For example, in the case of a circumnavigation of the Moon by a spacecraft, the problem of calculating swiftly the orbit of the vehicle in the Earth–Moon gravitational field can be adequately treated only by special perturbations and a computer.

The main disadvantage of this method is that it rarely leads to any general formulas; in addition, though they may be of no interest to the worker, the bodies' positions at all intermediate steps must be computed to arrive at the final configuration.

A further disadvantage is the accumulation of round-off error. During the numerical work for each step, a great many additions, subtractions, multiplications and divisions are carried out. Each is rounded off and this process is a source of error. If, for example, we worked to four significant figures and a number 1·2754 was, therefore, rounded to 1·275 then multiplied by two, the answer would be 2·550. If, however, we had retained five figures and multiplied by two, the answer would have been 2·5508. This number, rounded to four figures, is 2·551. This shows a difference of one in the fourth place when compared with the previous answer. Thus, in long calculation, where *millions* of operations are carried out, the number of significant figures gradually decreases until, by the end of the calculation, the answers may have lost any meaning.

Several long-term calculations of the positions of the bodies in the Solar System have been carried out by special perturbation methods. In recent years the positions of the five outer planets, Jupiter, Saturn, Uranus, Neptune and Pluto, have been computed over a period of 2×10^8 years. Such investigations show that the orbits, at least within that length of time, are stable. The orbits of Pluto and Neptune seem to be locked together so that no close approach of these bodies to each other can take place. Jupiter and Saturn's orbits also seem to be coupled, the two orbital planes rotating as if stuck together at a constant angle. Periodic disturbances occur, having cycles ranging from a few years to several million years.

Electronic computers, by writing suitable programmes for them, have also been used to prepare the general perturbation theories. The general perturbations are true analytical theories; the algebraic manipulations that formerly occupied *years* of a researcher's life in carrying out are now done by the computer in *hours*, so that problems of a much higher order of complexity in celestial mechanics can now be tackled.

14.5.2   Chaos and unpredictability

It would be a mistake to pin too much faith on the power of suitably programmed computers to predict to any desired future date the orbits in which *all* Solar System bodies must travel.

The concepts of **chaos** and the **predictability horizon** must be taken into account, concepts quite different from that of round-off error. It is a measure of the genius of the great French celestial mechanician Henri Poincaré that almost a century ago he grasped clearly the concept of chaos and its prevailing influence not only in dynamical systems but in human life. In an article published in 1907, he used the word *chaos* and associated it with large effects in the final outcome produced by slight differences in the initial conditions, with consequent long-term unpredictability. More recently, Sir James Lighthill (1986) published in a paper the concept of the *predictability horizon*, a future time beyond which we have no reliable knowledge of the positions and velocities of the dynamical system being studied.

For the major planets and satellites in the Solar System, the predictability horizon, using modern computers, is thousands of millions of years in the future. Even if we cannot know the precise masses, positions and velocities of these bodies at the present time, the absence of close encounters between pairs of these bodies avoids chaotic effects and drastically nearer predictability horizons. For the less massive bodies of the Solar System, comets, meteors, asteroids and small satellites, the picture is very different.

A simple example illustrates these concepts.

Suppose a billiard player can strike the cue-ball with an angular precision of 1 arc minute (the

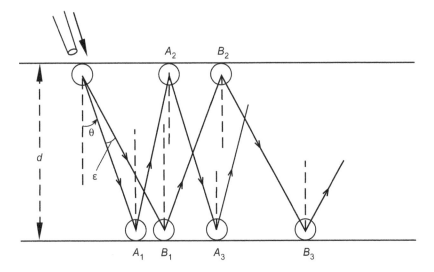

Figure 14.4. The effect of a small error ϵ in the desired angle θ in which a billiards player wants the cue-ball to travel. In n rebounds, the distance D between the intended and actual places of the ball on the opposite side of the table of width d, is given by $D \sim nd\epsilon / \cos^2 \theta$, still a small quantity.

limit of resolution of the human eye). Let the billiard table have width d and let the player try to push the ball, C, from one side of the table to the other at an angle θ degrees to the normal to that side the ball is resting against. Then the ball rebounds repeatedly from side to side of the table making impacts on alternate sides at points $A_1, A_2, \ldots, A_n$ (see figure 14.4). Suppose that the player has projected the ball at angle $(\theta + \epsilon)$ degrees instead of θ, ϵ being a small error of one arc minute. The ball's impact points would now be $B_1, B_2, \ldots, B_n$. It is obvious that the error between the directions of travel of the intended shot at θ and the actual shot at $(\theta + \epsilon)$ degrees is still ϵ after n rebounds. The distance between the 'intended' and 'actual' points where the ball strikes the table side $(A_n - B_n)$ at the nth rebound will be $\sim (nd\epsilon) / \cos^2 \theta$ which remains a small distance even for reasonably large n. For example, if $n = 10, d = 1 \cdot 5$ m, $\theta = 30°$ and $\epsilon = 1$ arc minute, $A_{10} - B_{10} \simeq 0 \cdot 0058$ m.

Consider now the situation as in figure 14.5(a) where the player, using the same cue-ball at the same table, tries to strike a second ball B_1 fixed against the opposite side of the table so that the cue-ball rebounds from B_1 and strikes a second ball B_2 fixed to the opposite side. It is easily shown that if the radius of the balls is R, an error ϵ in the direction the cue-ball is projected becomes an error after rebounding off the fixed ball B_1 of $(2d\epsilon)/R$ (see figure 14.5(b)). For $d = 1 \cdot 5$ m, $R = 0 \cdot 025$ m, $\epsilon = 1$ arc minute, the new error is almost two degrees. The linear error of the ball when it reaches the second fixed ball is about $0 \cdot 06$ m. If, indeed, it does strike the second ball and rebounds, the angular error can be larger than 90 degrees. One could safely wager money against even an expert billiards player in such circumstances managing to strike a third fixed ball B_3.

If we now consider a small body in the Solar System (the cue-ball) to pursue an orbit that does not involve a close encounter with a massive planet (the fixed ball), its orbit will be relatively stable and its predictability horizon will be measured in millions of years. But if the small body makes a close encounter, even a slight change in the circumstances of the encounter will produce large changes in the post-encounter orbit. A subsequent encounter with the same or another planet (which may not have taken place if the first close encounter had occurred under slightly different conditions) will again drastically alter the small body's orbit. It is obvious that the body's orbit is chaotic and has a predictability horizon far closer to the present than the body that suffers no close encounters.

A current concern of humanity is the possibility of comets and asteroids that cross the Earth's

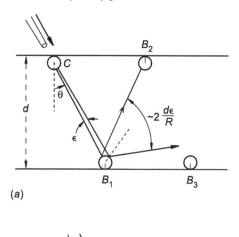

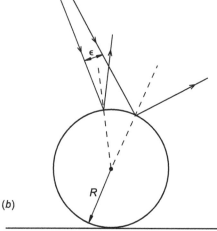

Figure 14.5. (*a*) The amplified effect of a small angular error ϵ in the desired angle θ if the cue-ball is to rebound from a fixed ball B_1 of radius R and strike a second fixed ball B_2, rebounding from it, in turn, to strike a third fixed ball, B_3. (*b*) Magnification of (*a*) to show the rebound of the cue-ball from the fixed ball B_1, the error is now of the order $2d\epsilon/R$.

orbit colliding with our planet. The consequences are now well understood. A collision of an asteroid only 1 km in diameter would cause world-wide catastrophe. One 10 km across would destroy about 90% of all living species. Since there are many such asteroids and comets and their orbits are such that they will have close encounters with one or more planets, their orbits are chaotic and, once found, require constant monitoring to separate the potential civilization destroyers from the others.

The collision of the 22 fragments of Comet Shoemaker–Levy with the planet Jupiter in 1994 is just one piece of evidence supporting the view that such a hazard exists. The object that is going to destroy humanity is out there now: we do not know where it is or when it is going to fall like Nemesis from the skies.

14.6 Dynamics of artificial Earth satellites

14.6.1 Forces acting on artificial satellites

Since October 4th, 1957, many hundreds of artificial satellites have been placed in orbit about the Earth. We have seen (section 13.3) that Newton himself showed that if the projectile was given a sufficient velocity outside the Earth's atmosphere, it would become a satellite of the Earth. But it was only by the development of the rocket during and after the Second World War that a means was provided of imparting to a payload of instruments the velocity necessary to keep it in orbit.

Artificial satellites are subject to Newton's laws of motion and the law of gravitation. They usually obey Kepler's laws very closely. If the Earth were a point-mass and no other force acted upon the satellite, a satellite would obey Kepler's laws exactly and remain in orbit for ever. Many forces, however, may act on the satellite. Among these forces are:

(1) the Earth's gravitational field,
(2) the gravitational fields of the Sun, Moon and the planets,
(3) the Earth's atmosphere and
(4) the Sun's radiation pressure.

In almost every case, the orbital changes produced by the Sun, Moon and the planets are so small that they can be neglected. Only in the case of those artificial satellite orbits that take the satellite many thousands of kilometres away from the Earth does the disturbing effect of the Moon have to be considered. Even then, it is still small.

As a result of the momentum associated with photons within any flow of radiation, the flux produces a pressure or force on any surface which intercepts the radiation (see section 13.7). For a satellite whose size is large (for example a balloon satellite) and whose mass is small, the Sun's radiation pressure can produce large changes in the satellite orbit over many months. For all other satellites, the orbital changes due to solar radiation pressure are negligible unless orbital positions are required to very high precision.

The two main causes of change in a satellite orbit are, therefore, the departure of the Earth's shape from that of a perfect sphere and the drag due to the Earth's atmosphere on those satellites low enough to experience it. We examine each in turn.

14.6.2 Effect of the Earth's shape on a satellite orbit

The Earth is not a perfect sphere. Even before the days of artificial satellites, it was known that the diameter of the Earth, measured from pole to pole, was about 43 km less than an equatorial diameter. Newton explained this equatorial bulge of matter as being a consequence of the Earth's rapid rotation. The departure of the Earth from a sphere, therefore, acts gravitationally as a disturbing force on the satellite. Indeed, it acts upon the Moon itself, producing perturbations in its orbit.

Many artificial satellites are so close to the Earth that other departures of the Earth's shape from a sphere produce observable changes in the satellite orbits. By predicting the types of orbital change due to particular departures and then observing them, our knowledge of the Earth's **figure**, as it is called, has increased enormously. For example, the equator is found to be elliptical rather than circular, the difference between longest and shortest equatorial diameters being about half a kilometre. The northern hemisphere is slightly sharper than the southern hemisphere, imparting a 'pear-shaped' aspect to our planet. The order of magnitude of this difference is only about 20 m. Other, even smaller, features of the Earth's figure have been measured.

The effects upon a satellite orbit due to these causes are best described by the changes that take place in the orbital elements. The semi-major axis, eccentricity and inclination of the orbital plane suffer purely periodic changes. This means that the Earth's departure from a sphere has no disastrous

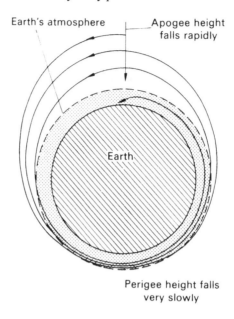

Earth's atmosphere   Apogee height
                            falls rapidly

Earth

Perigee height falls
very slowly

Figure 14.6. The evolution of a satellite's orbit due to atmospheric drag.

effects upon the satellite; it will neither spiral in to be burned up in the Earth's atmosphere nor recede indefinitely to be lost in interplanetary space.

The other three elements are the **right ascension** of the ascending node (since the equator in the case of an artificial Earth satellite is the most useful reference plane), the argument of perigee (again measured along the orbital plane from ascending node to the point in the orbit nearest the central body—in this case the Earth) and the time of perigee passage. All three suffer secular as well as periodic changes.

Thus, the ascending node precesses slowly backwards, at a few degrees per day for typical satellites, so that a revolution of the orbital plane takes some months in contrast to the satellite's own period of revolution about the Earth of an hour or two. It is also found that if the inclination is less than $63° 26'$, the orbit rotates slowly in the direction in which the satellite moves but precesses backwards within the plane, like the plane itself, if the inclination is greater than $63° 26'$.

Since a world-wide network of tracking stations keeps constant watch on many satellites, these secular rates, caused by the Earth's departure from a sphere, can be measured accurately. In turn, through the formulas relating these rates to parameters describing the Earth's figure, the mathematical description of the planet's shape can be evaluated

14.6.3   Effect of the Earth's atmosphere on a satellite orbit

If any part of a satellite orbit lies within the Earth's atmosphere, the satellite experiences a drag force that is a function of the cross-sectional area of the satellite, its velocity and the air-density. Even at the height of 200 km, where the air-density is less than that in a good laboratory vacuum, the drag will be sufficient to bring the satellite spiralling in slowly to the deeper and denser layers of the atmosphere. The process is then accelerated: the satellite loses energy (kinetic plus potential) more swiftly until it plunges into air dense enough to burn it like a meteor. A typical orbital evolution is shown in figure 14.6.

It is found that apogee height falls more rapidly than perigee height. Indeed, the apogee may change by hundreds of kilometres while the perigee changes by under ten kilometres. Because velocity in an elliptic orbit is greatest when the satellite is nearest its primary, and because drag is proportional to the square of the velocity, most orbital change in a revolution is caused at or near perigee. This, together with the near-constant perigee height in a satellite orbit, is an opportunity to measure and monitor the air-density at a particular height over many months. The many hundreds of satellites in orbits of different perigee heights, therefore, provide an excellent system for building up knowledge of the rate of change of air-density with height above the Earth's surface. Daily changes, seasonal changes and changes that occur due to complicated solar–terrestrial relationships can also be studied in this way.

Fortunately, orbital changes due to the air-drag are different in character to those caused by the Earth's departure from a sphere.

The inclination and right ascension of the ascending node are unaffected; the argument of perigee and the time of perigee passage suffer periodic changes. As we have seen before, not only do the semi-major axis and eccentricity vary periodically but they decrease secularly. The rate of decrease of semi-major axis, a, is best found by measuring the rate of decrease of the satellite's period of revolution T.

We have, from equation (13.31),

$$T = 2\pi \left(\frac{a^3}{\mu}\right)^{\frac{1}{2}}$$

where, in this case, $\mu = GM$, M being the mass of the Earth.

Then a knowledge of T gives a value for a.

Formulas exist relating the drag force to the rate of change in the semi-major axis. Knowing the satellite's velocity, size and mass, a calculation provides the air-density.

14.7   The geostationary satellite

An orbit of particular interest for an artificial satellite is a circular one at a height above the Earth's surface of some 35 000 km.

Every orbit has its own period of revolution, T, given by equation (13.31). For a circular orbit of an approximate height of 35 000 km, the period is $23^h 56^m$, in other words 1 sidereal day. A satellite placed in a circular equatorial orbit at such a height will, therefore, remain above a particular point on the Earth's equator. Such **geostationary** satellites are used for communications purposes, relaying radio and television signals over most of the Earth. These now numerous satellites are said to occupy the Clarke belt, so named after the science fiction writer, Arthur C Clarke, who first suggested the use of stationary satellites for communication purposes. It may be noted that because of the relative proximity of stationary satellites, their apparent position in the sky suffers from a substantial parallax.

14.8   Interplanetary transfer orbits

14.8.1   Introduction

The choice of orbits for most interplanetary probes may be readily understood using the concepts discussed in this and the previous chapter. We assume in what follows that the planetary orbits are circular and coplanar. As in many scientific problems, we begin with the simplest cases and gradually progress to more complicated ones.

With chemical rockets, burning times are short, being at most a few minutes. During the 'burn', therefore, it may be assumed that the change in the rocket's situation is one of velocity. The 'burn', in fact, is designed to change the orbit and does so by altering the velocity.

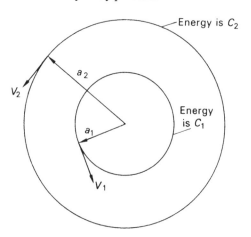

Figure 14.7. Description of two co-planar orbits.

One of the first men to study interplanetary orbits in which instruments could be sent from Earth to other planets was Dr Walter Hohmann. In the 1920s he showed that a particular type of orbit was the most economical in rocket fuel expenditure. We can best appreciate his argument by studying the situation presented in the next section.

14.8.2   Transfer between circular, coplanar orbits about the Sun

In figure 14.7, we have two circular, coplanar orbits of radii a_1 and a_2 astronomical units (AU).

Consider the energy C_1 of a particle in the orbit of radius a_1. It is given by using equation (13.20). We obtain

$$\frac{1}{2}V_1^2 - \frac{\mu}{a_1} = C_1$$

where V_1 is the velocity in the orbit. By equation (13.30),

$$V_1^2 = \frac{\mu}{a_1}.$$

Then,

$$C_1 = -\frac{\mu}{2a_1}.$$

Similarly, the energy C_2 in the orbit of radius a_2 is given by

$$C_2 = -\frac{\mu}{2a_2}.$$

Now $a_2 > a_1$, so that $C_1 < C_2$, in other words a change of orbit would be a change of energy. This change of energy is brought about by the rocket engine. By imparting a velocity increment $\Delta \mathbf{V}$ to the rocket, it changes its kinetic energy and, hence, its total energy. Hohmann studied how this could be most effectively done.

The kinetic energy in equation (13.20) is the term $\frac{1}{2}V^2$. In figure 14.8, we see the original velocity $\mathbf{V}$. The increment velocity $\Delta \mathbf{V}$ could be applied as shown in figure 14.8 in such a way that only the *direction* of the vehicle's velocity is changed (from $\mathbf{V}$ to $\mathbf{V}'$) but not the magnitude ($V' = V$). In that case, the new kinetic energy $\frac{1}{2}V'^2$ would be the same as the pre-burn kinetic energy. No change in total energy would be achieved.

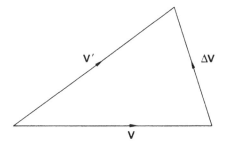

Figure 14.8. An incremental change ΔV in the velocity V.

Figure 14.9. An incremental change in velocity along the original velocity direction.

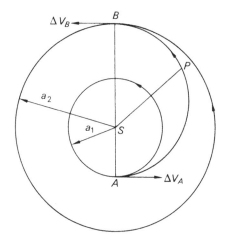

Figure 14.10. A Hohmann transfer orbit APB between two circular, co-planar orbits.

However, if the increment ΔV is applied as in figure 14.9, along the instantaneous velocity vector V, then the maximum increase in kinetic energy is achieved for a given burn, i.e. the full effect of ΔV is added to V. Obviously if it is desired to decrease the kinetic energy, the velocity increment ΔV would be applied in the opposite direction to V.

Hohmann showed that, in practice, the most economical transfer orbit between circular, coplanar orbits was an elliptical orbit cotangential to inner and outer orbits at perihelion and aphelion respectively. It is shown in figure 14.10 as ellipse APB. Only one-half of the transfer orbit is used. At A, the rocket-engine is fired to produce a velocity increment ΔV_A, applied tangentially to place the vehicle in the transfer orbit. The vehicle coasts round the half-ellipse APB, reaching aphelion at B. If no further change in energy took place, the vehicle would coast onwards along the mirror half of APB to return eventually to A. A second impulse is, therefore, required to produce a second velocity increment ΔV_B. This is again obtained by firing the rocket-engine tangentially and the rocket enters the outer circular orbit of radius a_2 AU.

Such transfer orbits are known as **Hohmann least-energy two-impulse cotangential orbits**.

We now show how to calculate various parameters associated with the transfer.

(i) The semi-major axis, α of the transfer orbit. In figure 14.10 it can be seen that

$$AB = 2\alpha = a_1 + a_2.$$

Hence,

$$\alpha = \frac{a_1 + a_2}{2}. \tag{14.1}$$

(ii) The eccentricity e of the transfer orbit.

$$SA = a_1 = \alpha(1 - e)$$
$$SB = a_2 = \alpha(1 + e).$$

Hence,

$$e = \frac{a_2 - a_1}{a_2 + a_1}. \tag{14.2}$$

(iii) The transfer time τ spent in the transfer orbit. This is the time interval spent in coasting from A to B. It must be half the period of revolution T in the transfer orbit. Then by equation (13.31),

$$\tau = \frac{T}{2} = \pi \left(\frac{\alpha^3}{\mu} \right)^{\frac{1}{2}}$$

or, using equation (14.1),

$$\tau = \pi \left(\frac{(a_1 + a_2)^3}{8GM} \right)^{1/2} \tag{14.3}$$

since the mass of the rocket is negligible compared to that of the central body.

If distance is measured in astronomical units, time in years and the unit of mass is the Sun's mass, then by section 13.7, $GM = 4\pi^2$.

We may write, therefore,

$$\tau = \left(\frac{(a_1 + a_2)^3}{32} \right)^{1/2}. \tag{14.4}$$

(iv) The velocity increments ΔV_A and ΔV_B. At A, the required increment ΔV_A is the difference between circular velocity V_{c1} in the inner orbit and perihelion velocity V_P in the transfer orbit. Then

$$\Delta V_A = V_P - V_{c1}.$$

By equations (13.26) and (13.30) respectively, we may write

$$\Delta V_A = \left[\frac{\mu}{\alpha} \left(\frac{1+e}{1-e} \right) \right]^{\frac{1}{2}} - \left(\frac{\mu}{a_1} \right)^{\frac{1}{2}}$$

$$= \left(\frac{\mu}{a_1} \right)^{1/2} [(1 + e)^{1/2} - 1].$$

Hence, using equation (14.2) we obtain

$$\Delta V_A = \left(\frac{\mu}{a_1} \right)^{\frac{1}{2}} \left[\left(\frac{2a_2}{a_1 + a_2} \right)^{\frac{1}{2}} - 1 \right]. \tag{14.5}$$

At B, the required increment ΔV_B is the difference between circular velocity V_{c2} in the outer orbit and aphelion velocity V_A in the transfer orbit.

Then $\Delta V_B = V_{c2} - V_A$, since $V_{c2} > V_A$.

Proceeding in the same way, we obtain, finally

$$\Delta V_B = \left(\frac{\mu}{a_1}\right)^{\frac{1}{2}} \left[1 - \left(\frac{2a_1}{a_1 + a_2}\right)^{\frac{1}{2}}\right]. \tag{14.6}$$

14.8.3   Transfer between particles moving in circular, coplanar orbits

In the first attempts to send packages of scientific instruments from Earth to Venus and Mars, the transfer orbits closely resembled Hohmann transfers. Only the first of the two impulses was required since there was no need to inject the payload into either the Venusian or the Martian orbits. Apart from that difference, there was a very important factor present that was not taken into account in the discussion of section 14.8.2.

In that problem, the first burn could be made at any time. In the present problem a timetable has to be kept, dictated by the necessity that when the vehicle reaches aphelion in the transfer orbit, the particle moving in the outer circular orbit should also be there. This problem, often called the **rendezvous problem**, arises because departure point and arrival point have their own orbital motions in the departure and destination orbits. When the transfer burn is initiated, the object in the second orbit is a little behind the point where it is due to cross the transfer orbit. During the time taken by the transfer, the object progresses on its orbit to be at the crossing point at the correct time.

Let two particles P_1 and P_2 revolve in coplanar orbits of radii a_1 and a_2 about a body of mass M. Let their longitudes, measured from some reference direction Υ be l_{10} and l_{20} at time t_0. The problem is to obtain the time conditions enabling a vehicle to leave particle P_1 and arrive at particle P_2 by a Hohmann cotangential ellipse.

The angular velocities of the two particles are n_1 and n_2, given by

$$n_1 = \left(\frac{GM}{a_1^3}\right)^{\frac{1}{2}} \qquad n_2 = \left(\frac{GM}{a_2^3}\right)^{\frac{1}{2}} \tag{14.7}$$

since

$$n = \frac{2\pi}{T} \quad \text{and} \quad T = 2\pi \left(\frac{a^3}{GM}\right)^{\frac{1}{2}}.$$

The longitudes of the particles at time t are, therefore, given by

$$l_1 = l_{10} + n_1(t - t_0) \tag{14.8}$$

$$l_2 = l_{20} + n_2(t - t_0). \tag{14.9}$$

The time spent by the vehicle in the transfer orbit must be the time taken by the particle P_2 to reach the touching point of transfer and destination orbits. This point C, say, therefore, lies ahead of the position of P_2, namely B, when the vehicle leaves P_1 at A (see figure 14.11).

Then, if $\angle BSC = \theta$, the transfer time is τ, given by

$$\tau = \frac{\theta}{n_2}. \tag{14.10}$$

But, by equation (14.3),

$$\tau = \pi \left(\frac{(a_1 + a_2)^3}{8GM}\right)^{1/2}$$

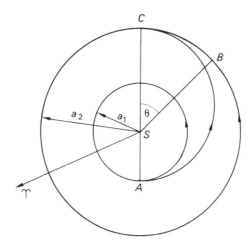

Figure 14.11. The rendezvous problem. The destination body (in the outer orbit) must be at B when the rocket leaves the inner body at A in order that rocket and outer body arrive together at C.

so that we have

$$\theta = \pi n_2 \left(\frac{(a_1 + a_2)^3}{8GM} \right)^{1/2}. \tag{14.11}$$

Now the longitude of P_1 when the vehicle leaves that body is π radians less than the longitude of P_2 when the vehicle arrives. Thus, the longitude of the particles at the vehicle departure time must differ by $(\pi - \theta)$ radians, or L_{12}, given by

$$L_{12} = \pi \left[1 - n_2 \left(\frac{(a_1 + a_2)^3}{8GM} \right)^{1/2} \right] = \pi \left[1 - \left(\frac{1 + a_1/a_2}{2} \right)^{3/2} \right] \tag{14.12}$$

using equation (14.7). Hence, L_{12} can be found.

But by equations (14.8) and (14.9), the difference in the longitudes of P_2 and P_1 at any time t is given by

$$l_2 - l_1 = l_{20} - l_{10} + (n_2 - n_1)(t - t_0). \tag{14.13}$$

If the value of L_{12} is inserted in the left-hand side of equation (14.13), the time t can be calculated. In fact, a number of values of t, both positive and negative, will be found to satisfy the equation. From these can be taken all future epochs (out-of-date timetables are never of much use!) at which the vehicle can begin a cotangential transfer orbit from P_1 to P_2. Obviously such epochs are separated by a time interval which is the synodic period of one particle with respect to the other, since it is the time that elapses between successive similar configurations of the particles and the central mass.

For a return of the vehicle from P_2 to P_1, the same period must elapse between successive favourable configurations for entry into a cotangential ellipse. The transfer time τ will be the same as on the outward journey and the angle θ' between the radius rector of P_1 when the vehicle departs and that of the arrival point in P_1's orbit must be given by

$$\theta' = n_1 \tau.$$

Then for a suitable configuration of bodies, the difference in longitudes of P_2 and P_1 must be L'_{21} where

$$L'_{21} = \pi \left[n_1 \left(\frac{(a_1 + a_2)^3}{8GM} \right)^{1/2} - 1 \right] = \pi \left[\left(\frac{1 + a_2/a_1}{2} \right)^{3/2} - 1 \right]. \tag{14.14}$$

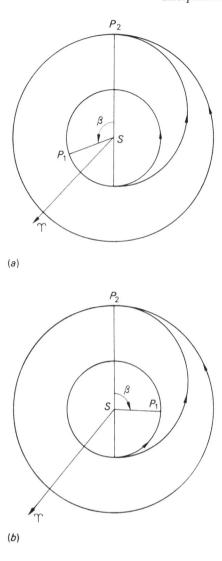

(a)

(b)

Figure 14.12. Transfer and return from one orbit to another, illustrating the term 'waiting time'.

Equation (14.14) may be used with equation (14.13) to compute the available epochs for the return journey. The waiting time interval t_W between the arrival time at C (figure 14.11) and the first available departure time from P_2 can then be found and can be added on to 2τ to give the round trip time or mission time t_M. Thus,

$$t_M = 2\tau + t_W. \tag{14.15}$$

From symmetry considerations the minimum waiting time t_W can be readily obtained. If P_1 is β radians 'ahead' of P_2 when a transfer from P_1 to P_2 has just ended, as in figure 14.12(a), the first available transfer back from P_2 to P_1 will begin when P_1 is β radians 'behind' P_2.

Hence, if S is the synodic period of the two particles,

$$t_W = \left(\frac{2\pi - 2\beta}{2\pi}\right) S = \frac{2\pi - 2\beta}{n_1 - n_2}. \tag{14.16}$$

Alternatively, if P_1 were β radians 'behind' P_2 when an outward transfer (P_1 to P_2) has just

Table 14.1. Transfer from Earth to the planets by Hohmann orbit.

Planet	Transfer time τ (yr)	Minimum waiting time t_W (yr)	Total mission time $t_M = 2\tau + t_W$ (yr)	Eccentricity of transfer orbit
Mercury	0·289	0·183	0·76	0·44
Venus	0·400	1·278	2·08	0·16
Mars	0·709	1·242	2·66	0·21
Jupiter	2·731	0·588	6·05	0·68
Saturn	6·048	0·936	13·03	0·81
Uranus	16·04	0·932	33·01	0·91
Neptune	30·62	0·766	62·01	0·94
Pluto	45·47	0·061	91·00	0·95

ended, as in figure 14.12(*b*), the first available return from P_2 to P_1 can begin when P_1 has reached a point β radians 'ahead' of P_2.

In this case,

$$t_W = \frac{2\beta}{2\pi} S = \frac{2\beta}{n_1 - n_2}. \tag{14.17}$$

To compute β, we note that by equations (14.8) and (14.9),

$$l_1 - l_2 = (l_{10} - l_{20}) + (n_1 - n_2)(t - t_0).$$

Suppose l_{10}, l_{20} were the longitudes of the particles at take-off time t_0 and l_1, l_2 were the longitudes at arrival time t. Then (figure 14.9)

$$t - t_0 = \tau \qquad l_{10} - l_{20} = \theta - \pi = n_2\tau - \pi$$

giving

$$l_1 - l_2 = n_1\tau - \pi.$$

Then β is given by

$$n_1\tau - \pi = 2\pi k + \beta \tag{14.18}$$

where $-\pi \leq \beta \leq \pi$ and k is a positive integer or zero.

If β is positive, equation (14.16) gives t_W.

If β is negative, equation (14.17) gives t_W.

When we apply the formulas derived in this and the previous section to travel in Hohmann orbits between Earth and the other planets, an extremely instructive table can be drawn up (see table 14.1).

The mission times for Venusian, Martian and Mercurian round trips are not impossible to contemplate for unmanned, or even for manned, voyages, the interesting fact emerging that the Mercurian mission lasts only about a third and a quarter as long respectively as the Venusian and Martian missions. The important factor in these cases is the long waiting times at Mars and Venus before the return journey can be begun. The decrease of such long waiting times by the use of transfer orbits other than the minimum energy expenditure Hohmann orbits, therefore, has a high priority in the list of factors involved in planning such voyages.

For the outer planets, even one-way outward transfers take a long time, certainly too long for manned voyages and possibly too long for instruments to function without breaking down. Exploration of these planets by interplanetary probes, however, has been made possible by utilizing the concept of velocity amplification by close encounter with Jupiter (see later).

Table 14.2. Sphere of influence of the planets.

	Radius (r_A) of sphere of influence		
Planet	Millions of kilometres	Fraction of planetary orbit's semi-major axis	Astronomical units
Mercury	0·112	0·001 93	0·000 747
Venus	0·615	0·005 69	0·004 11
Earth	0·925	0·006 19	0·006 19
Mars	0·579	0·002 54	0·003 87
Jupiter	48·1	0·061 9	0·322
Saturn	54·6	0·038 2	0·365
Uranus	52·0	0·018 1	0·348
Neptune	86·9	0·019 3	0·581
Pluto	3·59	0·000 61	0·023 95

14.8.4 Transfer between planets

In practice, the planets have their own gravitational fields and an interplanetary transfer made by a space-probe is a many-body problem. The concept of a planet's sphere of influence, first introduced in celestial mechanics, is useful in this type of problem.

The **sphere of influence** of the planet is the volume of space about the planet, within which it is more useful to consider a satellite or probe to be in orbit about the planet and perturbed by the Sun than to be in orbit about the Sun and perturbed by the planet.

The fact that the planet's gravitational field is so weak compared to the Sun's means that its sphere of influence is almost spherical and has a radius r_A much smaller than the planet's heliocentric radius vector r_P. It is found that

$$\frac{r_A}{r_P} \approx \left(\frac{m}{M}\right)^{2/5}$$

where m and M are the masses of the planet and Sun respectively.

Table 14.2 summarizes the relevant information about the planetary spheres of influence. The consequence of the fall-off in intensity of the Sun's gravitational field with distance from the Sun is evident on comparing the sizes of the spheres of influence of Earth and Pluto. The latter sphere is about four times as large as the former, even though the mass of Earth is about three hundred times that of Pluto.

A spacecraft destined for Mars, say, has a trajectory that begins in the Earth's sphere of influence. The spacecraft's rocket engines give it sufficient velocity to boost it to the limits of the Earth's sphere of influence. It enters interplanetary space with a certain geocentric velocity that, together with the Earth's own orbital velocity, will put it into the heliocentric transfer orbit required to take it out to the Martian orbit. During this transfer, its orbit will be perturbed only very slightly by the Earth and, as it draws nearer and nearer to Mars, by that planet. Once it enters the Martian sphere of influence, it can be said to be in a solar-perturbed orbit about that planet. In general, unless the vehicle fires its engines, it will not enter a closed satellite orbit about Mars but will swing round the planet in a hyperbolic orbit that will ultimately eject it from the Martian sphere of influence. Its subsequent solar orbit will be quite different from its pre-Martian encounter orbit because of its deep penetration into the Martian sphere of influence. The total energy of its solar orbit may have been increased or decreased by its flypast of Mars.

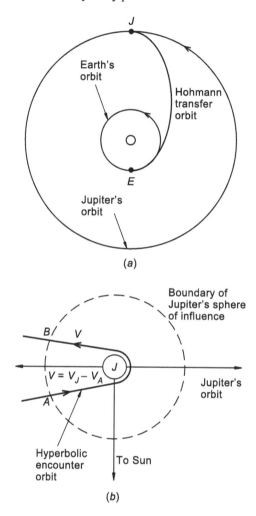

Figure 14.13. (*a*) Hohmann transfer orbit of a spacecraft from Earth to Jupiter, the spacecraft entering Jupiter's sphere of influence. (*b*) How, in the hyperbolic jovicentric orbit of the spacecraft within Jupiter's sphere of influence, the spacecraft's velocity V at A is effectively the exit velocity V at B.

This ability of a planetary flypast to change the energy of a small body in orbit and give it an entirely different heliocentric orbit is not a new phenomenon in celestial mechanics. Cometary orbits have been observed to be drastically changed by close encounters with Jupiter. For example, Comet Brooks (1889V) had its period of revolution shortened from 29 to 7 years by passing so close to Jupiter that it spent two days within the bounds of Jupiter's satellite system.

14.8.5 Interplanetary billiards

The ability to amplify a spacecraft's heliocentric energy or dramatically change the direction of its trajectory by a planetary encounter was used to send Mariner 10 via Venus to within 1500 km of Mercury. The same principle was used to make Pioneer 11 and Voyagers 1 and 2 pass by Saturn after a Jupiter flypast, ultimately enabling Voyager 2 to visit Uranus and Neptune. The principle, sometimes called *interplanetary billiards*, is illustrated as follows.

Let a spacecraft follow a Hohmann elliptic transfer from Earth orbit to Jupiter orbit (figure 14.13(a)). Let its aphelion velocity be V_A. The circular heliocentric velocity in Jupiter's orbit is V_J which is greater than V_A. Hence, from Jupiter's point of view, the planet effectively overtakes the spacecraft with a relative velocity $V = V_J - V_A$, the spacecraft entering Jupiter's sphere of influence at a point A close to Jupiter's orbit (figure 14.13(b)). Once within Jupiter's sphere of influence, the spacecraft makes a hyperbolic flight past Jupiter, exiting from Jupiter's sphere of influence at B with the same velocity V but now effectively in the opposite direction to which it entered at A. Its new velocity relative to the Sun is

$$V_\odot = V + V_J = 2V_J - V_A > V$$

thus enabling the spacecraft to travel into the outer reaches of the Solar System without any additional expenditure of rocket fuel. It is almost as if the spacecraft 'billiard ball' has rebounded from the moving Jupiter 'billiard ball', gaining speed in the process. Careful planning of the exact point of entry of the spacecraft into the Jovian sphere of influence can produce almost any desired subsequent trajectory even to sending the spacecraft far above and below the plane of the Solar System for detailed study of the Sun's polar regions.

This extreme sensitivity of the orbit of a spacecraft subsequent to slight changes in its Jovian flypast is another example of chaos (section 14.5.1) but one where the concept is used to produce a desired effect.

Problems—Chapter 14

Take the radius of the Earth to be 6378 km.

1. A communications satellite is to be injected into a circular equatorial orbit at such a height that the satellite remains permanently above a particular point in Brazil with declination zero. At present the satellite is in a circular equatorial parking orbit 320 km above the Earth surface. The transfer orbit is to be a two-impulse cotangential ellipse. Calculate (i) the radius of the stationary orbit, (ii) the semi-major axis and eccentricity of the transfer orbit, also the transfer time, given that the Earth's sidereal period of rotation is $23^h 56^m$ and the period of revolution of the satellite in its parking orbit is 90 minutes.

2. Two spacecraft, A and B, are in circular equatorial orbits about the Earth at geocentric distances of 7240 and 12 870 km respectively. If their right ascensions at a certain Greenwich Date are $0°$ and $64°$ respectively, calculate the time that must elapse before the first suitable occasion when A can begin a two-impulse transfer to rendezvous with B (i.e. enter B's circular orbit at B's position). Calculate the semi-major axis and eccentricity of the transfer orbit, also the time spent in the transfer orbit.
   (Take the period in a circular orbit 160 km above the Earth's surface to be 84·5 minutes.)

3. A lunar probe is put into an elliptical transfer orbit from a circular parking orbit (radius 6878 km) about the Earth. It is intended that the apogee of the transfer orbit should touch the Moon's orbit (assumed circular of radius 384 400 km). If the circular velocity in the parking orbit is 7·613 km s^{-1} calculate:
   (i) the semi-major axis and eccentricity of the transfer orbit,
   (ii) the time it takes the probe to reach apogee and
   (iii) the required velocity increment to put the probe into the transfer orbit.

4. Two asteroids move in circular coplanar heliocentric orbits in the ecliptic with the following elements:

Asteroid	Orbital radius (AU)	Longitude at epoch (degrees)	Epoch (Julian date)
A	2	79	
			244 1681·0
B	3·5	211	

Find the first available date after the epoch on which a Hohmann cotangential least-energy transfer from asteroid A to asteroid B can begin, also the minimum waiting time on the second asteroid if the return to the first is also a least-energy transfer.

PART 3

OBSERVATIONAL TECHNIQUES

Chapters 15–23

PROGRAMME: To complete the presentation of the astronomical 'software', some concepts related to radiation and the means whereby the radiation is collected and then analysed are set out.

In the first place, the generation of radiation from macroscopic bodies and from individual atoms is discussed. This is followed by the description of optical telescopes and their function in collecting radiation. The developing technologies associated with the quest of providing larger instruments and defeating the problems associated with the Earth's atmosphere are described. Detectors used to record data are presented with special terms and systems of measurement which are used internationally by astronomers. Other more special equipment and techniques are also described.

Finally, the technologies of radio and high energy astronomies are presented.

Chapter 15

The radiation laws

15.1 Introduction

Before we consider the nature of the measurements which are made in optical astronomy and their interpretation, it will be profitable to take a brief look at how our understanding of electromagnetic radiation has progressed, particularly in the early part of the 19th century and to summarize some of the laws of radiation.

The effects of the refraction of light when passing through transparent materials have been known from very early times. Experiments on refraction were recorded by Ptolemy in the 2nd century AD and effects of atmospheric refraction on the positions of stars (see section 10.2.2) were known at least as early as 1038 AD to the Arabian astronomer, Al-Hasen. However, it was not until the experiments of Sir Isaac Newton in the 17th century that the composite nature of 'white' light was discovered. He demonstrated that when the light from the Sun is passed through a prism and refracted, it is dispersed to reveal that it is made up of a range of colours, corresponding to those seen in a rainbow. He also demonstrated that this spectrum of colours could be reconverted to a beam of white light by the appropriate use of a second prism. Little further advance in the study of spectra was made until about a century later, when Melvill (1752) discovered that the spectrum of light emitted by a sodium flame could not be seen as a continuous range of colours but that only a strong yellow band was present, corresponding to what we know today as the D lines.

At the turn of the 19th century great advances were made in the understanding of optical radiation. The now classical interference experiments of Young and Fresnel revealed that light had a wave-nature and that for each point within the optical spectrum, there corresponded an exact value of wavelength. Young, in 1802, performed calculations on some of the original observations made by Newton of colours produced by thin films and assigned wavelengths to the seven colours designated by Newton. By proposing a wave-nature to light, Young was able to interpret observations which escaped explanation by Newton's corpuscular theory. When converted into nanometres, Young's calculations show that the solar spectrum extended from 424 to 675 nm.

In 1800 Sir William Herschel performed his experiments on the distribution of radiant heat through the solar spectrum and he found that his sensitive thermometer recorded its maximum value at a position just beyond the red end of the spectrum where no colour was apparent. At the time, he believed this unseen radiation was dissimilar to visible light. Unknowingly, he was the first astronomer to extend observations beyond the visible electromagnetic spectrum—in this case, to the infrared. Three years later, Ritter published the results of his experiments on the effectiveness of different parts of the spectrum to blacken silver chloride. He found that the strongest effects occurred at a position just beyond the violet end of the spectrum. In this way he detected the ultraviolet region of the electromagnetic spectrum.

During this same period Wollaston investigated the spectra of flames and sparks but gained no

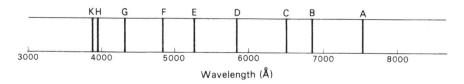

Figure 15.1. The positions of the chief absorption features in the solar spectrum as designated by Fraunhofer.

real clue to the relationships and identities of the spectral lines he observed. In the paper he published in 1802, describing his laboratory measurements of the refractive indices of various materials, he also makes comment on the spectrum of the Sun which he obtained. His description is vague but it appears that he saw dark bands which he supposed to be natural boundaries separating the zones of pure colour. He appears not to have attached much significance to this observation and dispenses with it in a few sentences. It may have been that he expected there to be natural divisions between the seven zones of the spectrum which had been designated by Newton and was, therefore, not surprised to see the dark bands. Newton himself made no mention of what we know as the solar absorption features, probably because of the poor quality of his glass—a point on which he himself complained.

Some years later in 1817, Fraunhofer, who had been experimenting in ways of defining the colours which are used to determine the refractive index of glasses, published the results of his observations of the solar spectrum made by using an improved spectroscope. He noted that the spectrum was crossed by many dark lines. By repeating his observations with a range of optical elements he proved that the lines were a real feature of the spectrum and were contained in the sunlight. He made a map of several hundred of the lines and designated the prominent ones with the letters A, B, C, . . . , by which they are still known (see figure 15.1). Although Fraunhofer was unable to offer an explanation for their cause, the positions of these lines provided the first standards of colour for use in measuring the refractive index and dispersion of different materials and in comparing the spectra from other luminous sources.

Fraunhofer also was the first to make visual observations of the spectra of the planets and of the brighter stars including *Sirius* by using an objective prism. He observed that all the spectra exhibited dark lines—those of the planets were similar to the Sun, those of the stars were different. Thus, it can be said that he was the first pioneer stellar spectroscopist. His work did not rest here, however, and he continued by inventing and manufacturing many diffraction gratings. With the aid of these devices he succeeded in measuring accurately the wavelengths, in terms of laboratory standards of length, of the spectral features and providing a basis whereby the results of different observers could be compared directly. In other words he was the instigator of observational spectrometry.

15.2   The velocity of light

It will already have been appreciated that the velocity of light *in vacuo* and indeed the velocity of all electromagnetic waves is a fundamental constant; its value appears in the description of and in formulas relating to many different physical processes. This fundamental constant of nature is one of the starting points for our understanding of the behaviour of matter from the extremes of processes involving individual atoms to those involving the whole conglomeration of matter within the universe. Indeed, it was because of experiments connected with the velocity of light that these new physical theories, in particular, Einstein's theories of relativity, arose.

We have already seen (section 11.3) that light has a finite velocity—this was first discovered by the astronomer Roemer. Confirmation that light had a finite velocity came from Bradley's discovery of aberration (see section 11.4).

Because of the magnitude of the velocity involved, laboratory measurements of the velocity of light proved to be very difficult. The first laboratory determination was achieved by Fizeau in 1849 and since that time many further experiments have been conducted with the aim of producing improved values. The modern value for the velocity of light is given by

$$c = 2 \cdot 997\,924\,58 \times 10^8 \text{ m s}^{-1}$$

and, by definition, its value is exact.

Establishment of the value for the velocity of light has had real practical importance in the determination of planetary distances by radar. In these experiments, radar pulses have been directed to the planets and by timing accurately the interval between transmission and reception of the reflected pulse, the planet's distance can be obtained with high precision. In particular, radar measurements of the distance of Venus have provided an accurate absolute scale for the dimensions of the Solar System and the astronomical unit (see section 10.6).

The advent of the laser has also allowed 'radar' measurements to be made in the optical region. By placing efficient reflectors on the lunar surface, it is now possible to detect laser pulses which are sent back by the reflectors. The time required for a pulse to travel from the Earth to the Moon and back again can be measured extremely accurately. By this new method, the distance of the Moon can be determined to within a metre.

15.3   Kirchhoff's law

The problem of making sense of the various spectra continued for some thirty years after the work of Fraunhofer. At this distance of time it seems obvious, at first sight, that Fraunhofer had all the evidence staring him in the face: an element like sodium, when introduced to a flame, produces the D lines in the spectrum—again, dark lines, at the position of the D lines, appear in the solar spectrum. Surely he should have seen the connection and arrived at the conclusion that individual spectra are produced for particular atoms when they are present in the source?

An understanding of the period in which he worked gives us some idea as to why he was unable to make what would eventually be a great leap forward. In the first place, he was already faced with the current theories; for example, Young believed that the colours of flames were caused by interference phenomena within the thin layers of the flame. In the second place, the experimental circumstances are likely to have thrown him and others off the scent. Perhaps the most important factor would have been the impurity of the salts which were added to the flame for observation. It turns out that sodium is a difficult contamination to remove—it is always present on the hands as a result of perspiration and is found in most animal and vegetable matter. The sodium D lines were, therefore, likely to occur in all flame spectra. Several other contaminations were also likely to cause confusion in the comparison of spectra.

It appears that several investigators after Fraunhofer were on the verge of identifying spectra with the presence of particular elements but no-one seems to have made an unequivocal statement to the effect that each type of atom or molecule has its own characteristic spectrum until Kirchhoff in 1859. This idea became clear finally as a consequence of the law proposed by Kirchhoff. In 1833, Ritchie had shown experimentally that the ratio of the total emissive power of a surface to its absorptive power was a constant. Kirchhoff extended this work and proposed the following law which states quite generally as follows:

At any given temperature, the ratio of the ability of the body to emit radiation and the ability to absorb it at a particular wavelength is constant and independent of the nature of the body.

Thus, if the emissive power of a body in the narrow wavelength range λ to $\lambda + \mathrm{d}\lambda$ is E_λ and its absorptive power is A_λ, then for a given temperature

$$\frac{E_\lambda}{A_\lambda} = \text{constant} \tag{15.1}$$

Table 15.1. The strong Fraunhofer lines.

Fraunhofer line	Å	Element
A	7594	Oxygen in Earth's atmosphere
B	6867	Oxygen in Earth's atmosphere
C	6563	Hydrogen ($H\alpha$)
D_1	5896	Sodium
D_2	5890	Sodium
E	5270	Iron
F	4861	Hydrogen ($H\beta$)
G	4340	Hydrogen ($H\gamma$)
H	3968	Calcium
K	3933	Calcium

the constant being independent of the type of body under consideration. The actual value of the constant evaded an evaluation by theory until the development of the quantum theory, which at the same time gave an explanation of how individual atoms radiate.

It follows from Kirchhoff's law that a body which is a good emitter of radiation must correspondingly be a good absorber. Simple confirmation of this is given by the appearance of the Sun. Being a good emitter of radiation, the Sun's material also absorbs radiation efficiently. Radiation which is generated at any depth inside the Sun is absorbed before travelling any appreciable distance. Consequently, we are only able to see radiation which is emitted in the Sun's outermost layers. Thus, the Sun appears to have a sharp edge and we cannot see to any depth into its interior.

Kirchhoff's law also opened the door to modern spectrometry whereby atoms and molecules are identified by their spectra. By the law's principle, a gas which produces bright spectral lines should also, at the same temperature, be able to absorb energy at these wavelength positions from a beam of light which would otherwise have produced a continuous spectrum. In his now famous experiment, Kirchhoff passed a beam of sunlight through a flame containing common salt (sodium chloride) and found that the yellow emission lines of the flame coincided exactly with the Fraunhofer D lines and that the sodium flame enhanced the absorption feature. He put forward the explanation that the Fraunhofer lines in the solar spectrum are caused by the absorption by the elements in the outer, cooler region of the Sun of the continuous spectrum emitted by its hot interior.

The idea that each element gives rise to a characteristic spectrum rapidly gained momentum from this point. In partnership with Bunsen, Kirchhoff continued his experiments by investigating the spectra of other metals in their purest form then available and was able to identify most of them in the solar spectrum. The elements corresponding to the strong Fraunhofer lines are presented in table 15.1.

Spectral analysis thus opened the way to performing analytical chemistry at a distance. In particular, the methods could be applied to all astronomical bodies and the problem of determining the composition of stars—a problem which had been said to be impossible by the philosopher, Comte, only a few years previously—was now realistically solved.

The science of spectrometry has now, in fact, developed further than this in that, by detailed study of the positions and shapes of spectral lines, it is possible to ascertain physical parameters of a source such as its speed and the temperature and density of its materials.

Very broadly, a laboratory spectrum may be considered to be one of three types, although, on some occasions, it cannot be categorized simply, i.e. there is no sharp division between the three types.

Many solids, liquids and very dense gases are opaque to all optical radiation and, therefore, according to Kirchhoff's law, provide spectra which are continuous. A **continuous spectrum** is one which provides a continuous distribution of energy through the observed wavelengths and, for example, such a spectrum is provided in the visible region by a tungsten electric lamp filament. No clue is given to the chemical composition of a body radiating a continuous spectrum. The energy–wavelength distribution, however, gives a means of determining the body's temperature.

Hot tenuous gases provide spectra with bright lines but with no continuous background. **Emission spectra**, as they are called, may be used to identify the atoms and molecules in the source. Between the emission lines, the spectrum generally displays gaps where there is no radiation present. Such a gap acts as a transparent window. If a spectrometer were to be tuned to a zone of wavelengths corresponding to a gap between emission lines, it would not receive radiation from the gas, which would remain undetected and transparent. Any radiation emitted in that zone of wavelengths by an object behind the gas would be detected by the spectrometer and the object would be 'visible' through the gas.

In the laboratory, emission spectra are usually readily available in the form of discharge lamps containing some selected gas at reduced pressure. Emission spectra of metallic elements can be obtained by using electric arcs.

Absorption spectra may be produced as in Kirchhoff's famous experiment. If light from a source which would normally present a continuous spectrum is made to pass through a tenuous gas, then energy may be removed from the wavelength positions which correspond to emission lines of the particular gas. The amount of energy removed in the absorption lines—and hence the strength of these features—increases if the gas is cooled.

The difference between a continuous spectrum and a line spectrum results from the circumstances under which the atoms emit or absorb the radiation. It will be seen later in this chapter that spectral lines are associated with the jumping of electrons within an atom from one discrete orbit to another. In the case of gases at low pressure, the atoms are able to radiate and absorb without the influence of neighbouring atoms and very sharp spectral lines are produced. An increase of the gas pressure causes the spectral lines to broaden. For emission line spectra, the broadening increases with pressure until lines overlap and merge to produce a continuous spectrum. In bodies which are liquid or solid, the atoms are packed even closer together than for a gas at high pressure and they are unable to radiate without being influenced by the neighbouring atoms. They, therefore, normally present a continuous spectrum.

The application of Kirchhoff's law and the principles relating to the understanding of laboratory spectra immediately provides a simple explanation for the interpretation of stellar spectra. The most common stellar spectra, as is the case of the solar spectrum, are *absorption spectra*. This suggests that the stars have structure. The general wavelength–energy distribution curve must originate from some body which must be a hot solid, liquid or dense gas and the absorption lines must be produced by some surrounding cooler tenuous gas, being in effect the stellar atmosphere. This idea is summarized pictorially in figure 15.2.

Thus, by correct interpretation of some of the possible basic measurements of solar light, we know that its spectrum is produced by a hot dense gas which is surrounded by a cooler more tenuous layer. From the general appearance of stellar spectra, it is not unreasonable to assume that the stars have this same basic structure.

Immediately after Kirchhoff's work, the problem of spectra identification was taken up by many investigators and it was soon evident that, because of the many thousands of lines which were being discovered, results by different workers could only be compared by using some definite standards. One of the first sets of standards to be accepted was as a result of Ångström's work on his *Normal Solar Spectrum*; his wavelength values were given in terms of a known grating spacing to six significant figures and expressed in units of 10^{-8} cm. This unit has since been known as the **ångström unit** (Å).

In the remainder of this chapter, the basic physical concepts associated with the radiation emanating from stars is presented with some indications as to how its collection and analysis leads to basic information on these bodies.

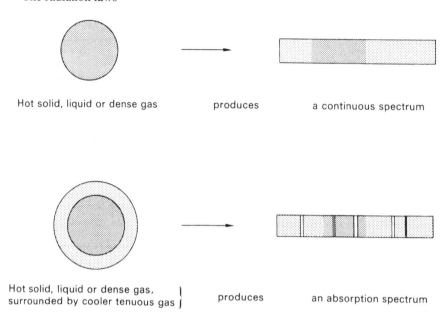

Figure 15.2. The production of a stellar absorption spectrum.

15.4   Solid angle

In order to apply fully the quantitative aspects associated with radiation flow, the concept of **solid angle** first needs to be appreciated.

When an assembly of atoms emits energy, the radiation flows in a range of directions and the dispersed power may not be isotropic. In order to describe such situations, the radiated power is described in terms of a flow into a cone with some specific angle. If at some distance d from the source, power is received in an area A (see figure 15.3), the associated solid angle is simply defined as

$$\Omega = A/d^2.$$

A solid angle is expressed in units of steradians (abbreviated to sterads [sr]). One steradian corresponds to a unit area placed at unit distance. Since the area of a sphere is given by $4\pi R^2$, where R is its radius, it is easy to see that the solid angle subtended by the sphere at its centre is equal to 4π steradians.

So far we have considered the measurement of angles along arcs of the celestial sphere. When objects of angular extent such as a nebula are seen on the celestial sphere, they subtend angles which are essentially two-dimensional. In terms of the measured power received from such an extended source, this obviously depends on the extent of the solid angle over which the radiation is collected. It is, therefore, important to know the strength of the source per steradian or in terms of some other two-dimensional measure such as 'arc seconds squared' ($\square''$). Figure 15.4 provides an illustration of an extended source but with only part of it with a small solid angle being examined by a telescope's field of view.

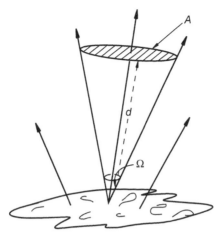

Figure 15.3. The outflow of radiation from a surface has a range of direction. For radiation collected by an area A at distance d from the surface, the received energy emanates from the solid angle $\Omega = A/d^2$.

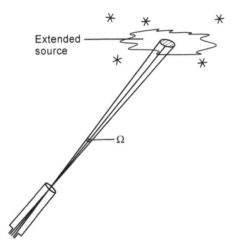

Figure 15.4. When directed to an extended source, the radiation received in the field of view of a telescope is limited to some solid angle, Ω.

15.5 Black body radiation

15.5.1 The basic behaviour

Laboratory experiments of the 19th century showed that the strengths of the various colours in a continuous spectrum depended on the temperature of the radiating body. For example, a body glowing at a 'dull red heat' obviously concentrates most of its energy into the red part of the spectrum and exhibits very little energy in the blue wavelengths. If the temperature of the body is made to increase, the wavelength at which the maximum of energy is radiated creeps further into the visible region of the spectrum and the amount of energy radiated at the blue wavelengths increases. By making quantitative measurements of the strength of the emitted energy at different wavelength positions, it is possible to build up the energy–wavelength distribution curve or **energy envelope** for any source and this can be done in the laboratory for a range of temperatures.

In order to be able to compare the emission efficiencies of radiating bodies, standardization is achieved by comparing their energy envelopes with those of a hypothetical body, known as a **black body**. This type of body is defined as one that absorbs all wavelengths with 100% efficiency and consequently, by Kirchhoff's law, it is able to radiate all wavelengths with the same perfect efficiency. Close approximation to a black body can be achieved in the laboratory in the form of a small aperture contained in a hollow sphere whose inside surface is blackened; it is the small aperture allowing radiation to pass from one side of the sphere to the other which constitutes the black body.

A mathematical description of the energy envelope of a black body defeated the 19th century classical physicists until it was postulated by Planck in 1900 that energy must be radiated in small discrete units or quanta (see section 4.4.2). The energy, E, contained in each quantum is given by

$$E = h\nu$$

where ν is the classical frequency of the emitted radiation and h is Planck's constant, which has been determined to be equal to 6.63×10^{-34} J s. By using the quantum concept, Planck showed that the energy envelope of a black body could be represented by a curve given by

$$B_{\nu b} = \frac{2h\nu^3}{c^2} \frac{1}{\exp(h\nu/kT) - 1} \, d\nu. \tag{15.2}$$

The quantity $B_{\nu b}$ is the **surface brightness** or **specific intensity** of radiation radiated per projected unit area per unit solid angle per unit time by a black body in the frequency range ν to $\nu + d\nu$. The dimensions are given by W m^2 Hz^{-1} sr^{-1} with $d\nu$ equal to unity. On the right-hand side of the equation h is Planck's constant, c is the velocity of light, k is Boltzmann's constant and T the absolute temperature. It can be seen from equation (15.2) that the surface brightness of black body radiation depends on the temperature of the body only and not on such parameters as size or shape. It can also be seen that no limits are set to the range around ν and that radiation of a body which has its maximum value of specific intensity somewhere in the visible region of the spectrum will also provide easily measurable radiation in the infrared and ultraviolet, as discovered by Herschel and Ritter respectively (see section 15.1).

Equation (15.2) is sometimes preferred in the alternative form, where the surface brightness is considered in terms of a wavelength range, λ to $\lambda + d\lambda$ in units of the metre, and in this form, Planck's curve can be expressed as

$$B_{\lambda b} = \frac{2hc^2}{\lambda^5} \frac{1}{\exp(hc/\lambda kT) - 1} d\lambda \qquad \text{W m}^2 \text{ sr}^{-1} \tag{15.3}$$

since by definition, $\nu = c/\lambda$ and hence $d\nu = c(d\lambda/\lambda^2)$.

Plots of the energy envelope described by equation (15.3) have been made in figure 15.5 for temperatures of a black body of 4000, 5000 and 6000 K.

By measuring the energy envelopes of stars, it is found that the majority of them correspond fairly closely to black body curves. According to the particular form of the measured curve, a temperature may be assigned to a star. In general, the term **effective temperature** is used to describe a source and it is defined as the temperature of the black body which would radiate with the same specific intensity, integrated over all wavelengths, as that of the considered source. It will be seen that the energy–wavelength distribution of the radiation from the source may have any arbitrary shape and is not unique in defining a particular effective temperature. The energy envelope of the source may be quite different in shape from the black body curve corresponding to the defined effective temperature.

The wavelengths which are analysed by radio astronomers are very much longer than those in the optical region. For most purposes in the radio region it is, therefore, reasonable to assume that $\lambda kT \gg hc$, which may also be written as $kT \gg h\nu$. With this approximation the expressions

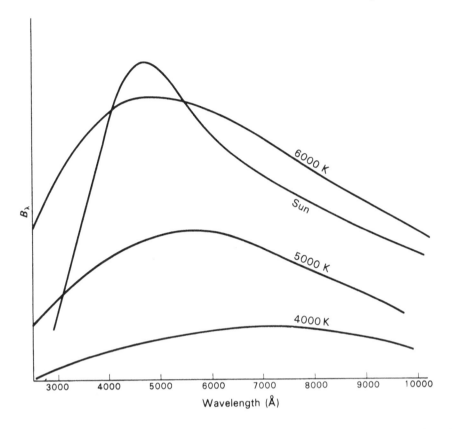

Figure 15.5. The black body radiation curves for temperatures 4000, 5000 and 6000 K. For comparison the Sun's energy envelope is drawn; its effective temperature is approximately 5800 K.

describing the black body curve simplify as the term $\exp(hc/\lambda kT)$ may be written as $[1 + (hc/\lambda kT)]$ and the term $\exp(h\nu/kT)$ may be written as $[1+(h\nu/kT)]$. Thus, for the radio region, Equations (15.2) and (15.3) may be written as

$$B_{\nu b} = \frac{2kT}{\lambda^2}\,\mathrm{d}\nu \quad \mathrm{W\ m^2\ sr^{-1}} \tag{15.4}$$

and

$$B_{\lambda b} = \frac{2ckT}{\lambda^4}\,\mathrm{d}\lambda \quad \mathrm{W\ m^2\ sr^{-1}}. \tag{15.5}$$

These two equations are known as the **Rayleigh–Jeans approximation** to the black body law. In fact, the Rayleigh–Jeans law had been proposed on the grounds of classical physics and it was because of its failure to explain the observed black body curves, particularly in the ultraviolet, that the new concept of quanta was introduced by Planck.

15.5.2 Stefan's law

Prior to Planck's revolutionary explanation of the black body curve, two other important laws of radiation had already been discovered. The first, known as **Stefan's law** or the **Stefan–Boltzmann law**, states that the total energy, $E_{b_{\mathrm{tot}}}$, radiated in all wavelengths by a black body per unit area per unit time is proportional to the fourth power of the absolute temperature, T, of the body. Hence,

$$E_{b_{\mathrm{tot}}} = \sigma T^4 \quad \mathrm{W\ m^{-2}} \tag{15.6}$$

where σ is known as **Stefan's constant**. It appears that Stefan made a lucky guess at the law in 1879; it was deduced theoretically by Boltzmann in 1884. The value of σ may be evaluated by integration of the black body curve and is given by

$$\sigma = \frac{2\pi^5 k^4}{15c^2 h^3} = 5.67 \times 10^{-8} \text{ W m}^{-2} \text{ K}^{-4}. \tag{15.7}$$

15.5.3 Wien's displacement law

The second important law concerning the black body radiation curves is that known as **Wien's displacement law**, proposed in 1896, which states that the product of the wavelength at which the maximum radiation is liberated and the temperature of the black body is a constant. This may be expressed by

$$\lambda_{\max} T = \text{constant} \tag{15.8}$$

the numerical value of the constant being equal to 2.90×10^{-3} m K. Wien's law, therefore, provides a means of determining the temperature of a body at any distance without measuring the absolute amount of energy that it emits. By investigating the spectrum and assessing the wavelength at which the maximum of energy is being liberated, knowledge of Wien's constant allows calculation of the temperature of the body. In astronomy, however, this method is limited to a fairly small range of temperature for which $\lambda_{\max}$ occurs in the visible part of the spectrum.

15.6   Magnitude measurements

15.6.1   The stellar output

If a star is considered simply as a spherical source radiating as a black body, its total energy output can be determined by equation (15.6), according to its surface temperature and its surface area. This total output is referred to as the **stellar luminosity**, L, and may be expressed as

$$L = 4\pi R^2 \sigma T^4 \quad \text{W} \tag{15.9}$$

where R is the radius of the star.

A typical value of stellar luminosity may be of the order of 10^{27} W, that of the Sun being 3.85×10^{26} W. The power received per unit area at the Earth depends on the stellar luminosity and on the inverse square of the stellar distance. If the latter is known, the flux provided by the source may be readily calculated and expressed in terms of watts per square metre (W m^{-2}). More usually, the flux density from a point source such as a star is defined as the power received per square metre per unit bandwidth within the spectrum, i.e. $\text{W m}^{-2} \text{ Hz}^{-1}$, with the bandpass expressed in terms of a frequency interval, or $\text{W m}^{-2} \Delta\lambda^{-1}$, with the selected spectral interval expressed in terms of wavelength (m). If an extended source is considered, then this pair of expressions would be rewritten as $\text{W m}^{-2} \text{ Hz}^{-1} \text{ sr}^{-1}$ and $\text{W m}^{-2} \Delta\lambda^{-1} \text{ sr}^{-1}$, respectively.

15.6.2   Stellar magnitudes

Certainly, in the optical region of the spectrum, it is not normal practice to measure stellar fluxes absolutely. In section 5.3.2, a preliminary description was given of the magnitude scale as proposed by Hipparchus whereby the brightnesses of stars are compared in a relative way. This scheme has perpetuated through the subsequent centuries.

In the late 18th and 19th centuries, several astronomers performed experiments to see how the magnitude scale was related to the amount of energy received. It appeared that a given difference in magnitude, at any point in the magnitude scale, corresponded to a ratio of the brightnesses which was virtually constant. Following his studies of apparent brightness variations of asteroids resulting from their changes in distance from the Earth, in 1856 Pogson proposed that the value of the ratio, corresponding to a magnitude difference of five, should be 100. Thus, the ratio of two stellar brightnesses, B_1 and B_2, can be related to their magnitudes, m_1 and m_2, by the equation

$$\frac{B_1}{B_2} = 2\cdot512^{-(m_1-m_2)} \tag{15.10}$$

since $(2\cdot512)^5$ equals 100. This is known as **Pogson's equation**. The negative sign before the bracketed exponent reflects the fact that magnitude values increase as the brightness falls. By taking logarithms of equation (15.10), we obtain

$$\log_{10}\left(\frac{B_1}{B_2}\right) = -(m_1 - m_2)\log_{10}(2\cdot512) = -0\cdot4(m_1 - m_2)$$

or

$$m_1 - m_2 = -2\cdot5\log_{10}\left(\frac{B_1}{B_2}\right) \tag{15.11}$$

thus showing that the early tabulated magnitudes were proportional to the logarithm of the stellar brightness. This fact is a result of the way that the eye responds, a fact embodied in the Weber–Fechner law. This states that the smallest recognizable change in stimulus is proportional to the already existing stimulus, giving rise to the notion that our senses to stimuli are logarithmic[1]. Hence, Pogson's equation is normally written in one of the three following ways:

$$\frac{B_1}{B_2} = 2\cdot512^{-(m_1-m_2)} \tag{15.10}$$

$$m_1 - m_2 = -2\cdot5\log_{10}\left(\frac{B_1}{B_2}\right) \tag{15.11}$$

$$\log_{10}\left(\frac{B_1}{B_2}\right) = -0\cdot4(m_1 - m_2). \tag{15.12}$$

More generally, Pogson's equation in the style of equation (15.11) can be presented in a simplified form as

$$m = k - 2\cdot5\log_{10}B \tag{15.13}$$

where m is the magnitude of the star, B its apparent brightness and k some constant. The value of k is chosen conveniently by assigning a magnitude to one particular star such as α Lyr, or set of stars, thus fixing the zero point to that magnitude scale. It should also be noted that the numerical coefficient of $2\cdot5$ in equation (15.11) is exact and *is not* a rounded value of $2\cdot512$ from equation (15.10).

Example 15.1. Two stars are recorded as having a magnitude difference of 5. What is the ratio of their brightnesses?

By putting $(m_1 - m_2) = 5$ in equation (15.12), the ratio of the

$$\log_{10}\left(\frac{B_1}{B_2}\right) = -0\cdot4 \times 5 = -2$$

leading to

$$\left(\frac{B_1}{B_2}\right) = 100.$$

[1] In actuality, the eye's response to different light levels does not behave exactly in a logarithmic way according to the Weber–Fechner law—see 1997 *Am. J. Phys.* **65** 1003 and 1998 *Mercury* p 8, May/June.

The brightness of a first magnitude star is, therefore, 100 times greater than that of a sixth magnitude star. ***This is a very useful figure to remember***.

Example 15.2. When a telescope is pointed to two stars in turn, the received power is 5.3×10^{-14} W and 3.9×10^{-14} W. What is the difference in apparent magnitude of these stars?

Now the energy received is proportional to brightness. Therefore, by equation (15.11),

$$
\begin{aligned}
m_1 - m_2 &= -2.5 \log_{10} \left(\frac{5.3}{3.9} \right) \\
&= -2.5 \times 0.13 \\
&= -0.325.
\end{aligned}
$$

Note that, in this case, the magnitude difference $(m_1 - m_2)$ is a negative quantity, indicating that $m_2 > m_1$, this being so since $B_1 > B_2$.

Example 15.3. At an observing site the brightness of the night sky background per $\square''$ is equivalent to a 21st magnitude star. In the search for a faint object a telescope scans the sky with a field of view limited to 200 $\square''$. Calculate the equivalent magnitude of the star matching the total effective brightness of the sky background.

The ratio of recorded energy in the experiment to that from 1 $\square''$ is 200:1. Using equation (15.11)

$$
m_{200} - 21 = -2.5 \log_{10} \left(\tfrac{200}{1} \right)
$$

giving $m_{200} = 21 - 5.75 = 15^{\mathrm{m}}25$.

15.7 Spectral lines

15.7.1 Introduction

Soon after the turn of the 20th century, experiments were performed which revealed that atoms had component particles. As a result of the investigations of electrical discharges through gases, a 'radiation' was discovered which caused a fluorescence on the walls of the glass discharge tube opposite the end at which the negative voltage was applied. The radiation was at first given the name **cathode rays**, as it appeared to emanate from the negative terminal or cathode. Experiments with electric and magnetic fields demonstrated that the rays consisted of negatively charged particles and the name **electron** was given to them. Determination of the ratio of their charge to their mass (e/m_e) showed that it was about 1840 times the same ratio (e/M) obtained by Faraday for the hydrogen ion. As the charges on the two types of particle were found to be of the same value (but different signs), it follows that the mass of the electron is only 1/1840 times the mass of the hydrogen ion. It was immediately obvious that with such a low mass, the electron could not take its place in the periodic table of chemical elements and it was suggested that it constituted one of the fundamental parts of an atom. An experiment by Millikan in 1905 gave a measure of electronic charge (1.6×10^{-19} C) and this allowed determination of the mass m_e of the electron (9.1×10^{-31} kg).

In a further experiment with the discharge tube, Goldstein punctured the cathode with small holes and discovered what he called **canal rays** which appeared to flow from the anode of the tube. Again these rays were found to have a particle nature but the e/M ratio for the particles depended on the gas contained in the tube. The highest value for e/M is obtained when the canal rays are produced in a hydrogen discharge tube. This was indicative of the hydrogen ion being the fundamental unit of positive charge and it was, therefore, called a **proton**.

With the discovery of two of the fundamental particles—the electron and the proton—the problem of understanding the nature of atoms was tackled. A further experiment by Lord Rutherford in 1911,

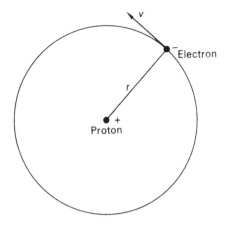

Figure 15.6. The Bohr model of the hydrogen atom.

his now famous α-particle scattering experiment, indicated that the positive charge in an atom is concentrated at its centre, the nucleus occupying only a small space in relation to the distances between atoms. In order to keep the atom electrically neutral, he proposed that the correct number of electrons should surround the nucleus and revolve around it in orbits in a similar way to the planetary orbits round the Sun. This explanation caused difficulties as it immediately contradicted the laws of classical physics. If the electrons are revolving in circular or elliptical orbits round the positive nucleus, they are subject to a constant acceleration along the line joining the electron and the nucleus. According to the classical laws of Maxwell and Lorentz, a charge which suffers an acceleration will radiate electromagnetic waves with energy which is proportional to the square of the acceleration. Classical laws predict that electrons could not occupy such stable orbits as suggested by Rutherford and these rapidly lose their orbital energies by radiation and spiral into the nucleus. However, this difficulty was overcome by Bohr in 1913 who applied a quantum concept based on the principle suggested by Planck. His theory was first applied to the hydrogen atom because of its apparent simplicity.

15.7.2 The Bohr hydrogen atom

According to Bohr's theory, the hydrogen atom consists of the heavy, positively charged nucleus around which the electron performs orbits under a central force provided by the electrostatic force which normally exists between charged bodies. (As the mass of the proton, M, is very much greater than the mass of the electron, m_e, it can be assumed that the proton is at a fixed centre of the electron's orbit). The orbit is illustrated in figure 15.6. For simplicity, we shall consider the simplest case of an electron orbit which is circular.

Suppose that the electron is at a distance r from the proton, that its velocity is v and that its energy is E. Now the electron's energy is made up of two parts: it has kinetic energy (KE) and potential energy (PE), i.e.

$$E = \text{PE} + \text{KE}. \tag{15.14}$$

According to classical dynamics, the kinetic energy of the electron is $m_e v^2/2$. Coulomb's law, relating the force which exists between charged bodies, shows that the electrostatic force, F, between the proton and electron is given by

$$F = \frac{e^2}{4\pi \varepsilon_0 r^2}$$

where ε_0 is the permittivity of free space and that for the orbit to be stable this must be balanced by the centrifugal force, $m_e v^2/r$. Hence,

$$\frac{m_e v^2}{r} = \frac{e^2}{4\pi \varepsilon_0 r^2}.$$

We, therefore, have

$$v = e(4\pi \varepsilon_0 m_e r)^{-1/2}.$$

The kinetic energy is, therefore, given by

$$\text{KE} = \frac{1}{2}\frac{e^2}{4\pi \varepsilon_0 r}. \tag{15.15}$$

The potential energy of the atom can be assessed by considering the electron not to be in motion but at a distance from the proton. If the separation is altered, work must be performed in taking the electron to a new distance from the proton: by increasing the distance, *positive* work must be done against the force of attraction between the two particles; by decreasing the distance, a *negative* amount of work must be performed. The amount of work applied in altering the distance corresponds to the change of potential energy in the system. In a stable orbit, the potential energy is given by the work done in taking the electron from infinity to the distance from the proton given by the radius of the orbit and, as the force between the particles is one of attraction, it will be seen that this is a *negative* quantity. Now the work done is the force multiplied by the distance moved and, as the force depends on the separation of the particles, the total work done in moving of particle from infinity to the orbital distance is evaluated by integration. Hence, the potential energy of the electron in its orbit is given by

$$\text{PE} = \frac{1}{4\pi \varepsilon_0} \int_\infty^r \frac{e^2}{r^2}\, dr$$

and, therefore,

$$\text{PE} = -\left(\frac{1}{4\pi \varepsilon_0}\right)\frac{e^2}{r}. \tag{15.16}$$

By using the equations (15.15) and (15.16), equation (15.14) reduces to

$$E = -\frac{1}{2}\left(\frac{1}{4\pi \varepsilon_0}\right)\frac{e^2}{r}. \tag{15.17}$$

The angular momentum, H, of the electron in its orbit is given by the classical expression

$$H = m_e v r$$

and, therefore,

$$H = e\left(\frac{m_e r}{4\pi \varepsilon_0}\right)^{1/2}. \tag{15.18}$$

Bohr suggested that the radii of the orbits could *not* take on any value as they might according to classical mechanics. He proposed that the electron could only revolve around the proton in preferred orbits and he proposed that the size of such orbits is given by allowing the angular momentum of the electron about the nucleus to have a value given by an integral multiple of the unit $h/2\pi$ where h is Planck's constant. His concept is, therefore, one of the **quantization of angular momentum**.

By this principle it was postulated that while the electron occupies a quantized orbit it does not radiate energy. As in the case of classical mechanics where a change in orbit corresponds to a change energy, a change in energy occurs when an electron jumps from one quantized orbit to another. At the time of a jump, the transfer is achieved by the emission or absorption of electromagnetic radiation but, as the orbits are quantized, so must be the amounts of radiation that are involved. In other words, the radiation emitted or absorbed by atoms is quantized.

Following Bohr's postulate, equation (15.18) may be written as

$$\frac{nh}{2\pi} = e \left(\frac{m_e r}{4\pi \varepsilon_0} \right)^{1/2}$$

where n has any integral value ($n = 1, 2, 3 \ldots$).

The radii of the Bohr orbits can, therefore, be expressed as

$$r = \frac{\varepsilon_0 n^2 h^2}{\pi e^2 m_e} \tag{15.19}$$

and by putting $n = 1$ and inserting the values of ε_0, h, e and m_e, the size of the smallest Bohr orbit is obtained, giving a value equal to 5.3×10^{-11} m or 0.53 Å.

The energies of the Bohr orbits are given by substituting equation (15.19) into (15.17) which gives

$$E_n = -\frac{e^4 m_e}{8\varepsilon_0^2 n^2 h^2}. \tag{15.20}$$

The energy of the first orbit, known as the **ground state**, is obtained by letting $n = 1$, giving a value equal to -2.17×10^{-18} J.

It is more convenient to describe the energies of electron orbits in units of electron volts (eV) rather than in joules. One **electron volt** is defined as the energy acquired by an electron after it has been accelerated by a potential difference of 1 volt. Thus,

$$1 \text{ eV} = \text{charge} \times \text{potential difference}.$$

Now the charge on an electron is 1.6×10^{-19} C and, therefore,

$$1 \text{ eV} = 1.6 \times 10^{-19} \text{ J}.$$

The energy of the ground state in the hydrogen atom in units of eV is given by

$$-\frac{2.17 \times 10^{-18}}{1.5 \times 10^{-19}} \text{ eV} = -13.6 \text{ eV}$$

and the energies of the other orbits or **excited states** are, therefore, given by

$$E_n = \frac{-13.6}{n^2} \text{ eV}. \tag{15.21}$$

Energy levels are frequently depicted in a pictorial way as a series of lines—the energy levels for the hydrogen atom are illustrated in figure 15.7.

15.7.3 The hydrogen spectrum

By setting up this model of the hydrogen atom involving the quantization of the orbital energies, Bohr succeeded in providing an explanation of emission line spectra, in particular the hydrogen spectrum which had been found by Balmer in 1885 to be positioned according to a fairly simple formula. Bohr suggested that a change of energy level, or **transition**, corresponds to the absorption or emission of a quantum of radiation whose energy, $h\nu$, is given by Planck's formula. Thus, if two energy levels are designated by E_m and E_n, their energy difference is given by

$$\Delta E = E_m - E_n = h\nu.$$

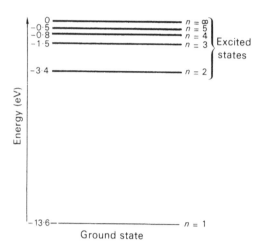

Figure 15.7. The energy levels of the hydrogen atom.

For a transition between the m and n levels, it is, therefore, expected that a spectral line would occur at the frequency v, the line being in emission for a downward transition (i.e. from an excited state to a less excited state) and being in absorption for an upward transition. Accordingly, spectral lines should appear at frequencies given by

$$v = \frac{E_m - E_n}{h}$$

or, in terms of wavelength, at

$$\lambda = \frac{ch}{E_m - E_n}. \tag{15.22}$$

By substituting values for E_m and E_n (see equation (15.20)), we have that

$$\lambda_{mn} = \frac{8\varepsilon_0^2 ch^3}{e^4 m_e} \frac{1}{(1/n^2 - 1/m^2)} \quad \text{m}$$

and by substituting for the constants this reduces to

$$\lambda_{mn} = \frac{9 \cdot 12 \times 10^{-8}}{(1/n^2 - 1/m^2)} \quad \text{m}.$$

or,

$$\lambda_{mn} = \frac{912}{(1/n^2 - 1/m^2)} \quad \text{Å}. \tag{15.23}$$

It will be seen from equation (15.23) that the spectral lines can conveniently be expressed in terms of series, the first corresponding to $n = 1$, $m \geq 2$ (Lyman series), the second corresponding to $n = 2$, $m \geq 3$ (Balmer series), $n = 3$, $m \geq 4$ (Paschen series), $n = 4$, $m \geq 5$ (Brackett series), $n = 5$, $m \geq 6$ (Pfund series), etc. Figure 15.8 illustrates the series which are particularly important in astronomy and the wavelengths of the chief spectral lines, in close agreement with those predicted by the simple Bohr theory (see equation (15.23)), are collected in table 15.2.

It will be seen from table 15.2 that the Lyman series is to be found in the ultraviolet. This series was discovered in 1906 just before Bohr's theory. The Balmer series occurs through the visible region and is an important feature in stellar spectra. Three of the lines of this series, Hα, Hβ and Hγ, correspond respectively to the C, F and G lines designated by Fraunhofer.

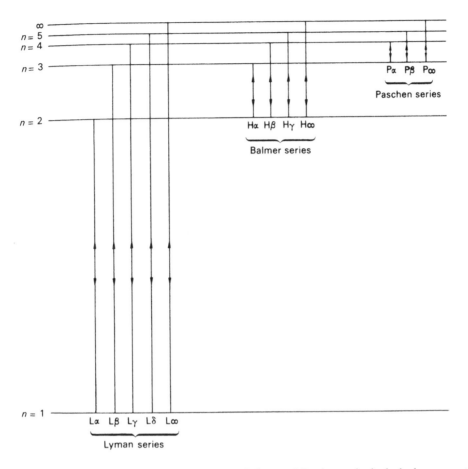

Figure 15.8. Energy transitions giving rise to the Lyman, Balmer and Paschen series in the hydrogen spectrum.

Table 15.2. The wavelength positions of spectral lines in the hydrogen spectrum.

$n = 1$ Lyman series			$n = 2$ Balmer series				$n = 3$ Paschen series		
		λ (Å)				λ (Å)			λ (Å)
$m = 2$	Lα	1215	$m = 3$	Hα	(Fraunhofer C)	6562	$m = 4$	Pα	18 751
$m = 3$	Lβ	1025	$m = 4$	Hβ	(Fraunhofer F)	4861	$m = 5$	Pβ	12 818
$m = 4$	Lγ	972	$m = 5$	Hγ	(Fraunhofer G)	4340	$m = 6$	Pγ	10 938
$m = 5$	Lδ	949							
.			.				.		
.			.				.		
.			.				.		
∞		912	∞			3646	∞		8 208

If a hydrogen atom is sufficiently excited to allow an energy level which is greater than the upper limit—a limit which does not exist in classical dynamics—the electron will leave the atom, which is then said to be **ionized**. A similar reverse procedure can also be imagined where an electron which is moving past the nucleus drops into one of the allowed orbits. In this case, the emitted photon will correspond to the difference between the energy of the electron in its free and closed orbits. As the range of free energy levels is continuous, the emitted photon can have a continuous range of wavelengths. Thus, it is expected that a continuous spectrum might exist, starting at a wavelength given by a series limit ($m = \infty$) and extending to shorter wavelengths; the spectrum may be one of continuous emission or continuous absorption, depending on whether the electron is being captured or ejected by the hydrogen nucleus. An absorption effect at the limit of the Balmer series is a frequent feature in stellar spectra.

Bohr's theory is useful in describing radiating atoms containing only one electron, i.e. the theory can be extended to describe the spectra of singly-ionized helium, doubly-ionized lithium and trebly-ionized beryllium, etc but it met with little quantitative success when it was applied to atomic systems containing two or more electrons. It has now been displaced by what is known as the **new quantum theory** or **wave mechanics**. It is by the mathematics of these theories that the complex patterns and fine structure within the spectra of the heavier atoms can be interpreted. However, a discussion on this topic is beyond the scope of this text.

15.7.4 Molecular spectra

It may be noted that molecules also provide spectral features and patterns that allow their identification. A simple diatomic molecule such as CO, for example, suffers vibration such that the axial distance between the components oscillates. Again the molecules have spin with components normal to and around the axis. The energies associated with the vibration and rotation are also quantized so that radiation at specific wavelengths is emitted or absorbed by a molecule according to the change of its energy state.

For example, the rotational energy, E, of a molecule depends on its moment of inertia, I, and the spin rate, ω, so that

$$E = \frac{I\omega^2}{2}$$

and, for a simple diatomic molecule,

$$I = \mu r^2$$

where r is the separation of the component atoms of mass M_1 and M_2 with μ the reduced mass of the molecule given by

$$\mu = \frac{M_1 M_2}{M_1 + M_2}.$$

By applying the quantization principle, the angular momentum can only have integer (J) values such that

$$I\omega = \frac{h}{2\pi} J.$$

The energy levels corresponding to the possible states of J may be written as

$$\Delta E = \left(\frac{h}{2\pi}\right)^2 J(J+1) \Big/ 2\mu r^2.$$

From knowledge of the structure of the molecule, i.e. the masses of the component atoms and their separation, the wavelength positions of spectral lines are readily calculated. The photon energies associated with the transitions are normally less than those generated by the electron jumps within atoms this being reflected by the fact that spectral features associated with molecules are likely to be more apparent in the infrared and millimetre region rather than at optical wavelengths. Many of the spectral features of identifiable molecules from the interstellar medium do, in fact, appear in the millimetre region. For example, one of the strong features associated with the CO molecule is the rotational line corresponding to $J \rightarrow 2$ to 1 at $\sim$230·5 GHz or $\sim$1.3 mm and this has been used to advantage to map out the distribution of this common compound in the molecular clouds within the interstellar medium.

An important adjunct to molecular spectroscopy is the possibility of undertaking measurements of isotope abundances. For the case of a diatomic molecule, if one of the component atoms, say M_1, has two isotopes, two possible moments of inertia ensue so generating spectral lines at differing wavelength positions. Hence, lines associated with $^{12}C^{16}O$ and $^{13}C^{16}O$ will differ in wavelength position in the ratio of the reduced masses of their associated molecules, i.e.

$$\frac{12 \times 16 \times (13 + 16)}{13 \times 16 \times (12 + 16)} \cong 4\%.$$

15.8 Basic spectrometry

15.8.1 Simple considerations

By attaching spectrometric instruments to telescopes, the collected radiation is dispersed to provide spectra of any observed source. Analysis of the records provides information on the overall spectral distribution of the received energy, so allowing temperatures to be assigned to the various objects. Within the overall energy–wavelength envelope, spectral features may be discerned corresponding say to the Balmer sequence of hydrogen as described earlier or to lines associated with other atomic or molecular species.

In addition to the detection of the various chemical elements from identification by their spectral fingerprints, some of the lines may correspond to excited or ionized states of the radiating or absorbing atoms and again such features lead to a determination of the source's temperature.

From laboratory measurements and theory, the wavelength positions of the lines associated with the various chemical elements are known to high accuracy. In the investigations of astronomical sources, it turns out that many of their velocities are not negligible in comparison with that of light. As a consequence, Doppler shift effects (see later) may be apparent whereby the position of a spectral line may be shifted with respect to its laboratory position. Measurement of the displacement can, therefore, lead to a determination of the line-of-sight velocity of the source.

Examination of the detailed shape of spectral lines, or **line profiles**, also holds much astrophysical information. For example, careful observation with subsequent data reduction of line profiles can again provide information on the rotation of the investigated body, be it a planet, star or galaxy. The rationale for undertaking spectrometric studies is amplified in the following sections by presenting a selection of mechanisms which influence the behaviour of line profiles in terms of their wavelength position and their shape.

15.8.2 The Doppler shift

It is an everyday experience that the pitch of sound depends on the speed of the object emitting it. The effect was studied by Doppler in the early 19th century and is named after him. In the case of an object which is moving relative to a fixed observer, the sound waves are either lengthened or shortened, depending on whether the object is receding or approaching the observer: the effect of the change of wavelength gives a change in the sensation of the pitch of the sound. It was suggested that the Doppler effect should also occur in connection with light waves but this could only be verified much later in the laboratory when improved optical equipment and techniques became available. It is of historical interest to recall that Doppler erroneously attributed the colours of stars to their motions relative to the Earth. The derivation of the Doppler law here is sufficient for cases where the velocities involved are small relative to the velocity of light.

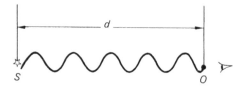

Figure 15.9. A source S at a distance d from the observer O.

Suppose that light is being emitted by atoms in a source, S, with a frequency, ν, and that the waves travel to an observer, O, at a distance, d, from the source (see figure 15.9). After 'switching-on' the source, the first wave arrives at O after a time d/c. During this time the source has emitted $\nu d/c$ waves and the apparent wavelength is obviously

$$\lambda = \frac{\text{Distance occupied}}{\text{Number of waves}} = \frac{d}{\nu d/c} = \frac{c}{\nu}.$$

Suppose that the source has velocity, V, away from the observer. During the same time interval as before (d/c), the source moved a distance equal to $V \times d/c$. Thus, the waves emitted during the same interval now occupy a distance equal to

$$d + \frac{Vd}{c}.$$

The apparent wavelength, λ', of the radiation is, therefore, given by

$$\lambda' = \frac{d + Vd/c}{\nu d/c}$$
$$= \frac{c}{\nu} + \frac{V}{\nu}$$
$$= \lambda + \frac{V\lambda}{c}. \tag{15.24}$$

Therefore,

$$\lambda' - \lambda = \Delta\lambda = \frac{V\lambda}{c}$$

so giving

$$z = \frac{\Delta\lambda}{\lambda} = \frac{V}{c}. \tag{15.25}$$

The difference, $\Delta\lambda$, between the observed wavelength and the wavelength that would have been observed from a stationary source is called the **Doppler shift**. It is *positive* ($\lambda' > \lambda$) for an object which is receding from the observer and *negative* for an object which is approaching the observer. Thus, lines present in a spectrum of the light from a moving source are shifted towards the *red* for a *receding* object and are shifted towards the *blue* for an *approaching* object.

Equation (15.25) is valid only for $V \ll c$. For velocities near that of light, such as for some galaxies and quasars, the concepts of relativity theory need to be applied giving a Doppler shift which is nonlinear in V. Under this circumstance, the apparent wavelength is written as

$$\lambda' = \gamma\lambda\left[1 + \frac{V}{c}\right] \tag{15.26}$$

where γ is the Lorentz factor:

$$\gamma = \left[1 - \left(\frac{V}{c}\right)^2\right]^{-\frac{1}{2}}.$$

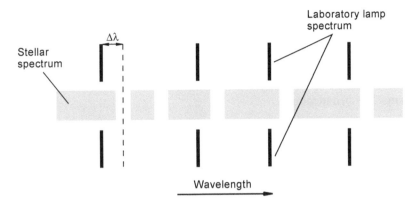

Figure 15.10. A schematic diagram displaying a stellar Doppler shift. Absorption lines within the stellar continuum are matched with displaced emission line spectra of a laboratory reference lamp imposed above and below. Note that the stellar spectrum is redshifted; note also that not all of the stellar lines have matching but displaced lamp lines.

Thus, the relativistic Doppler formula may be written as

$$z = \frac{\Delta\lambda}{\lambda} = \sqrt{\frac{1 + V/c}{1 - V/c}} - 1. \tag{15.27}$$

By comparing the wavelengths of spectral features in the light from celestial objects with standard laboratory wavelengths given by spectral lamps, the Doppler shifts can be measured and velocities deduced. The same optical Doppler effects occur, however, if the source is stationary and the observer moves. Doppler shifts, therefore, represent the **relative motion** between the source and the observer. What is more, the shifts represent only the components of velocity along the line joining the object to the observer. This component is known as the **radial velocity** component and so it can be said that the Doppler shift provides a means of measuring **relative radial velocities**. A schematic spectrum of a star exhibiting the effect of a Doppler shift is illustrated in figure 15.10.

15.8.3 Natural line width

The transitions from one energy level to another are not instantaneous and, as a consequence, the radiation absorbed or emitted does not occur at a unique frequency. The natural spread of any generated line depends on the 'lifetimes' of the excited states of the particular atom. A short lifetime produces narrow lines—as the lifetime increases the natural spread of the line increases.

If the Heisenberg uncertainty principle is applied, the energy of a given state cannot be assigned more specifically than

$$\Delta E = \frac{h}{2\pi \, \Delta t}$$

where Δt is the lifetime of the state. Consequently, any assembly of atoms produces a spectral line in emission or absorption with a spread in frequency such that

$$\Delta \nu = \frac{\Delta E}{h} \approx \frac{1}{\Delta t}.$$

A typical excited atomic state may have a lifetime $\sim 10^{-8}$ s and, in the visible spectral domain (~ 550 nm), this provides a wavelength spread $\sim 1 \cdot 6 \times 10^{-5}$ nm or $0 \cdot 016$ mÅ. (In order to check the calculation it will be remembered that $\Delta\lambda = \lambda^2/c \times \Delta\nu$.)

In most physical situations, practical determination of natural emission widths is compromised by a variety of physical causes affecting the behaviour of the radiating atoms. Certainly this is the case in the general astrophysical environments as highlighted in the next three subsections which serve as examples of the importance of recording the details of spectral line profiles.

15.8.4   Thermal line broadening

If atoms are contained in a hot gas cloud and are giving rise to emission lines in their spectra, then the profiles are likely to be broadened by thermal Doppler effects.

As a result of the local temperature, the atoms acquire a distribution of kinetic energies referred to as the Maxwell distribution. Within such a radiating cloud some atoms will be receding and others will be approaching the observer such that at the times of the emissions, individual contributions may be red or blue shifted as a result of the Doppler effect. The overall shape of the emission line is, therefore, broadened into a 'bell-shaped' profile with the halfwidth dependent on the temperature.

A typical kinetic energy of an atom may be obtained by taking the most probable value of the Maxwell distribution this being

$$\tfrac{1}{2}mv^2 = kT$$

where k is Boltzmann's constant. Hence,

$$v \sim \sqrt{\frac{2kT}{m}}$$

If, for example, a cloud of hydrogen is considered to be at a temperature of 10 000 K, a typical value for the velocity is

$$v \sim \sqrt{\frac{2 \times 1 \cdot 38 \times 10^{-23} \times 10^4}{1 \cdot 7 \times 10^{-27}}} \quad \text{m s}^{-1}$$

giving a value for $v \sim 1 \times 10^4$ m s^{-1}. The Doppler shift associated with such a velocity is readily detectable. It is obvious that thermal Doppler broadening would be less apparent for the heavier atomic species.

15.8.5   Collisional line broadening

For atoms which are radiating in an environment subject to a high pressure, the energy levels of the radiating or absorbing atom may be influenced by local charged particles (ions and electrons). In a gas, such perturbations are random and give rise to a broadening of the line—the Stark effect. There will be a high probability that the radiating atoms will suffer collisional impacts with adjacent atoms during the time interval of the individual emissions. As a result, the frequency associated with the emission will be blurred. Overall, the assembly of radiating atoms provides spectral line profiles which are broadened by the process according to the local density or pressure.

15.8.6   Line broadening by rotation

When a telescope–spectrometer combination is applied to a stellar radiation, the resultant spectrum results from the light received from the whole of the presented stellar disc.

For rotating stars, this means that the generated absorption lines from each location on the disc are subject to Doppler displacements according to their position and line-of-sight velocity. Radiation from the approaching limb will be blue shifted, while that from the receding limb will be red shifted. Integration of these effects over the presented disc produces an overall absorption line profile which is saucer-shaped. From observations of the line profile spread, the projected equatorial velocity, $V \sin i$, of the star can be determined.

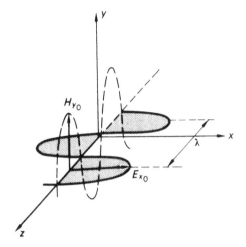

Figure 15.11. A section of simple electromagnetic disturbance.

15.9 Polarization phenomena

The early part of the 19th century also saw the development of our understanding of the polarization properties associated with electromagnetic radiation and of the reasons why an interaction between radiation and a material may depend on the polarization of the beam.

En route to the formalism by Clerk Maxwell of the wave equations which relate the nature of the electric and magnetic disturbances associated with the radiation, one of the major steps was the appreciation that electromagnetic waves have a transverse motion. It was Fresnel who used this notion in developing his now classical laws describing reflection and refraction by a dielectric material such as glass and providing formulas for evaluating the amount of radiation that is reflected by such a surface according to the angle of incidence of the radiation and the plane of the wave related to the plane of incidence.

Before looking at the forms the polarization property may take, it may be mentioned that it was Arago who first applied polarimetric observation to astronomy. He observed that moonlight was quite strongly polarized and also the light from comets. Further, he showed that the light from the limb of the Sun is not substantially polarized and proved that the Sun could not be considered as being solid or liquid and that it must be gaseous in nature.

Now the passage of electromagnetic energy through a point in space can be described simply when the energy is in the form of a plane monochromatic wave. For this case, the electric and magnetic disturbances may be considered to be in the xz and yz planes respectively and the equations describing the disturbances at any point along the z-axis may be written in the form:

$$E_x = E_{x0} \cos\left(\omega t - \frac{2\pi z}{\lambda} + \delta\right) \quad \text{and} \quad H_y = H_{y0} \cos\left(\omega t - \frac{2\pi z}{\lambda} + \delta\right) \quad (15.28)$$

where E_{x0}, H_{x0} are the amplitudes of the disturbances, $\omega(= 2\pi\nu)$ is the angular frequency of the radiation, λ is the wavelength of the monochromatic wave and δ is the phase whose value is controlled by the size of the disturbances when $t = 0$.

Thus, the electric and magnetic disturbances are orthogonal (at right angles) and in phase with each other and they are transverse to the direction of propagation. Figure 15.11 illustrates the disturbances as they might appear along the direction of propagation at some particular instance of time. The forms of the disturbances, therefore, have a similar relationship and the magnitude of the disturbances (i.e. E_{x0} and H_{y0}) are related by the optical properties of the material through which the radiation happens to pass. Consequently, a plane wave electromagnetic disturbance can be expressed simply by using either the electrical or magnetic disturbances alone. In the optical region, it is the electric vector which usually plays the dominant role in any interaction of the radiation and matter, and the equation describing the electric disturbance is normally used to express the form of the disturbance. Thus, provided that the optical properties of the medium are known, a plane wave electromagnetic disturbance may be summarized by any equation in the form:

$$E_x = E_{x0} \cos\left(\omega t - \frac{2\pi z}{\lambda} + \delta_x\right).$$

The strength of the radiation associated with such a wave is proportional to the square of the amplitude of the electric field disturbance and it is related to the brightness of the source.

At any instant in time, the disturbance passing through a point in space is the resultant of the radiation originating from atoms which happened to be emitting energy at exactly the same time. This resultant may be described by considering it to have components vibrating in the xz and yz planes, with amplitudes and phases which are unlikely to be the same. Thus, the resolved components of the disturbance may be written as

$$E_x = E_{x0} \cos\left(\omega t - \frac{2\pi z}{\lambda} + \delta_x\right) \quad \text{and} \quad E_y = E_{y0} \cos\left(\omega t - \frac{2\pi z}{\lambda} + \delta_y\right). \tag{15.29}$$

If it were possible to measure the electric disturbance at a point over one cycle of the wave (this would correspond to a time approximately equal to 10^{-15} s for visible radiation), the resultant electric vector would sweep out an ellipse. This ellipse is known as a **polarization ellipse** and for the short time of the measurement, the radiation can be considered to be **elliptically polarized**. The geometry of the polarization ellipse can be investigated by combining equations (15.29) and removing the time dependence. Thus, the equation describing the ellipse can be written as

$$\left(\frac{E_x}{E_{x0}}\right)^2 + \left(\frac{E_y}{E_{y0}}\right)^2 - \frac{2E_x E_y \cos(\delta_x - \delta_y)}{E_{x0} E_{y0}} = \sin^2(\delta_x - \delta_y). \tag{15.30}$$

A special form of polarization occurs when the phase difference $(\delta_x - \delta_y)$ between the components is zero. For this condition, equation (15.30) shows that the resultant oscillates along a line and the disturbance is known as **linear polarization**. Another special form occurs when the amplitudes are equal ($E_{x0} = E_{y0}$) and the phase difference is equal to $\pi/2$. For these conditions, the resultant has a constant magnitude and rotates at the frequency of the radiation and the disturbance is known as **circular polarization**.

In order to describe the general polarization ellipse, three parameters are required. They are the **azimuth** (i.e. the angle that the major axis makes with some reference frame), the **ellipticity** and the sense of rotation of the electric vector or its **handedness**. These three parameters, together with the strength of the radiation (i.e. the size of the ellipse), describe the characteristics of the radiation. The various polarization forms are depicted in figure 15.12.

The general polarization form described here was considered by measuring the electric vector over one cycle of the wave. However, during this small period of time, some of the wave-trains emitted by the collection of atoms will have passed through the point of observation and other wave-trains emitted by different atoms in the collection will have just started to contribute to the resultant disturbance. If the atoms are radiating in a random way, the amplitudes and phases of the resolved components will be changing continuously, thus causing the polarization to change. After a period of time equal to a few cycles of the wave, the 'instantaneous' polarization form bears no relation to the initial 'instantaneous' polarization form. If the radiation from a collection of atoms, with no special properties, were to be analysed over a time period required for normal laboratory experiments, no preferential polarization

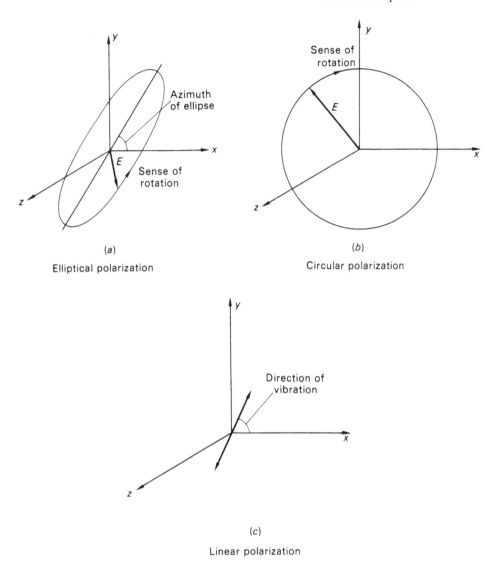

(a)
Elliptical polarization

(b)
Circular polarization

(c)
Linear polarization

Figure 15.12. The various polarization forms.

form would be detected. Under these conditions, the radiation is said to be **unpolarized**. If such radiation could be looked at with extreme time resolution, it would be possible, in principle, to record the probability of finding a particular polarization form at any instant. The probability distribution so obtained is unique to naturally occurring unpolarized light.

Under some special circumstances, a collection of atoms can give rise to radiation with a polarization form which persists over normal experimental time periods. The persistent polarization ellipse can be described by the same set of parameters which were used to describe the instantaneous polarization ellipse described earlier (i.e. azimuth, ellipticity and handedness). This type of radiation is said to be **perfectly** polarized. The general form of the polarization is **elliptical** and the extreme forms of **linear** and **circular** also occur.

Many of the interactions of radiation and matter (e.g. reflection, scattering, etc) give rise to polarization effects. The incident radiation may be unpolarized, but after the interaction, the resulting radiation is likely to be polarized, although the polarization may not be **complete** or **perfect**. This type of radiation is said to be **partially polarized**.

A beam of partially polarized radiation may be thought as being made up of two combined beams, one beam being perfectly polarized and the other being unpolarized. On a highly-resolved timescale, a partially polarized beam of light would show variations of the 'instantaneous' polarization form but, on average, one form, perhaps with a particular azimuth, would predominate. The degree of predominance would depend on how near to perfect the partially polarized radiation is. The ratio of the strength of the polarized component to the total strength of the beam defines the **degree of polarization**, p, of the beam.

The most frequently met case in astrophysics is the measurement of partially linearly polarized light. If a polarimetric analyser is rotated in such a beam, the transmitted intensity will change according to $I(1 + p\cos^2(\alpha - \beta))$ where α is the angle of the analyser's axis in the instrument and β is the vibration azimuth of the polarization relative to the instrumental frame. On a complete rotation of the polarizer, the recorded signal will oscillate between a maximum and minimum value. If these levels are recorded as I_{max} and I_{min}, it is readily shown that p can be determined from

$$p = \frac{I_{max} - I_{min}}{I_{max} + I_{min}}. \tag{15.31}$$

The degree of polarization, together with the parameters which describe the ellipse associated with the polarized component, summarize the polarizational properties of a partially polarized beam. This polarizational information, together with the brightness, gives a complete description of any beam of radiation.

All the parameters which are needed to describe a general beam of radiation are capable of carrying information about the condition of the radiating atoms which give rise to the energy, the nature of the matter which scatters the radiation in the direction of the observer or the nature of the matter which is in a direct line between the radiating atoms and the observer. If the observer wishes to gain as much knowledge as possible of the outside universe, all the properties associated with the electromagnetic radiation must be measured. Various scattering phenomena such as that associated with free electrons or dust in extended stellar atmospheres are important means of generating polarization. It also turns out that aligned dust grains in the interstellar medium give rise to linear polarization by differential extinction for the radiation's components resolved orthogonally to the axis of the grains. Generation of polarization by the synchrotron process is revealed in the supernova remnant known as the Crab nebula (see figure 15.13).

15.10 The Zeeman effect

Although it is not the only mechanism that gives rise to a particular spectral line behaviour accompanied by polarizational phenomena, the Zeeman effect serves as a classical example for the rationale of undertaking spectropolarimetric measurements.

In the middle of the 19th century, Faraday examined the emission spectrum of sodium vapour which had been placed in a magnetic field. His apparatus, however, was insufficiently sensitive to detect any effect caused by the field. By repeating these experiments with more sensitive equipment, Zeeman, in 1896, demonstrated that the emission lines were broadened. Lorentz, shortly afterwards, proposed a theory, based on the electron theory of matter, which suggested that if the broadening were to be looked at closely, a splitting of the lines should be observed. His theory also predicted that the light of the split lines should be polarized. When the magnetic field is in a direction along the line of sight (a longitudinal field), two lines should be observed at wavelength positions on either side of the position for the emission line when there is no field. These lines should be circularly polarized but with opposite handednesses. When the field is across the line of sight (a transverse field) three lines

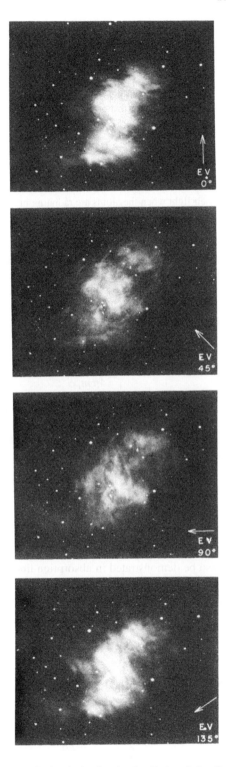

Figure 15.13. Evidence of strong optical polarization in the Crab nebula; four photographs taken at marked rotational settings of a polarizer. Polarized light is produced as synchrotron radiation by spiralling fast-moving electrons in a magnetic field (photographs from the Mt Wilson and Palomar Observatories).

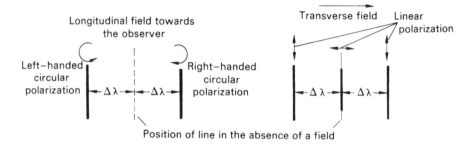

Figure 15.14. The classical Zeeman effect in an emission spectrum illustrating the positions of the observed lines and the polarizational qualities of their light when the atoms are radiating in a magnetic field.

should be observed: one line should be at the normal position and be linearly polarized with a direction of vibration parallel to the field; the other lines should be placed either side of the normal position and both be linearly polarized with vibration perpendicular to the field. The splitting of the lines and the polarization properties of their light were later verified by Zeeman. Figure 15.14 illustrates the important features of the classical Zeeman effect.

According to Lorentz's theory, the split lines should appear at distances from the position of the line without the field given in terms of a frequency shift by

$$\Delta \nu = \pm \frac{eH}{4\pi m_e c}$$

where H is the strength of the magnetic field. Thus, the degree of splitting depends on the strength of the field. By remembering that $\nu = c/\lambda$, it is easy to show that $\mathrm{d}\nu/\mathrm{d}\lambda = -c/\lambda^2$ and, therefore, that $\Delta \lambda = -\lambda^2/c$. Hence, in terms of the wavelength shift, the Zeeman splitting can be expressed as

$$\Delta \lambda = \pm \frac{eH\lambda^2}{4\pi m_e c^2}$$

and when H is expressed in tesla (T)

$$\Delta \lambda = \pm 4 \cdot 67 \times 10^{-17} \lambda^2 H \quad (\text{Å}). \tag{15.32}$$

The Zeeman effect may also be demonstrated in absorption lines and from measurements taken from spectra it has been found possible to determine the magnitude of magnetic fields of the Sun and some special stars. In practice, the splitting of spectral lines by magnetic fields is more complex than that predicted by the Lorentz theory.

Problems—Chapter 15

1. The following is a table of stars together with their listed magnitudes. Commencing with the brightest star, place them in order of decreasing brightness.

Star no	Magnitude
1	+3·1
2	+2·6
3	−0·1
4	+1·1
5	−0·9
6	+3·3

2. One star is three times brighter than another. What is the difference in magnitude between the two stars.

3. Two stars differ in magnitude by three. What is the ratio of their brightnesses?

4. The energy received per unit time from two stars is in the ratio 2:1. The brighter star has a magnitude of +2·50. What is the magnitude of the fainter star?

5. The apparent magnitude of Jupiter on a particular night is −1·3. Its brightness is compared with a star whose magnitude is +1·0. What is the ratio of the brightnesses of Jupiter and the star?

6. Two stars have magnitudes +4·1 and +5·6. The brighter star provides 5×10^{-8} W for collection by a particular telescope. How much energy would be collected from the fainter star?

7. Two stars are close together on the celestial sphere and measurements show that they have the same brightness. It is also known that one star has a magnitude of +8·5. What would be the magnitude of the pair of stars if they were seen as a single object?

8. The surface of a star is radiating as a black body with a temperature of 6000 K. If the temperature were to increase by 250 K, by what fraction would the energy liberated per unit area increase?

9. The constant in Wien's displacement law is $2·90 \times 10^{-3}$ m K. What is the wavelength corresponding to the maximum radiation output of a star whose surface temperature is 5000 K?

10. The wavelength associated with the hydrogen spectrum may be expressed in the form:

$$\lambda_{mn} = \frac{R}{(1/n^2 - 1/m^2)}.$$

For the lines in the visible part of the spectrum (Balmer series), $n = 2$ and the first line of the series, Hα is at 6562 Å. Show why observations of the lines of the Lyman series ($n = 1$) may only be made from rockets or satellites.

11. A spectrometer is just able to detect a wavelength shift of 0·01 Å. What would be the minimum strength of magnetic field that might be detected in a star if the spectral region around 4500 Å were to be used?

12. From the spectrum taken of a star, the Hβ line (4861 Å) exhibits a blue wavelength displacement of 0·69 Å. What is the relative velocity of approach of the star?

13. The equatorial radius of Saturn is 60 400 km and its rotation period is $10^h\ 14^m$. What is the maximum Doppler displacement of a spectral line at 5000 Å obtained from the light reflected by a selected area of the planet's disc?

Chapter 16

The optics of telescope collectors

16.1 Introduction

The invention of the telescope, as its name implies, allowed objects to be seen at a distance and it was inevitable that it would be used to have a look at some of the astronomical bodies. Ever since its application to astronomy at the dawn of the 17th century, it has become an essential part of the array of equipment which is applied to make the basic measurements. For the first two hundred years of the telescope's use, the eye and brain was the only means of recording the picture of the Universe which the telescope presented. In its ability to collect more radiation than the naked eye, the telescope allowed objects to be seen which were previously too faint to be detected. With the application of eyepieces, the property of the magnifying effect of the optical system could be put to purpose and structural details of some objects were recorded for the first time.

The telescope–eye combination no longer holds any place in the field of astronomical measurement. Most of the observations are made today by using more efficient and more reliable detectors such as the photoelectric cell or CCD cameras. The application of these detectors, however, has not made any basic change to the role that the telescope plays. The primary functions of telescope usage may still be summarized as follows:

1. to allow collection of energy over a larger area so that faint objects can be detected and measured with greater accuracy;
2. to allow higher angular resolution to be achieved so that positional measurements can be made more accurately and so that more spatial detail and information of extended objects can be recorded.

Since its invention, the telescope system has been improved progressively. Its development has been controlled by a variety of factors. At the beginning of telescope application, the eye and brain was the only way of recording and interpreting the images that the optics formed; consequently, the aim of early telescope design was to provide the sharpest images for visual inspection. The first form of the telescope was the **refractor** which uses lenses to produce the images. For the simplest of refracting systems, the telescope has the defect of the image position being dependent on the wavelength of light. For objects whose emitted or reflected light contains a spread of wavelengths, their images formed by such a system do not occur at the exact same position. If the image is viewed, its appearance is not sharp and it exhibits coloured tinges, particularly at its edges. This defect is known as **chromatic aberration**.

Many people, including Sir Isaac Newton, thought that this defect, which is inherent in simple lens systems, could not be overcome and indeed Newton himself introduced the alternative method of forming optical images by using a curved mirror. **Reflectors**, by their nature, do not suffer from chromatic aberration. Interest in refracting systems was aroused again when it was discovered by

Dolland in 1760 that the effect of chromatic aberration could be reduced considerably by the use of a compound lens system.

When the photographic process was introduced, its potential as a means of recording many star images simultaneously was immediately apparent. Optical design was, therefore, directed to the improvement of the telescope to perform as an astronomical camera. In the first instance, all-lens systems were used but now modern astronomical cameras may make use of a combination of both lenses and mirrors in the same instrument.

When the emphasis in astronomical research was directed to making use of telescopes with the maximum light-gathering power, the mechanical aspects of design dictated that large telescopes must be reflectors. In fact, large professional refractors are no longer constructed because of the technological problems that abound.

Thus, the designs of telescopes have been influenced by many factors, some of these being:

1. our knowledge and understanding of image formation,
2. the detector system available for appending to the telescope and
3. the technology of optical materials and methods of preparing optical surfaces.

It is usual practice to have a series of instruments of various purposes for attachment to a given telescope. As a particular telescope is normally designed for some special function, it is impossible to have a complete range of instruments which can be efficiently matched to that telescope. For example, a long-focus refractor is ideally suited for making measurements of the separations of double stars, either by eye or by photography—a long focus gives a good separation between images and this allows accurate measurement. However, this same telescope is unsuitable for the usual spectrometric techniques: any single star image is physically large, again as a result of the long focus, and it is usually impossible to pass the whole light from this image through the slit of the spectrometer. Any combination which wastes light in this way is inefficient. Whenever possible, efforts should be made to use the best telescope–instrument combination for the particular type of observation that is being undertaken. Because of the wide range of observations which are performed, it is not always possible to conduct the measurements with maximum efficiency and some compromise is normally made.

Novel telescope designs are emerging which rely on the support of recent technological developments. Some large telescope mirrors are being made as composite mosaics, with each of the elements being actively kept in position under computer control to maintain the sharpest image. Another concept features the idea of large multiple mirrors whose collected light might be combined by using fibre optic links. After reading the following sections, it will be appreciated that one of the advantages of such a system is that a large collection aperture can be achieved but with an overall system of relatively short length, thus reducing the costs of the supporting engineering framework.

16.2   The telescope collector

The principal part of a telescope consists of the **collecting aperture**, acting as a means of producing a primary image. This function is depicted in a simplified way in figure 16.1. It will be seen from this figure that the light entering the collector is arriving in parallel rays. This is general for all astronomical objects as they can all be considered to be at infinity in comparison with the dimensions of any telescope.

The quantity of energy which is collected per unit time by the telescope is proportional to the area of the collecting aperture. The larger the aperture of the telescope, the better it performs as a means of collecting energy and the more capable it is as a tool for investigating faint sources. As the aperture is usually circular in form, the light-gathering power is proportional to the square of its diameter, D. It is convenient, therefore, to express the size of a telescope in terms of its **diameter**.

A telescope would be working with perfect transmission efficiency if it were able to concentrate all the energy collected into the images. However, it is impossible to do this as some of the energy

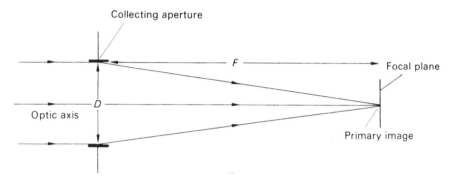

Figure 16.1. A schematic representations of the function of a telescope.

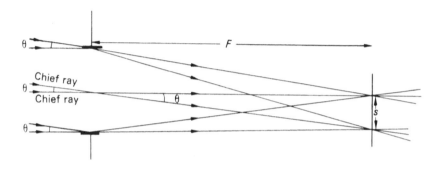

Figure 16.2. A simplified ray diagram illustrating the separation of images in the focal plane according to the angular separation of the point objects.

is lost either by **reflection** or **absorption**. From the design of a telescope it is possible to predict the fraction of the energy found in an image in relation to the amount of energy that is collected by the aperture. This fraction defines the **transmission efficiency** of the system and this is one of the factors which must be considered in the design of telescopes and which must be taken into account when brightness measurements are being determined absolutely.

As all the objects to be investigated are effectively at infinity, the distance of the primary images from the collecting aperture defines the **focal length**, F, of the telescope. A plane through this point and at right angles to the optic axis is defined as the **focal plane**. Telescopes of the same diameter need not have identical focal lengths. This latter quantity is engineered according to the purpose for which the telescope is primarily designed. Thus, the focal length is another parameter which is used to describe the optical properties of any telescope.

It is often convenient, especially when describing a telescope's ability to act as a camera, to describe a telescope in terms of its **focal ratio**, f, which is defined as the ratio of the focal length of the collector to its diameter. Thus,

$$f = \frac{F}{D}. \tag{16.1}$$

This definition is synonymous with that of the **speed** of an optical system when applied to general photography. (It may be remembered that, in everyday photography, the exposure time is proportional to f^2.)

When a telescope is directed to the sky, an image of a part of the celestial sphere is produced in the focal plane. The relationship between the size of this image and the angular field which is represented by it is governed by the focal length of the telescope. In figure 16.2, rays are drawn for two stars which are separated by an angle, θ. For convenience, one star has been placed on the optic axis of the telescope. Rays which pass through the centre of the system (chief rays) are not deviated. Thus, it can be seen from figure 16.2 that the separation, s, of the two images in the focal plane is given by

$$s = F \tan \theta.$$

As θ is normally very small, this can be re-written as

$$s = F\theta \tag{16.2}$$

where θ is expressed in radians. In order to be able to examine images in detail, their physical size or separation must be large and this can be achieved by using a telescope of long focus.

The correspondence between an angle and its representation in the focal plane is known as the **plate scale** of the telescope. By considering equation (16.2), it is obvious that the plate scale is given by

$$\frac{\mathrm{d}\theta}{\mathrm{d}s} = \frac{1}{F}.$$

It is usual practice, however, to express the place scale in units of arc seconds mm^{-1} and, in this case, the plate scale is given by

$$\frac{\mathrm{d}\theta}{\mathrm{d}s} = \frac{206\,265}{F} \tag{16.3}$$

where F and s are expressed in mm and θ in seconds of arc, the numerical term corresponding to the number of arc seconds in a radian.

16.3 The telescope and the collected energy

16.3.1 Stellar brightness

When any object such as a star can be considered as a point source, to all intents and purposes its telescope image can also be considered as a point, no matter how large a telescope is used. All the collected radiation is concentrated into this image. The larger the telescope, the greater is the amount of collected radiation for detection. Thus, the *apparent* brightness of a star is increased according to the collection area or the square of the diameter of the collector.

In allowing detection and measurement of faint stars, the role of the telescope may be summarized as being that of a **flux collector**.

Although stellar brightness measurements are usually expressed on a magnitude scale, they result from observations of the amount of energy collected by a telescope within some defined spectral interval over a certain integration time. A stellar brightness may be put on absolute scale and be described in terms of the **flux**, $\mathcal{F}$, or the energy received per unit area per unit wavelength interval per unit time. A measure of the star's brightness might be expressed in units of W m^{-2} Å^{-1}.

Estimation of the strength of any recorded signal and determination of the signal-to-noise ratio of any measurement is normally performed in terms of the number of photons arriving at the detector. In the first instance, the calculations involve determination of the number of photons passing through the telescope collector and this can be done by remembering that the energy associated with each photon is given by $E = h\nu$ or $E = hc/\lambda$ (see sections 4.4.2 and 15.5.1). Thus, the number of photons at the telescope aperture may be written as

$$N_T = \frac{\pi}{4}D^2 \times \Delta t \int_{\lambda_2}^{\lambda_1} \lambda \frac{\mathcal{F}_\lambda}{hc}\,\mathrm{d}\lambda \tag{16.4}$$

where λ_1 and λ_2 are the cut-on and cut-off points for the spectral range of the measurements and Δt is the integration time of the measurement.

If the stellar spectrum is relatively flat over the detected wavelength interval and if it is assumed that the photon energy is constant over this interval, then we may write

$$N_T = \frac{\pi}{4} D^2 \Delta t \frac{\lambda}{hc} \mathcal{F}_\lambda \Delta\lambda \tag{16.5}$$

where $\Delta\lambda$ is the spectral interval for the measurements.

The arrival of photons at the telescope is a statistical process. When the arriving flux is low, fluctuations are clearly seen in any recorded signal as a result and any measurements are said to suffer from **photon shot noise**. If no other sources of noise are present, the uncertainty of any measurement is given by $\sqrt{N_T}$. Hence, any record and its 'error' may be expressed as $N_T \pm \sqrt{N_T}$. In this circumstance, the signal-to-noise (S/N) ratio of the observation is given by

$$\frac{N_T}{\sqrt{N_T}} = \sqrt{N_T} \propto \sqrt{D^2 \Delta t} \longrightarrow D\sqrt{\Delta t}. \tag{16.6}$$

Although the light gathering power of a telescope depends on its collection area and hence on D^2, equation (16.6) shows that the signal-to-noise ratio of basic brightness measurements only increases according to D. It may also be noted that the S/N ratio improves according to the square root of the observational time illustrating a law of diminishing returns according to the time spent making any measurement. As will be seen later, in the real situation, the signal strengths depend on the photon detection rate, after taking into account the transmittance of the telescope and its subsidiary instrumentation and the efficiency of the detector. However, these additional factors do not alter the conclusions coming from equation (16.6).

In passing, it may also be mentioned that under some circumstances of detection of faint objects, the effectiveness of a telescope may not be proportional to D^2. For example, when faint stars are being detected photoelectrically against a background night sky which is also emitting light (no sky is perfectly black), measurements of the night sky brightness need to be made also so that this background signal can be subtracted and allowed for. In this circumstance, where observations are required of the star plus sky background and then sky background alone, the effectiveness of the telescope is then only proportional to D.

Example 16.1. At 6300 Å (630 nm) the flux from a source is 10^{-18} W m^{-2} Å^{-1}. Determine the photon rate passing through a telescope aperture with $D = 2\cdot2$ m over a wavelength interval of 100 Å. Calculate the best possible S/N ratio of a measurement with an integration time of 30 s.

Using equation (16.5)

$$N_T = \frac{\pi}{4}(2\cdot2)^2 \times 30 \times \frac{630 \times 10^{-9}}{6\cdot63 \times 10^{-34} \times 3 \times 10^8} \times 10^{-18} \times 100$$

this assuming that the flux exhibits no spectral variation over the wavelength interval of the measurements and that the photon energy is constant over this interval. By performing the arithmetic,

$$N_T \approx 7463.$$

If the measurements are limited purely by photon shot noise, the S/N ratio is simply $\sqrt{N_T}$ which equals 86·4. In other words, the accuracy is of the order of 1 part in 86·4 or just better than 1%.

16.3.2 Brightness of an extended object

The **brightness** of any extended object is defined in the same way as surface brightness (see section 15.5.1) as the flow of energy it provides through a unit area normal to the radiation per unit

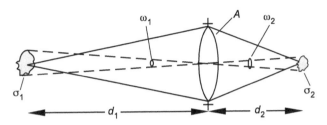

Figure 16.3. The relationship between the brightness of an extended object and the brightness of its image.

solid angle of the source per unit time. A convenient measure of the brightness might be expressed in power units as $J\,m^{-2}\,sr^{-1}\,s^{-1}$ or $W\,m^{-2}\,sr^{-1}$. Suppose that an extended source of brightness, B_1, at a distance, d_1, from an optical system of collecting area, A, presents an elemental area, σ_1, and that an image of this is formed (see figure 16.3). The amount of energy collected by the system per unit of time, or the **flux** collected by the aperture is given by

$$\mathcal{F}_1 = B_1 \sigma_1 \left(\frac{A}{d_1^2} \right) \tag{16.7}$$

where the bracketed term corresponds to the solid angle associated with the flux accepted by the aperture. The ratio of d_2 to d_1 sets the magnification of the system and the corresponding area of the image is given by

$$\sigma_2 = \sigma_1 \left(\frac{d_2^2}{d_1^2} \right). \tag{16.8}$$

Now for this same optical system, we can also consider the case with the radiation originating at the position of the first image from a source of brightness B_2, providing an elemental area σ_2, with the drawn ray paths following the opposite direction to the original situation. The flux now entering the aperture is now

$$\mathcal{F}_2 = B_2 \sigma_2 \left(\frac{A}{d_2^2} \right). \tag{16.9}$$

If the fluxes for the two scenarios are set equal, then by equations (16.7) and (16.9),

$$B_2 \sigma_2 \left(\frac{A}{d_2^2} \right) = B_1 \sigma_1 \left(\frac{A}{d_1^2} \right)$$

so that

$$B_2 = B_1 \left(\frac{\sigma_1}{\sigma_2} \right) \left(\frac{d_2^2}{d_1^2} \right).$$

By using equation (16.8),

$$B_2 = B_1 \left(\frac{\sigma_1}{\sigma_2} \right) \left(\frac{\sigma_2}{\sigma_1} \right)$$

with the result that

$$B_2 = B_1.$$

The brightness of an image of an *extended* object, therefore, matches that of the object. A telescope *does not* increase the brightness of an extended object—a fact not always appreciated. In fact, because of transmission losses in the optics, the brightness of any image of an extended source is slightly less than that of the object. *Brightness* is a fundamental parameter related to one of the properties of the radiating source. The application of a telescope cannot cause the source to alter the way in which it is radiating!

Relevant to the measurement of brightness is the amount of flux received from an object per unit telescope area per unit solid angle, sometimes referred to as the **luminance**, $\mathcal{L}$. Referring to figure 16.3 again, suppose that the flux received and measured by the telescope within the solid angle ω_1 is equal to $\mathcal{F}_1$. Thus, the luminance is given by

$$\mathcal{L}_1 = \frac{\mathcal{F}_1}{A\omega_1}. \tag{16.10}$$

By using equation (16.7) to substitute for $\mathcal{F}_1$ and noting that $\omega_1 = \sigma_1/d_1^2$,

$$\mathcal{L}_1 = \frac{1}{A} B_1 \sigma_1 \left(\frac{A}{d_1^2}\right)\left(\frac{d_1^2}{\sigma_1}\right) = B_1. \tag{16.11}$$

Thus, the simple measurement of luminance provides a direct value for the source's brightness. Without knowledge of the object's physical size and distance, a luminance measurement provides a determination of the energy emitted by the source per unit area per unit solid angle per unit time.

In relation to measurements of luminance, it is of interest to understand the role that the telescope plays. Suppose that a photoelectric photometer is attached to a given telescope. For the optical systems to be matched, the telescope and the photometer will have identical focal ratios. In order to make brightness measurements, a diaphragm within the photometer is placed in the focal plane of the telescope to limit the accepted angular field on the sky. Thus, the physical area of the diaphragm might correspond, say, to σ_2 in figure 16.3. The solid angle accepted by the photometer is ω_2 which equals ω_1, this being the solid angle on the sky over which the source is measured. According to the brightness of the extended object, the photometer will provide an output signal of a certain level.

If the photometer is now attached to a larger telescope but within identical focal ratio and directed to the same extended object, the output signal *does not* increase as perhaps might have been anticipated. The larger telescope collects more flux yet the photometer does not give a larger output. How does this come about?

Increasing the diameter of the telescope means that the collected flux increases according to D^2. But if the focal ratio is kept constant, then the larger telescope has a longer focal length and the solid angle on the sky, as limited by the physical size of the diaphragm of the photometer in the focal plane of the telescope, reduces according to F^2. Thus, if by going to a larger telescope the flux is increased by a certain proportion, the solid angle on the sky measured by the photometer decreases by the same proportion and the output levels are identical. The situation can be looked at in two ways:

1. For a certain output signal level, the spatial resolution of a photometered object may be increased by using a larger telescope.
2. If a larger telescope is used and the photometer's diaphragm is also enlarged to keep the field of view constant, the output signal *will* increase.

16.3.3   Illumination

If a screen is placed in the focal plane of a telescope, an extended source provides it with an **illumination**, $\mathcal{I}$, this being defined as the amount of collected flux that is concentrated into unit area of the screen. It is obvious that when the image is recorded on a two-dimensional detector (photographic plate or electronic detector), the length of exposure required depends on the strength of the illumination. Using the identities defined in the previous subsection, the flux collected by a telescope from an extended source is

$$B\omega A_T$$

and the area of the image is given by

$$F^2 \omega.$$

Hence, the illumination is given by

$$\mathcal{I} = \frac{B \omega A_T}{F^2 \omega} = \frac{B A_T}{F^2}.$$

If D is the diameter of the telescope collector, then the illumination $\mathcal{I}$ of the image is, therefore, given by

$$\mathcal{I} = \frac{B \pi D^2}{4 F^2}$$

and, therefore,

$$\mathcal{I} = \frac{\pi}{4} \frac{B}{f^2} \qquad (16.12)$$

where f is the focal ratio of the telescope. As mentioned briefly earlier, the correct exposure times for landscape photography can be adjusted by altering the focal ratio of the camera lens. When recording the image of an extended astronomical source with a fixed focal ratio telescope, the required exposure time is governed by the brightness of the source and the focal ratio, not the physical size of the telescope collection area.

One implication of this result is that for a range of telescopes of different sizes but with an identical focal ratio, although the flux collected increases with telescope size, so does the image size and the illumination to the image of a particular object is constant. The collected flux increases as the square of the telescope's size and the area of any image also increases in the same proportion. However, as the size of image increases by using a larger telescope, more detailed spatial studies may be made. Thus, we may summarize one of the uses of the telescope as being its ability to achieve angular resolution, allowing spatial analysis to be made of the radiation arriving from an extended source.

16.4 Telescope resolving power

The **resolving power** of a telescope may be defined as the ability of a telescope to separate objects with a small angle between them. Even if a telescope is made perfectly and does not suffer any aberrations (see later), there is a fundamental limit to the ability of any telescope to separate objects which are apparently close together. This limit is known as the **theoretical resolving power** of the instrument. If the telescope is of good design and in correct adjustment, it should be possible to achieve this theoretical value or even a slightly better one according to the nature of the observations.

Any radiating source can be considered to give rise to **wavefronts**, defined as surfaces over which the electromagnetic disturbances are in phase. As time proceeds, the wavefronts expand from the radiating source at a rate depending on the velocity of light for the medium through which they are travelling. The direction of propagation of the energy is at right angles to any wavefront as can be seen in figure 16.4, where the wavefronts are depicted for a point source. For a star, which can be considered to be at infinity, the wavefronts are in the form of parallel planes by the time they arrive at the telescope. The effect of the collector is to alter the shape of the wavefronts by inducing a differential phase change over the telescope aperture so that an image is produced. This is illustrated in figure 16.5. It can be seen from the figure that, because of its limited size, the telescope collector is able to use only a part of the wavefronts which arrive from space. As a consequence, some 'information' is lost and the resulting image does not correspond exactly to that which might be expected from a point object. It is, in fact, in the form of a **diffraction pattern**. The pattern is caused by the interruption of the plane wavefronts by the telescope aperture. In the construction of the focal plane image, it is necessary to take into account interference from the different points within the wavefront. As the collected wavefronts forming the

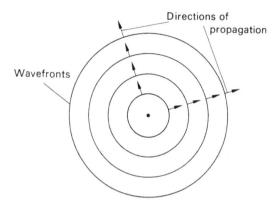

Figure 16.4. Wavefronts radiating from a point source.

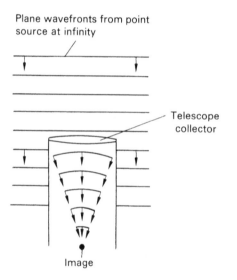

Figure 16.5. Wavefronts arriving at the telescope aperture.

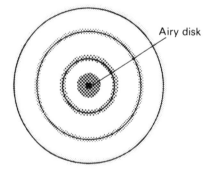

Figure 16.6. The diffraction pattern of a star image in the focal plane of an objective telescope.

image are no longer infinite in extent, the interference effects do not cancel out completely and will be present in the image. The form of the diffraction pattern can be predicted from theory.

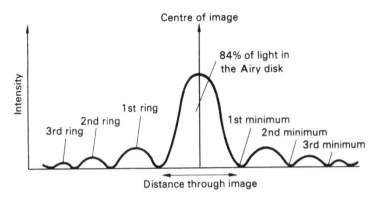

Figure 16.7. Intensity scan through a diffraction pattern.

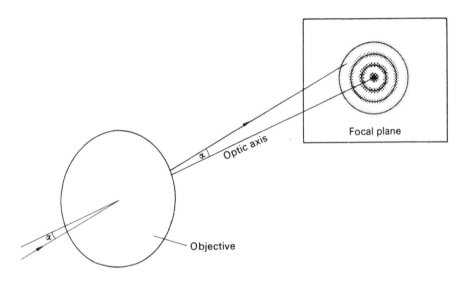

Figure 16.8. The diffraction pattern in the focal plane of the objective.

For a collector in the form of an objective lens, the resultant diffraction pattern produced from a point object appears as a spot at the centre of a system of concentric rings (see figure 16.6). The strength, or energy contained in each ring, decreases according to the number of the ring. However, after the third ring it is difficult to detect the presence of the outer rings. According to theory, 84% of the energy is concentrated into the central spot. As this type of diffraction pattern was first investigated by Airy, the central spot is sometimes referred to as the **Airy disc**. By making a scan along a line through the centre of the pattern, an intensity profile such as that illustrated in figure 16.7 can be obtained.

The sizes of the diffraction rings are also predicted by theory. By defining α to be the angle subtended at the collector by the centre of the Airy disc and a point in the diffraction pattern (see figure 16.8), the positions for minima of intensity are given by

$$\sin \alpha_n = \frac{m_n \lambda}{D}$$

where n is the number or order of the minimum, m is a numerical factor which is obtained by performing the integration of the interference from points over the telescope aperture, λ is the

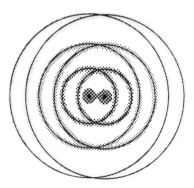

Figure 16.9. The combined diffraction pattern obtained from two point sources which are separated by an angle given by Rayleigh's criterion.

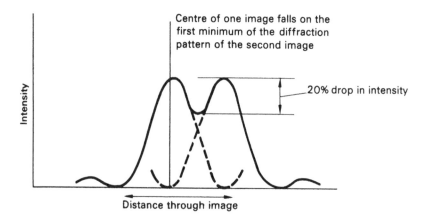

Figure 16.10. Intensity scan through a diffraction pattern obtained from two point sources which are separated by an angle given by Rayleigh's criterion.

wavelength of light and D the diameter of the telescope. Since the diffraction pattern is small, $\sin \alpha \approx \alpha$ and, therefore, minima are obtained at angles given by

$$\alpha_n = \frac{m_n \lambda}{D}. \tag{16.13}$$

The numerical factors are:

$$m = 1\cdot22 \quad \text{for } n = 1$$
$$m = 2\cdot23 \quad \text{for } n = 2$$
$$m = 3\cdot24 \quad \text{for } n = 3.$$

If two point sources are very close together, their images will result in a combination, being the superimposition of two diffraction patterns. It will only be possible to resolve the resultant image as being made of two components if the individual Airy disks are sufficiently separated. According to **Rayleigh's criterion** for resolution, the two images are said to be resolved when the centre of one Airy disc falls on the first minimum of the other diffraction pattern. This condition is illustrated in figure 16.9 and an intensity scan through such an image is illustrated in figure 16.10.

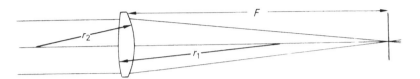

Figure 16.11. A beam of parallel light is brought to a focus by a double-convex lens (r is positive and r_2 is negative).

Thus, by considering Rayleigh's criterion in relation to equation (16.13), it should be possible to resolve two stars if they are separated by an angle (in radians) greater than

$$\alpha = \frac{1 \cdot 22\lambda}{D}. \qquad (16.14)$$

This value is known as the **theoretical angular resolving power** of the telescope. The physical separation of the resolving power in the focal plane can be obtained by multiplying the value of α from equation (16.14) by the focal length of the telescope. It can be seen that the resolving power is inversely proportional to the diameter of the objective. By taking a value of 5500 Å as being the effective wavelength for visual observations, the resolving power can be expressed in seconds of arc as

$$\alpha \approx \frac{140}{D} \qquad (16.15)$$

where D is expressed in mm.

In summary, the ability of a telescope to resolve structure in a celestial object simply depends on the ratio λ/D. The angle that can be resolved grows smaller according to the diameter, D, of the telescope. It may be noted too (section 16.3.1) that the S/N ratio of any point image photometry also improves according to the linear size of the telescope aperture.

We can now look at the main types of telescope system with particular reference to their ability to collect light and produce good primary images.

16.5 Refractors

16.5.1 Objectives

The lens-maker's formula expresses the focal length, F, of a lens in terms of the refractive index, n, of the material (usually glass) and the radii of curvature, r_1, r_2, of the two lens' surfaces. It can be expressed conveniently in the form

$$\frac{1}{F} = (n - 1) \left(\frac{1}{r_1} - \frac{1}{r_2} \right). \qquad (16.16)$$

(In this form, values of r are positive when light rays meet a convex curvature and are negative when they meet a concave curvature.) Thus, in general, if r_1 is positive and r_2 is negative, a simple lens has a positive, real focus. A beam of parallel light which falls on the lens is, therefore, brought to a focus as shown in figure 16.11. By applying this lens to a beam of light from a star which can be considered to be at infinity, an image of the star will be formed at the focus of the lens. This image is available for viewing with an eyepiece or for recording on a detector. The lens acts as light-collector and image-former. When it is used in this way in a telescope system it is commonly known as the **objective**.

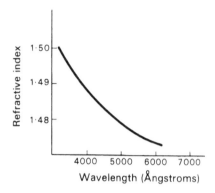

Figure 16.12. The dispersion curve of a typical crown glass.

A single-lens objective is unsatisfactory for astronomical purposes as the images produced suffer from defects or **aberrations** of different kinds. Considerable effort has been applied to the design of objectives to remove or reduce the aberrations and refractor telescopes show a variety of construction depending on their intended function and the way particular aberrations have been compensated.

Telescope objectives may suffer from the following defects:

(i)   chromatic aberration,
(ii) spherical aberration,
(iii) coma,
(iv) astigmatism,
(v)   curvature of field,
(vi) distortion of field,

and each of these effects is discussed in the following subsections.

16.5.2   Chromatic aberration

A closer look at the lens-maker's formula (equation (16.16)) reveals why the image produced by a single lens suffers from **chromatic aberration**. For a given lens of known shape, the lens-maker's formula may be re-written so that the focal length may be expressed by

$$F = \frac{K}{(n-1)} \tag{16.17}$$

where $K = r_1 r_2/(r_2 - r_1)$. (For a positive lens, K is also positive.)

The term $(n-1)$ is known as the **refractive power** of the material of the lens.

The refractive index and, hence, refractive power of all common materials used for lenses exhibits **dispersion**, i.e. its value is wavelength-dependent. A typical dispersion curve is illustrated in figure 16.12 and it depicts the fact that the refractive index progressively decreases as the wavelength of the light increases. Thus, the focal length of a single lens depends on the wavelength of light that is used and, for a simple positive lens, the focal length increases as the wavelength increases. If, therefore, a single positive lens is illuminated by a parallel beam of white light, a spread of images is produced along the optic axis of the lens. This is illustrated in figure 16.13. At no point along the optic axis is there a position where a point image can be seen: at the position where an image is formed for the extreme blue end of the spectrum, this image is surrounded by a red halo; similarly at the position where an image is formed for the extreme red end of the spectrum, this image is surrounded by a blue halo. Between these extreme positions, there is a plane which contains the smallest possible image

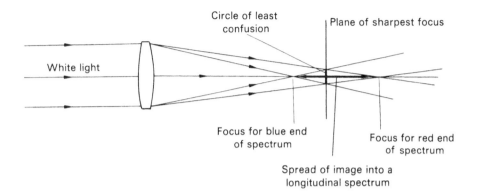

Figure 16.13. A ray diagram illustrating the effect of chromatic aberration resulting from the use of a single positive lens.

which can be obtained. The image in this plane of sharpest focus is not a point image. It is known as the **circle of least confusion** and its size can be determined from the physical properties of the lens. The spread of the image along the optic axis is known as **longitudinal chromatic aberration** and the spread of the image in the plane of sharpest focus is known as **lateral chromatic aberration**.

It will be remembered that it was first supposed that chromatic aberration could not be removed from lens systems but, eventually, Dolland proposed a method for **achromatism**. It is possible by using in combination with a positive lens, a negative lens of different material—and hence refractive power—to cancel the dispersion without complete cancellation of the refractive power. This is easily demonstrated as follows.

If two lenses of focal lengths F_1 and F_2 are fixed together by optical cement, the focal length of the combination, F, is given by

$$\frac{1}{F} = \frac{1}{F_1} + \frac{1}{F_2}.$$

Hence,

$$F = \frac{F_1 F_2}{F_1 + F_2}. \tag{16.18}$$

By expressing the individual focal lengths in terms of the shape of each lens and their refracting powers, equation (16.18) can be rewritten as

$$F = \frac{K_1 K_2}{K_2(n_1 - 1) + K_1(n_2 - 1)}. \tag{16.19}$$

Now the aim of the combination is to provide a system whose focal length is identical for the blue and red ends of the spectrum. If the refractive indices of the materials of the lenses are denoted by n_{1B}, n_{2B}, for blue light and n_{1R}, n_{2R}, for red light, the achromatic condition is achieved when

$$\frac{K_1 K_2}{K_2(n_{1B} - 1) + K_1(n_{2B} - 1)} = \frac{K_1 K_2}{K_2(n_{1R} - 1) + K_1(n_{2R} - 1)}$$

i.e.

$$\frac{K_1}{K_2} = -\frac{n_{1B} - n_{1R}}{n_{2B} - n_{2R}}. \tag{16.20}$$

Since $n_{1B} > n_{1R}$ and $n_{2B} > n_{2R}$ by the general properties of optical materials, K_1/K_2 must be negative. Hence, achromatization can only be achieved by combining a positive and a negative lens.

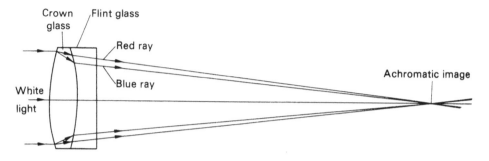

Figure 16.14. An achromatic doublet, made by cementing a positive crown glass lens to a negative flint glass lens, depicts how light rays of different colours are brought to the same focus.

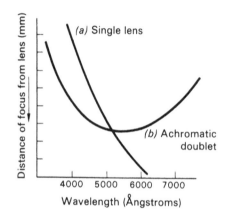

Figure 16.15. The variation of the focal length with wavelength of (*a*) a single positive lens and (*b*) a corrected achromatic doublet.

The choice of the types of glass available for the two lenses imposes a condition on the ratio K_1/K_2. According to the required focal length of the combination, K_1 and K_2 may then be determined individually: the particular values of the radii of each surface are usually decided from conditions set for reducing other aberrations.

An **achromatic doublet** or **achromat** usually consists of a positive lens made of a glass with a medium refractive index and dispersion followed by a negative lens made of glass with a higher refractive index and dispersion. A typical doublet is depicted in figure 16.14. The same figure illustrates the way in which the first lens introduces separation of the rays corresponding to the blue and red wavelengths present in the incident light and the second lens produces a closing of these rays so that they will cross at the same point along the optic axis.

A simple achromatic doublet does not completely remove all the chromatic aberration. There still remains a small spread of focus along the optic axis of the lens but this has been reduced considerably over the single lens. At each position within this reduced spread, light rays corresponding to a pair of wavelengths are brought to the same focus. Figure 16.15 illustrates how the focal length of an achromat might vary with wavelength. The same figure also illustrates how the focal length of a single lens would vary over the same spectral range.

There are, in fact, many different designs of achromatic objective. The wavelengths for which the best achromatism occurs can be chosen according to whether the telescope is to be used for visual or photographic purposes—the plates having a different spectral response to that of the eye—and the

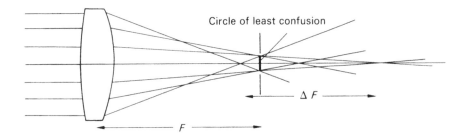

Figure 16.16. Spherical aberration produced by a single positive lens; F represents the focal length measured to the circle of least confusion and ΔF (exaggerated for clarity) represents the spread of the image.

objective is designed accordingly. Hence, some objectives are termed to be **visual** or **photographic**. Not all objectives are constructed in the form of a cemented doublet; there are some variations which have an air-gap between the two elements. One special objective, known as a **photovisual**, is constructed of three elements. The achromatism produced by this system is so effective that it can be used either visually or with photographic plates, as its name implies.

16.5.3 Spherical aberration

The simple theory for lens design is based on the effects of refraction produced by spherical surfaces and this is particularly convenient as a spherical surface is one of the easiest to obtain on an optical grinding and polishing machine. However, simple lens theory only takes into account **paraxial** rays, i.e. rays that are very close to the optic axis, allowing $\sin \theta$ to be written as θ. For a point source at infinity, lying on the axis of a single lens, the image produced by the lens does not retain a point-like appearance and is spread out into a disc. This effect results from the rays which cannot be considered as paraxial. The position of the focus for any incident ray depends on its distance from the optic axis. The defect of the image is known as **spherical aberration** and its effect is illustrated in figure 16.16. If the spread in focus is denoted by ΔF, the severity of any spherical aberration may be expressed by assessing the value of the ratio $\Delta F/F$. As in the case of chromatic aberration, there is one plane through the spread of focus which contains the smallest image, again known as the circle of least confusion. Also the spread of an image along the optic axis is known as **longitudinal spherical aberration** and the spread of an image in the plane containing the circle of least confusion is known as **lateral spherical aberration**. The size of any image can be predicted by the physical properties of the lens or by performing a ray-tracing analysis.

The amount of spherical aberration depends on the shape of a lens. It is, therefore, convenient to define what is known as the **shape factor** of a lens. By denoting the radii of the two lens surfaces as r_1 and r_2, the shape factor, q, is expressed as

$$q = \frac{r_2 + r_1}{r_2 - r_1}. \tag{16.21}$$

Typical shapes of lenses have been drawn in figure 16.17 with a range of shape factors running between $q < -1$ and $q > +1$.

By examination of lenses over the complete range of shape factors, it is found that spherical aberration has minimal effect when q is close to $+0.7$—it *never* goes to zero.

Spherical aberration can be overcome completely by figuring a lens so that the curvature of the faces is not constant. This process is known as **aspherizing** and is sometimes used in producing

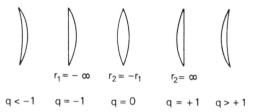

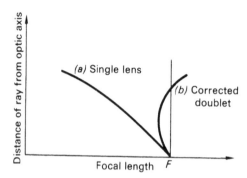

Figure 16.17. A range of lenses with different shape factors.

Figure 16.18. Longitudinal spherical aberration for (*a*) a single lens and (*b*) a corrected doublet.

telescope objectives. However, this process can be costly and it is not the only way to remove spherical aberration.

For a negative lens, the spherical aberration is also negative, i.e. the numerical value of the focal length increases as incident rays which are more distant from the optic axis are considered. A combination of a positive and a negative lens can be designed to provide a system which is free from spherical aberration. This is particularly convenient as we have already seen that chromatic aberration can be reduced by a two-lens system. Thus, achromatic lenses are designed to have minimal spherical aberration. The residual effect of spherical aberration for a corrected objective is illustrated in figure 16.18. In the same figure, the effect of spherical aberration of a single lens is drawn for contrast.

If effects of chromatic and spherical aberration have not been removed completely, they may be noticeable when the primary image is viewed with an eyepiece. However, an eyepiece normally limits the field of view to a small angle and these may be the only defects of the image that will be detected. By increasing the field of view, either by using a special wide-angle eyepiece or by placing a two-dimensional imaging detector in the focal plane of the telescope, other types of aberration may become apparent. Such aberrations that might be detected in this way are: coma, astigmatism, field-curvature and field-distortion. Any of these aberrations may be present in images which result from incident rays which arrive at an angle to the optic axis. The effects and causes of each of these aberrations will now be discussed briefly.

16.5.4 Coma

The effect of **coma** derives its name from the comet-like appearance that an image can have when a point object is off the axis of a lens. Its cause can be considered by treating a lens as being made of a series of annuli. Each annular zone gives rise to an annular image in the focal plane. The total aberrated image results from the combination of each component image. The effect of coma is illustrated in a simplified way in figure 16.19.

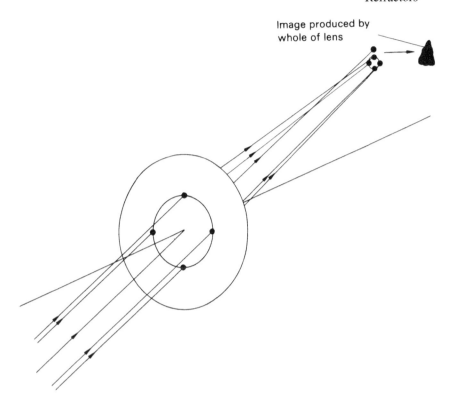

Figure 16.19. When coma is present, any annulus of the lens produces an annular image; the total aberrated image can be thought of as being made up of a series of such annular images, the sizes increasing as the outer zones of the lens are considered.

For any single lens, the size of the comatic image depends on the shape factor. In contrast with spherical aberration, it is possible for an image to be free of coma for a particular value of the shape factor. For objects at infinity, the coma-free condition is obtained when the shape factor is close to the value given by

$$q = \frac{(n-1)(2n+1)}{(n+1)} \qquad (16.22)$$

and for a lens made of glass with refractive index $n = 1 \cdot 5$, $q = +0 \cdot 8$. This value is close to the value which results in a lens having minimum spherical aberration. When an achromatic doublet is designed, it is possible to correct coma and spherical aberration at the same time. A telescope objective which is free from these defects is known as an **aplanatic** lens.

16.5.5 Astigmatism

A lens system which has been corrected for both spherical aberration and coma may not be free from aberration completely, especially when images are formed of objects which are at a considerable distance from the optic axis. Such images may suffer from **astigmatism**. The effect of astigmatism is illustrated in figure 16.20. The position of the focus depends on which section of the lens is used to form the image. The spread in the position of the image lies between two points which correspond to the image position for rays in the plane formed by the point object and the optic axis (i.e. the tangential plane) and to the image position for rays in the plane at right angles to this (i.e. the sagittal plane).

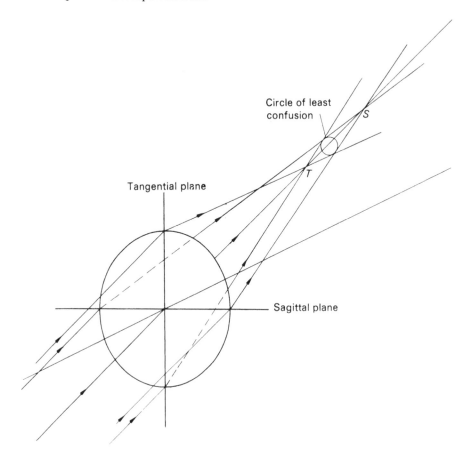

Figure 16.20. Points T and S represent the foci of rays passing through the tangential and sagittal planes of a lens; the spread of the astigmatic image lies between the points T and S.

Between these points there is a position where the smallest image can be found and again this image is known as the circle of least confusion.

By examining the images produced by objects with a range of distances from the optic axis, it is possible to record the surfaces on which the focus, T, of the rays of the tangential plane and the focus, S, of the rays of the sagittal plane lie. They are found to approximate to paraboloids of revolution.

A single achromatic combination is likely to exhibit appreciable astigmatism. However, the combination of two achromats at the correct spacing can provide a paraboloidal surface containing both the T and S images. This surface is known as the **Petzval surface**, after the investigator of this property.

16.5.6   Curvature of field

A system which is designed to remove completely astigmatism suffers from **curvature of field**, as the points of sharp focus lie on a curve rather than on a plane. Flattening of the field can be achieved at the expense of having incomplete removal of astigmatism and usually some compromise is made. The judicial placement of stops in the combination of lenses can also be used to reduce the effect of field curvature.

16.5.7 Distortion of field

Objectives that have been designed to remove aberrations which would show up by the blurring of images of point objects may still suffer from **distortion of field**. The effect occurs when the correspondence between the distance of an object from the optic axis and the distance of its image from the optic axis in the focal plane is not constant over the field of view, i.e. the plate scale varies over the focal plane. **Barrel** distortion is said to occur when the correspondence of the image position decreases with distance from the optic axis and **pincushion** distortion is said to occur when the correspondence increases with distance from the optic axis. In the design of compound objectives, the effect of field distortion can be reduced by the judicious placing of stops.

16.6 Transmission efficiency of the refractor

The refractor has imperfect transmission efficiency because of light losses which occur by reflection at each lens surface and by absorption in the material from which the lenses are made.

When a light beam crosses a dielectric surface such as glass, a fraction of the energy is not transmitted but is reflected by the interface. At normal incidence, the fraction, R, of light which is reflected is given by

$$R = \left(\frac{n-1}{n+1}\right)^2. \tag{16.23}$$

For a typical glass with $n = 1.5$, this amounts to about 0.04, i.e. 96% of the light is transmitted. Such a small loss hardly seems significant but when the losses from each lens surface in a multiplet lens system are summed, the total reflection loss may be quite high. For an air-spaced crown-flint doublet, reflection losses occur at each of the four lens surfaces. If the effects of multiple reflections are neglected, the transmission, T, can be written as

$$T = \left[1 - \left(\frac{n_c - 1}{n_c + 1}\right)^2\right]^2 \left[1 - \left(\frac{n_f - 1}{n_f + 1}\right)^2\right]^2 \tag{16.24}$$

where n_c and n_f are the refractive indices of crown and flint glass respectively, since transmission equals one minus reflection. By using typical values of $n_c = 1.5$ and $n_f = 1.6$, equation (16.24) predicts that the transmission of the doublet is equal to 83%. Both n_c and n_f exhibit dispersion and, hence, the transmission efficiency, as determined by reflection losses, will be slightly wavelength-dependent.

Most modern lenses are finished by vacuum-depositing an anti-reflection coating on the lens surfaces to cut down on reflection losses. This process, known as the **blooming of lenses**, is likely to be applied to small lenses, perhaps those used in an eyepiece system but it is unlikely that a telescope of a medium-to-large aperture would be bloomed.

When energy is transmitted by transparent materials, some of it is lost by processes of scattering and absorption. The fraction which is lost depends on the actual material and the wavelength; it is also proportional to the thickness of the material through which the light is transmitted. For most ordinary glasses, absorption is not uniform across the visible spectrum. The transmission falls noticeably towards shorter wavelengths, the change in absorption becoming apparent around 500 nm or 5000 Å. Transmission curves for two types of glass which might be used in a doublet are depicted in figure 16.21. A typical large objective may be constructed by the combination of two lenses, each a few centimetres thick. With modern materials, the resulting absorption losses may be lower than 10% over the central part of the visible spectrum. However, in the ultraviolet, the absorption is so severe as to prevent observations being made in the U band (see section 19.4).

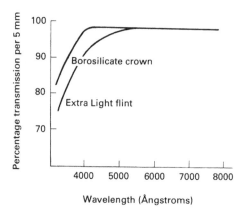

Figure 16.21. The percentage transmission per 5 mm as a function of wavelength for two types of glass which would be suitable for use in a doublet lens.

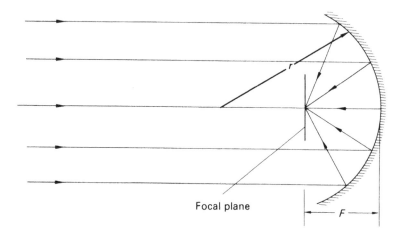

Figure 16.22. Simple ray diagram illustrating the ability of a spherical concave mirror to act as an image former. (All rays are paraxial.)

For old established refractors, the combination of reflection and absorption losses gives values for the overall transmission for typical objectives between 50 and 70% in the central part of the visible spectrum.

16.7   Reflectors

16.7.1   Principles

A spherical concave mirror of radius of curvature, r, has a focal point at a distance equal to $r/2$ from the mirror surface. When such a mirror is directed to objects which are at an infinite distance, images will be formed in a plane at right angles to the optic axis and passing through this focal point. A simple ray diagram illustrating this principle is drawn in figure 16.22. Thus, a large front-surfaced mirror can be used for astronomical purposes to act as a light collector and primary image former. Such a mirror would be made of rigid material with a low thermal expansion coefficient to prevent distortions in the reflecting surface as the ambient temperature changes. The figured surface is covered with an evaporated layer of aluminium to produce a high reflection coefficient.

By the method employed to produce images, reflector telescopes are inherently free from chromatic aberration. However, they are susceptible to all the other forms of aberration which have been discussed briefly in relation to refractor telescopes. The effects of spherical aberration, coma, astigmatism, curvature of field and distortion of field may be apparent in the images which are formed by the primary mirror.

Since the original application of mirror optics to telescope systems, there have been many improvements in design to eliminate the various aberrations. For example, the effects of spherical aberration can be removed by using a mirror which has been figured to be in the form of a paraboloid of revolution. Simple ray diagrams illustrating the difference between the spherical and paraboloidal mirror for on-axis objects are drawn in figures 16.23(a) and (b). It may be noted that if the molten blank from which the mirror is made to spin while cooling, its surface takes on a parabolic profile. This technique has been used to produce parabolic mirrors of large size. Although the paraboloid effectively removes spherical aberration, it suffers badly from astigmatism and, therefore, has a severely limited field. This aberration can be removed by using corrector plates prior to the primary mirror, but the discussion of this type of system (Schmidt telescope) is reserved until later.

In order to have access to the primary image, it is inevitable that the central part of the collecting area of the mirror will be ineffective. The fraction of the light which is lost depends on the size of the apparatus which is placed at the prime focus. For a telescope with a large aperture, the light-loss may be quite small if an imaging camera is placed in the focal plane. However, the telescope must be extremely large if the observer wishes to inspect an image by eye at the primary focus. In some designs, it has been found possible to use the collecting mirror off-axis, so that the primary image is produced outside the cylinder which contains the rim of the aperture. However, the more usual practice of producing the image relies on a secondary mirror in the system. This is normally placed along the optic axis, towards the focus of the primary mirror and, consequently, the small central part of the collecting area is lost. It is held in position by means of the thin frame (spider) which is attached to the telescope tube. There are several designs of two-mirror combinations, the most common being those of the **Newtonian** and the **Cassegrain** systems. The general principles of the systems are now described.

16.7.2   Newtonian reflectors

The Newtonian system provides access to the image formed by the primary mirror, without altering the effective focal length of the system. The essential optical parts are illustrated in figure 16.24. A flat front-surfaced mirror is placed at 45° on the optic axis of the primary mirror so that the image is formed just outside the cylindrical beam which the primary mirror collects. The rim of the secondary mirror describes an ellipse and the mirror is referred to as the **Newtonian flat** or **elliptical flat**.

Adjustment of the position and tilt of the Newtonian flat is fairly easy to control and, from this point of view, the system is very convenient. For some purposes the position of the focus, which is towards the open end of the telescope, is inconvenient for direct observation and the use of subsidiary equipment. For these reasons, large telescopes are not frequently used with a Newtonian focus.

16.7.3   Cassegrain reflectors

The Cassegrain normally consists of a large spherical or paraboloidal primary mirror with a secondary mirror which is complex or hyperboloidal. The optical arrangement is illustrated in figure 16.25, from which it can be seen that not only is the central portion of the primary mirror not used but it is also machined out to allow the converging beam to be brought to focus behind the primary mirror. It can also be seen that the effective focal length of the system is increased by the secondary mirror. If F_p is the focal length of the primary mirror (a positive value), F_s the focal length of the secondary mirror (a

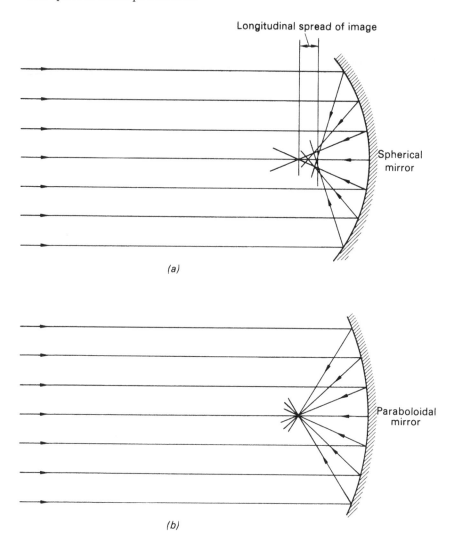

Figure 16.23. (*a*) Spherical mirror exhibiting the effect of spherical aberration. (*b*) Paraboloidal mirror removes spherical aberration; incident rays parallel to the optic axis are brought to the same focus, independent of their distance from the axis.

negative value) and if the secondary is placed at a distance d inside the focus of the primary mirror, the resulting focus will be found in the plane of the primary mirror when

$$\frac{1}{(F_p - d)} = \frac{1}{d} + \frac{1}{F_s}. \tag{16.25}$$

The equivalent focal length of the combination (defined as the focal length of a single imaging device which would provide the same size of image in its focal plane) is given by the focal length of the primary mirror multiplied by the magnification provided by the secondary mirror. If the equivalent focal length is denoted by F_e, then

$$F_e = \frac{(F_p - d)F_p}{d}. \tag{16.26}$$

For typical Cassegrain telescopes, the value of the magnification $(F_p - d)/d$ by the secondary

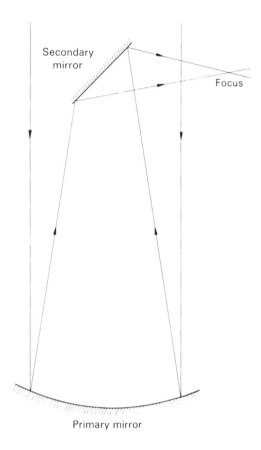

Figure 16.24. The Newtonian system.

mirror lies somewhere between two and five, according to the individual design. The secondary mirror must be sufficiently large to accept all the light of the convergent cone from the primary. The diameter needed and hence the fraction of the central part of the primary mirror which is not used, depends on the magnification of the secondary mirror. Some small adjustment of the position of the secondary mirror is usually made available by push-button control and the position of the final focus can be altered by this means. This is particularly convenient when a range of equipment is available for fixing to the base plate of the telescope.

One of the modified forms of the Cassegrain system is known as the **Ritchey–Chrétian** telescope. In this telescope, both coma and spherical aberration are removed. Astigmatism and field curvature are greatly reduced at the expense of having a larger diameter of secondary mirror than is usual for an ordinary Cassegrain telescope. Thus, although the light-grasp is not as great as the same size of Cassegrain telescope, the image quality is greatly improved.

16.8   Transmission efficiency of the reflector

Not all the light which is collected by the aperture of a reflecting telescope is able to pass through the system without loss, with the result that the images in the focal plane are less intense than might be expected. The causes of the loss of energy are the secondary mirror which blocks out the central cylinder of the light beam and the inability of mirror coatings to act as perfect reflectors.

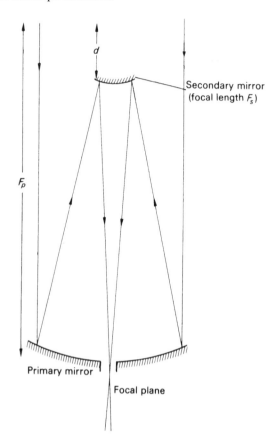

Figure 16.25. The Cassegrain system.

The fraction of light lost as a result of the obscuration by the secondary mirror is given simply by the ratio of the squares of the diameters of the secondary and primary mirrors. For the typical reflector telescope, the loss may be only a few per cent but may be as high as 20% for a design such as the Ritchey–Chrétian.

Modern reflecting telescopes use aluminium as the mirror surface. After completion of the optical figuring and polishing, the glass is given a uniform coating of aluminium by vacuum-deposition. Aluminium has a high reflectivity throughout the visible region. It is now used in preference to silver for, although silver has a higher reflectivity in the visible spectrum, it does not reflect the ultraviolet; in addition, silver films deteriorate more rapidly than aluminium coatings. Reflecting telescopes are now capable of making measurements in the ultraviolet down to the limit which is dictated by the transparency of the atmosphere.

As some designs of reflecting telescope use mirrors which are set at an angle to the incident rays, the overall transmittance of a system can only be estimated if this is taken into account. The curve of the reflection coefficient of freshly deposited aluminium as a function of the angle of incidence is illustrated in figure 16.26.

Thus, in the case of the Newtonian telescope, two reflections are involved and from values taken from figure 16.26, the light collected into the image is reduced as result of reflection losses by a factor given by

$$0.9 \times 0.86 = 0.77$$

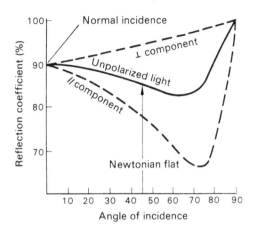

Figure 16.26. Reflection coefficients of freshly deposited aluminium as a function of the angle of incidence.

the first element of the product corresponding to the reflection efficiency at normal incidence while the second term is the reflection efficiency at 45° incidence for unpolarized light.

By taking a typical figure of 0·9 as being the fraction of light which is not cut off by the secondary mirror, the overall transmission efficiency of the telescope is given by

$$0.77 \times 0.9 = 0.69.$$

With an overall transmittance of 69%, the typical reflector system is little different in its efficiency from a typical refractor system. The curves depicted in figure 16.26 are only valid for a freshly coated aluminium surface. As the coating ages, its reflectivity deteriorates and the transmission efficiency of a reflector system cannot be maintained unless the mirrors are re-aluminized periodically.

It will also be noted on figure 16.26 that the reflectivity curves are different for the orthogonally resolved components of a beam. If polarization is present in any radiation collected by the telescope, the effective transmission coefficient of the overall Newtonian system will depend on the relationship between the polarization and the orientation of the Newtonian flat. When the brightnesses of sources with differing polarization properties are being compared, false assessments may be made unless the polarization of effects of the telescope are allowed for. The Newtonian system also makes polarization measurements themselves difficult to record without systematic errors.

16.9   Comparison of refractors and reflectors

Refractors range in focal ratio from about $f/4$ to $f/30$. Systems employing two achromatic lenses are in use to provide objectives of small focal ratio somewhere in the range $f/4$ to $f/8$. The system is specially corrected for astigmatism and will be used for star-field photography. Such telescopes are sometimes called **astrographs**. An objective of medium focal ratio ($f/8$ to $f/15$) would be used for a telescope intended for general observations by eye or with auxiliary equipment. A higher focal ratio objective, providing a large plate-scale, would be used when details of images are being investigated. Some telescopes are found with a focal ratio as high as $f/30$ but these are long, cumbersome instruments.

The prime focus of a reflector provides a focal ratio usually between $f/3$ and $f/7$, as such a system is frequently used for high speed imaging. Larger focal ratios are achieved using the Cassegrain system and such systems might provide focal ratios in the range $f/8$ to $f/20$. The coudé system (see later) can provide very big focal ratios, sometimes as high as $f/60$. Such a system is used almost exclusively for spectral studies at high resolution.

Figure 16.27. The 40-inch Yerkes refractor. (By courtesy of the Royal Astronomical Society.)

Figure 16.28. The 200-inch Hale reflector telescope at Mt Palomar, California. (Photograph from the Mt Wilson and Palomar Obervatories.)

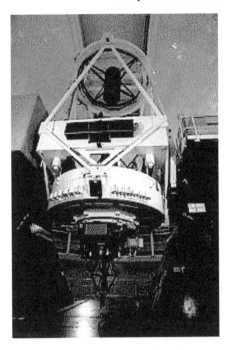

Figure 16.29. The William Herschel Telescope (WHT)—4·2 m. This telescope is driven on an alt-azimuth mounting. (Photograph by courtesy of the ING of Telescopes, La Palma.)

Figure 16.30. The Gemini (North) Telescope (8·2 m) prior to its commissioning. (Photograph by courtesy of NOAO/AURA/NSF.)

Table 16.1. A comparison between reflector and refractor systems.

Comparison	Reflector	Refractor	Comments
Optics	Fewer optical surfaces	More optical surfaces	Accuracy of reflecting surface needs to be better than that of refracting surface
	Material needs to be fairly homogeneous to prevent strains.	Material needs to be optically homogeneous to obtain high quality images.	This consideration favours reflecting systems for large telescopes.
Aberrations	No chromatic aberration.	Some chromatic aberration	Other aberrations will be present in both systems depending on the individual designs.
Theoretical image	Diffraction pattern is large because of the spider. The shape of the pattern may be detrimental to some kinds of observation.	Diffraction pattern has circular symmetry.	See section 16.4 on telescope resolving power.
Transmission efficiency	Losses by inefficient reflection and obscuration by secondary mirror.	Losses by reflection at lens surfaces and absorption within the lenses.	For small and medium-sized telescopes, the transmission efficiencies are comparable ($\approx$50 to 60%). Because of large absorption losses in big objectives, larger reflectors are superior to large refractors.
Wavelength range	Aluminium coated mirrors exhibit little or no change in reflectivity over the visible spectrum.	Absorption of glass is strongly wavelength dependent. Most of the ultra-violet light is absorbed.	Refractors are incapable of making measurements in the internationally accepted U (ultraviolet) band.
Method of mounting	Is supported across the back and round the rim.	Is supported round the rim only.	Large refractors suffer from strain—reflecting systems are used for very large telescopes.
Stability	Subject to temperature changes.	Less subject to temperature changes.	Visual observers prefer refractors.
Servicing	Optical surfaces need recoating periodically as the coatings deteriorate with age.	Once the objective has been aligned, the system requires little or no attention.	—

Thus, because of the wide range of telescope designs and their broad application, it is impossible to make a comprehensive comparison of the merits of refractor and reflector systems. However, some general comments can be made.

From a cursory consideration of the optical systems, it might be thought that a reflector system

Table 16.2. Some of the World's larger optical telescopes.

Type	Size	Designation and location
Refractors[a]	102 cm (40 in)	Yerkes, Williams Bay, WI, USA
	91 cm (36 in)	Lick, CA, USA
	83 cm (32·7 in)	Meudon, France
	80 cm (32 in)	Potsdam, Germany
	76 cm (30 in)	Allegheny, Pittsburgh, PA, USA
Reflectors	10·0 m	Keck I Mauna Kea, Hawaii—collector comprises 36 segments
	10·0 m	Keck II Mauna Kea, Hawaii
	9·2 m	Mt Fowlkes, Texas—segmented collector, fixed elevation, spectrometry
	8·3 m	Subaru—Mauna Kea, Hawaii—NAOJ
	8·2 m	Antu–Kueyen–Melopal–Yepun: Cerro Paranal, Chile—four independent telescopes that can be linked for multi-aperture synthesis
	8·1 m	Gemini North and South—Mauna Kea, Hawaii and Cerro Pachon, Chile
	6·5 m	Mt Hopkins, AZ, USA—multi-mirror telescope
	6·5 m	Magellan I—La Serena, Chile
	6·0 m	Nizhny Arkhyz, Caucasus Mts, Russia—altazimuth telescope
	5·0 m	Hale—Palomar Mt, CA, USA
	4·2 m	William Herschel—La Palma, Canary Islands—Obs del Roque de los Muchachos
	4·0 m	Victor Blanco—Cerro Tololo, Chile—CTIO
Under construction	4 × 8·2 m	Cerro Paranal, Chile—interferometry
	2 × 10 m	Keck Interferometer—Keck I + II with several other collectors
	2 × 8·4 m	Large Binocular Telescope—two separated mirrors with resolution ≡23 m
	10·4 m	La Palma, Canary Islands—segmented collector, based on Keck design
	9·2 m	SALT—South African Astronomical Obs, based on HET design
Project Studies	100 m	OWL—OverWhelmingly Large Telescope
	50 m	Swedish Optical Telescope
	30–50 m	MaxAT
	30 m	CELT—California Extremely Large Telescope

See also: W 16.1.

[a] All the large refractors were constructed before the beginning of the 20th century.

would produce better quality images than a refractor. Its immediate advantage is that it does not suffer from chromatic aberration. The quality of the images produced by any system depends on the accuracy of the optical surface which form them and the system with the fewest surfaces might be expected to perform best. A conventional reflector has two optical surfaces while an achromatic refractor has four surfaces within the objective and it might be expected on this account that the reflector would be a superior instrument. This argument cannot be applied fully, however, as in order to provide the same quality of image, a reflecting surface has to be made more accurately than a refracting surface. For an objective to be of good quality, the glass of the components must be optically homogeneous, whereas a reflector system needs only to supply well-figured surfaces. The selection of the block (blank) of glass,

(a)   (b)

Figure 16.31. (*a*) One of the four 8·1 m telescopes making up the VLT array on the Paranal Mountain. (*b*) The panoramic disposition of the VLT array. (Photographs by courtesy of ESO.)

prior to the optical working, is more critical for an objective than for a mirror. Large homogeneous disks of glass are very difficult to manufacture. This is one of the reasons why all the very large telescopes are reflectors.

Despite these advantages, some experienced amateur visual observers prefer to use refractors for their measurements and it appears that refractors do perform better under the actual working conditions in the telescope dome. Perhaps the chief reason for this is that, in general, the optics of the refractor are less sensitive to the changes in temperature which occur during the night. A fall in temperature causes the usual optical materials to contract, thus causing the optical surfaces to change shape. In the case of refractors, any warping produced in the front surface is largely cancelled out by the changes in the rear surface of the lens. The images produced by an objective are usually affected only slightly by changes of temperature within the dome. However, in the case of a mirror, its optical and rear surfaces are not exposed in identical ways and, therefore, suffer different rates of temperature changes. Some of the smaller reflectors are susceptible to heat which is generated by the observer at the telescope. Strain is set up within the mirror with consequent warping of the optical surface. Under severe circumstances, some reflecting telescopes produce multiple images, each image being produced by a different zone in the optical surface. The mirror material must have a small thermal coefficient of expansion. Ordinary plate glass is poor in this respect. Pyrex, which has been used in the past, has an expansion coefficient of about one-third of that for plate glass. Many of the new larger telescope mirrors are made of materials such as Cer-Vit, with an extremely small thermal expansion.

All large telescopes, i.e. greater than 1 m ($\sim$40 inches) in diameter, are reflectors. The mechanical and optical problems involved in the design of a large telescope are more easily overcome in a reflector than in a refractor. It is obvious, for example, that the weight of the collector increases as the telescope

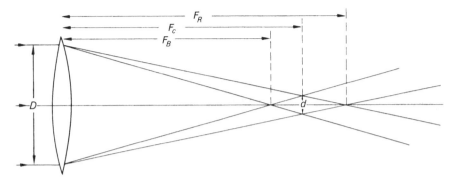

Figure 16.32. Example 16.2—circle of least confusion of a chromatic change.

size increases. The mechanical strains, due to the collector's own weight, will vary according to the point in the sky to which the telescope is directed; for large telescopes the strains are sufficient to distort the optical surfaces and spoil the images. Very little can be done in the case of the refractor as the objective can be supported only by its rim. However, in the case of a reflector, the collector can be supported over all of the rear surface. In fact, it is possible to design a system of pressure pads in the support to relieve the strain in the mirror, according to its position.

From the point of view of light-gathering power, it is not profitable to construct large refractors. As the diameter of the objective increases, so does its thickness and the amount of absorption. The ratio of the transmission to the diameter decreases as the telescope aperture increases and, thus, the law of diminishing returns applies. For reflectors with the same focal ratio but ranging in size, the fractional loss in transmission efficiency is constant. Other properties of telescope systems are discussed later in their relevant sections. Those which can be compared have been summarized in table 16.1.

Example 16.2. A 400 cm diameter lens has focal lengths in the blue and red regions of the spectrum given by: $F_B = 2995$ mm, $F_R = 3000$ mm.

(i) What is the value of the focal length corresponding to the position of the circle of least confusion?
(ii) What is the linear size of the image of a star at its focal position?

(i) By similar triangles (see figure 16.32):

$$\frac{D}{F_B} = \frac{d}{F_c - F_B} \quad \text{and} \quad \frac{D}{F_R} = \frac{d}{F_R - F_c}.$$

Dividing these identities,

$$\frac{F_R}{F_B} = \frac{F_R - F_c}{F_c - F_B}.$$

giving

$$F_c = \frac{2 F_B F_R}{F_R + F_B}.$$

Inserting the values gives

$$F_c = \frac{2 \times 3000 \times 2995}{5995}$$
$$F_c = 2997 \text{ mm}.$$

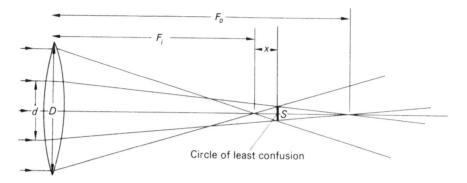

Figure 16.33. Example 16.3—circle of least confusion of a spherically aberrated image.

(ii) Again by similar triangles,

$$\frac{d}{(F_c - F_B)} = \frac{D}{F_B}$$

Hence,

$$d = \frac{D(F_c - F_B)}{F_B}$$

$$d = \frac{40 \times 2}{2995}$$

$$d = 0\cdot33 \text{ mm}.$$

Example 16.3. A lens suffers from spherical aberration. Its diameter is 200 mm and it has focal lengths of 990 mm for rays incident at the edge of the lens and 1000 mm for rays 50 mm away from the optic axis.

(i) What is the position of the plane which contains the circle of least confusion?
(ii) What is the size of the image in this plane?

(i) By similar triangles (see figure 16.33):

$$\frac{D}{F_i} = \frac{S}{x}$$

and

$$\frac{d}{F_o} = \frac{S}{F_o - (F_i + x)}.$$

Eliminating S, we have

$$\frac{d}{F_o} = \frac{Dx}{F_i[F_o - (F_i + x)]}.$$

Rearranging we have

$$x = \frac{dF_i(F_o - F_i)}{DF_o + dF_i}$$

$$= \frac{100 \times 990 \times (1000 - 990)}{(200 \times 1000) + (100 \times 990)}$$

$$= 3\cdot3 \text{ mm}.$$

The plane containing the circle of least confusion is at 99·33 mm from the lens.
   (ii) Now

$$s = \frac{xD}{F}$$
$$= \frac{33 \times 200}{990}$$
$$= 0 \cdot 67 \text{ mm}.$$

Details of many of the world's optical telescopes can be found from links listed at W 16.2. The large list also includes sites giving information on 'Telescopes' for spectral domains other than the optical region.

Problems—Chapter 16

1. The Moon has an angular diameter of 30 minutes of arc and is photographed using a telescope of 5000 mm focal length. What will be the size of the lunar image on the plate?
2. A 500 mm diameter $f/12$ telescope is used for double star measurements. What is its plate scale? Two stars are separated by 2 arc sec. What is their linear separation in the focal plane of the telescope?
3. A comet displays a tail of $15°$ and it is planned to take a photograph with a regular 35 mm camera with a picture format having dimensions 24 mm $\times$ 36 mm. Calculate the focal length of the lens that would be appropriate.
4. A 2·2 m Cassegrain telescope has a focal ratio of $f/12$. Calculate the diameter in the focal plane of the field stop which limits the field of view to 15 arc sec.
5. A star at a zenith distance of $70°$ is photographed using a reflecting telescope of effective focal length 8000 mm. Supposing that the photographic plate responds to the wavelength range 3600 Å to 6500 Å, what is the length (mm) of the image of the star? (The refractive indices of air at these wavelengths are respectively 1·000 301 and 1·000 291.)
6. The energy output from a star is such that at the surface of the Earth it provides 5×10^{-19} W m^{-2} in the visible part of the spectrum. How many photons are collected per second by a 500 mm diameter telescope?
7. A star provides 4×10^{-17} W m^2 of Hα radiation at the bottom of the Earth's atmosphere. Calculate the number of Hα photons per second entering a telescope of 500 mm diameter.
   Assuming the chief source of noise is the random nature associated with photon counting, estimate the signal-to-noise ratio of an Hα brightness measurement of the star using an integration time of 50 s.
8. A simple positive lens made of a glass with a refractive index 1·55 at a particular wavelength has a focal length of 5000 mm. What is the focal length of the lens for the wavelength for which the refractive index is 1·54?
9. A 500 mm diameter primary mirror has a focal ratio of $f/4$. It is used in a Cassegrain telescope where the secondary mirror is placed 400 mm inside the primary focus, the final image being formed in the plane of the primary mirror.
   (i) What is the effective focal length of the system?
   (ii) What is the effective focal ratio of the system?
   (iii) What is the focal length of the secondary mirror?
   (iv) If the whole area of the secondary mirror is used, what is the fraction of light lost to the collector by obscuration?
10. A Cassegrain telescope has a diameter ratio of the secondary to primary of 1/20. Assuming the mirrors to have an 75% reflection efficiency, calculate the effective area or overall optical transmittance of the system.
11. A 700 mm diameter $f/6$ primary mirror is used in a Newtonian telescope. The flat is placed so that the images are formed 100 mm outside the collected beam. What percentage of collected light is lost by obscuration due to the flat?
12. A 2 m diameter telescope is used in the infrared, the passband of the detector having a wavelength of 1·2 μm. What is the theoretical angular resolution of the telescope?

Chapter 17

Visual use of telescopes

17.1 Magnifying power

In order to make a visual inspection of the images which the telescope collector provides, some kind of eyepiece must be used. The effect of the use of an eyepiece is illustrated in a simplified way in figure 17.1, where the virtual image is at a position on or beyond the near point of the eye. (The actual final image is formed on the retina of the eye.) Under this circumstance, it can be seen that the image produced by the collector lies inside the focus of the eyepiece. The figure illustrates that, with this simple eyepiece, the viewed image is inverted. The magnifying power, m, of the overall optical system is defined as the ratio of the angle subtended by the virtual image at the eye, α_e, and the angle, α_c, subtended by the object at the collector. Thus,

$$m = \frac{\alpha_e}{\alpha_c}. \tag{17.1}$$

For astronomical observations, it is usual practice to adjust the position of the eyepiece so that the virtual image appears to be at infinity. It is obvious that in this adjustment, the eye lens must be at the exact distance of its focal length from the image produced by the collector in its focal plane. This situation is represented in figure 17.2. The collector aperture acts as the **entrance pupil** and the image of the collector aperture formed by the eyepiece acts as the **exit pupil**. The distance from the eye lens (or the last lens of a compound eyepiece) to the exit pupil is known as the **eye relief**. As all the rays from all parts of the field which can be viewed by the telescope pass through the exit pupil, it is at this position that the eye should be placed. The ray passing through the centre of the collector is known as the **chief ray**. It is not deviated at the collector aperture and it can be seen from figure 17.2 that it passes through the centre of the exit pupil.

From figure 17.2, it can be seen that

$$\tan \alpha_e = \frac{h}{v}$$

and

$$\tan \alpha_c = \frac{h}{u}$$

where h is the distance of the chief ray from the optic axis when it arrives at the eyepiece, u and v are, respectively, the object and image distances corresponding to the entrance and exit pupil distances from the eyepiece. As both α_c and α_e are small angles,

$$\alpha_e \approx \frac{h}{v} \tag{17.2}$$

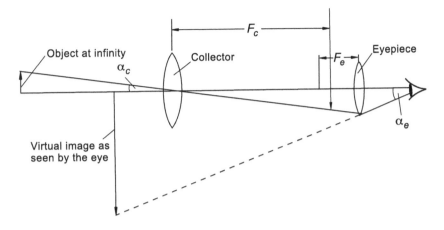

Figure 17.1. Visual use of a telescope. The collector is depicted as an objective; it could equally well be a reflecting system.

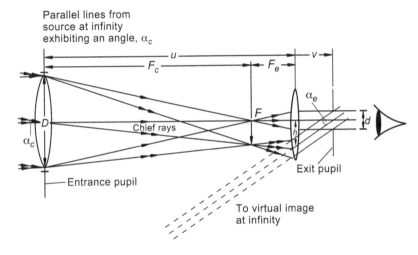

Figure 17.2. A schematic diagram of the astronomical telescope, illustrating the positions of the entrance and exit pupils.

and

$$\alpha_c \approx \frac{h}{u}.$$ (17.3)

By applying the general lens formula to the eyepiece lens,

$$\frac{1}{u} + \frac{1}{v} = \frac{1}{F_e}.$$

Therefore,

$$\frac{1}{v} = \frac{u - F_e}{u F_e}$$

and, since $u = F_c + F_e$,

$$\frac{1}{v} = \frac{F_c}{F_e(F_c + F_e)}.$$

By substituting for v in equation (17.2) and u in equation (17.3),

$$\alpha_e = \frac{h F_c}{F_e (F_c + F_e)}$$

and

$$\alpha_c = \frac{h}{F_c + F_e}.$$

Hence, the magnifying power of the telescope is given by

$$m = \frac{\alpha_e}{\alpha_c} = \frac{F_c}{F_e} \tag{17.4}$$

which is the ratio of the focal length of the collector to that of the eyepiece. Thus, in order to alter the magnifying power of the system, it is only necessary to change the eyepiece for another of a different focal length.

A further inspection of figure 17.2 shows that by considering the rays which enter the collector parallel to the optic axis, the triangle formed by the diameter of the collector, D, and the primary image at F is similar to the triangle given by the diameter of the exit pupil, d, at the distance of the eye lens, and the primary image. Thus,

$$\frac{D}{d} = \frac{F_c}{F_e} = m. \tag{17.5}$$

An alternative way to express the magnifying power of a telescope is to evaluate the ratio of the diameters of the collector (entrance pupil) and the exit pupil.

If an object subtends a certain angle to the unaided eye, the use of a telescope increases this angle at the eye by a factor equal to the magnifying power. In some cases, detail within an object cannot be seen by the naked eye, as the angles subtended by the various points within the object are too small to be resolved by the eye. By using a telescope, these angles are magnified and may be made sufficiently large so that they can be resolved by the eye at the eyepiece. A simple example of the use of the magnifying power of the telescope occurs in the observation of double stars. Many stars which appear to be single to the unaided eye are found to be double when viewed with the aid of a telescope. To the naked eye, the angular separation of the stars is insufficient for them to be seen as separate stars. By applying the magnification of the telescope, the angle between the stars is magnified and, in some cases, it becomes possible to see the two stars as separate bodies.

17.2   Visual resolving power

As described in section 16.4, the angular resolving power of a telescope may be defined as the ability of a telescope to allow distinction between objects which are separated by only a small angle. For the case of a telescope used visually, the theoretical resolving power based on Rayleigh's criterion may be written as (see equation (16.15))

$$\alpha \approx \frac{140}{D} \tag{17.6}$$

where α is in arc seconds when D is expressed in mm.

A skilled observer, however, can resolve stars which are, in fact, closer than the theoretical resolving power. In other words, the observer is able to detect a drop in intensity at the centre of the combined image which is less than 20% (see figure 16.10). In other words, if the observer is able to detect a dip in intensity smaller than 20% in a dumbbell-like image, then the two stars have been resolved at an angular separation smaller than the Rayleigh criterion. Other more practical criteria for resolving power have been proposed according to particular observers' experiences. The **Dawes'**

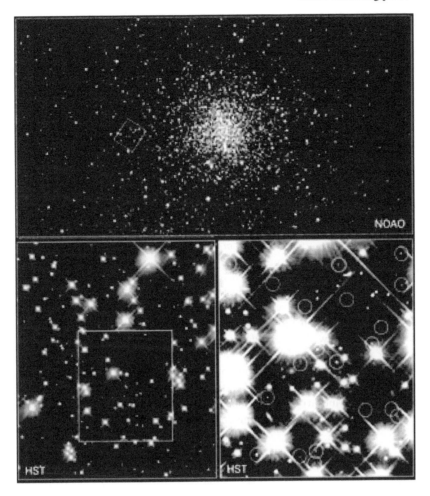

Figure 17.3. The upper frame is by a ground-based telescope of a star field. A section of this has been observed by the HST telescope. The lower right frame is further magnified and, without the effects of atmospheric seeing, clearly shows the telescope diffraction pattern imposed on the brighter stellar images. The circles indicate the positions of very faint stars. (By courtesy of STScI.)

empirical criterion gives resolving powers which are typically about 20% better than Rayleigh's theoretical criterion. Thus, to a very good approximation, Dawes' resolving power can be written as

$$\alpha = \frac{115}{D}$$

where α is in seconds of arc when D is expressed in mm.

In practice, the clarity of separation of two stars depends on many factors including the seeing conditions, the amount of scattered or background light, the relative brightnesses of the two stars and the apparent colours of the stars.

For a reflector telescope, the diffraction pattern corresponding to a point object is influenced and made more complicated by the central obscuration and the spider which supports the secondary mirror. The effects are sometimes apparent on long exposed photographs of star fields, where the very bright stars exhibit a cross-like pattern as depicted in figure 17.3.

17.3 Magnification limits

There are both lower and upper limits to the magnification which can be usefully applied with any given telescope. The first consideration for a lower limit is that all the light which is collected by the telescope should be made available for viewing by the eye; the magnification must be sufficiently great to make the exit pupil equal to or smaller than the entrance pupil of the eye (see figure 17.2). Thus, by using the definition of magnification given by equation (17.5), this condition can be written as

$$m \geq \frac{D}{d}$$

where d is the diameter of the pupil of the eye and D the diameter of the telescope aperture. Under normal observing conditions, a typical value of d is 8 mm and, thus, the lower limit of magnification is set by

$$m \geq \frac{D}{8} \tag{17.7}$$

where the diameter of the telescope is again expressed in mm. If the magnification is less than the value given by equation (17.7), some light will be lost and the full collecting power of the telescope is not being utilized. This consideration is very important, except perhaps when the Moon is being viewed.

A further lower limit to the magnification is set by the resolving power of the eye. If any detail in a complex object is to be viewed, the angular size of its image must be larger than the eye's resolving power. This latter quantity depends, to a great extent, on the observer but a typical value can be taken as one minute of arc. If a certain resolution has been achieved by using a particular telescope, the magnification must have a value which is sufficient to allow the resolved details to be seen by eye. Thus, by matching the resolution of the telescope (equation (17.6)) to that of the eye (≈ 60 arc seconds) by means of magnification, a lower limit of magnification is set by

$$m \geq \frac{60D}{140}$$

or

$$m \geq \frac{3D}{7}.$$

As the value taken for the resolving power of the eye favours an extremely good eye, the lower limit of magnification can be conveniently approximated to

$$m \geq \frac{D}{2} \tag{17.8}$$

where the diameter of the telescope, D, is again expressed in mm.

The useful magnification of a telescope cannot be increased indefinitely. The upper limit is set by the impracticability of making eyepieces with extremely short focal lengths, by the quality of the optics of the collector and by the fact that the ability of the eye to record good images deteriorates when the beam that it accepts becomes too small. For the image to be seen without loss in the quality of the eye's function, the exit pupil must be larger than 0·8 mm. By using equation (17.5) for the expression defining magnification,

$$m \leq \frac{D}{0 \cdot 8} \tag{17.9}$$

where, again, the diameter of the telescope is in mm. Comparison of equations (17.7) and (17.9) shows that from a consideration of the size of exit pupil, the magnifying power of any telescope has a useful range covered by a factor of ten.

According to an empirical relation known as **Whittaker's rule**, deterioration of a viewed image sets in when the value of magnification exceeds the diameter of the telescope, D, expressed in mm. Using this rule, $m \leq D$.

The lower and upper limits discussed earlier do not pretend to be hard and fast rules but serve only as guides. The limits of magnification also depend on the type of object that is being viewed. It may be possible, for example, to use a magnification given by $2D$ when the object is a double star. Magnifications greater than D (Whittaker's rule) can be used occasionally, especially with telescopes of small aperture. In fact, it has been found that in the case of double star observations with telescopes of small-to-medium aperture, the upper limit for magnification does not have a linear relationship with the telescope's diameter. According to Lewis, the upper limit is given by

$$m \leq 27\cdot 8\sqrt{D}. \tag{17.10}$$

Inspection of equation (17.10) shows that for telescopes of small aperture, magnifications close to $2D$ can be used but for medium-sized telescopes, the upper limit of magnification is reduced to values which are closer to D. The nonlinear dependence of the upper limit of magnification on the aperture of the telescope probably results from the way in which the telescope image is distorted by the seeing effects of the Earth's atmosphere. The appearance of an image distorted by seeing conditions depends on the aperture of the telescope (see section 19.7.3). In many cases, the upper limit of magnification is set by the seeing conditions and these vary greatly from site to site and from night to night.

17.4   Limiting magnitude

The amount of energy collected by any aperture is proportional to its area and, therefore, to the square of its diameter. In the case of the eye, the sensitive area responds to the energy which is accepted by the pupil and, accordingly, there is a limit to the strength of radiation that can be detected. For starlight, the limit of unaided eye detection is set at about sixth magnitude. By using a telescope, with its greatly increased aperture over the pupil of the eye, it should be possible to record stars which are much fainter than sixth magnitude.

The brightness of the sixth magnitude star corresponds to the arrival of a certain amount of energy per unit area per unit time. Thus, if the star of naked eye brightness B_e is viewed by a telescope, its apparent brightness, B_t, will be given by

$$\frac{B_t}{B_e} = \frac{D^2}{d^2} \tag{17.11}$$

where D and d are the diameters of the telescope and eye pupil respectively, provided that the telescope is used with a sufficient magnification so that all the collected light enters the eye. If m_e and m_t correspond to the magnitude of a star as seen by the naked eye and by the telescope, respectively, the apparent difference in magnitude can be obtained by using Pogson's equation (see equation (15.12)). Thus,

$$\log_{10}\left(\frac{B_t}{B_e}\right) = -0\cdot 4(m_t - m_e)$$

and, by using equation (17.11), this becomes

$$\log_{10}\left(\frac{D^2}{d^2}\right) = -0\cdot 4(m_t - m_e)$$

which may be re-written as

$$m_t = m_e - 5\log_{10}\left(\frac{D}{d}\right). \tag{17.12}$$

By letting m_t correspond to the faintest star that can be detected by the telescope–eye combination (i.e. $m_t = 6$), then the value obtained for m_e via equation (17.12), corresponds to the original magnitude of the star. This value is known as the **limiting magnitude**, m_{lim}, of the telescope and, if a typical value of $d = 8·0$ mm is used, it can be expressed as

$$m_{lim} = 6 - 5\log_{10}(8) + 5\log_{10} D$$
$$m_{lim} = 1·485 + 5\log_{10} D. \tag{17.13}$$

Thus, for a telescope of 500 mm, the theoretical limiting magnitude is approximately 15·0.

In this discussion it has been assumed that the transmission efficiency of the telescope is perfect. A typical figure for the efficiency is about 0·65 (including losses in the eyepiece) and, to allow for this, equation (17.11) should be corrected so that it becomes

$$\frac{B_t}{B_e} = (0·65)\frac{D^2}{d^2}.$$

Consequently, equation (17.12) should be corrected to

$$m_t = m_e - 5\log_{10}(0·81)\frac{D}{d}$$

and, by rounding off the figures, the equation expressing the limiting magnitude of the telescope may be written as

$$m_{lim} = 6 + 5\log_{10}\left(\frac{D}{10}\right) \tag{17.14}$$

where D is expressed in mm. Thus, in practice, the limiting magnitude for a 500 mm telescope is likely to be about 14·5. Half a magnitude has been 'lost' due to the telescope's imperfect transmission. Equation (17.14) is again not a hard and fast law, however, as each telescope must be treated individually and also the limiting magnitude will depend on the observer to some extent.

17.5   Eyepieces

In the previous discussion on the magnification of the telescope, it has been assumed for simplicity that the eyepiece, or ocular, is in the form of a single positive lens. It is obvious that one of the drawbacks of using a single lens for an eyepiece is that the viewed image would suffer from chromatic aberration. In practice, an eyepiece is generally in the form of a combination of lenses and there are several types which are frequently used; and perhaps the most traditional are the **Huygens** eyepiece and the **Ramsden** eyepiece. These eyepieces are constructed of two simple separated lenses made of the same material. The two-lens system serves two purposes: it allows a field of increased size to be viewed and also allows compensation for chromatic aberration. The ability of such a system to overcome chromatic aberration is demonstrated here.

If two lenses of focal length F_1 and F_2 are separated by a distance, x, the focal length of the combination, F, is given by

$$\frac{1}{F} = \frac{1}{F_1} + \frac{1}{F_2} - \frac{x}{F_1 F_2}.$$

Hence,

$$F = \frac{F_1 F_2}{F_1 + F_2 - x}. \tag{17.15}$$

By expressing the individual focal lengths in terms of the shape of each lens and their refractive powers (see equation (16.14)) and by letting the lenses be made of the same type of glass so that $n_1 = n_2 = n$,

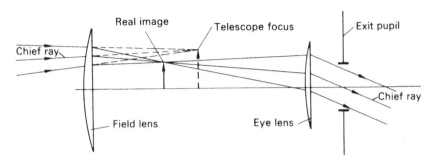

Figure 17.4. The Huygens eyepiece.

equation (17.15) can be re-written as

$$F = \frac{K_1 K_2}{(n-1)(K_1 + K_2) - x(n-1)^2}. \qquad (17.16)$$

The focal length of the combination, F, is independent of wavelength, i.e. free from chromatic aberration, if $dF/d\lambda$ is zero. For a given pair of lenses, K_1 and K_2 are constant so that this condition can only be obtained if $dF/dn = 0$. By differentiating equation (17.16) and putting $dF/dn = 0$, it follows that

$$x = \frac{F_1 + F_2}{2}. \qquad (17.17)$$

Thus, for an achromatic system involving two positive lenses made of the same material, their separation must be equal to half the sum of their focal lengths.

The optical arrangement of the Huygens eyepiece is depicted in figure 17.4. The first lens of the system is known as the **field lens** and the second is known as the **eye lens**. The ratio of the focal length of the field lens to that of the eye lens is usually in the range 1·5:3·0. It can be seen from figure 17.4 that the field lens lies inside the focal plane of the telescope and that it forms a real image between the two lenses of the eyepiece, at the focus of the eye lens. The purpose of the field lens, as its name implies, is to increase the field of view. When the eyepiece is in the correct position, the light rays emerge from the eyepiece in a parallel bundle and the viewed image appears to be at infinity.

If it is desired to view the field against a cross-wire or graticule, it should be located between the lenses at the focus of the eye lens. Thus, the observed images and graticule will be in focus simultaneously. Although the combined lens system corrects for chromatic aberration in the observed images, the image of the graticule is obtained by using only the eye lens and, consequently, this image suffers from coloured effects. The Huygens eyepiece also suffers from spherical aberration, astigmatism and pincushion distortion. These aberrations are very noticeable if the eyepiece is used with a telescope of focal ratio smaller than about $f/12$. The exit pupil corresponds to the image of the collector which is produced by the lens combination and is close to the eye lens. The small eye relief (see section 17.1) sometimes makes this type of eyepiece difficult to use.

In the Ramsden eyepiece, the two lenses have the same focal length and to compensate for chromatic aberration (see equation (17.17)), their separation, therefore, should be equal to the value of this focal length. If this were to be done rigorously, the eye lens would only produce a virtual image at infinity when the field lens is placed in the focal plane of the telescope. Any graticule would also be required to be placed on the field lens and dust on the same lens would be in focus with the images under inspection.

At the expense of introducing a small amount of chromatic aberration, the two lenses of the Ramsden eyepiece are brought together so that their separation is just less than their focal lengths. The arrangement of the two lenses is depicted in figure 17.5. It can be seen that the field lens is situated

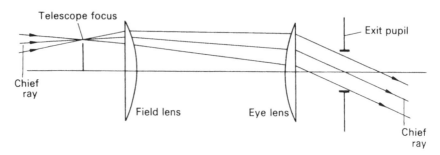

Figure 17.5. The Ramsden eyepiece.

outside the focal plane of the telescope and it is an easy matter to arrange to have a graticule in this same plane. The Ramsden eyepiece is superior to the Huygens in that the aberrations are much smaller and it is more convenient to use because of its greater eye relief.

By using a doublet instead of a single lens for the eye lens, it is possible to utilize the desirable qualities of the Ramsden eyepiece and, at the same time, remove the chromatic aberration. This type of ocular is known as a **Kellner eyepiece**.

In the case of the single element eyepiece, it is a simple matter to measure its focal length and the expected magnification, when used with a particular telescope, can be obtained from knowledge of the focal length of the telescope. It is usually more difficult to make a direct measurement of the effective focal length of a compound eyepiece and the magnification that it produces is best measured when it is used on the telescope. A simple way to do this is to point the telescope at the daytime blue sky and measure the diameter of the parallel beam (exit pupil) emerging from the eyepiece. Application of equation (17.5) allows the magnification to be determined.

By joining simple ray diagrams, illustrating the effect of each of the optical components, it is easy to demonstrate that an astronomical telescope, using one of the previously described eyepieces, provides an inverted image. A more complicated eyepiece giving an upright field (a terrestrial eyepiece) is of no real advantage to the astronomer. In fact, its extra lenses lose more light.

17.6   Micrometer eyepieces

By measurement of the physical separation of images in the focal plane of a telescope, it is a simple matter to convert the records into angular measure on the celestial sphere (see equation (17.2)). One way of doing this directly at the telescope was by the use of specially designed eyepieces. Although no longer in commission, such micrometer eyepieces had their main application in the measurement of the separations of visual binary stars.

The micrometer eyepiece would perhaps be of Ramsden design but, in the plane that coincides with the telescope focal plane, a system of adjustable cross-wires was placed. There are many different arrangements in the design of the cross-wires. However, the main feature is that one of them, or sometimes two, could be adjusted relative to the others. The adjustment was made by a micrometer screw and the amount of movement recorded on the scale or drum which was attached to the side of the eyepiece. The eyepiece, together with the cross-wire system, was fitted into a rotatable tube and the orientation of the cross-wires could be altered. The essential components of the micrometer eyepiece are shown schematically in figure 17.6, which also shows the positions of the cross-wires when they have been adjusted for a determination of the separation of two stars.

In a typical arrangement, as shown in figure 17.6, two wires are fixed exactly at right angles to each other with their intersection at the centre of the field of view. The orientation of the two wires is conveniently displayed by placing a pointer along the direction of one of the wires and allowing this

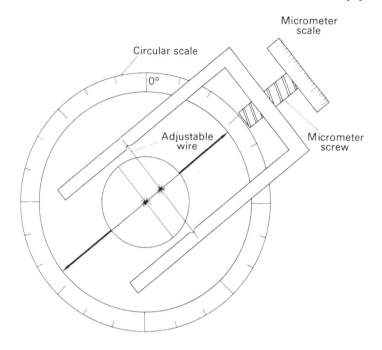

Figure 17.6. A micrometer eyepiece set on two stars.

pointer to be swept over a circular scale as the eyepiece is rotated. A third wire is set into a yoke so that it is parallel to the wire which is not carrying the pointer. Its position is made adjustable by allowing the yoke to be driven by a micrometer screw.

The measurement of a double star could be performed as follows. After the position of the telescope has been adjusted so that one of the stars corresponds to the intersection of the fixed wires, the eyepiece is rotated so that the wire which is attached to the orientation pointer is made to pass through the two stars. In this way, the angular relationship between the stars and the coordinates on the celestial sphere are determined. The adjustable wire is first placed on the star of the centre of the field and then displaced so that it lies on the image of the second star. The difference in these positions of this wire is known in terms of the movement of the micrometer screw and, by means of a calibration, the separation of the two stars is determined in angular measure.

17.7 Solar eyepieces

Although they are no longer in general use, special eyepieces have been designed for viewing the Sun.

It cannot be emphasized too strongly that the Sun should NEVER be viewed by an ordinary telescope system, no matter how small the collector is. Without special precautions, PERMANENT DAMAGE can be inflicted on the eye, even with only the briefest glimpses of the solar image.

It is also *insufficient* and *dangerous* to reduce the intensity of the image by placing a filter over the eyepiece; the heat that is generated in the filter can cause it to crack.

One of the special arrangements relies on partial reflection to reduce the brightness of the solar image. In such a system, only about 5% of the light is allowed to enter the eyepiece; the remaining 95% is usually reflected out of the system. In some designs, the solar image is viewed by the eyepiece in a direction at 90° to the telescope's optic axis and this attachment is called a sun diagonal.

For small telescopes, coated mylar disks may be fitted over the telescope aperture and these are commercially available. An alternative and convenient way of viewing the solar image produced by a small telescope, using a conventional eyepiece, is that of **projection**—see section 24.5.2.

Problems—Chapter 17

1. An $f/10$ telescope with a focal length of 3000 mm is used with an eyepiece of 2 mm focal length. What is the magnifying power of the system? Would the magnification allow all the collected light to enter the pupil of the eye?

2. The star with the largest apparent diameter subtends an angle of the order of 10^{-7} radians. What size of telescope would be required to resolve the star as a disc? (Assume an effective wavelength of 5500 Å.)

3. An $f/10$ telescope has a diameter of 500 mm. What is the resolving power of the telescope and what is the size of the Airy disc in the focal plane? (Assume an effective wavelength of 5500 Å.)

4. An $f/10$ achromatic telescope has a diameter of 500 mm. The images of a star corresponding to the blue and red ends of the spectrum are separated by 0·05 mm along the telescope's axis. How does the size of the circle of least confusion compare with the theoretical size of the Airy disc? (Take the effective wavelength as being 5500 Å for the Airy disc calculation.)

5. An $f/12$ reflecting telescope with a diameter of 300 mm suffers from spherical aberration. The longitudinal spread of the image is given by $\Delta F/F = 0·001$. How does the size of the circle of least confusion compare with the theoretical size of the Airy disc? (Assume an effective wavelength of 5500 Å.)

6. Given two refracting telescopes A and B, the diameter of the objective of A being 1000 mm, calculate the diameter of the objective of B if the limiting magnitudes of A and B are $+16^m40$ and $+15^m40$ respectively. Neglect any light losses. Calculate the limiting magnitudes of A if one-fifth of the incident light is lost.

7. Calculate the diameter of a telescope in which the star ζ CMi (5^m11) would appear as bright as *Sirius* -1^m58 to the unaided eye. (Assume that the diameter of the pupil of the eye is 8 mm and that 30% of the light incident on the telescope collecting area is lost.)

8. Tycho's supernova of 1572 attained its greatest brilliance at magnitude -4 on approximately October 1st and thereafter began to fade, finally becoming invisible to the unaided eye on approximately March 1st 1574. If the apparent brightness b at a time t after October 1st 1572 is related to the apparent brightness b_0 on that date by the equation

$$b = b_0 10^{-t/\tau}$$

where τ is a constant to be determined, find the last date (to the nearest month) on which observers could have seen the nova if they had had a 150 mm telescope available.

9. A telescope with a focal length of 6000 mm is used to make double star observations. A turn of the micrometer screw causes the cross-wires to alter their separation by 0·1 mm. When a particular double star is observed, the setting of the cross-wires corresponds to 3·72 turns of the micrometer. What is the angular separation of the two stars?

Chapter 18

Detectors for optical telescopes

18.1 The optical spectrum

We have seen earlier that the Earth's atmosphere has a window which allows transmission of electromagnetic radiation with a range of frequencies, the centre of the band being close to the peak sensitivity of the eye. The eye is not sensitive to all of the frequencies which arrive at the bottom of the atmosphere in this band. Energy in the form of near ultraviolet and infrared radiation is arriving from space and penetrates down to ground-level but is not sensed by the eye. It can, however, be collected by ordinary optical telescopes and made available for measurement by other detectors. There are also telescopes which are designed with high efficiency for infrared measurements.

For convenience in the following discussion, we shall call this window which includes the visible wavelengths the **optical window** and the range of frequencies which it covers the **optical spectrum**. The term **visible spectrum** is reserved to describe the range of frequencies which can be detected by eye.

Before discussing three important types of detector which are used for making measurements in the optical spectrum, it will be useful to consider the concepts of spectral sensitivity.

18.2 Spectral sensitivity

Any detector is not equally sensitive to energy of all wavelengths. By this we mean that if a series of recordings is made whereby the detector is illuminated by radiation from different spectral wavebands but each containing equal amounts of energy, its recorded response varies according to the position of the waveband in the spectrum. The way in which the detector responds to equal amounts of energy at different wavelengths is described by its spectral sensitivity curve, $S(\lambda)$. This curve of **spectral response** is frequently depicted in a relative way by setting the maximum value of $S(\lambda)$ equal to unity.

18.3 Quantum efficiency

Besides knowing the spectral response of any detector in a relative way, it is important to know how efficient it is in producing a response in relation to the quantity of radiation which falls on to it. This can be conveniently expressed in terms of its quantum efficiency.

In many cases where light interacts with matter, such as in a detector, the interaction can only be described by assigning a 'particle' nature to the radiation. This description requires that a beam of radiation be made up of discrete packets of energy called **photons** or **quanta**. As presented in earlier chapters, the energy, E, carried by each quantum is given by

$$E = h\nu$$

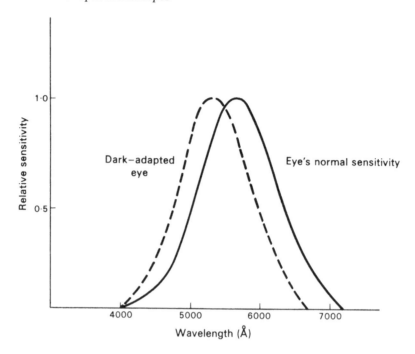

Figure 18.1. The relative spectral sensitivity of an average human eye; the dotted curve illustrates the Purkinje effect, showing how the sensitivity curve is modified when the eye is dark-adapted.

where h is Planck's constant, equal to 6.63×10^{-34} J s and ν is the frequency of the electromagnetic wave. It is easy, therefore, to perform the conversion between the amount of energy in a beam, given in watts for example, to the number of photons s^{-1} which are travelling with the beam—as done, for example, in section 16.3.1.

From the size of response of a detector to a beam of light whose energy is known in terms of photons s^{-1}, it is possible to evaluate the fraction of the photons which are effectively used to provide the response. The ratio of the number of photons present in the beam to the number which contribute to the detector's response defines the **quantum efficiency** of the detector.

18.4   The eye as a detector

Very briefly, the main elements of the eye consist of a pupil which controls the amount of light which enters it, a lens and a photosensitive surface called the **retina** on to which the lens focuses the images. The brain converts the sensations of the retina to give the observer the impression of there being objects set out in space.

There is wide variation in the different properties of the eye according to the individual and the discussion here is related to that of an average observer. The daytime spectral response or visibility curve of an average eye is illustrated in figure 18.1. It can be seen that the eye is capable of covering the optical range from 4000 to 7000 Å and this range is known as the **visible spectrum**. The eye is unable to store or make a permanent record of any image. The image it produces can be considered to be instantaneous. If the light-level is below a certain **threshold**, then the eye does not respond. When the eye has been working in conditions of normal illumination and is then removed to a darkened environment, such as might be expected for night observations, its sensitivity undergoes changes, lowering the threshold of visibility. These changes take about half an hour before the sensitivity settles to its improved value and, in this condition, the eye is said to be **dark-adapted**.

There are two causes which give rise to dark-adaption. One of them is due to the automatic expansion of the pupil, so allowing a larger collecting area for the incoming radiation. The other, providing the greater effect, is due to the biochemical changes occurring in the retina itself.

Under dark-adapted conditions and with a little practice, it is apparent that the eye's sensitivity depends on the direction of the object. The minimum of sensitivity occurs when the eye is looking straight ahead and, by using **averted vision**, it may be possible to detect faint objects which disappear when looked at directly.

Dark adaption also gives rise to changes in the spectral sensitivity of the eye. The peak in sensitivity moves by approximately 500 Å towards the blue end of the spectrum and the eye loses its sensitivity to red wavelengths. This change is known as the **Purkinje effect** and the dotted curve in figure 18.1 illustrates an average spectral sensitivity under dark-adapted conditions.

Prior to the application of the photographic plate, photoelectric detectors and CCD devices, the eye was used to judge brightness differences between astronomical objects. The values of differences which might be detected depend on several factors such as the colours of the objects for the comparison and their absolute brightnesses. With practice, brightness differences of a few per cent can be detected. It is not really convenient to assign a quantum efficiency to the eye but by considering its ability just to be able to detect a sixth magnitude star, it can be said that it is necessary to receive a few hundred photons per second to register a star image. The limiting magnitude of the telescope–eye combination has already been discussed in section 17.4.

It is difficult to give a hard and fast rule for the resolving power of the eye as it depends critically on the type of observation which is being attempted. However, under ordinary circumstances, the average eye is able to resolve angles of one minute of arc. This figure corresponds (see equation (17.6)) to the resolving power which might be expected by an aperture of the size of the pupil ∼2·5 mm, a typical diameter when operating in an everyday environment, and to the sizes of the detector elements which make up the retina. Under special circumstances, the eye has special properties of high **vernier acuity** and **symmetry judgment** allowing even smaller angles to be resolved. It is capable of resolving a break in a line which might correspond to an angular difference of ten seconds of arc. This ability is put to use in measuring instruments whose settings are read off a vernier scale. Although such instruments are now no longer used on a telescope, they may still be found on ancillary equipment which might be employed to analyse astronomical data recorded, say, on a photographic plate. The continuing advance of new techniques and automation, however, will eventually displace the use of the eye for all data reductions. As well as taking advantage of the eye's symmetry judgment in general measuring instruments, it was this ability which made the eye useful for making classical double-star measurements on the telescope.

18.5 The photographic plate

18.5.1 Introduction

As soon as the photographic process was developed, it was immediately applied to the skies and has been one of the chief ways of recording brightness and positional information associated with starfields, galaxies and spectra since the middle of the 19th century. With the advent of solid state detectors (see later), its role has diminished by a large degree in the last 20 years but continues to be used in some areas of study particularly when large areas of the sky are being surveyed (see the Schmidt telescope—section 20.4). It may be noted that research on the production of improved emulsions continues.

Photographic material consists of an emulsion of silver halide (mainly bromide) crystals in gelatin, which is attached to a glass plate or celluloid sheet in a thin uniform layer. The exact mechanism occurring within the emulsion when it is illuminated is not fully understood but the interaction results in the production and accumulation of silver ions wherever the light was incident on the emulsion. Exposure of the emulsion to optical radiation does not alter its appearance to a visual inspection. The distribution of the illumination of the exposure over the surface of emulsion is not immediately apparent. Any images which have been recorded are, in the first instance, latent. The emulsion must be processed before the image is available for analysis.

In order to produce and retain any image in the emulsion, the exposed plate must undergo two processes: the images must be first **developed** and then **fixed**. This is carried out away from the telescope in a photographic darkroom. Development is performed by immersing the emulsion in a liquid chemical reducer which converts the silver ions to individual silver grains. After development, the emulsion contains a distribution of silver grains, matching the images that were imposed on it, and the remaining silver halide crystals which have been unaffected and which are, of course, still capable of responding to any further illumination. Fixation of the images is, therefore, obtained by the removal of the remaining sensitive silver halide from the emulsion, and this again is achieved by using chemicals in solution. Before the emulsion is dried, all traces of the fixer material are removed by washing in water.

There are many kinds of photographic plate available and it is important that the correct type should be chosen according to the image information which is to be recorded. The chief parameters that have to be considered are those of **spatial resolution, speed** and **spectral sensitivity**.

18.5.2   Spatial resolution

If an exposed plate is examined under a microscope, it will be seen that the images are made up of individual blackened grains. Any seemingly sharp edge in the image is fuzzy on the scale which depends on the grain size. Microscope views of star plates reveal that each image is made up of an agglomeration of grains with a central condensation; the edges of images are not well defined. It is the grain sizes in the emulsion which normally determine the limit of any spatial resolution. The large choice of emulsion types embraces plates covering a wide range of graininess and, hence, **resolving power**. Quantitatively, the resolving power of any particular type of plate is determined by forming the image of a grid which has numbers of black and white ruled lines on it at different separations. The resolving power, R, is defined as the highest number of ruled lines per mm that can be resolved. The various emulsion types give a range in R from about 50 to 1000. In any photographic observation, it is important that the value of R is chosen so that all the detail which is provided by the telescope and other optics is recorded without any deterioration.

Because of the large plate formats that are available, the photographic process scores well when large amounts of spatial information need to be recorded. Each collection of grains within an area equivalent to $\sim 1/R^2$ acts as a resolved picture element or **pixel**. Thus, for a plate measuring 100 mm × 100 mm with a value of $R = 100$, the plate contains

$$\frac{100 \times 100}{(\frac{1}{100})^2} = 10^8 \text{ pixels}$$

an exceedingly good number in comparison with other detectors.

18.5.3   Speed

The time of exposure required to obtain an image of a certain strength of a particular astronomical object with a given telescope varies according to the emulsion type that is used. A plate or film is said to have a certain **speed**, designating the amount of energy that needs to fall on to the emulsion to produce a good image.

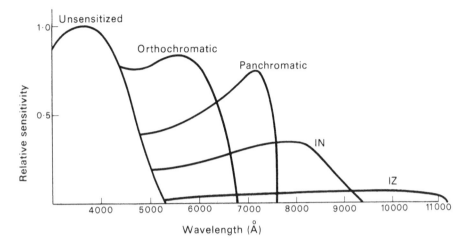

Figure 18.2. Relative spectral sensitivity of typical unsensitized, orthochromatic, panchromatic and infrared (IN and IZ) emulsions.

Obviously, it is important that any observation should be recorded in the shortest possible time by choosing a plate with a fast emulsion. However, the choice of speed has to be considered in relation to criteria which may be set by other requirements of the measurement.

It is usually the case that as the speeds of emulsion increase, so does the graininess—it is impossible to have a film with a high resolving power and fast speed. If, therefore, a particular measurement requires an emulsion with a high resolving power, then any notion of making the observation with a short exposure time may need to be sacrificed.

Practically all astronomical photography is performed by using long exposures. During the exposure, the whole 'picture-area' is receiving the general scattered light from the sky background and from the optical system at the same time as the image of importance is being recorded. Thus, any image is recorded against a background which is generally termed the **fog**. Any image can only be seen well when its strength is significantly above the fog-level. The ability of plates to have low fog-levels varies according to the type of emulsion but it is usually the case that fast emulsions are susceptible to fog. If a particular measurement requires a faint object to be recorded, it would be thought that this could be done most quickly by using the fastest film. If, however, the object is to be recorded with maximum contrast against the fog, it may not necessarily follow that the fastest emulsion is the correct choice. A slower emulsion and longer exposure time may produce a better result.

18.5.4 Spectral sensitivity

Photographic emulsions are available for the whole of the optical spectrum except the infrared region beyond 12 000 Å. The basic sensitivity of the silver bromide runs from beyond the ultraviolet to about 5000 Å but, by adding various dyes to the emulsion, the spectral coverage can be improved. The relative spectral sensitivities of the major classes of emulsion types are illustrated in figure 18.2.

18.6 Photographic photometry

Besides its obvious role of being effective in providing records of the relative positions of celestial objects and providing a means of measuring these positions, the photographic process enables objective photometric studies to be undertaken. The strengths of the images on a processed plate depend in some way on the original brightnesses of the objects that are recorded. For photometric work, it is the astronomer's task to decipher the strengths of the images in terms of the original energy which caused the production of the images. The process is carried out by measuring the strength of an image and relating it to an amount of incident energy by means of a calibration curve.

The most convenient way of measuring the strength or blackening of a photographic image is to pass a beam of light through the processed plate and measure the intensity of the beam after its passage through the plate. The laboratory instrument designed especially to make such measurements is usually called a **microphotometer** and the transmitted beam's strength is normally monitored photoelectrically. If I_0 is the original intensity of the beam and I its intensity when it is passing through a particular area in the plate, then the **transmittance**, T, of the plate at the selected area is given by

$$T = \frac{I}{I_0}.$$

The **opacity**, O, of the plate is defined as

$$O = \frac{I_0}{I} = \frac{1}{T}.$$

When relating the strengths of photographic images, it is usual practice to use the term **density**, D, which is defined as the logarithm of the opacity. Hence,

$$D = \log_{10} O$$

and, therefore,

$$D = -\log_{10} T.$$

The calibration curve relating the densities of areas on the plate to the original energies producing the blackening is normally represented by plotting D against the logarithm of some known exposures, where the exposure, E, is defined as the illumination of the image, $\mathcal{I}$, multiplied by the time, t, for which the energy falls on to the plate. This form of calibration curve is known as the **characteristic curve** of the emulsion. A typical characteristic curve is illustrated in figure 18.3.

It can be seen from figure 18.3 that unless the exposure is above a threshold, any image cannot be seen against the fog-level. After the toe of the curve, the characteristic curve becomes linear, making it relatively easy to describe density values in terms of incident energy. The relationship can be expressed in the form

$$D = A + \gamma \log \mathcal{I}t = A + \gamma \log E. \tag{18.1}$$

The slope, γ, of the linear part of the characteristic curve is called the **gamma** of the emulsion and it is a measure of the way points within an extended image are contrasted. Beyond the linear part of the curve, the slope begins to decrease and the curve levels out. In this region, exposure to further amounts of energy gives rise only to small increases of blackening and the plate is said to be **over-exposed**.

Between the limits of under- and over-exposure, the photographic plate has a useful **dynamic range** giving scope to cover illuminations differing by a factor of just over 100. What this means is that the magnitude range that can be covered on any single exposure is of the order of five to six. If, for example, an exposure is made such that 12th magnitude stars are just detectable, stars brighter than the sixth will be overexposed. Fainter magnitudes might be recorded by extending the exposure, but the zone of over-exposure will also move to fainter stars.

Comparisons of characteristic curves allow the best emulsions to be chosen for particular observations. We can consider a simple example of their use by examining figure 18.4, where the toes of two characteristic curves, Y and Z, are illustrated. If it is required to use the least exposure to detect an object, then it is easy to see that emulsion Y would be chosen. If it is required that a density of *one* or over should be recorded using the least exposure, then it is obvious that emulsion Z would be chosen.

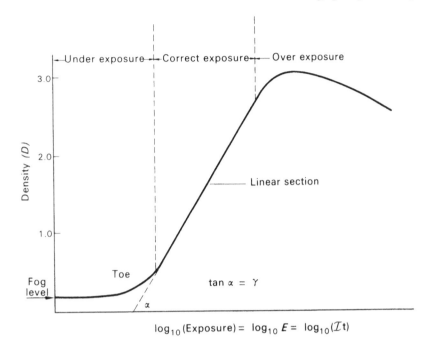

Figure 18.3. The characteristic curve of a typical emulsion, illustrating the regions of under exposure, correct exposure and over exposure.

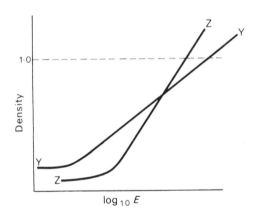

Figure 18.4. A comparison of the characteristic curves of two emulsions.

When photographic photometry is being undertaken, it is important that the plates should have calibration exposures imposed on them so that they receive exactly the same treatment in the development process as the images of interest. Because of the faintness of most celestial objects, their measurement by photography requires long exposure times and, in contrast, the calibration exposures are normally made by using a relatively bright illumination and short exposure times. It is possible, with care, to arrange that the total energy collected in a certain part of an astronomical plate is matched exactly by that which is used to provide one of the calibration exposures. If the densities of these images are compared, however, it is very likely that they will not be the same. The photographic plate is said to suffer **reciprocity failure**—the degree of blackening depends not just on the total amount

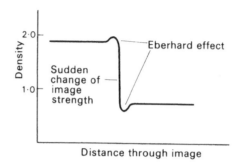

Figure 18.5. A density scan through a recorded image with a sharp change in strength illustrates the Eberhard effect.

of incident energy but on the rate at which it arrives. In effect, reciprocity failure means that there are different characteristic curves available according to how the calibration exposures are imposed so that, before any photometry can be attempted, the characteristic curve, as obtained via the calibration exposures, must be corrected for reciprocity effects. On this account alone, photographic photometry is open to systematic error. By a similar argument, the brightnesses of astronomical objects recorded on two different plates cannot be compared directly if the exposure times are different.

It is also important to consider the effects of reciprocity failure when an emulsion is being chosen for a particular type of photograph. For example, it might be thought that the fastest possible plate should be chosen to record faint stars in the shortest possible time. Now the speed of a film is normally determined according to how it behaves under the circumstances of everyday photography with very short exposures. If the reciprocity failure is severe, as it usually is for everyday fast emulsions, the effectiveness of the plate with long exposures can be so reduced as to be inferior to other plates with small reciprocity failures but with lower everyday photographic speeds.

There are several other effects besides reciprocity failure which can lead to distorted measurements unless they are allowed for. The following two result from the development process itself, the first affecting photometry and the second, positional measurements.

If an object has detail in it which is recorded on the photographic plate, during the process of development the chemical reactions will be under way at different rates, according to the position in the image. In regions where the image is most dense, the reaction is most speedy; however, in other regions, the development will proceed at a lower rate. Because of this, there will be localized variations in the rate at which the developer is being used and, therefore, in its strength. Unless they are disturbed adequately by agitation, these variations will control the development rate to some extent. The effect of unequal development rates is usually manifest where there are sudden changes in the density of an image, causing distortions of density values. It is known as the **Eberhard effect** and it is illustrated in figure 18.5.

Positional measurements are obviously affected if there are geometric distortions in the gelatin of the emulsion and this can occur in the development process. During development, the gelatin becomes tanned and the extent of the tanning process increases where the images are strongest. The effect of tanning is to make the emulsion dry quicker and this sets up strains as the tanned areas shrink. In the simple case of a double star measurement, the shrinkage causes the star images to move together and, if the effect is not allowed for, systematic errors may result.

18.7   Photographic efficiency

Perhaps the chief advantage of the use of the photographic plate in astronomy is its ability to record many picture points simultaneously. It also has the ability to integrate the energy which is incident on it and to build up images of objects which would otherwise remain unseen. In these senses, it can be said to be efficient.

The photometric accuracy which can be obtained depends on many factors but it is rarely better than 5%. Most photographic measurements provide photometry with an accuracy in the range 5–10%.

A major drawback of photography is that its quantum efficiency is very low. The interaction of photons with the silver halide crystals in the emulsion provides blackened grains on development of the plate and their distribution gives rise to the image. Each blackened grain corresponds to the response of emulsion to energy which is arriving at a particular point. The response, therefore, behaves in a quantized way: a certain number of recorded events (blackened grains) are obtained according to the amount of energy or number of photons which are incident on the plate. The quantum efficiency, defined as the ratio of the average number of blackened grains to the number of incident photons, is typically about 0·001 or 0·1%.

In order to have some appreciation of the absolute sensitivity of photographic emulsion, it is useful to remember the adage that when using some particular equipment,

**what the eye can see can be photographed
with an exposure of a few minutes.**

Now a star image can be recorded with certainty if its image is the result of an accumulation of a few tens of blackened grains, say fifty. By using the value of the quantum efficiency given earlier, the number of photons required to produce this faintest detectable star image is equal to 5×10^4. We have already seen that the eye can detect a star image if it is receiving of the order of 200 photons s^{-1}; therefore, the exposure time required to match the detectivity of the eye is given by

$$\frac{5 \times 10^4}{200} \text{ s} \approx 4 \text{ min}$$

so confirming the adage. Exposure times longer than a few minutes will, in general, allow stars to be recorded by a telescope which are too faint to be seen by eye using the same telescope.

18.8   Limiting magnitude

To a first approximation, a star image on a photographic plate can be considered to be a point. The total energy which is collected into the point is proportional to the apparent brightness of the star, the area of the collecting aperture and the integration time of the exposure.

We can, therefore, estimate the magnitude of the faintest star which can be recorded photographically in the following way.

The eye needs to receive approximately 200 photons s^{-1} in order for it to sense an image. By assuming that the pupil of the dark-adapted eye has a diameter of 8 mm, the energy arrival rate, E_e, is given by

$$E_e = 200 \times \frac{4}{\pi \times 8^2} \text{ photons s}^{-1} \text{ mm}^{-2}.$$

We have already seen that a photographic star image requires an accumulation of 50×10^3 photons and this can be achieved by having an energy arrival rate at the image of $(50 \times 10^3)/t$ photons s^{-1}, where t is the exposure time expressed in seconds. Thus, the energy arrival rate, E_t, per unit area of telescope aperture is given by

$$E_t = \frac{50 \times 10^3}{t D^2} \frac{4}{\pi} \text{ photons s}^{-1} \text{ mm}^{-2}$$

where D is the diameter of the telescope expressed in mm. If we consider that E_e and E_t are produced by stars of magnitudes m_e and m_t respectively, Pogson's equation (see equation (15.11)) allows us to form the relation

$$m_t - m_e = -2 \cdot 5 \log_{10} \frac{E_t}{E_e}$$
$$= -2 \cdot 5 \log_{10} \frac{50 \times 10^3 \times 4 \times \pi \times 8^2}{t D^2 \pi \times 200 \times 4}$$
$$\approx 2 \cdot 5 \log_{10} t D^2 - 2 \cdot 5 \log_{10} 1 \cdot 6 \times 10^4.$$

Now, E_e corresponds to the energy arrival rate of a star which can just be seen by the naked eye and, hence, by putting $m_e = 6$, the value obtained for m_t corresponds to the limiting magnitude, m_{lim}, of the star which can be recorded by a telescope of diameter, D (mm), and exposure time, t (s). Hence,

$$m_{\text{lim}} = 6 + 5 \log_{10} D + 2 \cdot 5 \log_{10} t - 10 \cdot 5$$
$$= -4 \cdot 5 + 5 \log_{10} D + 2 \cdot 5 \log_{10} t. \qquad (18.2)$$

As an example, consider the use of a 500 mm telescope and an exposure time of 10^3 s. The faintest star recorded as given by equation (18.2), is approximately

$$m_{\text{lim}} = -4 \cdot 5 + 13 \cdot 5 + 7 \cdot 5$$
$$= 16 \cdot 5.$$

In practice, however, this simplified approach of evaluating the limiting magnitude cannot be applied generally because of the wide range in the choice of telescopes, photographic materials and the site conditions. The ultimate magnitude limit is, of course, set by the fog of the plate as a result of the sky background. Equation (18.2) may also give values which are too optimistic, as no allowance has been made for the transmission efficiency of the telescope or for the reciprocity failure of the emulsion. It was also derived on the assumption that the star images are point-like and this is not strictly true.

It has been found, in practice, that the detectability of star images on plates which have been exposed to the limit of the fog-level depends on the sizes and the images: a larger image can be detected more easily. Now the size of any image depends on the focal length of the telescope and, hence, for telescopes of the same aperture, the one with a longer focal length will allow fainter stars to be detected. In other words, the right-hand side of the equation expressing the value of the limiting magnitude should also contain a term whose increase is related to the focal ratio of the telescope.

If the image has a truly extended form, such as a nebula, then its strength is again proportional to the area of the collector, i.e. $\propto D^2$, and it is also inversely proportional to the square of the focal length, F^2, of the telescope. The latter dependence results from the area of the image being proportional to the square of the focal length. The image illumination is, therefore, inversely proportional to the square of the focal ratio, i.e. $\propto 1/f^2$—see equation (16.12). Hence, for a given emulsion, an extended object can be photographed by a range of telescopes with the same exposure, provided that the focal ratios of the instruments are identical. The actual size of each image depends on the plate scale which in turn depends on the focal length. Accordingly, a nebula of small apparent angular size, such as a galaxy, might be photographed by a large reflector at its prime focus in order to obtain a detailed image, while a nebulosity with larger apparent angular size might be photographed by a wide-field camera of small focal length. If the apparent brightnesses of the two objects are the same and the focal ratios of the telescopes identical, then the exposure times will be the same for the two photographs.

18.9   Unsharp masking

Much fine detail is often lost in an astronomical photograph of an extended field because it varies little with respect to the background brightness. **Unsharp masking** is an image-processing technique that enhances the fine detail. The original picture is processed to obtain a negative mask, such a mask being of low density and slightly out of focus. This mask is then superimposed on the original, resulting in a decrease in the brightness range of the original but without decreasing the fine detail. The tiny brightness variations over the original can then be seen more clearly. The Australian astronomer, David Malin, has produced many beautiful pictures of celestial objects such as the Orion Nebula using unsharp masking, revealing by this method fine detail that otherwise would effectively be lost. A selection of such images is available from the Anglo-Australian Observatory (AAO)[W 18.1]. This processing technique can also be applied to images recorded in electronic form with the operation undertaken by software rather than by using a physical mask.

18.10   Photoelectric devices

18.10.1   Introduction

When optical radiation falls on certain alkali materials, electrons are liberated and ejected from the material's surface. By encapsulating the photosensitive material in a vacuum and providing a plate with a positive potential, the ejected electrons, or **photoelectrons** as they are called, can be collected and their flow can be measured as an electric current. This type of detector provides the means of performing extremely accurate photometry. One of the chief reasons for this is that the size of the current is exactly proportional to the amount of energy which falls on the sensitive surface and this holds for the whole of the range of energies which are likely to be presented to the device. The device is said to have a **linear response**.

18.10.2   Spectral sensitivity

A range of photocathodes is available covering the spectrum from the ultraviolet to about $10\,000$ Å. The cut-off in the ultraviolet is controlled by the glass envelope of the cell and, by using a quartz window above the cathode, the ultraviolet sensitivity can be extended. The various spectral responses are designated by a number (e.g. S–1) and three such sensitivity curves are depicted in figure 18.6.

18.10.3   Quantum efficiency

The quantum efficiency of the response is typically of the order of 0·1 or 10%, meaning that, on average, one photoelectron is emitted for every ten photons which are incident on the sensitive surface. Using this information, it is very easy to predict the current to be measured according to the energy arrival rate. For example, if the photoelectric cell is receiving 200 photons s^{-1} (i.e. the photon rate just detectable by eye), the number of photoelectrons emitted is 20 s^{-1}. Now the charge on an electron is equal to $1·6 \times 10^{-19}$ coulomb (C) and, therefore, the total charge liberated is

$$20 \times 1·6 \times 10^{-19} \, C \, s^{-1} = 3·2 \times 10^{-18} \, A.$$

This, of course, is an extremely small current and it would require large amplification before it could be registered by some recording device.

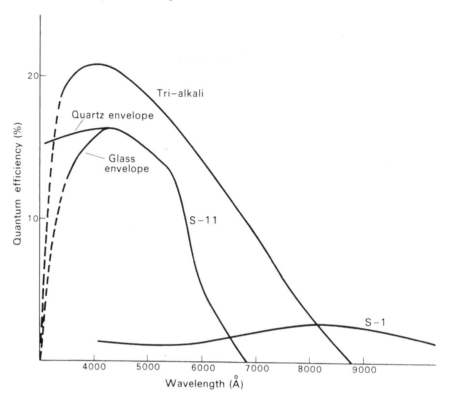

Figure 18.6. The spectral sensitivity of various photocathodes.

18.11   The photomultiplier

18.11.1   Its principle

The simple photoelectric cell is rarely used in astronomy as a detector because of the small currents which are generated. Devices have been developed, however, which rely on the photoelectric effect but which supply their own amplification of the current; and these are now one of the standard detectors for astronomical measurement. They are known as **photomultipliers**. Each photomultiplier has a **photocathode**, acting as the basic detector, and this is followed by a series of **dynodes** which provide the amplification or multiplication process. The dynodes are arranged in series so that each one is at a higher positive potential than its immediate previous neighbour. When a single photoelectron is liberated from the photocathode, it is accelerated towards the first dynode as a result of its positive potential. By the time it arrives there and impinges on the dynode, it has sufficient energy to cause liberation of several (about three or four) further electrons. These are then accelerated towards the second dynode where further multiplication takes place. After leaving the last dynode, the large bunch of electrons—corresponding to one single photoelectron—is collected by the **anode**. Measurement of the current from the anode allows the amount of energy falling on to the photocathode to be determined.

Photomultipliers take on a variety of geometric forms. In one of the main types, the photocathode is set at an angle to the incoming radiation and the dynodes take the form of cups which help to focus the secondary electrons towards the following dynode. Another type has a semi-transparent cathode, allowing its surface to be normal to the radiation. The photoelectrons in this case are ejected from the opposite face of the photocathode to that on which the radiation falls. The dynode chain for this type of

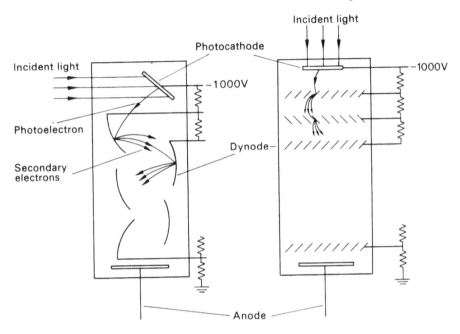

Figure 18.7. Two arrangements of the dynodes within a photomultiplier. The high voltage is generated by a stabilized power pack and the resistor chain provides successive dynodes with an increased positive potential.

photomultiplier is in the form of a series of venetian blinds. Both types of photomultiplier are depicted in figure 18.7.

The amplification, or gain, provided by a photomultiplier depends upon two factors: one is the number of dynodes or stages; the other is the number of electrons ejected at each stage by the arrival of a single electron from the preceding stage. Hence, for a typical cell with twelve dynodes, each giving an application of four, the gain, G, is given by

$$G = 4 \times 4 \times 4 \times 4 \times \cdots$$
$$= 4^{12}$$
$$\approx 1.68 \times 10^{7}.$$

By using the same figures given in the example of the estimate for the current produced by an ordinary photocell, the output current provided by a typical photomultiplier receiving 200 photons s^{-1} is given by

$$3.2 \times 10^{-18} \times 1.68 \times 10^{7} \text{ A} \approx 5 \times 10^{-11} \text{ A}.$$

A current of this magnitude can be measured easily using subsidiary electronic systems.

The currents likely to be obtained from various telescope systems can be estimated as follows if it is assumed that the spectral sensitivities of the eye and the photocathode are approximately the same. Now the pupil of the eye, with a diameter of 8 mm, is capable of detecting the arrival of 200 photons s^{-1}, this energy being that which is provided by a sixth magnitude star. This is equivalent to an energy arrival rate of

$$\frac{200 \times 4}{(8)^{2}\pi} \text{ photons s}^{-1} \text{ mm}^{-2}.$$

For a telescope of diameter, D (mm), the energy collected by it is given by

$$\frac{200 \times 4 \times D^{2}\pi}{(8)^{2}\pi \times 4} = \frac{200D^{2}}{64} \text{ photons s}^{-1}.$$

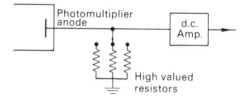

Figure 18.8. The DC amplification technique.

By assuming a quantum efficiency of 10%, a photomultiplier gain of 10^7 and taking the value of the charge of an electron to be $1\cdot6 \times 10^{-19}$ C, the output current is given by

$$\frac{200D^2}{64} \times 0\cdot1 \times 10^7 \times 1\cdot6 \times 10^{-19} \text{ A} = \frac{D^2}{2} \times 10^{-12} \text{ A}.$$

This current is produced when the telescope is directed to a sixth magnitude star. Using Pogson's equation (equation (15.10)), the current obtained according to the brightness of the star observed is, therefore, given by

$$A = \frac{D^2}{2} \times 10^{-12} \times 2\cdot512^{-(m-6)} \tag{18.3}$$

where m is the magnitude of the star. As an example, consider the observation of an 11th magnitude star, using a telescope of 500 mm diameter. The output signal is, therefore, given by

$$A = \frac{(500)^2}{2} \times 10^{-12} \times 10^{-2}$$
$$= 1\cdot25 \times 10^{-9} \text{ A}.$$

Because of the wide range in spectral sensitivity and gains for photomultipliers, equation (18.3) should not be taken as a hard and fast rule.

It may also be noted that the photomultiplier detector simply responds to all the radiation illuminating its cathode irrespective of the direction of arrival or the presence of any image structure. Unlike the photographic plate, it can only record the brightness of one field of view at a time and is essentially a 'single pixel' device. Consequently, when the photomultiplier provides the basis of a photometric instrument, a field stop in the form of a diaphragm is placed in the focal plane of the telescope so that only one star at a time is measured.

There are two basic principles applied to the output signal for its registration. These are referred to as DC amplification and pulse-counting or photon-counting photometry. The two schemes are sketched out below.

18.11.2   DC amplification

In this technique, the current is made to flow through a high-valued resistor and the voltage developed across it is then amplified. A simplified scheme of the arrangement is depicted in figure 18.8. By using Ohm's law, the magnitude of the voltage developed for any desired further amplification can be estimated. Thus, if the current from the anode is of the order of 10^{-12} A and the value of the resistor is 10^9 Ω, the voltage developed is given by

$$V = 10^{-12} \times 10^9$$
$$= 10^{-3} \text{ V}.$$

It is usual practice to keep the gain of the DC amplifier fixed and to allow for a wide range in signal levels by having a series of high-valued resistors and a selector switch. Accurate photometry can only be performed if the gain of the DC amplification is drift free. The final output signal can be displayed on a pen recorder or integrated and displayed digitally. With these types of system, the accuracy limit of the photometry is set by the electronic circuitry or the mode of display and it is usually possible to obtain a good value to 1%.

18.11.3   Photon-counting photometry

The current which is measured by DC amplification represents the average flow of electrons from the anode of the photomultiplier. The flow is made up of bursts or pulses of electrons, each one being the result of the liberation of a single photoelectron from the cathode. These pulses can be seen easily if the output from a photomultiplier is fed to an oscilloscope. Since the number of primary photoelectrons is directly proportional to the intensity falling on to the photocathode, so will be the number of pulses which is available at the anode. Hence, by measuring the rate at which the pulses appear or by counting the number of pulses in a given time, the intensity of the radiation can be determined. Since the output is digital, it is already in suitable form for acceptance by computer.

Each of the pulses carries a quantity of charge which is given simply by the gain of the detector. They behave as if generated across a capacitance and given by the stray capacitance of the cables: the lower this is, the higher is the voltage of the pulse. Hence, the voltage of the pulses may be estimated as follows. Using the value of the charge of a single electron and assuming the value of 10^7 for the gain of the photomultiplier, the charge carried by each pulse is given by

$$1{\cdot}6 \times 10^{-19} \times 10^7 = 1{\cdot}6 \times 10^{-12} \text{ C.}$$

Now the voltage, V, across a capacitance, C, when holding a charge, Q, is given by

$$V = \frac{Q}{C}.$$

If the stray capacitance can be held to a value as low as 10^{-11} farad (10 picofarad), then the voltage appearing across it is given by

$$V = \frac{1{\cdot}6 \times 10^{-12}}{10^{-11}}$$
$$= 0{\cdot}1 \text{ V.}$$

A further amplification, by means of a wide-band amplifier, is, therefore, normally required before the pulses can be counted by a scaler (see figure 18.9). It must also be remembered that the production of secondary electrons by the dynodes is a statistical process and that there is a range in the voltage heights of the pulses.

Following the discussion in section 16.3.1, photon-counting photometry lends itself to immediate assessments of the quality of any data. Successive counts over a fixed integration time will be distributed about some mean value, the photon count rate being subject to a noise sometimes referred to as **shot noise**. The distribution of values can be checked against that predicted by the statistics of photon counting. If the match is good, then the observations are at the limit of the best possible accuracy, with no other noise sources degrading the measurements. Under this circumstance, the fractional uncertainty (the inverse of the signal-to-noise ratio) of any photometric measurement is given by

$$\frac{\sqrt{N}}{N} = \frac{1}{\sqrt{N}}.$$

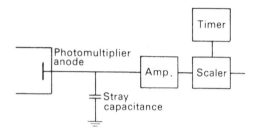

Figure 18.9. A simple pulse counting photometer.

Hence, any record may be expressed as $N \pm \sqrt{N}$. It is easily seen, therefore, that if measurements with an accuracy of 1% are planned, it is necessary only to choose an integration time which provides a total count of $N > 10^4$; and this will depend on the brightness of the object and the aperture of the telescope. Higher accuracies can be achieved by recording greater total counts.

18.11.4 Dark background

If a photomultiplier is made to see no light at all, a small output current persists. This is known as the **dark background** or **dark signal** and its source is chiefly from electrons which have sufficient thermal energy to allow them to be spontaneously emitted from the photocathode. When photon-counting techniques are used, this background signal from the detector is referred to as the **dark count**. The magnitude of the dark background can, therefore, be reduced by enclosing the photomultiplier in a cooled housing: dry ice (solid CO_2) is frequently used as a coolant. A recorded signal is, therefore, the combination of the detector's response to light plus the contribution from the dark background. When observations are made of faint objects, the dark signal may be sufficiently significant to require separate measurement and then subtraction from the measurement of the combined signal. If the magnitude of the dark signal were known perfectly, its effect would not cause any deterioration in the measurement obtained from the faint object. However, as in the case of the photoelectrons, the emission of thermal electrons is a statistical process and the dark background shows shot noise fluctuations. Any measurement of the background carries with it an uncertainty and this, in turn, affects the accuracy of any determination of the weak light-level.

18.12 Limiting magnitude

We have seen that the accuracy in any photometry is ultimately determined by the number of photoelectrons which are recorded. By extending this idea, it is obvious that any object can only be considered to have been detected when the counted number of photoelectrons produced by its light has reached some significant value above the background counts. The time that it takes to reach this value depends on the number of photons arriving from the source and the size of the telescope employed. In principle, then, extremely faint sources could be detected by using very long integration times. In practice, the limit is set by some disturbance in the experimental conditions such as a temperature change, so altering the level of the dark signal or a change in the brightness of the sky background.

When a photoelectric photometer is applied to detecting weak sources, the background signal is not just that of the detector alone but that from the small patch of sky which is also inevitably included in the field of view. The level of the sky signal is likely to change during the course of a night and, on top of this, it will exhibit photon shot noise. It is the sky background which sets the ultimate limit to the detection of faint stars. Results obtained some years ago for the 150-in (3·81 m) telescope at Kitt Peak, Arizona, show that a patch of sky, 10 seconds of arc across, is equivalent to the brightness of

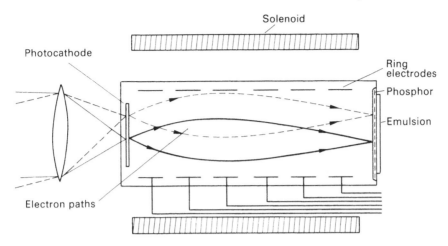

Figure 18.10. A simple image intensifier. Electrons ejected from the photocathode are focused on the emulsion-backed phosphor by uniform electric and magnetic fields.

a 17th magnitude star and gives rise to 2000 photoelectrons s^{-1}—a 23rd magnitude star gives rise to only 6 photoelectrons s^{-1}. In order to detect such a star against the sky background and measure it to an accuracy of $\pm 0 \cdot 1$ magnitude, the total observation time is of the order of 90 minutes.

The integration time required to detect faint objects would be decreased for detectors with higher quantum efficiencies. Such devices are now available in the form of solid state semiconductors. The absolute limit to be aimed for is the production of a detector with a quantum efficiency of 100%, i.e. every collected photon produces a recordable event. When this is reached, the uncertainty in the measurements of very faint objects would be due to the statistical fluctuations in the photoelectron production rate which would be related directly to the statistical fluctuations in the arrival of the quanta themselves.

It is axiomatic that the aim of any observation is the extraction of the total information content from any collected radiation, dependent on the number of photons carried by the radiation.

18.13 Image converters

An **image converter** or **image intensifier**, now disappearing from regular use, is a device which uses a combination of photoelectric and photographic processes (see figure 18.10). Primary detection of an image is achieved by using a photocathode, giving the benefits of its relatively high quantum efficiency. By a system of accelerating electrodes, the photoelectrons gain energy and are made to impinge on a phosphor which then liberates light according to the number of electrons it receives. Immediately behind the phosphor is placed a photographic plate which records the intensity distribution across the phosphor. The light to which the final exposure is made is much stronger than the original amount falling on the photocathode, the amplification depending on the energy which the accelerated electrons have gained in transit through the tube and on the efficiency of the phosphor.

The accelerating electrodes also provide a means of focusing the trajectories of each electron which is emitted by the photocathode. By this means, any point in the phosphor where an electron arrives corresponds to a particular single point on the surface of the photocathode. This, in turn, corresponds to a single point in the primary image. On the plates behind the phosphor, there is a one-to-one correspondence between the recorded image and the detail in the original image. The image converter is, thus, capable of obtaining images of extended objects efficiently by using the good quantum efficiency of photocathode material and the spatial recording ability of a photographic plate.

18.14   TV systems and other detectors

Television sensors of various kinds have also found application as detectors. The many advantages that they have include a high quantum efficiency, broad spectral coverage, a large dynamic range, no reciprocity failure for long exposures and the data are in a convenient form for processing by digital computer. As data collectors, they have been superseded by CCD technology but they still remain in use on telescopes for field identification and guidance systems.

One detector system that had great success over a couple of decades was the Image Photon Counting System (IPCS) designed to detect single photoelectrons and assign their positions of origin from within a two-dimensional image, so allowing a picture to be built up. The complete apparatus used a television camera to look at the phosphor screen of a multistage image intensifier. The intensifier gave an amplification sufficient for a single photoelectron to be detected above the readout noise of the television camera. The frame rate of the camera was set to be sufficiently high so that the probability of more than one photoelectron arriving from a resolved picture element during a single frame is extremely low and the decay time of the phosphor was sufficiently short so that the same photoelectron was not counted twice in consecutive frames. A computer controlled the system and the photoelectron events were fed into memory locations which opened and closed in synchronization with the position of the scanning television raster. The IPCS was developed by Boksenberg and had particular success as a detector for the spectrograph on the Anglo-Australian Telescope recording the weak spectra of faint quasars.

18.15   Charge coupled devices

Based on developments of solid state technology, the invention of charge-coupled devices (CCDs) in the 1970s has revolutionized practically every aspect of observational optical astronomy. Although photography and photoelectric photometry are still practised, nearly all current optical imaging, stellar photometry and spectrometry is now performed using CCDs. A basic detector is small, looking like any familiar semiconductor 'chip'.

The CCD chip comprises a matrix of pixels typically $\sim 10^4 \times 10^4$ in number. Each pixel in the detector comprises a light-sensitive element with an associated charge storage capacitor similar to that found in a metal oxide semiconductor (MOS) transistor. A schematic diagram of a basic photosite is given in figure 18.11. By giving the gate a positive bias relative to the substrate, the majority of carriers or holes are repelled from the Si–SiO_2 junction and a depletion layer forms with increased depth according to the value of the voltage. As the bias voltage increases to just more than a couple of volts, the Si–SiO_2 junction becomes sufficiently positive with respect to the bulk substrate so that any free electrons in the vicinity are attracted to the junction and form an inversion layer. Prior to exposure, a positive potential is applied, so charging the photosite capacitor. During the exposure, any photons impinging on the site enter the silicon crystal lattice and raise electrons from a low-energy valence-band state to a high-conduction-band state so that they can migrate and be captured in the potential well, effectively discharging the capacitor. The degree of the discharge at each site is proportional to the number of photons impinging on the pixel, the detector essentially having a linear response. If the light levels are too high or the exposure time too long, the pixels saturate and quantitative information is lost. Typical full-well capacity is of the order of 5×10^5 photoelectrons with the value dependent on the chip design. It is sensible observational practice to plan any imaging or other photometry so that the photoelectron accumulations in the pixels are no more than $\sim 80\%$ of their full-well capacity.

The spectral sensitivity covers the whole of the optical range from about 400 to 1000 nm. The quantum efficiency is typically $\sim 40\%$ but can be higher at the longer wavelengths according to the basic photo response of silicon. The sensitivity can be further enhanced if the chip has been 'thinned' in the manufacturing process and is used with back illumination.

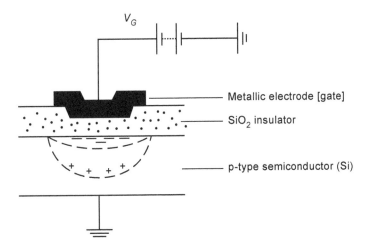

Figure 18.11. A basic photosite within a CCD showing the depletion volume beneath the electrode when a positive potential V_G is applied to the gate.

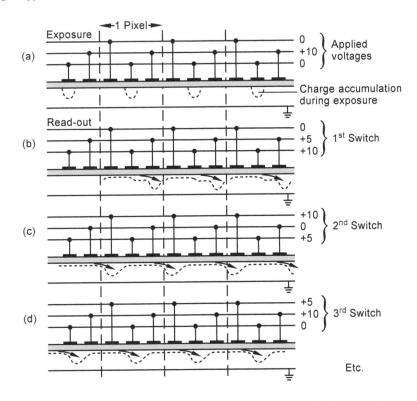

Figure 18.12. A section of a row within a CCD detector is shown in (*a*) with the gate voltages configured for exposure. The 'bucket brigade' principle whereby the pixels are read out, in turn, by a repetitive voltage switching cycle is illustrated in (*b*), (*c*) and (*d*).

Complete CCD detector chips are made with a variety of architectures. In the example depicted in figure 18.12, each pixel comprises three basic capacitors or gates within a space ~20 μm, formatted in rows. There are common connections between the elements comprising each pixel. Figure 18.12(*a*) indicates typical voltages that might be applied to the components of each pixel during the exposure phase. During this time, the generated photoelectrons accumulate under the right-hand well of each pixel according to the local illumination. The operation of the detector is controlled by computer. This includes the clocking of the voltage levels, the interrogation of the pixels and the organization of files to record the mapped illumination levels.

At the end of the exposure, the accumulated charges are successively transferred along the columns of the detector matrix by cyclically switching the voltages as indicated in figures 18.12(a)–(d). The technique is referred to as the 'bucket brigade'. The contents of each pixel arrive in turn at the extreme column, the charge flow is amplified and re-transformed to a digital value by an analogue-to-digital (A-to-D) converter and temporarily stored in the computer's memory, according to the location of the originating pixel on the chip.

Inevitably, during the transfer process, charge is left behind from one stage of the register to another. In the operation of the detector, it is very important that the efficiency of the transfer process is extremely high. Suppose that

$$N_0 = \text{number of electrons under a gate}$$

and

$$N_t = \text{number of electrons in the adjacent gate after transfer.}$$

The **charge transfer efficiency** (CTE) can be defined as

$$\text{CTE} = 1 - \left(\frac{N_0 - N_t}{N_0} \right).$$

To see just how important it is to have extreme high values of CTE, consider what happens to a starting charge value of $1000e^-$ under a gate following 100 transfers through the detector columns with a CTE of 99%. After the transfers, the readout of the charge would be

$$1000e^- \times (0.99)^{100} \longrightarrow 370e^-.$$

Thus, any image would be severely degraded, with the electron accumulations at one side of the chip being very much weakened with respect to the pixel columns which are immediately closer to the readout column. In practice, CTEs are $\sim 0 \cdot 999\,990$ but this depends on temperature and on the speed or clocking at which the transfers are made. Rather than quoting the CTE, a figure of merit which is important in calculating the noise associated with high precision photometry is the pixel readout noise. Typical values for this are about $\pm 10e^-$ per pixel. As CCD technology has continued to improve, this figure has been progressively reduced and is now generally lower than the photon shot noise associated with the number of photons collected in each pixel.

A regular required calibration procedure arises because the individual pixels have differing responses to unit intensity or, in other words, the quantum efficiency is not uniform over the pixel array. This is addressed by subjecting the detector to a source of even illumination and, according to the responses, all other records are adjusted by a direct ratio, pixel by pixel. This process is referred to as flat-fielding and it is common practice to point the telescope with a CCD camera attached to the twilight sky at the beginning and end of the nightly observations. In order for this process not to introduce its own noise, it is important that the flat-field frames are repeated a large number of times to obtain accurate values for the relative sensitivities of each pixel.

CCD detectors suffer also from background signals which accumulate during the exposure. For short to medium exposures, these can be reduced effectively to zero by housing the chip within a sealed unit and cooling it to liquid nitrogen temperatures. It is standard practice to record frames with a shutter in the optical path so that any effects of the thermal background can be subtracted from any target frames. Occasionally, a cosmic ray passes through the chip to produce a glitch in the image. According to the type of study, it is sometimes possible to clean the data of the defect.

It may be noted that the final outputs of the signal levels are expressed in 'analogue-to-digital' units or ADUs. In assessing the noise levels of any observational exercise, it is important to know the relationship between the number of electrons (related to incident photons) and the ADU number and this can be done by simple calibration experiments. The gradation of intensities or 'grey levels' depends on the number of *bits* used in the digitization process within the chip. For professional CCDs, the number of bits used is normally 2^{16} so providing a dynamic range of 2^{16} or 65 536 grey levels. With such a discrimination in the recording of intensity, the potential photometric accuracy is 1 part in 2^{16} per pixel or 1 part in 65 536 or $\pm 0.0015\%$. Some of the CCDs used by amateurs may provide only a 2^8 bit conversion, so providing a lower potential photometric accuracy of 1 part in 256 or $\pm 0.39\%$.

Approximate values for the number of photoelectrons corresponding to each ADU can be obtained from the manufacturer's specification of the pixel full-well capacity and the number of bits used in the analogue-to-digital conversion. If, for example, the full-well capacity $\sim 5 \times 10^5 e^-$ and the chip uses 2^{16} bits, each ADU corresponds to

$$\frac{5 \times 10^5}{2^{16}} \approx 8 \text{ electrons.}$$

With this figure, the accuracy of any single pixel measurement based on the ADU value and photon-counting statistics can be easily assessed. Quite generally, brightness assessments involve the combination of signals from several pixels, so improving the overall signal-to-noise ratio of any measurement.

One of the considerable advantages of the two-dimensional detector is its ability to allow for the night-sky background component, whether it be a direct image of a star field or a record of the spectrum of a source. Those pixels not forming the target area of the main measurements may be taken to provide a reference for the general illumination of the light from the night sky (for example, scattering from distant urban lighting, atmospheric airglow spectral lines or the integrated extra galactic background). Using an extrapolation from the pixels surrounding the target areas, the contribution of the night-sky background may be subtracted.

Suppose that over an exposure a star provides N_* photons spread over p pixels and the background sky signal provides N_{sky} photons to the same pixels, the signal-to-noise ratio of the stellar brightness measurement is given by

$$\text{S/N} = \frac{N_*}{\sqrt{N_* + N_{\text{sky}}}}.$$

For bright stars, the contribution of N_{sky} to the S/N calculation can usually be neglected.

For the exploration of very faint stars, in addition to the noise from the sky background subtraction process, noise associated with dark-signal subtraction and readout noise need to be included in S/N determinations. For an exposure time of t with a background signal of N_d s^{-1} per pixel and with a readout noise of $\pm \sigma$ per pixel, the S/N ratio obtained from p pixels may be determined from

$$\text{S/N} = \frac{N_*}{\sqrt{N_* + N_{\text{sky}} + pN_d t + p\sigma^2}}. \tag{18.4}$$

Unlike the photographic plate with its random scatter of small detectors with positions which are different for each used plate, the CCD has a regular structure of pixels and, providing that the registration can be maintained, the rows and columns of the matrix can be used to provide a scale for repeated positional measurements. This has advantages in such areas as the measurement of the separation of visual binary stars or in radial velocity measurements (see section 15.8.2).

The main advantages of the CCD are the high quantum efficiency of the photo-detection and the linearity between the output signals and the illumination of the pixels. The device also only requires low voltages to control it and the power consumption is low. Following the readout procedure, images can be displayed for immediate assessment. At a later time, any frame can be reduced to obtain quantitative data. Frames may also be processed to provide the best quality image. Unsharp masking and other techniques can be readily applied from image-processing software suites

During the last 50 years, the sizes of the world's largest telescopes have made a significant move forward allowing us to 'see' fainter and fainter objects. The fact that they are more efficiently productive and can see deeper than at the time they were originally commissioned, bears testimony to the simultaneous advances in detector systems. Telescope/detector systems are still not 100% efficient and progress in the field of detector technology no doubt has further directions to take.

Problems—Chapter 18

1. A telescope has a diameter of 300 mm and is used to investigate a star field. Compare the limiting magnitudes when it is used (i) visually and (ii) photographically with an exposure of 15 minutes.

2. When performing photoelectric photometry with a particular system, it is noted that a fourth magnitude star generates 10^4 primary photoelectrons per second. What is the integration time required to achieve a photometric accuracy of 1% on a ninth magnitude star? (Assume that the measurement noise simply arises from photon-counting statistics.)

3. For a photographic plate with a $\gamma = 0.6$, it is found that reasonable exposures cover a density range of 2. If a star of eighth magnitude is the limit for over-exposure, calculate the brightness of the faintest star that can be recorded satisfactorily, allowing a good measurement.

4. A CCD camera with 10 μm square pixels is used at the focus of a 1 m, $f/12$ telescope. If stellar images are out of focus and spread over an angle corresponding to 2 arc sec on the sky, estimate the number of pixels used in measuring stellar magnitudes.

5. A CCD camera records a star image using 30 pixels, each providing a signal of 250 ADU. If 1 ADU corresponds to $20e^-$, estimate the best percentage accuracy that can be achieved for the measurement of stellar brightness.

6. The image of a faint star fills ten pixels on a CCD giving rise to a combined signal of N photoelectrons. An adjacent section of the detector records the surrounding sky and 100 similar pixels provide a total signal of N_s photoelectrons. Following a sky subtraction procedure, show that the signal-to-noise ratio for the detection of the star can be estimated from

$$\frac{N - N_s/10}{(N + N_s/10)^{1/2}}.$$

Chapter 19

Astronomical optical measurements

19.1   Introduction

We have seen earlier that the main task of the observational astronomer is one of detecting electromagnetic radiation which arrives from space. The basic measurements involve the determination of the direction of arrival of the radiation, its strength and its polarization. The radiation is measured over as wide a range of wavelengths as is possible. Time-dependent variations in the radiation's parameters are also investigated.

Most measurements are referred to internationally accepted scales which have evolved partly as a result of the nature of the sources being observed and partly as a result of the way in which the analysing and detecting equipment has developed. In this chapter, some of the basic optical measurements will be discussed in relation to the methods and equipment that are employed, the methods by which effects of the Earth's atmosphere can be allowed for and the scales that are adopted to describe the measurements.

19.2   Positional measurements

Perhaps the most obvious kind of positional measurement involves recording the coordinates of celestial objects. Such determinations may lead to the evaluation of the orbit of a planet or binary star, the distances of stars by parallax, the distribution of stars in space, etc. The basic data are first recorded on two-dimensional detectors such as photographic plates or CCDs attached to the most appropriate telescope or camera that is available. A few reference stars with accepted coordinates are required and the positions of these stars and the ones of special interest are measured.

If conventional photography is used, the resulting plates are reduced using a two coordinate microscope or measuring engine. The photographic plate is clamped to a table whose position can be adjusted along two directions (X and Y) by means of highly accurate micrometer screws. A small area of the plate can be viewed by the microscope which is fixed to the mainframe. Movement of the micrometers allows any part of the plate to be inspected, and by setting the screws so that the cross-wire in the microscope bisects any image, the image's position may be determined in terms of X and Y. By measuring the X, Y coordinates of the reference stars, the measurement of the position of any other object in terms of X and Y can be converted to the astronomical (equatorial) coordinates, right ascension and declination. As the photographic plate usually records the curved celestial sphere on a plane, special formulas derived by spherical trigonometry are used to perform the conversion.

There are various degrees of automation which are applied to star plates. Where the plate carries images of only a few celestial objects, the astronomer may be able to readout the XY micrometer settings, or they may be automatically recorded each time a star is centred on the graticule of the eyepiece. This is obviously impossible in the case of large field surveys which may register one million images of many kinds of celestial object on each plate. In recent years sophisticated devices have been designed which automatically scan the plate, hunting for each of the star images, even distinguishing between a star and a distant galaxy. Highly automated instruments of this kind have been developed at the Royal Observatory, Edinburgh. Its Super COSMOS scanning system can operate on 20 in × 20 in glass plates and 'celluloid' films up to 14 in × 14 in. The information from the plate or film is digitized with a spatial resolution of 10 μm with 10^{15} bits (37 768 grey levels) for the intensity values. Special software processes the resulting data to produce a catalogue of all the objects detected on each plate, with 32 parameters associated with each object.

Prior to computerized systems, other special types of microscope for identifying moving objects or variable stars were sometimes used. One instrument, known as a **blink microscope**, has two carriages which support two plates taken of the same region of the sky at different times. An arrangement is provided so that the identical star field on each plate can be viewed in quick succession. Any object which has moved against the star background between the two exposures will appear to jump between two positions across the fixed star background. Any variable star will appear to pulsate as the microscope is operated.

Some types of positional measurement require only one coordinate to be measured and, in this case, the microscope need not have both X and Y movements. For example, for the measurements of radial velocity, comparison of the stellar spectral lines and laboratory spectral lines of a photographic plate can be obtained by a single coordinate-measuring microscope.

The accuracy to which any positional measurements can be made depends on several parameters, one of which is obviously the accuracy of the microscope, usually controlled by the quality of the micrometer screw. The uncertainty of any position may be typically $\pm 1 \times 10^{-3}$ mm and such a figure may then be used to determine the uncertainty of the parameter being measured.

Because of the smaller size of the CCD chip relative to a photographic plate, large surveys of stellar positional measurements have not been applied to ground-based surveys. However, very accurate differential positional measurements are readily obtained from CCD frames under direct computer control without the subjectivity of any eye-ball assessment of the XY positions of the image seen either through a microscope or on a VDU screen. For example, in the case of stellar images, the centroid of the distribution of the photons as collected by the local pixels is readily determined in terms of the structure of the light-sensitive chip, the rows and columns providing the reference grid. The positions are readily determinable as fractions of the basic pixel size, making for very accurate measurements of angular separations according to the focal length of the telescope.

19.3   Broadband spectral photometry

The response of any detector system to energy falling on to it obviously depends on the strength of the radiation received. It also depends on the frequency distribution of the energy and on the spectral sensitivity of the detector. In many instruments, rather than using the full spectral range of the detector, optical elements, such as colour filters, may be applied to select specific spectral zones for measurement. The variation with wavelength of the relative response of the system may be written as $S(\lambda)$ and may be displayed graphically (see figure 19.1). The width of the embraced spectrum, $\Delta\lambda$, may be defined as the spectral interval between points on the $S(\lambda)$ curve where the response is reduced to a half of the maximum value and this is termed the **full-width half maximum** or FWHM.

Each passband can be ascribed an **equivalent wavelength**, λ_0, defined by

$$\lambda_0 = \frac{\int_0^\infty \lambda S(\lambda)\,\mathrm{d}\lambda}{\int_0^\infty S(\lambda)\,\mathrm{d}\lambda}$$

this being the mean wavelength, weighted according to the response curve. To a first order, brightness measurements made using passbands which are narrow in comparison to the broader energy spectrum of a source can be considered as the monochromatic brightness value at the equivalent wavelength.

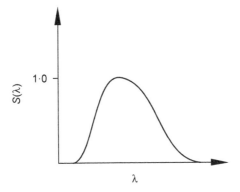

Figure 19.1. An example of the relative spectral response of a detector system normalized so that its maximum equals unity.

Let us suppose that we have been given some equipment and we use it to measure the brightnesses of stars. By defining a standard star, or a set of standard stars (see discussion relating to Pogson's relationships (see equation (15.10)–(15.13)), a magnitude scale can be set up and by comparing the response of the system to the standard star and all of the stars in turn, each star can be assigned a magnitude. If the same stars now have their brightnesses measured by a second detector system, providing a different passband and different equivalent wavelength, it will be found, in general, that the stars do not appear on this second scale at the identical positions at which they appeared on the first scale. This is a reflection of the fact that each star has an energy envelope defined by its temperature and that the signal from the detector is a function of the amount of energy that falls on it within the spectral range to which it responds. Thus, each detector system has its own magnitude scale.

19.4 Standard magnitude systems

In order to allow direct comparisons between observing stations, internationally accepted detector systems and magnitude scales are used. An early system was obviously one related to the sensitivity of the eye and is known as the **visual system** (m_v). It has now been virtually replaced by the **photovisual system** (m_{pv}). By using orthochromatic plates which are sensitive up to about 5900 Å and by using a yellow filter to cut out the response below 5000 Å, the spectral response roughly matches that of the eye and the two magnitude scales correspond closely.

The **photographic system** (m_{pg}) is obtained by using ordinary plates whose spectral sensitivities have not been deliberately extended and cut off at about 5000 Å. The sensitivity is, therefore, limited to the blue part of the spectrum down to the ultraviolet, where a cut-off is provided by the Earth's atmosphere or by the telescope optics. The zero point of the photographic scale is chosen so that stars of a certain temperature and colour, corresponding to certain types of star (spectral type A5), have identical magnitudes on the visual and photographic scales when the visual magnitude is equal to 6·0. Stars chosen as standards are found in a catalogue known as the *North Polar Sequence (NPS)*.

More modern magnitude systems depend on the combination of the spectral sensitivity of the cathodes of photoelectric multipliers and specially chosen filters. In the U, B, V system developed by Johnson and Morgan, three bands are used with equivalent wavelengths of 3650 Å (U), 4400 Å (B) and 5500 Å (V), corresponding to the ultraviolet, blue and yellow regions of the spectrum (see figure 19.2). Magnitudes measured in the V system correspond closely to those measured on the photovisual scale. Similar passbands to UBV have been established using CCDs and selected colour filters. Many other

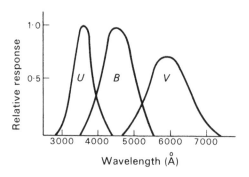

Figure 19.2. The international UBV system.

Table 19.1. Standard photometric passbands.

Band	λ_0	$\Delta\lambda$
U	365 nm	70 nm
B	440 nm	100 nm
V	550 nm	90 nm
R	640 nm	150 nm
I	790 nm	150 nm
J	$1 \cdot 25 \ \mu$m	$0 \cdot 12 \ \mu$m
H	$1 \cdot 66 \ \mu$m	$0 \cdot 16 \ \mu$m
K	$2 \cdot 22 \ \mu$m	$0 \cdot 22 \ \mu$m
L	$3 \cdot 45 \ \mu$m	$0 \cdot 35 \ \mu$m
M	$4 \cdot 65 \ \mu$m	$0 \cdot 46 \ \mu$m
N	$10 \cdot 3 \ \mu$m	$1 \cdot 0 \ \mu$m
Q	$20 \cdot 0 \ \mu$m	$5 \cdot 0 \ \mu$m

systems are in use, including extensions to six or even more passbands using detectors with sensitivity in the infrared. Some of the more regular systems are given in table 19.1.

19.5   Colour indices

The earlier simple discussion on black body radiation showed that the emitted energy from a star has a wavelength dependence according to its temperature. By applying Wien's law (section 15.4.3), the temperature of a body could be obtained by measuring the wavelength at which the energy received has a maximum value. However, many of the stars are so hot as to place the energy peak somewhere in the ultraviolet region, beyond the position which can be investigated at the bottom of the Earth's atmosphere. A more flexible technique involves temperature determination by measuring the ratio of received fluxes at two different wavelength positions on the energy–wavelength curve. This is illustrated in figure 19.3 by the black body curves depicted for two temperatures. It is obvious from an inspection of this figure that

$$\left(\frac{B_{\lambda_x}}{B_{\lambda_y}}\right)_{T_1} \neq \left(\frac{B_{\lambda_x}}{B_{\lambda_y}}\right)_{T_2}.$$

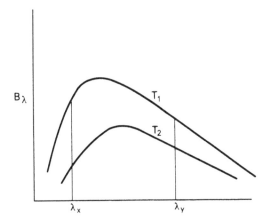

Figure 19.3. The black body curves for two temperatures.

It is, therefore, possible to express the ratio of energies received at two different wavelengths in the form

$$\frac{B_{\lambda_x}}{B_{\lambda_y}} = f(T)$$

where $f(T)$ is a known function that can be derived from the Planck formula describing the black body curve.

By using equation (15.13), the recorded magnitudes corresponding to these fluxes may be written as

$$m_{\lambda_x} = k_{\lambda_x} - 2\cdot5 \log_{10} B_{\lambda_x}$$

and

$$m_{\lambda_y} = k_{\lambda_y} - 2\cdot5 \log_{10} B_{\lambda_y}$$

where k_{λ_x} and k_{λ_y} are the appropriate constants giving the zero points for the two different magnitude scales. Therefore,

$$m_{\lambda_x} - m_{\lambda_y} = k_{\lambda_x} - k_{\lambda_y} - 2\cdot5 \log_{10} \left(\frac{B_{\lambda_x}}{B_{\lambda_y}} \right)$$
$$= k_{\lambda_x} - k_{\lambda_y} - 2\cdot5 \log_{10} f(T). \tag{19.1}$$

The difference in magnitude $m_{\lambda_x} - m_{\lambda_y}$ is known as the **colour index** of the star and its determination leads to an assessment of the star's effective temperature. In practice, many different bandpass systems corresponding to λ_x and λ_y are used. A colour index system can be applied to any pair of such magnitude scales with the difference in the values of the constants $k_{\lambda_x} - k_{\lambda_y}$ being chosen to be zero for some particular temperature. Thus, colour indices may be derived from photographic magnitude measurements—m_{pg}, m_{pv}. More modern systems are based on UBV magnitudes and are usually written as $B - V$ and $U - B$. Table 19.2 shows how the colour index defined by $B - V$ depends on the surface temperature of the star.

A note of caution is perhaps due here in respect of the 'signs' of the values of colour indices. Suppose that in figure 19.3 the wavelengths λ_x and λ_y correspond to B and V respectively. For the lower temperature (T_2), the flux at B is lower than at V. A quick assessment might suggest that $B - V$ would be negative. However, the values of B and V are expressed in magnitudes which carry an inverse scale and hence $B - V$ is, in fact, positive. For objects with higher temperature (T_1), the flux at B is

Table 19.2. Colour index as a function of temperature (main sequence stars).

$T(K)$	$B - V$
35 000	−0·45
21 000	−0·31
13 500	−0·17
9 700	0·00
8 100	+0·16
7 200	+0·30
6 500	+0·45
6 000	+0·57
5 400	+0·70
4 700	+0·84
4 000	+1·11
3 300	+1·39

greater than at V and, hence, $B - V$ (measured as a magnitude difference) is negative. The higher the temperature, the more negative $B - V$ becomes.

If theoretical black body curves are used to represent the radiation from stars, the $B - V$ colour index can be expressed to a good approximation as

$$B - V = -0 \cdot 71 + \frac{7090}{T}. \tag{19.2}$$

Using equation 19.2, temperatures may be estimated from measurements of the $B - V$ colour index. In practice, the expression needs modification to allow for the fact that stars do not behave as perfect black bodies and care must also be taken for the effects that absorption by interstellar material can have on the apparent colours of stars.

19.6 Bolometric magnitudes

When the properties of stars are being compared, it is advantageous to use a magnitude system which is independent of any particular detector system with its associated wavelength-selective property. It is useful to be able to know the brightness or magnitude of stars as they would be determined by a system which is equally sensitive to all wavelengths. For this purpose, a system known as the **bolometric magnitude scale** is frequently used. From a knowledge of the energy envelope of a star and the spectral response curve of the selective detector, it is possible to perform calculations so that the total amount of energy available for measurement by a bolometer can be predicted. In principle, any magnitude determination by a particular system can be converted to a bolometric magnitude (m_{bol}) by the appropriate correction, known as the **bolometric correction** (BC). Thus, in general,

$$m_{\text{bol}} = m + \text{BC}$$

and, therefore,

$$\text{BC} = m_{\text{bol}} - m. \tag{19.3}$$

Consider the bolometric correction in relation to the visual magnitude scale. By using equation (15.3), namely

$$m = k - 2 \cdot 5 \log_{10} B$$

Table 19.3. Bolometric corrections to V band magnitudes.

$T_e(K)$	BC
35 000	−4·6
21 000	−3·0
13 500	−1·6
9 700	−0·68
8 100	−0·30
7 200	−0·10
6 500	0·00
6 000	−0·03
5 400	−0·10
4 700	−0·20
4 000	−0·58
3 300	−1·20

the bolometric magnitude scale can be represented as

$$m_{bol} = k_{bol} - 2 \cdot 5 \log_{10} B_{bol}. \tag{19.4}$$

Similarly, the visual band scale can be represented as

$$m_V = V = k_V - 2 \cdot 5 \log_{10} B_V. \tag{19.5}$$

In equation (19.4), k_{bol} is an arbitrary constant while k_V in equation (19.5) has been chosen according to some suitable standard stars. Subtraction of equation (19.5) from (19.4) gives

$$m_{bol} - m_V = BC = k_{bol} - k_V - 2 \cdot 5 \log_{10} \left(\frac{B_{bol}}{B_V} \right). \tag{19.6}$$

The ratio B_{bol}/B_V has a minimum value for a star with a temperature providing the maximum in its energy envelope at the same wavelength as the peak sensitivity of the selective detector. For the V-band system, this occurs when the temperature is close to 6500 K. The value of k_{bol} is chosen so that the bolometric correction is zero for this temperature. Since B_{bol}/B_V has a minimum value at this temperature, it is easy to demonstrate, using equation (19.6), that the bolometric correction is *negative* for stars with temperatures on either side of 6500 K. The bolometric corrections to V-band magnitudes are listed in table 19.3.

19.7 Disturbances caused by the atmosphere

19.7.1 Introduction

We have already seen in section 10.2 that atmospheric refraction causes displacement of the apparent positions of stars but, by investigating the amount of refraction, allowances can be made for these effects.

The atmosphere, however, is the source of two other effects giving problems to photometry and spectrometry. In the first place, **extinction** affects all measurements and systematic errors may be introduced unless careful compensations are applied. Second, turbulence in the atmosphere may compromise the quality of the basic measurements. The causes of the problems are independent and we will deal with them in turn.

19.7.2 Extinction

When light passes through any medium, its strength is continuously weakened by absorption and by scattering which deflects the radiation out of the path. The combination of the two effects results in an extinction along the line of sight. For a medium of uniform extinction, the flux in any beam decreases exponentially with the distance travelled through the medium. The rule is very general and is known as **Bouger's law**. The amount of extinction depends on the nature of the medium and the wavelength of light which is transmitted. In the case of a point source with a flux F_0 on arrival at the medium, the measured flux, F, at any distance, x, within the medium is given by

$$F = F_0 e^{-\tau x}$$

where τ is defined as the **absorption coefficient** or **optical depth** of the medium. The decay of the flux can also be expressed in terms of a **decimal** extinction coefficient, a, such that

$$F = F_0 \, 10^{-ax}.$$

The Earth's atmosphere continuously absorbs and scatters starlight out of its original path on its passage to the telescope. Stars observed at the bottom of the atmosphere appear to be less bright in comparison to their brightness at the top of the atmosphere. The flux is continually reduced by extinction. The thickness of the atmosphere through which the light from the star travels depends on the altitude of the star and, consequently, its apparent brightness will vary during the course of a night as it rises, reaches culmination and sets. In order to be able to make sense of the observations so that the brightnesses of stars can be compared in a reliable way, it is important that the effects of atmospheric extinction should be removed from the basic measurements.

Up to 60° zenith distance, the following simple method is satisfactory. As in the simplified treatment of the effect of refraction, the atmosphere can be taken as a series of thin plane parallel layers. Each layer needs to be considered as having a well-defined extinction coefficient which increases as the starlight penetrates deeper and deeper into the atmosphere. Figure 19.4 depicts one such layer of thickness, dz, at a height, z, and shows that the actual path length for a star at a zenith distance, ζ, depends on the secant of that angle and is given by $dz \sec \zeta$.

If F_0 is the flux of a star above the first thin layer of the atmosphere and F the flux after its light has passed through the layer, then according to the decimalized form of Bouger's law

$$F = F_0 10^{-a \, dz \, \sec \zeta}$$

where a is the optical depth of the first layer. The flux at the bottom of the atmosphere is determined by integrating the extinction over the whole path length of the atmosphere through which the starlight has travelled. Thus, for a star at zenith distance, ζ, the recorded flux is given by

$$F(\zeta) = F_0 10^{\int_h^\infty -a(z) \sec \zeta \, dz}$$
$$= F_0 10^{-\sec \zeta \int_h^\infty a(z) \, dz} \tag{19.7}$$

where $a(z)$ represents the decimal extinction coefficient as a function of the distance into the atmosphere and h is the observer's altitude. By expressing this equation in terms of logarithms it may be written as

$$\log_{10} F(\zeta) = \log_{10} F_0 - \sec \zeta \int_h^\infty a(z) \, dz. \tag{19.8}$$

If it were possible to observe the star exactly at the zenith, equation (19.8) reduces to

$$\log_{10} F(0) = \log_{10} F_0 - \int_h^\infty a(z) \, dz$$

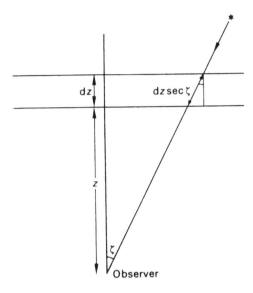

Figure 19.4. The path length of starlight passing through a thin layer.

and, therefore,

$$\int_{h}^{\infty} a(z)\, dz = \log_{10}\left(\frac{F_0}{F(0)}\right). \tag{19.9}$$

By expressing the right-hand side of equation (19.9) in terms of a magnitude difference (see Pogson's equation (15.11)), we may write

$$m(0) - m_0 = -2 \cdot 5 \log_{10}\left(\frac{F_0}{F(0)}\right)$$

$$= -2 \cdot 5 \int_{h}^{\infty} a(z)\, dz = \Delta m. \tag{19.10}$$

Thus, the integrand in equation (19.9) represents the extinction of the zenith column giving rise to a magnitude change, Δm. By substituting for this in equation (19.8) and, by using Pogson's Equation again, we may write

$$m(\zeta) = m_0 + \Delta m \sec \zeta \tag{19.11}$$

where $m(\zeta)$ is the observed magnitude at zenith distance ζ and m_0 corresponds to the magnitude as would be recorded above the Earth's atmosphere without any extinction.

According to this simple model, equation (19.11) shows that the observed magnitude of a star has a linear dependence on the secant of its zenith distance. The relationship is conveniently illustrated in figure 19.5 where typical empirical magnitude measurements of a star are plotted against zenith distance. The value of $\sec \zeta$ has its minimum value of unity when the star is at the zenith ($\zeta = 0°$) but if we assume that it can behave like any variable and take on any value, we can extrapolate the plot shown in figure 19.5 and consider the magnitude value which would have been obtained if it had been practically possible to allow $\sec \zeta$ to be equal to zero. By putting $\sec \zeta$ equal to zero in equation (19.11), it can be seen that the value of the deduced magnitude is equal to the magnitude that would have been obtained from measurements above the Earth's atmosphere. It will also be seen that the gradient of the line is equal to the increase in magnitude between a measurement above the Earth's atmosphere and the same measurement at the observing site if the star were at the zenith.

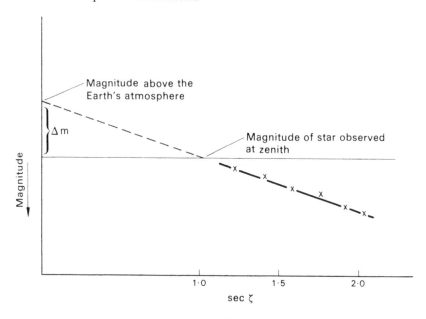

Figure 19.5. The variation of stellar magnitude with zenith distance.

Once the value of Δm has been evaluated for a particular site from a series of observations on standard stars, corrections for atmospheric extinction can be applied to all other observations according to the zenith distance at which they are made. As the value of Δm may have short-term fluctuations over periods of a few minutes and, as it may vary from night to night according to conditions within the atmosphere, a series of magnitude determinations needs to include frequent measurements of standard stars. For photoelectric photometry, this involves the speedy movement of the telescope from star to star as target and standard objects are observed in turn. For CCD photometry, differential magnitudes can be obtained more readily by arranging for the target and standard stars to be in the same field of view and recorded simultaneously on the same frame. It may be noted that there are seasonal or annual changes in the value of Δm influenced by volcanic activity that might be thousands of kilometres away.

It is obvious that individual observatories will suffer different amounts of extinction—for example, by being situated at different altitudes. In order to make magnitude determinations meaningful, it is necessary practice **to reduce the values to above the atmosphere**. Lists given in star catalogues of stellar magnitudes present data corrected in this way.

The need for reduced determinations is further emphasized by considering magnitude determinations in two colours. Most of the extinction is a result of scattering by the air molecules. According to Rayleigh's law (see section 5.3), the efficiency of such scattering is strongly wavelength dependent and inversely proportional to λ^4. Thus, it would be expected that the extinction is very much greater in the blue spectral bands than in the red. This difference markedly shows up in practice; the typical effect is illustrated in figure 19.6 where magnitude measurements in the B and V bands are plotted against sec ζ for a particular star. If the colour index $B - V$ of the star is to be determined, it is obvious that the direct values taken at specific zenith distances will depend on the particular value of ζ at which the measurements are made. In order for the colour index to be meaningful, the two magnitude values must be reduced to above the Earth's atmosphere before the subtraction is performed (see figure 19.6). For each magnitude system, a particular value of Δm may be found empirically at each observing station. Typical values of Δm are of the order of $0^{\mathrm{m}}6$, $0^{\mathrm{m}}3$ and $0^{\mathrm{m}}2$ for the U, B and V bands respectively, this reflecting the behaviour of Rayleigh's law.

For extremely accurate photometry, improved formulas are necessary for the reduction procedure.

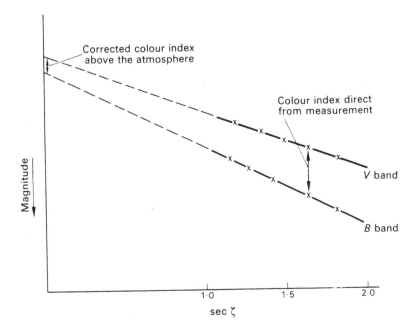

Figure 19.6. Variation of magnitudes *B* and *V* with zenith distance for a particular star.

These formulas take into account, for example, the fact that the light path through the atmosphere is curved and they may be applied to observations which are made at zenith angles greater than 60°.

19.7.3 Atmospheric turbulence

The Earth's atmosphere is never still with airflows and turbulence on a variety of scales. As a consequence, the wavefronts associated with the progression of the radiation through the atmosphere are subject to having random phase delays imposed on them. The effects of this are twofold. One noticeable outcome, even visible to the naked eye, is that stars display rapid fluctuations in apparent brightness, the effect referred to as **intensity scintillation**. It should be noted that scintillation is not caused by rapid fluctuations in atmospheric extinction. The second effect is referred to as **seeing**, affecting telescope image quality. Visual inspection of an image reveals that it wanders in the telescope focal plane and comes in and out of focus.

The simple underlying explanation of both phenomena is as follows. In the turbulent eddies in the atmosphere, there are minute temperature differences between the individual pockets of air and, consequently, there are small differences in the refractive index within the eddies. As a result of the small refraction differences that are introduced, the plane wavefronts become corrugated (see figure 19.7(*a*)).

With respect to intensity scintillation, the energy is transmitted in a direction at right angles to a wavefront. Consequently, there will be times when the corrugations effectively increase the flux entering the telescope and there will be times when they decrease the amount of energy collected. Thus, the flux collected is subject to random fluctuations. The noise on the recorded intensities depends on the size of the telescope and on the stellar zenith distance. For a large telescope, the aperture collects, at any one time, wavefronts which have many corrugations. The following wavefronts carry different corrugations but, providing the telescope is sufficiently large, the amount of energy collected fluctuates only by a small degree. However, if the aperture of the telescope is reduced to a size of the order

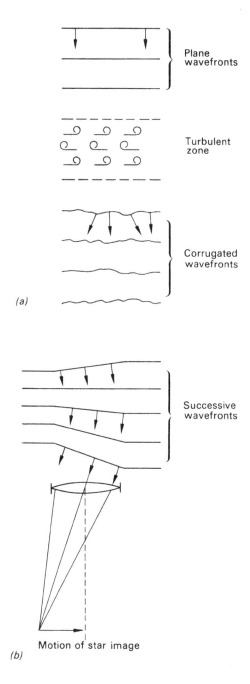

Figure 19.7. (*a*) The effect of atmospheric turbulence. (*b*) The dancing effect of a star image when viewed with a small telescope.

of a single corrugation, then the energy collected does show marked fluctuations and scintillation is exhibited.

For a telescope of the order of 1 m, the noise of intensity scintillation of a star may be a few per cent, with time-scales of a few milliseconds up to ~1 s. Unlike photon shot noise, the scale of the fluctuations is directly proportional to the brightness of the source, the constant of proportionality, κ, depending on the quality of the site, on the telescope size and on the direction of the star (altitude and azimuth). Thus, for bright stars with the photometric noise dominated by intensity scintillation, a single brightness measurement would be recorded simply as $B \pm \kappa B$. Without scintillation, with the noise dominated by photon-counting statistics, a measurement of the same star would be noted as $B \pm \sqrt{B}$. Clearly, there is a range of brightness values for which the accuracy of any photometry is dominated by scintillation and a range dominated by photon counting statistics. The demarcation occurs when $\kappa B = \sqrt{B}$ or $\kappa = 1/\sqrt{B}$. If, for example, $\kappa \sim 1\%$, over a measurement time of 1 s, the photometric noise is dominated by scintillation if the photon count rate is greater than 10^4 s^{-1}.

It is well known that, to the naked eye, planets generally do not appear to twinkle. This is because they have a greater apparent angular size than the stars. Any small component within an extended object exhibits intensity scintillation but if the object has a sufficient angular size, then there are sufficient elements for the overall scintillation to smooth out.

Experiments have shown that scintillation begins to be apparent when the object subtends an angle of 10 arc seconds or less and that the magnitude of the scintillation increases as the angular size of the object decreases, until it cannot be differentiated from stellar scintillation when the angle subtended is about 3 arc seconds. In effect, it can be said that the apparent angular size, α, of the corrugations in the wavefront is of the order of 3 arc seconds and their physical size, r_0, obtained from experiments investigating the relationship between scintillation and telescope diameter, is of the order of 100 mm. Thus, the typical distance in the atmosphere where the deforming turbulence is taking place is given by

$$\frac{r_0}{\alpha} = \frac{100}{3} \times 206\,265 \text{ mm}$$
$$\approx 7 \times 10^6 \text{ mm}$$
$$= 7 \text{ km}.$$

By considering the theory associated with turbulent media, Fried characterized the fluctuations with a scale over which the parts of the corrugated wavefront can be considered as being essentially plain. This dimension is referred to as the Fried parameter, being synonymous with the r_0 given here.

The visual behaviour of the 'seeing', as it affects the appearance of a star image, also depends on the value of r_0 in relation to the telescope diameter. For a small telescope of a few centimetres diameter, in addition to rapid fluctuations in apparent brightness, the expected diffraction pattern is likely to be seen but with the whole structure dancing around in constant agitation. When a telescope of a few tens of centimetres or larger is used, the image loses its sharpness and appears as an ill-defined blob without scintillation. This image is known as the **seeing disc**. When a starfield is recorded on a photograph or on a CCD frame, the images are recorded with a spread which is dominated by the seeing disc. By making an intensity scan through any stellar image, the profile of the seeing disc may be compared with that of the theoretical Airy diffraction pattern (see figure 19.8). The size of the disc is usually expressed in terms of arc seconds, one definition being the angular diameter between the positions at which the intensity has dropped by a factor of two relative to the peak intensity. The seeing disc size and shape gives an indication of the quality of the observing site—a good site will have many nights in the year when the seeing disc is 1 arc second or smaller. There are many observatories, however, where such good quality seeing is not usually available.

Another important aspect to the description of the deteriorated stellar image is the way the flux is redistributed relative to that expected in the diffraction pattern. The more the image is concentrated into an area of a detector, the more easily a faint object might be detected; the more the concentration of the image, the more radiation might be accepted into the narrow slit of a spectrometer. This aspect is described by comparing the peak intensity of the seeing disc, I_S, relative to that of the peak of the

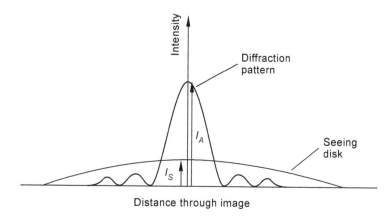

Figure 19.8. The seeing disc of a star is superposed in the theoretical diffraction pattern in the image plane. The ratio of the peak intensities, I_S/I_A is referred to as the Strehl index.

Airy disc, I_A. The ratio, $S = I_S/I_A$, is referred to as the **Strehl index** and it is not uncommon for it to be no greater than a few per cent.

For the small telescope, the entering wavefronts are essentially plain but arriving at the aperture with a range of tilts relative to the optic axis. Each plain section provides an image—the diffraction image produced by the telescope—which is sharp. Its position in the focal plane, however, depends on the tilt angle. As the orientation of the deformations in the successive wavefronts change, so does the position of the sharp image. This is illustrated in figure 19.7(*b*).

When a larger telescope is used, an image is formed which is the resultant of many simultaneous corrugations in the accepted wavefront. Each deformation produces a displacement of the sharp image with the result that their combination is a blurred-out patch. It should be obvious that if a photograph is taken of a star using a small telescope, the dancing of the sharp image during the exposure will result in a blurred image, the effect being the equivalent of a visual inspection of the star image provided by a larger telescope.

With the introduction of CCD detectors with quantum efficiencies $\sim$100 times higher than the photographic plate, it is possible to record frames of bright objects with sufficiently short exposure times such that the wavefront disturbances at the end of the exposure maintain correlation with those at the beginning. The time scales for this condition are of the order of 5–50 ms. Under this circumstance, each frame provides a record with distortions corresponding to particular patterns of wavefront disturbances. The images are recorded with 'frozen' seeing with some pictures being of better quality than others, the sharpest occurring when the whole wavefront over the aperture is essentially plain.

Any imaging system relies on the principle of the application of phase delays of differing degree across the surface of the accepted wavefronts (see section 16.4). Following manufacture, the curvatures of the final optical surfaces should follow contours according to the design specification with local departures which are no greater than about $\lambda/8$. This can be translated into phase departures by remembering that $\lambda \equiv 2\pi$. Acceptable manufacturing phase errors over the surface may be of the order of $2\pi/8 \approx 0.75$ rad.

Atmospheric turbulence is constantly introducing additional phase delays and advances over the telescope aperture. If these are small, i.e. some fraction of a radian, little or no deterioration of the image would be noticed. They are, however, usually larger than this with a very apparent blurring of the image. As the phase structures are constantly changing, over short time intervals there may be occasions when their effect is small and an 'instantaneous' sharp image is produced. According to the theory of atmospheric turbulence, the probability of having phase variations less than 1 radian over the telescope aperture of diameter D at any given instant is given by

$$5 \cdot 6 \exp \left(-0 \cdot 1557 \left(\frac{D}{r_0} \right)^2 \right).$$

(19.12)

This essentially defines the probability of obtaining a lucky snapshot with a non-distorted or sharp image. As might be expected, such a probability depends on the ratio of D/r_0. According to the probability formula with a telescope-seeing combination with $D/r_0 \approx 5$, there is of the order of a 1 in 10 chance of obtaining a sharp picture. It is for this reason that amateurs with small telescopes using fast CCD cameras are able to obtain very respectable pictures of planetary disc features fulfilling the potential resolution of their telescopes. For a telescope of medium size, such as the Nordic 2·5 m instrument on the island of La Palma, the D/r_0 ratio is ~ 7 giving the chance of about one good exposure out of 350 and such odds have been used to advantage in recording the close separations of some bright stars. For reference, it may be mentioned that the Hubble Space Telescope also has a diameter of 2·5 m but, as it is in orbit above the Earth's atmosphere, it is not subject to the problems of seeing. For very large telescopes with $D/r_0 \approx 10$, the probability of getting a sharp picture falls dramatically to ~ 1 in 10^6.

Following the simple calculation of the atmospheric height at which 'seeing' is generated, it is obvious that observations made above this zone should be free from any major disturbance. This is confirmed by the brilliantly sharp pictures obtained by balloon-borne imaging. It goes without saying that all seeing problems are completely eliminated by observations from orbiting satellite platforms.

The effect of seeing does not always cause significant deterioration in measurements and, in the cases where it does, it is sometime possible to design the recording equipment so that the effects can be overcome or compensated. The fact that all recorded star images are in the form of a seeing disc does not prevent positional measurements being made with an uncertainty which may only be a small fraction of the size of the disc. It is perhaps in connection with analyses which involve the measurement of one star which is apparently close to another, or those where the analysing equipment has a very small acceptance angle, that the size of the seeing disc is inconvenient. However, this can also be addressed by using adaptive optics (see section 20.2) in the telescope design.

Scintillation may also not always be important since, for example, a long exposure of a photographic plate or CCD chip smooths out the rapid intensity fluctuations. It is only when an 'instantaneous' stellar intensity measurement is obtained, say, from a photoelectric cell that scintillation noise appears in the output. In photoelectric scanning spectrophotometry, the departures from a mean intensity level may be on the same order as the depths of spectral lines and, in this case, it would be impossible to detect them against the scintillation. The remedy for this particular problem is to have a second detector which accepts a fixed band of the spectrum close to the section which is to be scanned and allow this to monitor the scintillation noise. By using the second channel to divide out the noise in the scanning channel, the recorded spectrum is, to all intents and purposes, freed from scintillation noise.

19.8 Image photometry

19.8.1 Photographic photometry

One of the purposes of recording a star field on a photographic plate is to allow determination of stellar brightness. The usual procedure is to place the photographic plate in the focal plane of the telescope so that in-focus images are produced. The spectral range may be limited by placing colour filters prior to the emulsion. Exposures may last from a few minutes to a few hours.

An inspection of any star field plate immediately reveals that the images vary in size and are in some way related to the brightnesses of the stars. It should be noted that the distribution of image

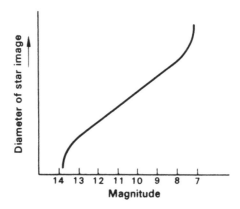

Figure 19.9. A typical magnitude–diameter calibration curve.

diameters *does not* reflect the distribution of the physical diameters of the stars. An inspection of the star images with a microscope reveals that any image is made up of a collection of blackened grains, the image being dense at its centre, becoming weaker at the edges. Simple measurements reveal that the sizes of the images are much greater than the Airy disc predicted by diffraction theory. There are two main reasons for this.

The first is caused by scattering within the emulsion of the plate. Any light falling on to the surface of an emulsion may be scattered by the grains before it is finally absorbed and this effect obviously spreads out the size of any pinpoint image. The second is caused by the seeing. During the exposure, any telescopic star image dances and wanders over the emulsion. The recorded image represents the blurred patch over which the image has been in motion and the distribution of the density within the image reflects the amount of time the star image has spent at particular positions during the exposure.

When the observer looks at the image in the developed plate as a disc, the eye is effectively deciding the positions in the image where the plate threshold has been reached, i.e. positions where there has been just sufficient energy to cause perceptible blackening on the plate. No matter what mechanism is responsible for the image-spreading, it is obvious that more energy will be available to spread the image for bright stars. Consequently, the emulsion threshold will be achieved at greater distances from the centre point of the image according to the star's brightness. Thus, the apparent size of the recorded image depends on the brightness of the star and on the length of the exposure. Without the image-spreading mechanisms, it would be very difficult to make assessments of the brightness of the star by an examination of a point-like image.

Determinations of stellar magnitudes can, therefore, be obtained by measuring the diameters of images recorded photographically, with some kind of microscope. Any measured diameter is converted to a magnitude by means of a calibration curve. This curve is first obtained by measuring the image diameters of the standard stars which have also been recorded on the plate. Over the useful range of the plate, a range usually allowing a coverage of the order of six magnitudes between the brightest and faintest stars (see section 18.6), the relationship between magnitude and image diameter is approximately linear. A typical calibration curve as obtained by microscope measurements is illustrated in figure 19.9.

An improvement to the method is obtained by using a specially designed laboratory instrument known as an **iris diaphragm photometer**. This machine allows measurements of star images to be made accurately and quickly.

With this special instrument, the plate is held on a moveable carriage which may have provision for reading its position in terms of coordinates, X and Y. By means of a lamp and projection system, a small area of any selected part of a plate can be illuminated uniformly. The size of this circular patch is controlled by an iris diaphragm. After the beam has passed through the plate it is directed to a photomultiplier. A chopping system allows the strength of the beam to be compared with a standard beam which is obtained from the original lamp. The strength of the beam, after passing through the photographic plate, depends on the size of the diaphragm and the density distribution of the area of the plate which has been isolated by this aperture. When a star image has been adjusted by movement of the carriage to be at the centre of the aperture, a balance between the compared beams can be achieved by controlling the size of the diaphragm.

It should be obvious that an image produced by a bright star will be very dense at its centre and will have a comparatively large disc. In order to allow a certain amount of light through this image to provide the balance, the diaphragm must have a large diameter. At the balanced or null condition, the diameter of the diaphragm is read off a scale. A calibration curve must first be obtained, by using standard stars, before the magnitudes of other stars may be determined. The uncertainties of any determined magnitudes are typically ±0.05 mag but, in some cases, it is possible to improve on this. The ease of working the photometer is improved in larger instruments when the part of the plate under investigation and the image of the diaphragm are projected on to a screen, allowing the star field to be identified quickly. The X, Y and diaphragm diameter scales may also be projected and an electronic record of these parameters may be available for automatic dispatch to a computer. Automatic reduction instruments, as already described in section 19.2, may have a facility for determining brightness values as well as positional coordinates.

19.8.2 Photoelectric photometry

In the discussion of the photomultiplier (section 18.11), it was pointed out that this detector, in its simplest form, responds only to the total amount of energy which falls on to its sensitive area. If a brightness picture is to be built up of an extended image or series of images, it is necessary to look at each picture point in turn. If the stars in a field are to have their brightnesses measured photoelectrically, it is necessary to look at each star one by one and, for this purpose, a diaphragm is placed in the focal plane of the telescope. This allows any single star to be isolated and keeps to a minimum the amount of background sky light entering the photometer.

Whereas in the case of photographic observations the plate is placed in the focal plane of the telescope, an in-focus image must *not* be allowed to fall on the photocathode. Any photocathode has a detecting area which is non-uniform in sensitivity and any image wander over its surface caused by poor telescope following or image motion due to seeing will generate fluctuations or **noise** in the output signal of the cell. To prevent such noise, a **field lens** or **Fabry lens** is provided so that the collector aperture is focused on to the cathode. A lens is chosen so that the whole of the sensitive area of the cathode is filled with the light for detection. It can be seen from the ray diagram of the simple photometer depicted in figure 19.10 that by imaging the collector aperture on the sensitive area, movement of the telescope's direction related to the star or agitation of the star's position caused by seeing does not produce movement of the patch of light over the cathode. In fact, there should be no change in the detector's output until the deviation of the telescope's direction is such that the diaphragm in its focal plane begins to cut off the star image. Intensity scintillation, giving fluctuations of the illumination of the telescope aperture, is not removed by the expedient of the field lens and is apparent in the cell's output.

Provision is usually made for the insertion of colour filters which limit the spectral passband. For ease of operation, retractable viewers are usually provided before and after the diaphragm so that the star field can be studied and particular stars chosen for measurement. As we have seen in section 18.11, the output from the photomultiplier is usually monitored by DC amplification or photon counting. A series of measurements involves recording the output signal over short but regular integration times

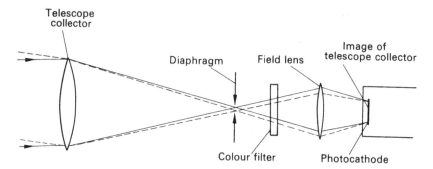

Figure 19.10. A simple photoelectric photometer. The solid lines represent light rays when the star is in the centre of the diaphragm; the dashed lines represent light rays when the star image is at the edge of the diaphragm; the field lens prevents movement of the illuminated patch on the photocathode.

1. of the target star under investigation,
2. of some adjacent standard stars—and, to obtain a background level,
3. of a patch of sky without any stars present in the field.

As the response of the photomultiplier to the amount of light falling on it is linear, no further calibration is required and it is an easy matter to convert the measured responses into stellar magnitudes after subtracting the background signals from the sky background and the dark signal from the detector itself. The accuracy to which any measurement is made may depend on factors such as the quality of the observing site or the length of time which is spent in integrating the detector's response but it is not unusual to be able to achieve an accuracy which is 1% or better, far superior to photographic photometry.

19.8.3   CCD photometry

As a result of the high quantum efficiency of these devices and their linearity of response to light, CCDs have excellent scope for undertaking differential photometry of stars. Unlike photometry using a photomultiplier, which is, of course, also differential in that target stars and standard stars are observed in a cyclic sequence, CCD measurements require the target and standard object to be in the same field of view covered by the chip. Because of the small area of the detector, this can sometimes be a restriction not usually encountered in photography. Unlike the photographic process, however, there is virtually no delay in obtaining the data as each picture or frame is read directly into a computer with the potential of fast analysis of the images.

When stellar photometry is being undertaken, the required differential measurements are best obtained if the targets and calibration stars are in the same field covered by the chip so that the required information is contained in one individual frame. In determining the brightness values, one star is selected to determine the **point spread function** (PSF), so exploring the way by which a would-be point image is spread by the seeing conditions during the exposure. Using the determined constants associated with the particular PSF of the individual frame, the function is fitted to the intensity distributions of the other stellar images and the appropriate magnitude values are assigned from which magnitude differences between the various stars can then be calculated. This is done using a suite of software. The process is fairly automatic requiring the minimum of keyboard interaction. Star images can be searched for automatically with the centroid of their positions given in terms of the pixel position by row and column numbers. Photometric accuracies better than ± 0.01 are readily achieved according to the number of photons that are accumulated within the point spread function.

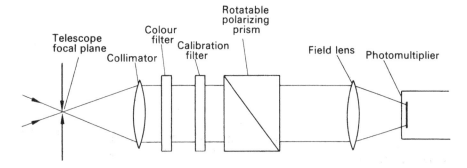

Figure 19.11. A single-beam polarimeter.

19.8.4 Polarimetry

In order to make measurements of polarization in the optical region, the necessary related photometric determinations (see equation (15.31)) need to be made with high accuracy. Most precision polarimeters, therefore, make use of photoelectric detectors and, as a consequence, they are able to make measurements of only one star at a time. There are several designs of photographic polarimeter which allow star field surveys to be undertaken but the results from individual stars are of inferior quality in comparison with the photoelectric measurements.

In the optical region, the amount of polarization in the light from an astronomical object is usually very small and great care is needed to avoid the introduction of systematic errors. In principle, the polarization can be measured by rotating polarization-sensitive optical elements in the beam, prior to the detector. The simplest design would use a piece of Polaroid in the beam and would require the signal output to be recorded as the rotational setting of the Polaroid is adjusted. A more elegant version would use a more efficient polarizer in the form of a birefringent prism. The improvement might require the use of an extra lens to provide a collimated beam for the prism. The basic elements of this type of polarimeter are illustrated in figure 19.11. Non-polarizing colour filters may be used to limit the spectral passband. Calibration filters may also be available so that known amounts of polarization may be added and the instrument's response checked. From the experiences of various observers, catalogues of stars exhibiting zero polarization and others with accurately determined values have been established for reference measurement and these are constantly under review.

19.9 Spectrometry

It is sufficient for some spectrophotometric purposes to use colour filters to isolate spectral regions. These may be simple dye filters, by which a passband is achieved by spectrally selective absorption, or interference filters, whereby a passband is achieved by interference in thin layers of materials, with chosen refractive indices and thicknesses, vacuum-deposited on a glass substrate. Dye filters have transmission efficiencies which are typically 10%, while the efficiency of an interference filter is much higher, being typically 50%. However, for other spectrophotometric studies, where, for example, detailed measurements are made on the profiles of individual spectral lines, filters cannot be used to isolate the wavelengths of interest. Other equipment must be employed, which is capable of looking at a continuous range of wavelengths, and with spectral passbands much narrower than those provided by colour filters.

The conventional way of doing this is by using a spectrometer. With this type of device, the light contained in an image in the focal plane of a telescope is dispersed and a spectrum obtained. By allowing the spectrum to fall on a photographic plate or a CCD chip, a broad range of wavelengths may

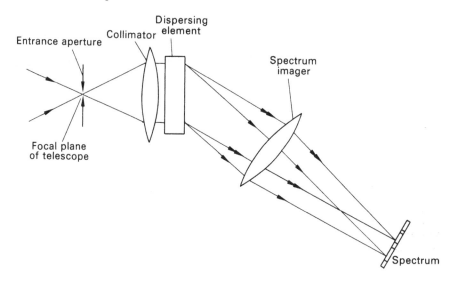

Figure 19.12. The essential elements of an astronomical spectrometer.

be recorded simultaneously. If a narrow exit slit is placed in the spectrum a small element can be made available for detection by a photoelectric cell; a progressive movement of the slit through the spectrum allows the spectrum to be scanned.

The basic elements of a conventional astronomical spectrometer comprise an **entrance aperture**, usually in the form of a slit when the spectrum is recorded photographically, a **collimator**, the **dispersing element** and an **imaging device** or **camera** to bring the spectrum to a focus. A simple spectrometer is illustrated in figure 19.12. The collimator and spectrum imager may either be in the forms of lenses or mirrors. In a photographic or CCD spectrometer, the spectrum imager is frequently referred to as the **camera** optics. The instrument is usually provided with retractable viewers so that any star field can be identified and checks can be made that the light from the star under investigation is falling centrally on the entrance aperture or slit.

The dispersing element is normally one of two types, being either a glass prism or a diffraction grating.

The glass prism causes the collimated beam to be deviated and, as its refractive index is wavelength-dependent, the amount of deviation is also wavelength-dependent. For a collimated beam of white light incident on the prism, a continuous range of collimator beams emerges from the prism with a spread in the angle of emergence. By focusing these beams into a plane, a spectrum is produced. Prism spectrometers are now very rarely used, however.

The diffraction grating consists of a series of accurately ruled slits or grooves. When a collimated beam is incident on it, the light is diffracted by each of the rulings. In some special angular directions, given by the spacing of the rulings and the wavelength of the light concerned, constructive interference takes place. In these particular directions, waves of a given wavelength from the rulings are exactly in phase. For this condition, the optical path length of the emerging light from one ruling to the next differs by an integral number of wavelengths. The general grating equation states that

$$d(\sin i + \sin \theta) = m\lambda \tag{19.13}$$

where d is the space between adjacent slits or grooves, i the angle of incidence of the collimated beam, θ the angle of emergence, λ the wavelength and m the order of interference, which for a conventional spectrometer would have a value of either one or two.

Two of the important properties of any spectrometer are given by the size of the spectrum that is produced and the ability to record fine detail within the spectrum. The first may be described by the term **angular dispersion** and the second by the term **spectral resolving power**.

The angular dispersion, AD, of any spectrometer is defined as the rate of change of the dispersed collimated beams with wavelength. Hence,

$$AD = \frac{d\theta}{d\lambda}.$$

This may be converted to a **linear dispersion**, LD, i.e. the length of spectrum produced representing a certain wavelength interval according to the focal length, F, of the camera lens (see equation 16.2). Hence,

$$LD = \frac{d\theta}{d\lambda} F. \tag{19.14}$$

It is usual practice to express the relationship between the wavelength range covered in the spectral plane by the **reciprocal linear dispersion** (RLD), usually in units of Å mm^{-1}.

The expression for the angular dispersion of a prism is complicated and will not be given here. It depends on the angle of the prism and the refractive index of the glass. As the second parameter is wavelength-dependent—the very property which is being made use of to produce the spectrum— the angular dispersion and, hence, the reciprocal linear dispersion will vary with wavelength. Equal wavelength intervals are, therefore, not covered by equal distances within the spectrum.

In the case of the diffraction grating, the angular dispersion may be obtained by differentiating equation (19.13). Thus,

$$\frac{d\theta}{d\lambda} = \frac{m}{d\cos\theta}.$$

By designing a grating spectrometer so that θ is close to zero, $\cos\theta$ will not differ significantly from unity and the angular dispersion will be practically constant for all wavelengths. The spectrum produced will have a scale which is linear, this being just one of the advantages that the grating has over the prism.

The **spectral resolving power** of any spectrometer is defined as its ability to allow inspection of elements of a spectrum which are close together. If a part of the spectrum centred at a wavelength, λ, is being investigated and $\lambda + \Delta\lambda$ is the closest wavelength to λ which can be seen distinctly as being separate from λ, then the spectral resolving power, R, is defined as

$$R = \frac{\lambda}{\Delta\lambda}. \tag{19.15}$$

The largest absorption features in a stellar spectrum are the order of 10 Å wide and it will be seen from equation (19.15) that a resolving power greater than 10^3 would be required to record them. A resolving power greater than 10^4 is needed to investigate spectral details which cover a range of 1 Å or less. If stellar absorption line profiles are to be recorded to real purpose, the resolving power must, in general, be at least of the order of 10^5. It may also be pointed out that the *reciprocal* of the spectrometer's resolving power is equal to the expression describing the Doppler shift (see equation (15.25)) and, hence, the value of R expresses the ability of a spectrometer to detect such shifts. Unless the Doppler shift is larger than a resolved spectral element, it obviously would not be detected.

There is a theoretical limit to the spectral resolving power of any spectrometer. Assuming that the aberrations of the optical elements within the spectrometer in no way cause deterioration in the instrument's performance, the resolving power is limited by the size of the dispersing element—prism or diffraction grating.

For a prism, the theoretical resolving power is given by

$$R = t\frac{dn}{d\lambda} \tag{19.16}$$

where t is the thickness of the base of the prism and $dn/d\lambda$ is the dispersion of the glass. By considering a prism with a base of 10 mm made from a typical glass which changes its refractive index by 0·02 over a spectral range of 2000 Å (see figure 16.12), it can be seen, using equation (19.16), that the resolving power is given by

$$R = \frac{10 \times (0·02)}{2 \times 10^{-4}}$$
$$= 10^3.$$

Prism spectrometers with resolving powers greater than this may be constructed by using prisms made of glass with greater dispersions (flint glasses) and by using larger prisms or a train of prisms.

For the diffraction grating, the resolving power is given by

$$R = Nm \tag{19.17}$$

where N is the total number of lines used across the grating and m is the order of interference. A typical grating may have 500 lines mm^{-1} and, consequently, a 10 mm grating, used in second order, would have a resolving power given by

$$R = 500 \times 10 \times 2$$
$$= 10^4.$$

It will be seen immediately that, size for size, the grating gives a resolving power which is typically an order of magnitude greater than that of the prism.

It is important to note that if full use is to be made of the dispersing element, its entrance face must be fully illuminated by the beam from the collimator. The cone of light provided by the telescope must also be accepted by the collimator. These conditions dictate the size and focal ratio of the collimator. A spectrometer is, therefore, said to have a certain focal ratio and, in practice, its value should match that of the telescope.

Any resolved spectral element within a spectrum, in effect, corresponds to an image of the entrance aperture or slit of the spectrometer in the monochromatic colour of that region of the spectrum. Thus, in order to achieve a given spectral resolution, it is important that the slit should be limited to a particular size. This prevents the spectrometer accepting light which lies outside a defined small angle. For a telescope–spectrometer system to be matched efficiently, it is important that the star image in the focal plane of a telescope should be smaller than or equal in size to the slit width so that all the light collected by the telescope is accepted by the spectrometer. Because of the practical limits to the sizes of the conventional dispersing elements, high resolving power can only be maintained by having small slits which, in some cases, are much smaller than the star images they are trying to accept. For example, it has been estimated that the slit of the spectrograph at the coudé focus of the 200-in (5·08 m) Hale telescope at Mt Palomar accepts only about 5–10% of the light in any star image.

There are, however, other spectrometric techniques which provide high spectral resolution with the advantage of having more reasonable angular acceptances. Such an instrument is based on the Fabry–Pérot interferometer. This comprises a pair of very reflective circular plates with a controlled space, the plate surfaces being parallel. Multiple beam interference occurs within the cavity so that only radiation of selected wavelengths is allowed to pass (see figure 19.13). Those allowed to pass correspond to the gap optical path length being at an integer number of wavelengths, i.e.

$$m\lambda = 2n\tau \cos\theta$$

where m is an integer (the order of interference), τ is the physical space between the plates, n its refractive index and θ the angle of the beam inside the cavity (see figure 19.13).

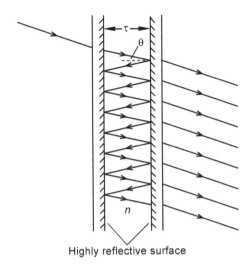

Highly reflective surface

Figure 19.13. An idealized Fabry–Pérot system depicting the generation of multiple beams within the optical cavity which constructively interfere on emergence when $2n\tau \cos \theta = m\lambda$.

The spectral resolution depends on the quality of the reflection of the plate surfaces and this is made to be very high by vacuum deposition of multilayer films on the cavity surfaces. The value of R may be written as $\pi m F$, where F is a coefficient depending on the quality of the reflection coatings. The value of m is high, usually a few thousand, and R can be in the range 10^3–10^6. It can be made to scan through the spectrum by changing the optical gap thickness and this can be done by adjusting the refractive index of the gas within the gap or by adjusting the physical gap thickness by piezoelectric pads. One of the advantages of this kind of spectrometer is its circular symmetry, making it efficient in accepting the collimated beam that is produced, following the focus of the telescope. The field stop is also circular rather than a slit and the spectral resolution is less compromised by seeing relative to grating instruments.

When a dispersed spectrum is recorded on a photographic plate, it is obvious that a suitable emulsion must be chosen with sufficient resolution to at least match the resolving power provided by the spectrometer. A similar consideration must be given to the matching of resolved spectral elements to the pixel size when CCD detectors are used. In order to allow the record to be examined easily, the star image is usually made to travel to and fro along the length of the spectrometer slit, by guiding the telescope, thus giving height to the spectrum.

For CCD records, the chip is orientated in the spectrum plane so that the dispersion runs parallel to the pixel rows. By allowing drift of the star image on the spectrometer slit, the spectrum is given height in the direction of the pixel columns. By averaging the recorded intensity values of the illuminated columns, a better S/N ratio is obtained for the various recorded spectral features.

For general survey purposes, it would be an exhausting task to attempt to record a spectrum of each star in turn. Such surveys were carried out by the use of an **objective prism**. This device simply comprised a small-angled prism placed over the telescope objective. Each star in any field was then recorded photographically as a small spectrum in the focal plane of the telescope, the latter simply acting as the camera of the spectrometer. The main application was the simultaneous recording, identification and classification of the different types of stellar spectra.

In comparative photoelectric spectrophotometry, the data are directly available on chart or in digital form and no calibration procedures are usually necessary. When the spectrum is recorded on a photographic plate, however, laboratory time is needed to perform an analysis. The variations of photographic density must first be measured and then, via the calibration on the plate, these must

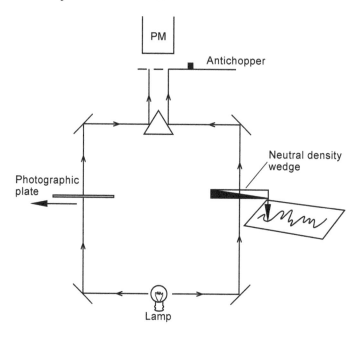

Figure 19.14. The schematics of a microdensitometer shows two light beams, one passing through the photographic plate, the other passing through a neutral density wedge. The beams are alternately detected by a photomultiplier. As the plate is scanned, any difference in intensity of the two beams causes movement of the neutral density wedge to restore the balance. The movement of the wedge is drawn out as a graph with the horizontal axis matching the movement of the plate.

be converted to intensity values. There are several laboratory instruments which enable analysis of spectrograms to be made and the brief description of one of them, the **automatic microdensitometer**, is given here.

The microdensitometer has an adjustable carriage onto which the photographic plate is placed (see figure 19.14). A projector lamp, followed by an optical system, illuminates a small area of the plate and light passing through the plate is collected by a microscope. The image produced by the microscope is projected on to a slit and light accepted by it is made to fall on to a photomultiplier. The current produced by the cell, therefore, corresponds to the transparency of the area of the plate which has been selected by the slit.

The same projector in the instrument also provides a beam of light which is made to pass through part of a neutral density wedge. This wedge is made of glass having a density which increases linearly along its length. After passing through the plate, this beam is made to fall on to the same photomultiplier as above.

By a chopping mechanism, the photomultiplier is allowed to see each beam in turn. If the density of the part of the wedge through which the second beam is passing does not match that of the investigated area of the photographic plate, an error signal is generated. This is made to control a servo-system which drives the position of the wedge until a match is achieved between the two beams. Thus, the density of the investigated part of a plate can be read off according to the position of the density wedge.

Automatic spatial surveys, such as scans through the length of the spectrum, can be effected by driving the carriage supporting the photographic plate and linking the motion to a second table which is made to pass under the density wedge. A pen attachment to the wedge allows the density variations to be drawn out automatically on a graph carried by the second table.

The availability of CCD technology is rapidly replacing photography as a means of recording spectra. The greatly improved quantum efficiency, the linearity of response and direct input to computer for immediate reduction and assessment makes the process very compelling. Again, because of the pixel structure of the detector, an effective reference scale for wavelength calibrations is automatically available.

Problems—Chapter 19

1. The apparent V band magnitude of a star is 8·72 and for its temperature the bolometric correction is $-0·48$. What is the apparent bolometric magnitude of the star?

2. A star has a colour index $(B - V)$ of $+1·0$ and its apparent B band of magnitude is 6·4. For this value of colour index the bolometric correction is $-0·5$. What is the apparent bolometric magnitude of the star?

3. Above the Earth's atmosphere a star has a magnitude of 7·5. If at an observing site the zenith magnitude gain is $+0·6$, what would be the apparent magnitude of the star at a zenith distance of 30°? What is the transmission factor of the Earth's atmosphere at this zenith distance?

4. After correction for the atmosphere, a star is found to have a colour index $(B - V) = -0·13$. If at a particular observatory the zenith magnitude gain for the B band is 0·29 and for the V band 0·17, at what zenith distance would the star have an *apparent* colour index of zero?

5. At a zenith distance of 45° a star has an observed colour index $(B - V)$ of $+0·0$; at 30° the measured value is $-0·1$. What is the value of the colour index corrected for atmospheric absorption?

6. While on the launch-pad, an astronaut observes a star at a zenith distance of 54° and estimates its apparent magnitude to be 2·50. Some time later when the star's zenith distance is 40°, he/she estimates its apparent magnitude to be 2·30. What would be his/her estimate of its apparent magnitude when he/she is placed in orbit above the Earth?

7. From a calibration study using standard stars in the field, it is found that a measured image diameter, d (in μm) on a photographic plate can be related to a magnitude by

$$m = 10 - \log_{10} d.$$

Two stars are measured to have image diameters of 100 and 150 μm, respectively. What is their magnitude difference?

8. Using a photon-counting photometer, measurements show that a second magnitude star provides $\sim 10^6$ counts s^{-1} with scintillation noise $\sim 2\%$. At what stellar magnitude would photon shot noise begin to dominate the accuracy of brightness measurements?

9. A spectrometer system using a camera lens with a focal ratio of $f/6$ provides a reciprocal linear dispersion of 40 Å mm^{-1}. If the camera lens is changed to work at $f/3·5$, what would be the new reciprocal linear dispersion?

10. A diffraction grating with a ruling of 500 lines mm^{-1} is 50 mm in length and is used in second order as the disperser in a spectrometer. What is the theoretical resolving power of the instrument? What is the minimum radial velocity that could be detected by Doppler shift? If the linear dispersion of the spectrum is 40 Å mm^{-1}, what is the physical size of a resolved element in the spectrum?

11. A spectrometer with a diffraction grating with a ruling of 600 lines mm^{-1} is used in second order. The grating is illuminated at an angle of 330° and the dispersed spectrum emerges along the direction of the normal to the grating. Calculate the reciprocal linear dispersion (RLD) when a camera of 0·5 m focal length is used.

    If the grating is 70 mm × 70 mm square, calculate the theoretical spectral resolving power of the instrument. If the system is used to record stellar spectra under atmospheric conditions such that the seeing disc is ~ 10 arc sec, estimate the wavelength spread of each resolved spectral element if all the light from the star image is allowed to pass through the system. How does this compare with the wavelength spread of the theoretical resolved element say at 500 nm?

Chapter 20

Modern telescopes and other optical systems

20.1 The new technologies

It might be thought, with the ability to put telescopes such as the Hubble Space Telescope into orbit above the disturbing effects of the Earth's atmosphere (see section 19.7), and with the well-known difficulties experienced in the past in building ground-based telescopes larger than the 6 m Hale Telescope at Mt Palomar, that the creation of larger ground-based telescopes would not have been undertaken. This is by no means the case. Two recent developments have enabled new observatories to be established with telescopes far larger than the Hale Telescope and rivalling the Hubble Space Telescope in resolving power. The developments are **active optics** and **adaptive optics**.

The current quest for providing new telescopes is chiefly the provision of larger diameters allowing the collection of more and more flux from the very faint objects and the utilization of the potential angular resolution that such diameters provide. Modern technologies involving new optical materials, new concepts in achieving large imaging apertures and the introduction of computer support have revolutionized the approach of producing new large telescopes.

20.1.1 Active optics

The previous generation of large telescopes involved the production of large monolithic primary mirrors to collect the flux from celestial sources and to produce images for detection. In order for the image quality to be maintained as the telescope is oriented to different parts of the sky, it was important to have mirrors with good mechanical stability. To achieve this, the mirror design required a thickness-to-diameter ratio of the order 1:9. As telescopes grow in size, the requirement of producing thick blanks from which the mirror is figured causes difficulties in their production. In addition, primary mirrors with large mass require very heavy engineering to allow the telescope to be readily manoeuvrable. As the telescope alters its orientation in moving from one target object to another, the mirror tends to flex under its own weight. Such flexure allows the optical surface to deform with the consequent deterioration of image quality.

However, by using 'active' supports on the underside of the mirror in the form of a distribution of pistons, it is possible to readjust the shape of the optical surface of the mirror and maintain the required figure. By continuously monitoring the image quality of some reference object in the field, the adjustments may be applied continuously by computer control. Because of the inertia of the large optical system, the response to the feedback is relatively slow ($\sim$ tens of seconds) but easily sufficient to allow 'continuous' adjustment as the telescope tracks a celestial object. Thus, it is now possible to use lighter mirrors with smaller thickness-to-diameter ratios. Such mirrors are obviously more likely to suffer flexure but can also be corrected more easily. Many modern monolithic mirrors are now manufactured with material removed from their underside to form a honeycomb pattern with the correcting pistons in contact with the rib structure.

An additional development involving active optics is the idea of producing a large collection aperture by constructing the main mirror from a mosaic of smaller ones. Such a system is referred to as a **multiple mirror telescope** (MMT). Small mirrors are so much cheaper and easier to make that the cost of the resulting MMT is still considerably less than it would be to construct a single mirror equivalent in size. The idea of using mirror mosaics is not new but the concept has only recently come to fruition with the advent of inter-active computer control systems to maintain the mosaic alignments while the telescope is being guided.

Perhaps the most famous MMTs are those of the Keck Observatory[W 20.1] on Mauna Kea, Hawaii, at a height of 4160 m. Two identical 10 m reflectors have been built, each with 36 hexagonal segments. The telescopes are only 85 m apart and can be used as an interferometer (see later). The Observatory, owned and operated by the California Association for Research in Astronomy, provides an unprecedented resolving power at visual and near-infrared wavelengths. It may be noted too that the James Clerk Maxwell telescope[W 20.2] designed for use in the mm spectral region and also sited on Mauna Kea is an MMT system. It comprises a primary dish of 15 m diameter made up of 276 aluminium panels, each of which is adjustable to keep the surface as near to perfection as possible.

Another interesting approach is that of the combination of images provided by individual telescopes which are linked to combine their foci either by superposition or in some other way. Technically, such a system should be referred to as a multi-aperture telescope (MAT) but the term of MMT is more frequently applied. The best known example of a multi-telescope combination is that sited at the Whipple Observatory[W 20.3] in Arizona at an altitude of 2382 m under the auspices of the Steward Observatory and the Smithsonian Astrophysical Observatory. Fully operational in 1980, it consisted of six 1·8 m mirrors arranged in a hexagonal array on an alt-azimuth mounting. This MAT produced a collecting area equivalent to one 4·5 m single mirror telescope. Star images of less than 0·5 arc sec were achieved. The individual images from each of the mirrors were sent to a six-sided hyperbolic secondary mirror system which produced a single image. Computer-controlled repositioning of the secondary mirrors provided a correcting system to enhance the final image. One of the engineering advantages of this approach is that the length of the system depends on the diameter and focal ratio of each of the component telescopes. For a single telescope with a diameter equal to that of the effective combination but working with the same focal ratio, the length of the system would be that much greater. More recently, however, the six primary mirrors in the mount have been replaced by a single 6·5 m mirror.

Currently, the largest multi-telescope system is operated by the European Southern Observatory (ESO) in Chile at Cerro Paranal. This system comprises four individual and independent 8·2 m telescopes that can be linked to provide interferometry and aperture synthesis (see later) across baselines according to their separation.

All the plans for future large optical telescopes are based on expansion of current MMT technology. The grandest project yet, the OWL (overwhelmingly large) telescope proposed by ESO, is for a diameter of 100 m, using 1600 segments. Each piece will need to be polished to an accurate shape and positioned with nanometre precision and kept in place by a system of sensors and activators, this being the essence of active optics.

The new telescopes with active optics satisfy one of the main purposes of using collectors with large diameters, i.e. the collection of larger amounts of flux to boost the strengths of recorded signals from celestial sources and improve the measurement signal-to-noise ratio. However, their potential of achieving improved angular resolution according to the telescope diameter, D, is spoiled because of seeing. This problem is being addressed through the principle and technology of adaptive optics.

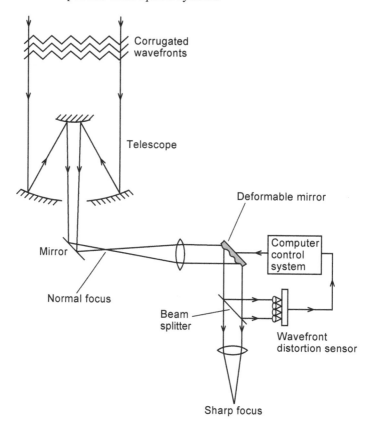

Figure 20.1. A schematic layout of an adaptive optics system.

20.2   Adaptive optics

As described in section 19.7, the atmosphere in the light path above a telescope distorts the incoming wavefronts with the image definition fluctuating at a rate of about 100 times a second. At any one instant, to a first order, over a large telescope diameter, the wavefront corrugations can be considered to comprise a number of plane sections with dimensions of the order of the Fried parameter, r_0, but with a random set of tilts. Substantial correction of the blurring that this introduces could be achieved by sensing the deformations, taking note of each of the locations of the plane sections, and applying compensating tilts so that the whole of the wavefront becomes planar again. Such fast correcting optics are referred to as adaptive optics (AO) systems.

One possible way of doing this would be to consider making the elements of any mosaic telescope to be of a size no larger than r_0. If the locations and tilts of the disturbed sections of the original plane wavefront could be continuously monitored, elaborate feedback systems might be designed to control the individual pieces of the mosaic so that elemental tilt corrections could be applied.

Although the concept is sound, its implementation cannot be effected directly in this way. The behaviour of seeing requires the response times of the mosaic elements to be no longer than a few milliseconds. Because of the substantial weight of each mosaic cell, their large inertias prevent adjustments on such timescales.

The problem can be addressed, however, by designing the AO with lightweight elements. One such scheme is shown in simplified form in figure 20.1. Following the normal focus of the telescope, the light beam is collimated and then reflected off a deformable mirror at a position such that the collimator produces an image of the telescope aperture on it. At this position the effective size of r_0 is reduced or demagnified by the ratio of the diameter of the telescope to that of the collimated beam. Subsequent to the reflection, part of the beam is deflected to a wavefront distortion sensor but with the greater part of the flux brought to a focus where the astronomical data are collected. One form of sensor is a fast readout CCD with lenslets prior to the pixels, so producing an image of the field on the chip. On successive readouts, a computer control system constantly provides signals to the activators controlling the shape of the small deformable mirror. This mirror may be constructed from small plane mirrors such that their individual sizes are no larger than the demagnified value of r_0, with each element controllable in tilt by piezo-electric pistons. Thus, the plane sections within the original broad diameter beam illuminating the telescope are realigned to produce a collimated beam with the greater part of the wavefront disturbances corrected. Much sharper images are then found at the chief focus for applying long exposure cameras or other analysing instruments for data collection. By improving the Strehl index (see section 19.7.3), for example, more of the flux collected by a large telescope diameter can be made to pass through the entrance slit of an attached spectrometer. Alternative schemes for sensing the seeing involve the production of 'artificial stars' by application of powerful lasers to illuminate the regions of the atmosphere where the turbulence occurs and to assess the deformations by recording the returned scattered light.

For most kinds of observation, it is important to have adaptive optics as a standard feature on all large telescopes. AO systems were automatically considered in the design of the twin 8 m Gemini telescopes[W 20.4]. Gemini North is located on Mauna Kea, Hawaii, at a height of 4200 m, Gemini South is on Cerro Pachon, Chile, at an altitude of 2715 m. While the performances of both giant telescopes depend upon their being at such great heights that much of the Earth's atmosphere is below them, their success in producing images as sharp as the Hubble Space Telescope's images also depends on their AO operating systems.

20.3 Measurements at high angular resolution

20.3.1 Michelson's stellar interferometer

The distances of stars are so great in relation to their diameters that, even with the world's largest telescopes, it is extremely difficult to make measurements of their apparent diameters from conventional imaging. By extending an interferometric method which he had developed for measuring the diameters of such objects as Jupiter's satellites, Michelson in the 1920s provided the first direct measurements of the diameters of a few stars. His interferometer was never in general use but it is worthwhile describing it here, as it represents a milestone in providing us with the basic knowledge of the sizes of stars.

Michelson's original interferometer consisted of two collecting apertures which were fixed to a beam so that their separation could be adjusted. The length of the beam was 6 m. A very stable platform was provided for the beam by attaching it to the 100-in (2·54 m) Mt Wilson telescope. The essential elements of the optical system are depicted in figure 20.2. Its principle is similar to that of Young's classical double-slit experiment.

Suppose that the interferometer is in exact adjustment so that the distances M_1, M_2, M_5, M_6, eye and M_4, M_3, M_5, M_6, eye provide the same optical path length. The wavefronts arriving perpendicularly to the system are intercepted by the two apertures M_1 and M_4 and the interference pattern set up from the two rays can be viewed in the eyepiece. For the second series of wavefronts arriving at an angle θ to the first set, a second interference pattern is set up but, in general, this is slightly displaced so that it does not overlap the first. For the second set of wavefronts, the ray leaving M_4 lags in phase behind that from M_1 as it has had to travel an extra distance given by $d \sin \theta$, where d is the separation of the mirrors. If the value of $d \sin \theta$ is equal to $\lambda/2$, then a bright fringe from the pattern set up by the second set of wavefronts will fall exactly on a dark fringe which was produced

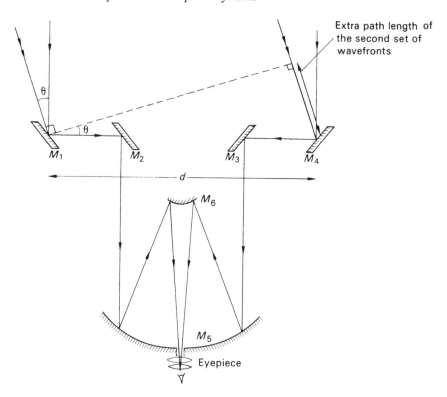

Figure 20.2. The Michelson stellar interferometer. Mirrors M_1 and M_4 collect the light from the star and act in the same way as the slits in Young's classical interference experiment. Wavefronts are drawn corresponding to those arriving from points separated by an angle θ.

from the first set of wavefronts. The overall effect at this condition is that the two fringe patterns are cancelled out by each other. Thus, for two point sources separated by an angle θ, a separation of the mirrors M_1 and M_4 can be found so that no interference pattern is present and, at this condition,

$$d \sin \theta = \frac{\lambda}{2}.$$

Hence,

$$\sin \theta \approx \theta = \frac{\lambda}{2d}. \tag{20.1}$$

When the object being investigated is a single star, the interference from each point within the optical disc needs to be considered. Mathematical procedures which are similar to those needed to evaluate the diffraction pattern produced by a circular aperture are required to determine the condition when the interference pattern is destroyed. For a circular object of uniform brightness, the interference pattern should not be detectable when

$$\theta = \frac{1 \cdot 22\lambda}{d} \tag{20.2}$$

where θ now corresponds to the angular diameter of the star. If the star is known not to be uniformly bright, allowances can be made which alter the numerical factor contained in equation (20.2).

The original Michelson interferometer was only able to investigate a few nearby supergiant stars. One of its limitations was that the interference fringes had to be assessed by eye. The experiment requires the separation of the mirrors to be adjustable but, for the white light fringe condition, the difference in path length for the two beams needs to be maintained to the order of *one* wavelength of light—a difficult mechanical problem at the time. With the application of modern equipment such as lasers to monitor the mirror separation and TV cameras to record fringe patterns, there has been a resurgence of interest in the principle of Michelson interferometry by several research groups around the world.

20.3.2 Aperture synthesis

The larger the telescope, the greater is the potential angular resolving power to investigate finer detail in an image. Unfortunately, the recorded sharpness is degraded by a large factor as a result of atmospheric seeing.

One approach to redress the situation is by the technique of **aperture synthesis**. If the telescope aperture is fitted with a mask so that only a few small patches of the primary mirror are used, the light is still brought to a focus but the image will be in the form of an interference pattern. In the experimental arrangement, the sizes of the smaller apertures should be no larger than the Fried parameter. By altering the structure of the mask, so that the smaller apertures are redistributed over the main mirror, the resultant image will again be an interference pattern but with a different form. From a combination of the various recorded patterns, a resultant image can be constructed as though achieved by the complete large aperture. Thus, the resolution potentially available from the large aperture telescope is synthesized from the smaller element combinations used to produce the 'component' images.

To make such a technique more powerful and to allow the synthesizing of an effective telescope aperture of some tens of metres, i.e. much larger than the currently available optical telescopes, it is possible to use an array of independent small telescopes. This notion was first developed by radio astronomers (see section 21.5) but the technological requirements for the optical region are more complicated because of the much smaller wavelengths in the radiation relative to the physical sizes of the engineered apparatus. The approach of a multi-telescope array is incorporated in COAST[W 20.5] (Cambridge Optical Aperture Synthesis Telescope). The instrument comprises five 40 cm telescopes arranged in a 'Y' configuration, the maximum separation of the elements being 22 m. Rather than by moving the individual elements to synthesize an aperture of this order, a range of effective configurations is achieved by taking records at different times during the night, the Earth's rotation changing the orientation of the array relative to the observed star.

A notable success of the system was the very clear imaging of the close binary star, α Aur (*Capella*), with a separation $\sim 0.''05$. Records taken weeks apart revealed movements of the components as they orbit their common centre of gravity with a period of 104 days. The COAST group are now planning an array with a baseline of 400 m, with the potential of 0·25 milli-arc sec angular resolution.

20.3.3 The intensity interferometer

Again, following a concept developed for radio astronomy by Hanbury Brown and Twiss, the principle of the intensity interferometer was considered in the 1950s for the optical region and an experimental arrangement was established at Narrabri in Australia. Rather than utilizing the classical concept of measuring the interference of waves based on their amplitudes and phases, the method involves the measurement of low-frequency beat signals that are generated when incoherent intensity signals from the two collectors are combined. In the Narrabri arrangement, the two light collectors (light buckets!) were large compound mirrors (6·5 m in diameter) but with little effort being made to obtain images of the quality that could theoretically be achieved by a telescope of this diameter.

The current arrangement at Narrabri is shown in figure 20.3. The experiment involves an investigation of the way in which two collectors receive the radiation from the star according to their separation. The fluctuations of the arrival of energy are compared or correlated. A determination of the behaviour of the degree of correlation according to the separation of the two collectors allows the

Figure 20.3. An aerial view of the current stellar interferometer, Narrabri, Australia. (By courtesy of the University of Sydney.)

diameter of the star to be calculated. As the strength of the correlation signal is very weak, some tens of hours of integration are required for each star.

Over the years, the Narrabri experiments have provided diameter measurements for a representative sample of stars across the whole spectral range.

20.3.4 Lunar occultation method

The natural phenomenon of lunar occultations has provided a method of measuring the diameters of a few of the very bright stars. When a star is occulted, the Moon's limb acts as a sharp edge producing a Fresnel diffraction pattern at the Earth's surface. A simplified investigation of the scale of the pattern is illustrated in figure 20.4 where x corresponds to the distance between two points on the Earth at which the optical path length differs by $\lambda/2$. At the position corresponding to the path length, $d_{\mathbb{C}}$, the star will appear at full brightness. At exactly the same time, on sampling the brightness westwards from this point, there will be a progressive fall to a minimum where the path length is $d_{\mathbb{C}} + \lambda/2$. From that point, the brightness will increase again and display further oscillations corresponding to the diffraction pattern. To a first order it can be seen from figure 20.4 that

$$d_{\mathbb{C}}^2 + x^2 = (d_{\mathbb{C}} + \lambda/2)^2.$$

By expansion and ignoring the very small terms, the typical fringe size may be written as

$$x = (d_{\mathbb{C}} \lambda)^{\frac{1}{2}}. \tag{20.3}$$

By taking the lunar distance as 384 405 km and the wavelength of light as 550 nm, the value of $x \simeq 14 \cdot 5$ m.

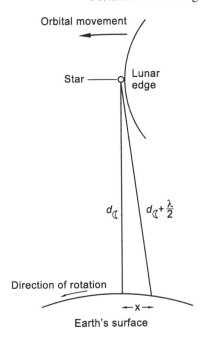

Figure 20.4. The basic geometry of the Fresnel pattern associated with the lunar occulting edge.

For any telescope/photometer system recording the disappearance of the star, the diffraction pattern is made to sweep across the aperture by the Earth's rotation and also by the diffracting edge moving according to the lunar orbital motion, the former being the dominant factor in determination of the timescales. A simple calculation of the timescale for the passage of a fringe is given by

$$\Delta t \approx \frac{x P_\oplus}{2\pi R_\oplus}$$

where $R_\oplus$ and $P_\oplus$ are the radius and period of rotation of the Earth respectively. By inserting the necessary values with their appropriate units

$$\Delta t = \frac{(14 \cdot 5) \times 24 \times 3600}{2\pi (6370 \times 10^3)} \approx 0 \cdot 030 \text{ s}.$$

Thus, the diffraction pattern may be recorded simply by letting it sweep across the telescope aperture and undertaking the brightness sampling with sufficient time resolution. Figure 20.5 displays the pattern according to the time of its progression across the photometer.

Because the stellar source has a finite angular size, the recorded pattern does not correspond to that expected of a point source, the oscillations being not as well defined (see figure 20.5). Differences between the record and that of a theoretical point source allows determination of the diameter of the stellar disc. The number of stars that offer diameter determinations in this way is of course limited to the very bright ones which happen to lie in the zone of sky through which the Moon travels.

20.3.5 Speckle interferometry

When a very large telescope is used with high magnification to observe a star, the image may be seen to have a rapidly changing grainy structure or exhibit a speckle pattern. Such effects are produced by the turbulence in the atmosphere above the telescope. Any bright portion in the pattern results from

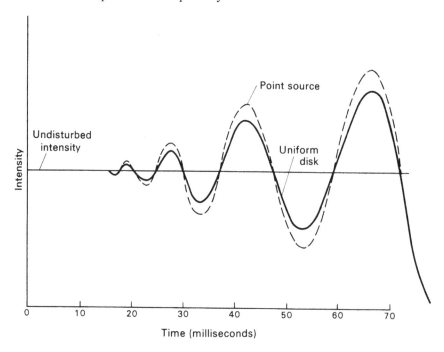

Figure 20.5. The rapid fluctuations of intensity of a star as it is occulted by the Moon. (A reflection of this curve is obtained at the reappearance of the star.)

interference of essentially plane portions of the beam which, at any instant, happen to be in phase but originate from widely separated areas of the primary mirror. An instantaneous speckle pattern, therefore, contains high spatial resolution information down to the diffraction limit of the telescope. Sketches of an instantaneous pattern of a single star and one for a close double star are depicted in figure 20.6.

The term **speckle interferometry** has been coined to cover the technique whereby a series of speckle patterns are recorded by photography, video camera or CCD, combined and then spatially analysed to investigate any difference from what would have been obtained from a point source. In the first place, the speckle patterns must be obtained using a large plate scale and with a short exposure. In order to detect whether a star has a noticeable diameter, a second star needs to be in the field of view at an angular distance such that the seeing affecting both objects is essentially coherent. The speckle pattern of the second star is then used as a reference to detect any difference that the target star's diameter has had on its recorded speckles. The technique has been most successful in determining the separations of double stars simply by noting the pairing of elements within the recorded speckle patterns and measuring their separations and angles of the lines joining the pairs.

The first successful speckle interferometric work was done with the 200-in (5·08 m) Mt Palomar telescope. Using a microscope objective at the Cassegrain focus, the focal ratio was converted to be about $f/500$ giving a plate scale of 1 arc sec per 25 mm and short exposures of the order of 1/100th of a second were made possible by employing image intensifiers. The disks of *Betelgeuse, Antares* and *Aldebaran* have been resolved and, for the first time, *Capella* was resolved as being a double star (see the result from COAST described earlier).

The techniques of speckle interferometry are continuously under development and, no doubt, there will be improvements, particularly in reducing the time between recording the speckles and determining the spatial content of the stellar disc.

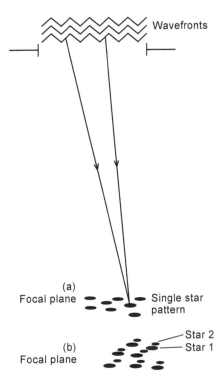

Figure 20.6. The production of speckle patterns for (*a*) a single and (*b*) a close double star.

20.4   The Schmidt telescope

Improvements in wide-field photography over astrographic refractor systems were contemplated by considering the use of specially designed reflector cameras. The simplest system would consist of a single spherical mirror with an aperture stop at its centre of curvature. Such a camera has no unique axis and any point in the field would be imaged equally well—the focal surface would be curved, with a radius given by the focal length of the mirror.

When such simple camera systems are designed to have smaller focal ratios than those of the fastest astrographic refractors, the images suffered severely from spherical aberration. Many ways were suggested whereby this aberration might be reduced but it was not until the 1930s that a design was successfully manufactured by Schmidt in Hamburg. The system made by Schmidt was similar, in principle, to an optical device for a good quality parallel light source suggested and patented by Kellner some 20 years previously.

The **Schmidt telescope** is basically a two-element system, whereby the spherical aberration produced by the spherical collecting mirror is compensated by using a thin aspherical correcting lens or plate at the centre of curvature of the mirror. There is a range of shapes that the corrector lens can take. The basic parts are depicted in figure 20.7 in which the departure of correcting lens from being a parallel plate is greatly exaggerated. It is usual practice to make no effort to correct for the curvature of the focal surface and specially produced photographic plates or film are required to match the curvature of the surface of best focus of the camera. Since the introduction of the basic design, improvements have been suggested and investigated which use more elements in the optical system but these will not be discussed here. Schmidt cameras can have focal ratios of $f/1\cdot0$ or smaller and can provide field

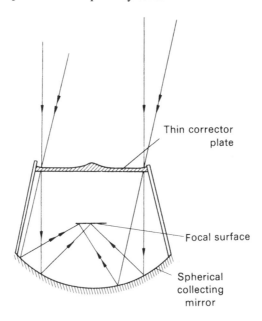

Thin corrector
plate

Focal surface

Spherical
collecting
mirror

Figure 20.7. A Schmidt telescope or camera.

coverage as wide as 25°. They are primarily intended to be used as high-speed cameras rather than as conventional telescopes.

One of the obvious uses of Schmidt telescopes is to allow star surveys of large areas of the sky. Perhaps the most famous survey undertaken is that of *The National Geographic Society—Palomar Observatory Sky Survey* whereby 935 pairs of plates from the blue and red parts of the spectrum have been obtained covering the whole of the sky north of declination −33°, each plate having a field diameter a little larger than 5°. The Palomar Schmidt camera used in the survey has a corrector plate aperture of 1·25 m (the telescope is known as the 48-in Schmidt), a main mirror diameter of 1·83 m (72 in) and the system is used at $f/2\cdot44$.

A complementary survey of the southern hemisphere skies involves the UK 1·2 m Schmidt telescope at Siding Spring Observatory[W 20.6], Australia (see figure 20.8). With the newer emulsions now available, this survey records objects one magnitude fainter than the Palomar survey. The photographs are taken on 14 in square glass plates, each covering a wide field of 6°5. On each exposed plate, there may be a million images. The task of measuring their brightness and position is a colossal one and special scanning systems have been developed to extract the enormous quantity of information recorded on such plates (see section 19.2).

The plates are stored at the Royal Observatory Edinburgh, UK, with high quality copies being made available to astronomers throughout the world. Together with other astronomical plates, they form an extremely valuable archive.

Another use of the Schmidt camera is for recording the positions of artificial Earth satellites.

Although the basic Schmidt system provides good quality images, they are not perfect. It is obvious that some chromatic aberration is introduced by the correcting plate. The plate also only has a correct shape for removing spherical aberration for one particular wavelength and, thus, the residual spherical aberration is wavelength dependent. Coma is also present in the images away from the centre of the field.

Figure 20.8. The UK Schmidt telescope at Siding Spring, Australia. (By courtesy of the Science Research Council.)

A development, since the original design of the Schmidt telescope, is the **Maksutov–Bouwers camera**. In this design, the aspheric correcting lens is replaced by a negative long-focus meniscus lens with spherical surfaces. By the correct choice of the curvatures of the lens surfaces, the amount of spherical aberration produced by the lens can be made to match that of the collecting mirror but with the opposite sense. The single lens can also be designed to be achromatic, again by the correct choice of the radii of curvature of the lens surfaces in combination with the correct choice of lens thickness, thereby providing a lens–mirror combination which is free from chromatic aberration. Whereas each corrector plate of Schmidt telescopes needs individual figuring, the simplicity of using spherical surfaces in the corrector lens allows easier production of Maksutov–Bouwers cameras. A simple Maksutov–Bouwers camera is depicted in figure 20.9.

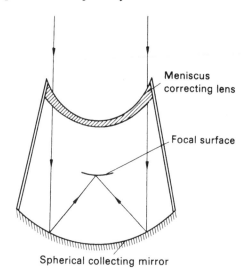

Figure 20.9. A Maksutov–Bouwers camera.

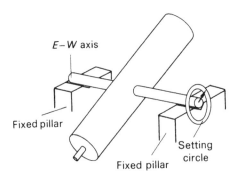

Figure 20.10. The transit telescope.

20.5   The transit telescope

The **transit telescope**, or **meridian circle**, usually takes the form of a small refracting telescope but it is mounted on only *one* axis which is horizontal, lying along the east–west direction. In this way, the telescope is restricted to movement in one fixed plane so that it points in directions which all lie on the observatory's meridian. The rotational axis is provided with a declination setting circle. The essential parts of the transit telescope are illustrated in figure 20.10 and an automated operational system is depicted in figure 20.11.

In its basic form, an eyepiece is fitted with a special graticule, the central vertical wire corresponding to the meridian. By setting the telescope to the correct altitude, the passage, or transit, of a star across the meridian can be timed accurately. This may be done by using the times of passage across each of the graticule wires in turn and taking a mean value for the transit of the central wire. At one time, the transit telescope's function was to supply checks for master clocks which were, in turn, used to provide civil time. Nowadays, they are used to make regular checks on the rotational period of the Earth. They are also used to provide fundamental data for the purpose of determining the coordinates of celestial objects.

As with all experimental procedures, current measurements are made with electronic detectors

Figure 20.11. The Carlsberg Meridian Telescope with CCD detector system attached for the automatic monitoring of stellar transits. Both positions and brightnesses of over 100 000 stars are regularly recorded each night.

under computer control. The operation of the Carlsberg Meridian Telescope[W 20.7] has progressively improved since its establishment on La Palma in 1984, at that time being one of the world's first automatic telescopes. The instrument was programmed to automatically set itself to the correct altitudes of about 500 stars in turn and measure the time of passage and their magnitudes. More recently, the photoelectric scanning-slit micrometer has been replaced by a CCD detector operating in a drift-scan mode. Between 100 000 and 200 000 stars are now measured each night and the limiting magnitude of the system is ~17.

20.6 Zenith tubes

Very accurate positions of the latitude of an observatory are obtained by using **zenith tubes** of which there are several designs. Their advantages over the transit telescope for this particular purpose are that the instrument does not suffer from flexure, as it is always used in a near vertical position, and the effects of atmospheric refraction are very small. Regular positional measurements by this type of instrument allow the motions of the Earth's poles to be studied.

The zenith tube or telescope consists of a small refractor which is limited in movement so that only the zenith field can be observed. In its basic form, the measurements are made using a micrometer eyepiece but there are several more modern versions using photographic and electronic recording techniques which provide improved accuracy.

A single determination of an instrument's position requires measurements of the difference in zenith distance of two stars when they are placed on the observer's meridian—absolute zenith distances are not required. The telescope is first directed to a position which is just south of the zenith and the micrometer eyepiece adjusted so that the transit of a star is made to occur along the horizontal wire. By turning the telescope through 180° about a vertical axis, transits occurring just north of the zenith can be observed. When a suitable star appears in the field, the horizontal wire is adjusted by the micrometer

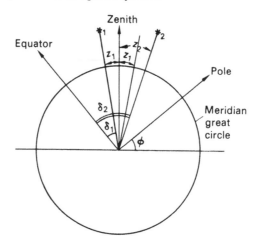

Figure 20.12. Measurements of two stars on the meridian using the zenith tube.

screw so that the transit is made along the same wire. The difference between the settings of the wire is read off the micrometer and this is then converted to angular measure.

Suppose that two stars have declinations δ_1 and δ_2 and appear on the meridian at a particular site when the zenith distances are z_1 and z_2 (see figure 20.12). The measurements made by the zenith tube provide a value for $(z_2 - z_1)$.

By considering sections 8.2 and 8.3, it is readily seen from figure 20.12 that the latitude, ϕ, of the observing station is given by

$$\phi = \delta_1 + z_1. \tag{20.4}$$

It can also be seen from the same figure that

$$\delta_2 = \delta_1 + z_1 + z_2 = \delta_1 + 2z_1 + (z_2 - z_1).$$

Therefore,

$$z_1 = \frac{(\delta_2 - \delta_1)}{2} - \frac{(z_2 - z_1)}{2}. \tag{20.5}$$

By substituting equation (20.5) into (20.4), we have

$$\phi = \delta_1 + \frac{(\delta_2 - \delta_1)}{2} - \frac{(z_2 - z_1)}{2}$$

or

$$\phi = \frac{(\delta_1 + \delta_2)}{2} - \frac{(z_2 - z_1)}{2}. \tag{20.6}$$

Thus, the latitude of the observing station is the mean of the stars' declinations, corrected for half the difference of their zenith distances.

Accurate positions of longitude may be determined if the transit times are also recorded.

20.7 Portable positional instruments

20.7.1 The theodolite

For some purposes, such as surveying, it is useful to have some kind of portable instrument which is capable of determining the position of an observer from measurements made of astronomical objects. Such an instrument is the theodolite.

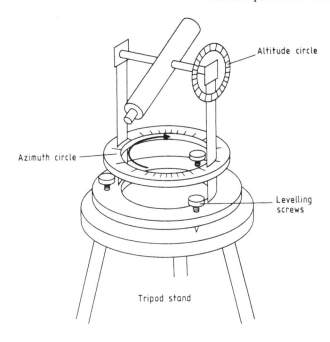

Figure 20.13. The theodolite.

The **theodolite** is essentially a small telescope, with an erecting eyepiece and graticule, which is mounted on an alt-azimuth stand. The basic design is illustrated in figure 20.13. A horizontal axis is set up between two pillars so that the telescope can be set to any value of altitude. The pillars are secured to a rotatable ring so that the horizontal axis can be directed to any azimuth. Levelling screws and spirit levels allow the instrument to be set up at any site. Graduated circles which are attached to the two axes allow measurements to be made of the azimuth and altitude of any celestial object. Filters are usually provided to cover the objective so that positional measurements can be made of the Sun. Positional measurements can be made with an uncertainty which is typically of the order of a few seconds of arc. By taking readings of the positions of different objects at known times or by taking readings of the position of one object over a known interval of time, the geographic position of the observer can be determined, after applying the relevant corrections for atmospheric refraction, semi-diameter (if the Sun or Moon is observed), etc.

For satellite tracking, special quickly-adjustable theodolites are required and such a system is depicted in figure 20.14.

20.7.2 The sextant

The **sextant** is primarily an instrument designed for navigation at sea. When it is used in its simplest way, it allows measurements of the altitudes of celestial objects to be made. From altitude measurements of objects at known times, the geographic position of the observer can be determined. The sextant is sufficiently light and compact so that it can be held by hand. The essential features of the instrument are illustrated in figure 20.15.

Two mirrors are mounted on to a frame which is provided with a handle so that it can be held by the observer. The first mirror, known as the **index mirror**, I, is fixed to a lever which is pivoted through an axis passing through the mirror's silvered surface. At the opposite end of the lever arm a vernier is attached so that the angle of the index mirror can be read off the graduated arc forming the lower part of the sextant frame. The second mirror, known as the **horizon mirror**, H, is set at a

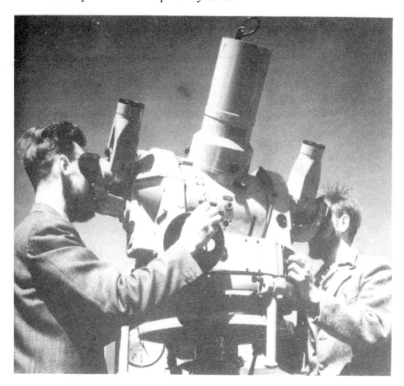

Figure 20.14. A satellite-tracking kinetheodolite of the Royal Observatory, Edinburgh. (By courtesy of the Royal Astronomical Society.)

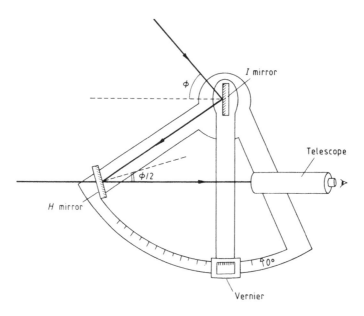

Figure 20.15. The sextant.

fixed position on the frame. It is divided into halves. The top half is transparent while the bottom half is silvered. When the field is viewed by the telescope, the split horizon mirror allows the horizon to be viewed simultaneously with the field which can be seen via the two reflections given by I and the lower half of H.

To permit solar observations, filters of various density are available to cover both the I and H mirrors.

It is obvious that when the mirror faces of I and H are parallel, I and the transparent half of H provide identical views of the horizon. By adjusting the lever which carries I, the two images can be made to overlap, thus allowing the fiducial point or index point to be determined on the graduated scale.

When the altitude of a celestial object is being determined, the lever carrying I is adjusted until the object is made to 'appear' on the horizon. Since a reflection by a mirror alters the direction of travel of a beam of light by twice the angle of incidence, an angle read off the graduated scale corresponds to twice the apparent altitude of the object. This is allowed for in the graduation of the scale, each $10°$ of rotation being registered as $5°$. Before an accurate figure of the true altitude is available, corrections need to be made for any error in the index position, for the observer's height above the horizon and for atmospheric refraction. The uncertainty in any one determination is likely to lie in the range between 5 and 10 seconds of arc.

20.8   The coelostat

The coelostat is a special collector system, usually applied to solar observations, providing an image which is fixed in space. Such a system allows heavy subsidiary analysing equipment to be used. Most coelostat systems are placed at the top of a tower with the analysing equipment below the level of the ground where it is more easily stabilized against temperature fluctuations. The positioning of the coelostat at a height above the ground is optimum in obtaining the best seeing conditions and the collected radiation's path to the analysing equipment can be made virtually free from disturbing air currents by enclosing it in a thermally insulated tube.

The collecting system consists of two circular flat mirrors as illustrated in figure 20.16. The first mirror is set in a cradle so that the polar axis is parallel to and passes through the centre of the front reflecting face. The second plane mirror reflects the beam into some imaging device, such as a long-focus objective, although, in some systems, reflection optics are used throughout. Rotation of the first mirror about the polar axis, at a rate equal to half the rotational speed of the Earth, allows celestial objects to be followed.

The range of declinations is achieved by adjustment of the relative positions of the two mirrors. As in the case of a solar tower, the image-forming optics are in a fixed position and, consequently, the movement of the second mirror of the coelostat can only allow adjustment by tilt. The first mirror and its drive are placed on a carriage which can be positioned along the north–south line. For each declination setting, the position of this carriage and the tilt at the second mirror need to be adjusted.

In order to overcome the problem of the second mirror blocking-off some parts of the sky from the view of the first mirror, two sets of runners are usually provided for the carriage which holds the first mirror. These runners lie on either side of the meridian which passes through the second mirror. For solar observations, it is easy to visualize that the runners on the east side would be chosen to hold the carriage for pre-transit (morning) observations and that the carriage would be moved to the runners on the west side for post-transit (afternoon) observations.

There are also two other collector systems which are sometimes used to provide a solar image. These are the **heliostat** and the **siderostat**. The heliostat has the disadvantage that, although the centre of the solar image is kept in a fixed position, the image rotates slowly. A siderostat is basically a single-mirrored system but the driving mechanism required to achieve a stationary image is complicated.

The McMath solar telescope, which is a heliostat design, is shown in figure 20.17. The mirror has a diameter of 1·5 m and the system provides a solar image 80 cm across (see figure 20.18).

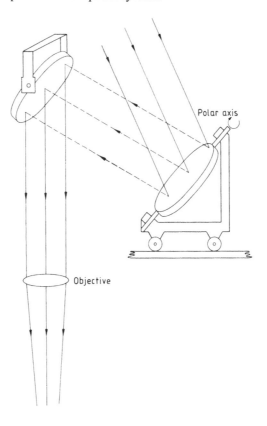

Figure 20.16. A simple coelostat.

20.9   The coronagraph

Until the invention of the coronagraph, the Sun's corona could only be seen on the occasion of a total eclipse of the Sun. The brightness of the corona is generally much less than the brightness of the daytime sky which surrounds the Sun. However, by choosing a good observing site at a height in the mountains, the amount of scattered light seen round the Sun is reduced and the chances of observing the corona are greatly improved.

An ordinary telescope gives rise to scattering in the optical system and this would again prevent the corona being seen. A specially designed instrument, called a coronagraph, has been developed which provides an obscurator for the solar disc and keeps any scattered light down to a minimum. The essential parts of a coronagraph are depicted in figure 20.19.

The objective is a single-element lens, specially chosen for its freedom from blemishes such as air-bubbles and scratches which give rise to scattering. A metal cone is fixed to the field lens so that the Sun's disc is obscured and its light reflected out of the optical path. A diaphragm with a small central obscurator is placed prior to the camera objective to eliminate light which is diffracted by the coronagraph objective. Although the camera lens can provide correction for the chromatic aberration which is produced by the single element objective, the obscurator, in any one position, is effective only for a narrow band of wavelengths. This is not too much of a disadvantage as narrow-band filters are frequently applied in coronal measurements. Use of the instrument over a range of wavelengths is achieved by adjusting the position of the obscurator.

With the aid of the coronagraph, a much longer observation time is made available than at an

Figure 20.17.   The McMath Pierce solar telescope, Kitt Peak National Observatory, Arizona.   (NOAO/AURA/NSF.)

eclipse, enabling faint coronal spectral lines to be investigated and providing a means for recording the development of solar prominences by time-lapse photography.

Example 20.1. Two stars of declinations $56° 30' 10''$ and $56° 30' 50''$ are used for observation with a zenith tube. After observation of the first star, the instrument is rotated through $180°$ and in order to observe the transit of the second star, the micrometer needs to be turned by 0.4 of a thread in a direction moving the wire away from the zenith. If the plate scale of the telescope is 30 seconds of arc per mm and the pitch of the thread 1 turn per mm, what is the latitude of the observing station?

$$\text{The movement of the micrometer} = 0.4 \text{ threads}$$
$$\equiv 0.4 \text{ mm}$$
$$\equiv 0.4 \times 30 \text{ arc sec}$$
$$= 12 \text{ arc sec.}$$

By considering equation (20.6),

$$\frac{(z_1 - z_2)}{2} = 6 \text{ arc sec}$$

and, therefore,

$$\phi = 56° 30' 30'' - 6''$$
$$= 56° 30' 24''.$$

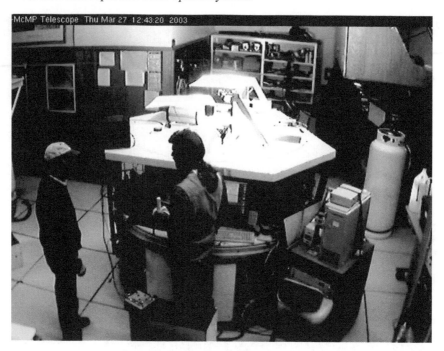

Figure 20.18. The main observing room of the McMath Pierce solar telescope on March 27, 2003. The current status of the observing room may be viewed via http://nsokp.nso.edu/mp/mpcam.html.

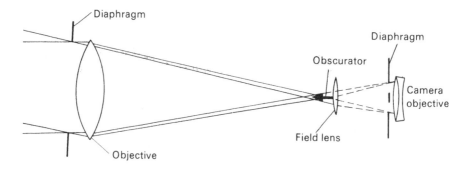

Figure 20.19. A schematic representation of a coronagraph.

Example 20.2. A double slit of variable separation is placed over an objective to facilitate the measurement of two equally bright stars which are separated by 0·1 arc sec. At what separation of the slits would the fringe pattern disappear?

By using equation (20.1), the fringe patterns produced by each star will cancel each other when

$$d = \frac{\lambda}{2\theta}.$$

Now,

$$\theta = 0 \cdot 1 \text{ arc sec} = \frac{1}{2\,062\,650} \text{ rad.}$$

By assuming a value of $\lambda = 5 \times 10^{-4}$ mm,

$$d = \frac{5 \times 10^{-4}}{2\left(\frac{1}{2\,062\,650}\right)} \quad \text{mm}$$
$$\approx 500 \text{ mm}.$$

Example 20.3. A Michelson stellar interferometer is used to determine the apparent diameter of a star. The fringe pattern disappears when the adjustable mirrors are at a separation of 5 m. What is the angular diameter of the star?

By using equation (20.1),

$$\theta = \frac{1 \cdot 22\lambda}{d}.$$

Again by assuming a value of $\lambda = 5 \times 10^{-4}$ mm,

$$\theta = \frac{1 \cdot 22 \times 5 \times 10^{-4}}{5 \times 10^3} \text{ rad}$$
$$= 1 \cdot 22 \times 10^{-7} \text{ rad}$$
$$\theta \approx 0 \cdot 025 \text{ arc sec}.$$

Chapter 21

Radio telescopes

21.1 Introduction

Developments in the technology of electronics, such as aeroplane detection by radar, accelerated by the Second World War, provided the means for the establishment of radio astronomy. The discovery that energy in the form of radio waves was arriving from space was made a few years prior to this war by Jansky. Other discoveries of importance to astronomers were made by chance during the war but it was not until its end that the new techniques could be applied to astronomical research for its own sake.

Basic measurement of the incoming energy is achieved by using a directional **antenna** or **aerial**. The brightness of a source, as in the optical region, is measured as the **flux density** or the energy, E, arriving per unit aperture of the telescope per unit frequency interval and, for example, may be expressed in units of $W\,m^{-2}\,Hz^{-1}$. For convenience, measurements of flux density may be expressed in **flux units** or janskys (Jy), with

$$1\,Jy = 10^{-26}\,W\,m^{-2}\,Hz^{-1}.$$

For a source with angular extent, its surface brightness is expressed in units of $Jy\,sr^{-1}$.

By using the Rayleigh–Jeans law (see equation (15.4)), any source can be assigned a brightness temperature, T_B, given by

$$\frac{2kT_B}{\lambda^2} = E \tag{21.1}$$

where k is Boltzmann's constant and λ the observed wavelength. However, as radio sources do not, in general, have black body energy envelopes, the brightness temperature may vary considerably with wavelength.

The disturbances set up in the antenna may be fed into a receiver, tuned to the frequency of interest, which amplifies them before rectification to provide the final output signal. In order to remove the effects of drifts in the equipment, a stabilized calibration source is usually provided. This is switched periodically to the receiver system and response of the equipment to this known input is monitored. Such a system, forming the essential part of a **radio telescope**, is illustrated in figure 21.1. The output from the receiver can be recorded on magnetic tape or, more likely, is digitized in readiness for direct computer processing with data storage on disc. These records provide the radio astronomer with measurements of the strength of the radio radiation and its polarization over the celestial sphere.

As in optical astronomy, one of the aims of the radio telescope is to be able to collect energy from well-defined regions on the celestial sphere. Because of the longer wavelengths which are involved, this is not easy and high angular resolution can be achieved only by having large antenna complexes

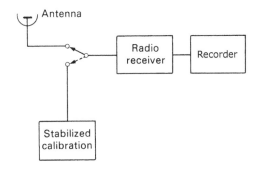

Figure 21.1. The basic elements of a radio telescope.

and collecting areas extending over many hundreds of square metres rather than by using single dish-type collectors. An appropriate substitution of the value of λ in equation (16.14) shows why this is so.

With the introduction of special techniques of very long base-line interferometry, in which telescopes are used thousands of kilometres apart, the angular precision of measurement is now of the order of fractions of milli-arcseconds.

21.2 Antennas

The purpose of antennas used in radio astronomy is to collect waves arriving from particular directions and provide at their terminals a disturbance which can be detected by a radio receiver. Because of their fundamental nature, antennas are selective to a fairly narrow range of frequencies and are also sensitive to the polarization of the radio waves. The range of frequencies to which the antenna responds is, however, usually much wider than that accepted by the receiver. The broad range of radio waves available for measurement is covered by using antennas of different sizes in combination with appropriately tuned receivers. Some particular frequencies are, indeed, a special interest as they correspond to energy transitions of identifiable atoms and molecules. The monitoring of these frequencies has led to investigations of the distribution in interstellar space of such atoms and molecules. Study at these radio spectral lines, e.g. the hydrogen line at 21 cm (1427 MHz), has provided a major contribution to our knowledge of the structure of our own galaxy.

It is important in most types of observation for the radio telescope to have high sensitivity and directional discrimination. When a radio telescope is directed to a particular source, a signal appears at the output of the receiver. If now the telescope is allowed to drift relative to the source's direction, the output signal does not fall to zero as soon as the telescope is just off the source's position. The signal may only fall to zero when the telescope is directed away from the source by several degrees. The rate at which the signal falls with angle describes the directional quality of the telescope. A simple representation of a directional sensitivity of a radio telescope can be given by drawing a polar diagram. In this representation, the length of a vector is used to represent how the magnitude of the output response would vary according to the apparent direction of a distant point source of constant radio brightness. The polar diagram may be said to be the instrumental profile of the radio telescope. A two-dimensional polar diagram for a typical telescope is depicted in figure 21.2. The diagram may not necessarily be symmetrical about the *z*-axis and, in general, a three-dimensional polar diagram is necessary to describe the directional sensitivity of a telescope.

It will be noted from this same diagram that a radio telescope has some sensitivity outside the main beam. These **side-lobes** may sometimes be troublesome by picking up sources which are way

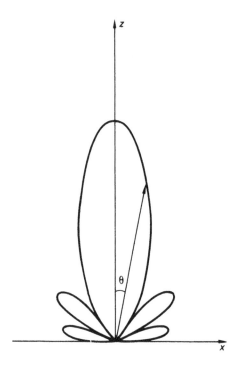

Figure 21.2. A radio telescope's sensitivity polar diagram in two dimensions.

out of the main beam—a strong source picked up by a side-lobe could contribute significantly to the telescope's output if its main lobe is directed to a relatively weak source.

The **beamwidth** of a telescope is sometimes defined by the angular extent of the central lobe or by the points in the primary lobe where the sensitivity has dropped to one-half of its maximum value. A telescope with a narrow main lobe is said to have a narrow beam.

As well as using the term beamwidth to describe a telescope's ability to pinpoint discrete sources, it is also sometimes convenient to use the concept of **power gain**. This concept is perhaps better understood by considering the ability of any antenna to act as a transmitter. Suppose that instead of an antenna acting as a receiver, it is made to perform as a transmitter radiating a certain amount of power (energy per unit time). At some point at a distance from the antenna, a certain quantity of energy will arrive per unit area per unit time. Consider a point lying on the line whose direction is given by the maximum of the polar diagram. Let the rate of arrival of energy at this point be equal to P_1, usually measured in W m^{-2}.

Consider that the power radiated by the first antenna is now radiated by a second of different design. If again the preferred direction is found, then at a point along this direction, at a distance identical to that chosen for the first antenna, the rate of arrival of energy will have a different value, being equal to P_2. The ratio of P_1 to P_2 is defined as the power gain, G, of the first antenna in relation to the second. Thus,

$$G = \frac{P_1}{P_2}. \tag{21.2}$$

The value of the power gain of an antenna can be put on an absolute basis by considering the second antenna to be one which accepts or transmits radiation equally well in all directions, i.e. its polar diagram is that of a sphere centred on the antenna. Although such an isotropic antenna can never exist in practice, it is a useful concept for defining absolute power gain.

Thus, the concept of the power gain of a transmitting antenna is one which defines the ability of an antenna to direct power into a narrow beam. When the antenna is allowed to act as a receiver, it is obvious that the value of the power gain reflects its competency to concentrate its collecting ability along a particular direction. The concept of power gain may be extended to consider point sources in directions not along the direction of maximum sensitivity. The variation of power gain with the source's direction allows the sensitivity polar diagram to be plotted absolutely.

The performance of any antenna system may also be described by its response to the rate of arrival of energy (the power) at a given frequency and with a given polarization from a point source. If P is the power developed by an antenna and E is the energy available per unit area per unit bandwidth from the point source, then these parameters can be related by

$$P = A_e E$$

where A_e has the dimensions of an area and is defined as the **effective area** of the telescope. Its value need not correspond directly to the geometric area of the antenna or telescope.

By defining A_e in this way (i.e. $A_e = P/E$), it is obvious that its value will vary according to the direction of the antenna or telescope in relation to the position of a point source. A plot of this variation is, therefore, equivalent to the polar diagram representing a telescope's directional sensitivity. The effective area of an antenna is, in fact, related to its gain by the expression:

$$G = \frac{4\pi A_e}{\lambda^2} \tag{21.3}$$

where λ is the wavelength for which G and AE are compared. The definition of the gain (equation (21.2)) involves the concept of an isotropic antenna and the factor of 4π in equation (21.3) corresponds to the solid angle of the sphere surrounding its centre. Alternatively, therefore, it may be noted that an antenna's effective area may be written as

$$A_e = \frac{\lambda^2}{\Omega_A} \tag{21.4}$$

where Ω_A is the solid angle of the telescope's beam.

As is the case with any kind of experimental measurement, radio astronomy measurements are subject to uncertainties. The output signal from any telescope does not remain at a fixed value but is seen to fluctuate being subject to noise. The signal-to-noise ratio of the records indicates the meaningfulness of any measurement (see figure 21.3). The strength of the noise indicates the value of the smallest change in the signal that can be detected as being real. In other words, the magnitude of the noise sets a limit to the strength of the weakest source that can be detected. In most cases, the signal-to-noise ratio of measurements is very much less than unity but, by using long integration times for the observations or by using sophisticated techniques, the radio astronomer is able to detect extremely weak sources.

The noise in radio astronomical measurements arises from the sky background or from the preamplifier following the antenna. In addition, the signals may suffer from interference generated by local terrestrial sources such as lightning or man-made disturbances such as electrical motors, car spark plugs, etc.

It is well known from Nyquist's theorem that any resistor generates noise by thermal agitation of the conducting electrons. The power, P, generated across any resistor can be shown to be given by

$$P = kT\Delta\nu \tag{21.5}$$

where k is Boltzmann's constant, T the resistor's absolute temperature and $\Delta\nu$ the frequency interval associated with the measurements. Suppose that a resistor with the same value as the antenna and its transmission lines were to be placed across the terminals, power would be generated across it according

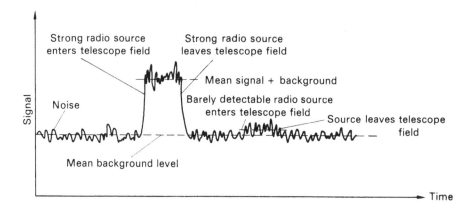

Figure 21.3. A radio telescope output illustrating the difficulty of detecting a weak radio source.

to equation (21.5). If both the antenna and the matching resistor are held in thermal enclosures set at identical temperatures, each system would generate identical amounts of power and there would be no net flow of energy between the two containers.

In terms of the signal generated across the antenna's terminals, any celestial source that fills the antenna beam can be considered as being equivalent to fluctuations across the matched resistor but with a temperature appropriate to the measured power. The more power received from the source, the higher would be the required temperature of the resistor. Thus, according to the flux density of the source, an **effective temperature** can be assigned to the antenna.

Quite generally, the antenna temperature can be defined as

$$T_A = \frac{P_R}{k\,\Delta\nu} \tag{21.6}$$

where P_R is the power received from the source, with the provisos that the source fills the antenna beam and that the source is thermal, behaving as a black body with a Planck spectrum. Typical antenna temperatures range from a few degrees at wavelengths of a few centimetres to several millions of degrees at wavelengths of tens of metres. It goes without saying that these temperatures do not represent the physical temperature of the antenna!

21.3 Antenna design

21.3.1 The basic dipole

The simplest antenna or dipole, sometimes referred to as a **Hertzian dipole**, is sketched in figure 21.4. The length between the ends of the pair of conducting rods must be smaller than the wavelength of the radiation to be detected.

According to the radiation impinging on the system, the associated oscillating electric fields set up currents in the antenna which, in turn, produce fluctuating voltages across the resistor. The strengths of these are determined to provide measurements of the power received from the radio source. It may be noted that the primary currents in the rods are only generated by the electric fields in the radiation which are parallel to the lengths of the rods. The electric field disturbances within the radiation which are orthogonal to the rods' axis have no effect in producing currents. Sensitivity only to electric field disturbances in the parallel direction to the rods means that the dipole is automatically sensitive to 'one polarization' associated with the received radiation.

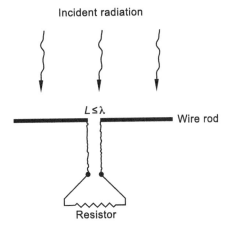

Figure 21.4. A simple Hertzian dipole.

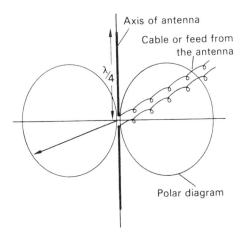

Figure 21.5. The polar diagram of a half-wave dipole antenna.

21.3.2   The half-wave dipole

A frequently met form of antenna is a **half-wave dipole**, consisting of two metal rods or wires which are held in line with each other but separated by means of insulated supports. Each rod has a length which is one-quarter of the wavelength to be monitored by the radio receiver. The feed from the antenna to the receiver is taken by cable from the ends of the rods at the centre of the antenna. A typical dipole is illustrated in figure 21.5 which also shows its far-field polar diagram of directional sensitivity.

The half-wave dipole polar diagram in three dimensions can be obtained by rotating the two-dimensional diagram about the antenna's axis. It can be seen that the dipole is insensitive to radiation which arrives from a direction along its axis. However, it can also be seen that it has poor directional sensitivity and, consequently, its absolute power gain is low, being 1·64. By itself, this antenna is of little use to act as a radio telescope.

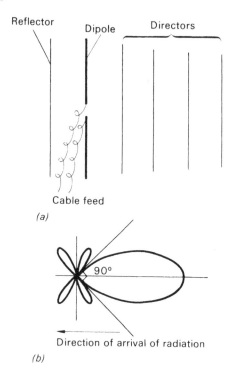

Reflector   Dipole   Directors

Cable feed

(a)

90°

Direction of arrival of radiation

(b)

Figure 21.6. The parasitic or Yagi antenna and its polar diagram

21.3.3   The Yagi antenna

The directivity of a dipole antenna is greatly improved by applying extra elements. They are not electrically connected to the antenna or the receiver and are said to be **parasitic**. Their purpose is to give radiative reaction with the dipole so that the polar diagram achieves greater directivity. The **Yagi antenna** usually has one reflecting rod and a few directors. It is schematically illustrated in figure 21.6(*a*). The polar diagram for this type of antenna is illustrated in figure 21.6(*b*) and it should be compared with figure 21.5. It will be seen that it has little response in the backward direction. The small side lobes may sometimes be troublesome by picking up signals which are out of the main beam. Its absolute power gain depends on the number of directors but a law of diminishing returns keeps this number to be about a maximum of 12, when G is about 50—its beamwidth is of the order of 20°. Again, however, the parasitic antenna by itself has little application to radio astronomy.

21.3.4   Antenna arrays

By connecting antennas in the form of an array, high absolute power gains can be obtained with values which are sufficient for radio astronomical purposes. There are two basic forms of array, the **collinear** and the **broadside** array.

   In the collinear array, as the name implies, the dipoles are arranged to be in line and are connected together so that their individual contributions are brought to the receiver in phase. A four-element collinear array with its polar diagram projected in the yz-plane is depicted in figure 21.7.

   In order to produce the polar diagram in three dimensions, the diagram as illustrated in figure 21.7(*b*) must be rotated about the axis of the array (i.e. the y-axis).

   In the broadside array, the dipoles are arranged in parallel formation as in figure 21.8. Each dipole is separated by a distance equal to half a wavelength. They are again connected so that their

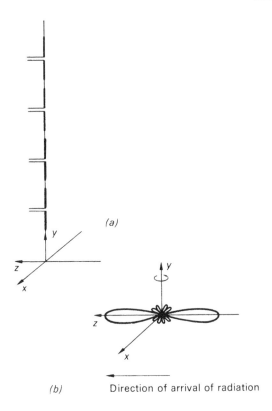

(a)

(b)   Direction of arrival of radiation

Figure 21.7. A four-element collinear array and its polar diagram. (The axes of the antennas are collinear with the *y*-axis.)

contributions arrive in phase at the receiver. The polar diagram drawn in figure 21.8(*b*) is that which is found in the *xz*-plane. In the *yz*-plane, the polar diagram is the same as that of a single half-wave dipole.

By a combination of the collinear and broadside arrays, the directivity of the two-dimensional array is reduced in both the *x* and *y* directions, leaving the main narrow lobes pointing in either direction along the *z*-axis. One of these directions can be eliminated by a wire mesh reflector put at the correct distance in a plane parallel to the two-dimensional array. An array of 64 half-wave antennas gives a beamwidth of about 14°. The absolute power gain is equal to the number of dipoles in the array multiplied by 2×1.64 (the factor of two is obtained by using the back reflector) and, in this case, the array has an absolute power gain equal to 210.

By mounting the whole system on an adjustable rig, the highly-directional antenna can be pointed to any source on the celestial sphere. Such a system is one of the forms of a radio telescope.

Similar highly-directional antennas can be devised by using an array of parasitic antennas. Again, the absolute power gain is obtained by multiplying the absolute power gain of a single antenna by the number of antennas in the array.

It may be noted that the contributions from the elements in an array may be connected in different ways and are referred to as **phased arrays**. In the simplest design of collinear array, each signal flows into a single line whose centre feeds the receiver (see figure 21.9(*a*)). In the **Christiansen interferometer**, or **Christmas Tree** array, the signals are added together in pairs in a progressive way so that the total length of feed between each antenna and the receiver is the same. This is illustrated in figure 21.9(*b*). The collinear system has smaller power losses in the feed than the Christmas Tree array

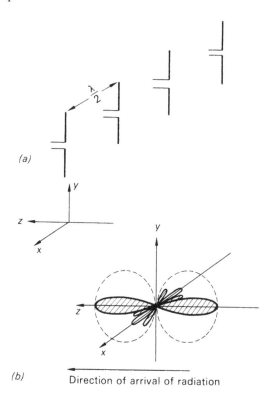

(a)

(b)   Direction of arrival of radiation

Figure 21.8. A four-element broadside array and its polar diagram. (The antennas are parallel to the y-axis and in a plane parallel to the xy-plane.)

but the latter has an advantage in being able to operate with a much broader band of frequencies as a result of the symmetry of the path lengths of the connections. The original Christiansen arrangement used for resolving areas of the Sun at a wavelength of about 200 mm consisted of 32 steerable parabolic collectors, each having a diameter of 2 m, spaced equally along an east–west line 217 m long. The angular resolution achieved by this arrangement was of the order of 3 arc min.

Under the heading of telescope arrays, mention should be made of the Square Kilometre Array (SKA[W 21.1]). The development of this project is most ambitious with the aim of obtaining a collecting area of 1 km^2 (10^6 m^2) which is about 100 times that of the largest present day instruments. In order to achieve this, a very large international collaboration is required. Various designs are under consideration with some 20 countries participating.

21.4   Parabolic dishes

Directional antennas such as the Yagi antenna and the two-dimensional array all suffer from the disadvantage that the wavelengths that they can monitor are at once fixed by the choice in the size of the half-wave dipoles. A more versatile system is that which uses a large parabolic collecting dish, bringing the collected radiation to a focus where a dipole or collecting horn and waveguide system is placed. A range of dipoles and horns can be made available according to the wavelengths that are to be investigated. To obtain even a modest angular resolution, the collecting dish must be at least several wavelengths in diameter. By applying the formula for the resolution of a telescope (equation (16.14)), it can readily be shown that, in order to be able to resolve an angle of 7°, the ratio of the diameter of the

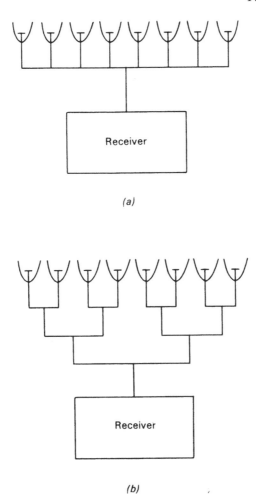

Figure 21.9. An eight-element interferometer as (*a*) a collinear array and (*b*) in the Christiansen arrangement.

dish to the observing wavelength must be of the order of at least ten to one. Higher angular resolution at the same wavelength would require an even larger collecting area and it is for this reason that radio telescope dishes run in size from a few metres to a few hundred metres. The world's largest steerable dish telescope is the 100 m National Radio Astronomy Observatory (USA) sited at Green Bank, West Virginia. Referred to as the Green Bank Telescope, or GBT[W 21.2], it was commissioned in 2000. The non-steerable radio telescope at Arecibo, Puerto Rico, is approximately 300 m in diameter and is a terra-formed structure.

A typical collecting dish has a small focal ratio. This allows the receiving antenna to be at not too great a distance above the bowl and the reflecting surface which collects the radio radiation then acts as a shield against locally generated signals and interference. The actual reflecting surface need not be made of metal plate; it can be constructed from wire mesh in which the size of the holes determines the lower limit of the wavelength which can be observed effectively. The contours of the parabolic dish must also be made to an accuracy of about one-eighth of the smallest wavelength to be observed if the telescope is not to lose its potential angular resolution.

The absolute power gain of a radio telescope dish is given by inserting the value of the geometric aperture in equation (21.3). This value is approximately the same as that of a two-dimensional array

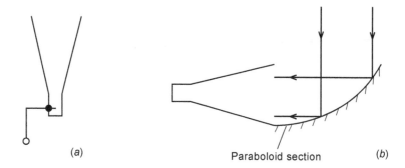

Figure 21.10. (*a*) A horn antenna with feed output. (*b*) A horn antenna with a reflection collector.

of antennas covering the same area. However, the calculation of power gain usually requires an adjustment to allow for the fact that the feed horn (see later) is designed to prevent 'spill-over' to the ground with the result that the dish is effectively under illuminated. A typical loss is of the order of a factor of two so that $A_e \approx A/2$, where A is the full geometric area of the dish. If the absolute power gain is expressed in terms of the diameter of the dish, it is given by

$$G \approx \frac{\pi^2 D^2}{2\lambda^2}. \tag{21.7}$$

Thus, if a 10 m dish is used at a wavelength of 1 m, the absolute power gain is close to 10^3.

As with the case of optical telescopes, some radio dishes are constructed to support a secondary reflector prior to the primary focus, so operating as a Cassegrain system. With this configuration, the position of the feed is more accessible being just behind the dish and the system's length is reduced.

21.5   Horn collectors

For frequencies $\gtrsim$GHz, horns are commonly used to act as collectors. Apertures are normally $\lesssim 15\lambda$. A basic design is depicted in figure 21.10(*a*). An alternative design might have a reflective paraboloid section attached to the horn lip to increase the effective area (see figure 21.10(*b*)). One of the features of horns is that their beams can be calculated accurately which, in turn, ensures good calibration of signal strengths and absolute measurements of bright sources. In some telescope designs, a collection of horns might be used and their signals combined.

A horn is sometimes placed at the focus of a radio dish to accept the collected radiation. Obviously it is important that the horn only 'sees' the dish collection aperture and not beyond its perimeter otherwise it would respond to illumination from the ground. The design is, therefore, engineered so that its beam tapers rapidly at its edges.

21.6   Interferometry

21.6.1   A basic interferometer

By using interferometric techniques, analogous to those used in optics, it is possible to obtain higher angular resolutions than those which result from telescopes in the form of two-dimensional arrays and reflecting dishes. The simplest form of radio interferometer consists of two antennas which point in the same direction but are separated by a considerable distance in relation to the radio wavelength which is being used. This system is equivalent to Young's double-slit interference experiment in optics. The

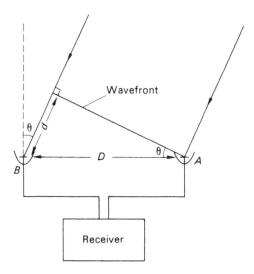

Figure 21.11. A double-beam radio interferometer.

radio version of this type of interferometer is illustrated in figure 21.11, where the incoming radiation arrives in a parallel beam, inclined at an angle, θ, to the normal direction of the antennas.

Consider a wavefront which arrives at the antenna A. This wavefront has an extra distance, d, to travel before arriving at antenna B. Thus, the phase of the radio wave arriving at B lags behind that at A. If the two antenna signals are combined at the receiver, they will interfere according to their relative phase. If the distance, d, is such that it corresponds to a whole wavelength, the antenna signals will be exactly in phase and will, therefore, reinforce each other—constructive interference. This condition is monitored by the radio receiver. From figure 21.11,

$$\frac{d}{D} = \sin\theta$$

where D is the separation of the two antennas. For constructive interference, $d = n\lambda$ (i.e. a whole number of wavelengths) and, therefore,

$$\sin\theta = \frac{n\lambda}{D}. \tag{21.8}$$

Thus, if a radio object sweeps across the sky under diurnal motion over the fixed interferometric system, a large output will be registered each time this condition is achieved for the range in values of n. The signal is reduced to its background level when the phase between the antennas is exactly an odd number of half-wavelengths—destructive interference. In this way, the position of the object can be determined accurately. The polar diagrams for a simple and double-beam interferometer are illustrated in figure 21.12. Obviously, some ambiguities may arise in the interpretation of the output from the radio interferometer if two or more objects happen to pass at the same time through the beam and other side-lobes but it is usually not very difficult to allow for this circumstance.

21.6.2 A phase-switched interferometer

In practice, the output from a simple interferometer has other contributions that reduce the effectiveness of the device. There is a background from the receiver that raises the mean level of the output, its power being designated as P_B. There is also a contribution from the very broad structures in the sky, i.e. the galactic background, being represented as P_S. This latter contribution will not be constant, depending

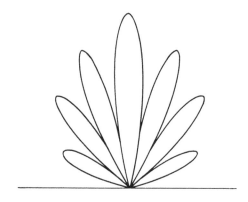

Figure 21.12. The polar diagram of a two-beam radio interferometer.

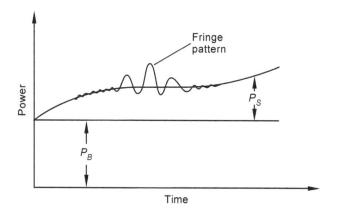

Figure 21.13. The output of a simple two-antenna interferometer illustrating the effects of the background signals P_B and P_S.

as it does on the area of the sky which is in the beam. The combination of these two sources within the signal is displayed in figure 21.13 with an interferometric pattern superimposed. The presence of the signals from backgrounds raises problems as follows.

(1) The final output needs correction by subtraction of the background contributions.
(2) As a result of variations in the gain of the receiver, the strengths of both P_B and P_S will appear to drift.
(3) The background signals are themselves noisy and cause fluctuations on the output signal.

The problems associated with (3) are fundamental and unavoidable but can be reduced by using long time constants within the recording circuitry to smooth out their effects. The problem under (2) can be reduced by having high stabilization of the electronic components. However, by the principle of **phase-switching** the effects of both (1) and (2) are removed. This technique involves the introduction of a phase delay, equivalent to half a wave path length, which is switched in and out of one of the arms of the interferometer. A schematic diagram of the modified interferometer is shown in figure 21.14. The phase-sensitive detector measures two levels, one corresponding to the normal situation without the phase delay, the other to that with the phase delay added, the latter effectively shifting the pattern by half a fringe. The switching frequency is set to be much faster than the frequency associated with the fringe record.

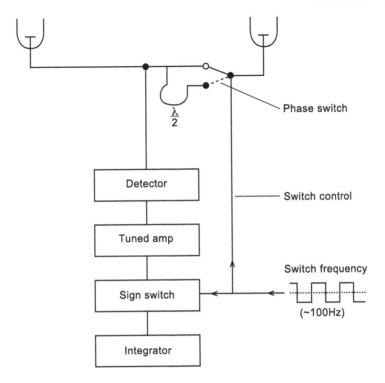

Figure 21.14. The basic elements of a phase switched two-antenna interferometer.

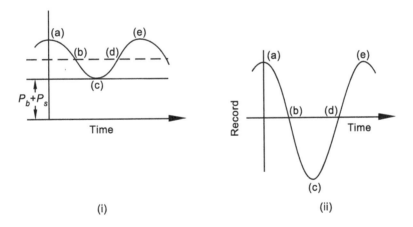

Figure 21.15. (i) A schematic interference pattern obtained from a two-antenna interferometer showing the background offset. (ii) The recorded pattern as for (i) but showing the effects of phase switching.

For reference, the fringe pattern from a simple two-element interferometer is displayed in figure 21.15(i) with maxima indicated at (*a*) and (*e*), with a minimum at (*c*) and with two mid-points at (*b*) and (*d*). According to the phase-switching process,

At situation (*a*): Output → $P(a) - P(c) = 2 \times$ fringe amplitude.
Similarly at (*b*): Output → $P(b) - P(d) = 0$.
And at (*c*): Output → $P(c) - P(e) = -2 \times$ fringe amplitude.

The record from the arrangement is depicted in figure 21.15(ii) displaying that the fringe amplitude has doubled and that the effects of P_B and P_S have been removed.

Problems of source identification, occurring with the two-beam interferometer as a result of the strong secondary lobes, are generally reduced by using multi-element interferometers. The designs normally allow the separations between the elements to be adjusted from day to day. For example, the Ryle Telescope[W 21.3] at the Mullard Radio Astronomy Laboratory of the University of Cambridge, is an eight-element interferometer operating at 15 GHz. The components are 13 m Cassegrain antennas on an east–west baseline. Four antennas are mounted on a 1·2 km rail track with the others fixed at 1·2 km intervals. Baselines between 18 m and 4·8 km are, therefore, available in a variety of configurations.

Long baseline interferometry (LBI) is also possible by linking the outputs from individual elements located at separated operating stations in such a way that the phase information is preserved. An LBI array known as MERLIN (Multi-Element Radio-Linked Interferometer Network) uses microwave communication links to send signals from its seven radio observatories in England to a computer system at Jodrell Bank, Cheshire, to make detailed maps of radio sources. With some 15 baselines ranging from 6 to 217 km in length, it can achieve a resolution of 0·05 arc sec.

21.6.3   Very long baseline interferometry

In principle, and practice, very large distances between two or more radio telescopes can be used to obtain high resolution in measuring a radio source's coordinates without the stations having direct or real time links. This is done by the principle of *Very Long Baseline Interferometry* (VLBI). Baselines between the radio telescopes have been extended in some arrays to thousands of kilometres by employing simultaneously several radio astronomy observatories.

With very accurate timekeeping methods available, the radio telescopes' signals are recorded individually but synchronously with accurate time markers added as the sky moves across each of the apertures. By adding the recorded signals together at some future convenient time, keeping their recorded times accurately matched, an interference pattern is produced, corresponding to the one that would have been produced if the signals had been combined at the actual time of the observations.

The Australia Telescope National Facility (ATNF) uses eight radio antennas located at three observatories in New South Wales. It has a baseline of 300 km. Like MERLIN, the antennas can be used individually or in various combinations and combines the principles of both VLBI and aperture synthesis.

The Very Long Baseline Array (VLBA) operated by the US National Radio Astronomy Observatory with headquarters at Socorro, New Mexico, consists of an array of ten radio telescopes situated with all but two in the continental USA, the others being on Hawaii and the Virgin Islands. With a maximum baseline of 8000 km, the VLBA can achieve a resolution of 0.001 arc sec, some 50 times better than the Hubble Space Telescope.

In principle, putting a radio telescope in a spacecraft or on the Moon to be used in conjunction with one or more situated on Earth should give baselines of hundreds of thousands of kilometres with resolutions orders of magnitude better than those obtained by the VLBA.

21.7   Aperture synthesis

Any large telescope aperture can be considered as being made up of a grid of small elements, each collecting energy and transferring it to a common receiver. At this point, all the contributions are added according to the phase.

Suppose that an aperture is taken as comprising N elements of equal area. The contribution from each element will produce an interference pattern with each of the other $N - 1$ elements. The overall effect is equivalent to the resultant of $N(N - 1)/2$ superposed interference patterns from pairs of elements. The spacing of the interferometers ranges from adjacent elements to a maximum across the

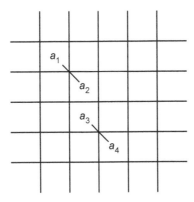

Figure 21.16. The interferometric elements $a_1 - a_2$ and $a_3 - a_4$ within a large array carry identical information.

full aperture—all orientations of linking axes are possible. There may be a fair degree of redundancy in the information. For example, in figure 21.16, the combination of a_1 and a_2 supplies the same pattern and, hence, identical information as the combination of a_3 and a_4, their separations and orientation being identical.

At the expense of not achieving the flux gathering power of the single large aperture, it is possible to devise an array with the right spacings and axis orientations so that it is equivalent to a single large aperture in terms of its angular resolving power. Such a notion is the basis of the **skeleton radio telescope**.

Consider an array configuration in the form of a 'T' with the row comprising $2\sqrt{N}$ elements and the column with $\sqrt{N}$ elements. With this pattern, there are approximately $2N$ different interferometer pairs that can be made by combining one element from the vertical strip with an element from the horizontal strip. These $2N$ interferometer pairs represent all the possible spacings and distances found in the $N(N-1)/2$ equivalent pairs in the complete filled aperture given by $\sqrt{N} \times \sqrt{N}$ shown in figure 21.17.

On the assumption that a radio source does not vary from day to day, the brightness distribution over the source can be derived from the elementary interferometer patterns taken one at a time. It is not necessary to have all the elements present simultaneously—their equivalent contributions can be generated sequentially.

In the 'T' formation, it is usual to have two elemental telescopes on the top arm running east–west, acting as a simple interferometer, with a third telescope on the north–south track. After setting the dishes for a particular declination, a series of recordings is made, with the separation of the interferometer and moveable telescope adjusted every 24 hours. From analysis of a complete series of recordings, it is possible by computer to synthesize the observations as though they had been obtained by an aperture given by the area described by the 'T'. There are $2N$ separate patterns containing all the information that would be present in the field aperture. Hence, if each of the recordings are made for $2N$ times longer, observations can be made with the same sensitivity and resolution as though they were made by a single large aperture. By progressively changing the declination, the whole of the available sky can be mapped at a particular frequency. The now-famous *Third Cambridge Catalogue* (3C) of 328 sources measured at 178 MHz was produced by the aperture synthesis method using a 'T' configuration.

Another well-established aperture synthesis system is the Very Large Array (VLA) at the US National Radio Astronomy Observatory at Socorro, New Mexico. It comprises 27 antennas, each in the form of a 25 m diameter dish. The elements are set out in a 'Y' configuration and the arrangement provides a resolution equivalent to a single antenna 36 km across with a sensitivity equivalent to a dish

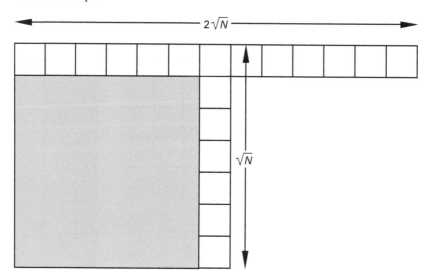

Figure 21.17. A 'T' aperture synthesis arrangement, with $2\sqrt{N}$ elements on the east–west row and $\sqrt{N}$ elements on the north–south column, provides interferometric information to give an angular resolving power equivalent to that of a single large aperture with area N.

of 130 m diameter. Four basic arrangements are used in the separations of the antennas within the 'Y' and the telescopes are switched about every four months. From 'snapshots' taken at each of the configurations, images are synthesized as though recorded by the equivalent filled aperture. For each of the four configurations, the declinations of the individual elements may be progressively adjusted so that high-resolution sky maps are generated.

The European VLBI network (EVN) comprises 18 radio telescopes operated by 14 different institutions. The individual telescopes are spread across Europe and into China. The shortest baseline is 198 m and the longest is 9169 km. EVN often co-observes with MERLIN and VLBA, both mentioned earlier, to form a 'global' VLBI array. It is also committed to supporting VLBI observations with the Japanese Orbiting Radio Telescope (HALCA). This latter instrument of 8 m diameter is part of the VSOP (VLBI Space Observatory Programme) mission and, as a result of the telescope's orbit, provides baselines three times longer than those achievable on Earth.

Currently under development and expansion is the Atacama Large Millimetre Array (ALMA) project. It involves the merger of a number of major millimetre arrays into one global system. Part of the system is being constructed at high altitude (5000 m—Zona de Chajnantor in Chile) with some 64 telescopes of 12 m, with baselines extending to 10 km. The receivers will operate in the frequency range 70–900 GHz.

It may also be noted that the process of aperture synthesis can also be effected by using the rotation of the Earth. If the individual elemental telescopes are made to track a source, the geometry of apparent array configuration presented to the source effectively changes according to the location of the source in the sky. The result of tracking a source is, therefore, equivalent to a continuous change of the telescopes' physical distribution on the ground. One of the consequences of this is that all VLBI systems can be used as imaging instruments using aperture synthesis.

A listing of some of the larger radio telescope facilities is given in table 21.1.

21.7.1 Lunar occultations

Accurate positions and the angular extent of some radio sources have been obtained from observations at *lunar occultations*. As a radio source is occulted, the Moon's limb acts as a diffracting edge. During

Figure 21.18. Parkes steerable 210 ft radio telescope, Australia. (By courtesy of the Royal Astronomical Society.)

Table 21.1. Some of the larger radio astronomy observatories. Links related to the descriptions of most of the world's radio telescope facilities including the VLBI Networks can be found in a catalogue of Radio Telescope Resources[W 21.4].

Name	Location	Major equipment
Arecibo Observatory	Puerto Rico	305 m (1000 ft) fixed spherical dish.
NRAO GBT	Green Bank West Virginia	100 × 110 m steerable dish.
Effelsberg Radio Telescope, Germany	Max-Planck-Institut	100 m (330 ft) steerable parabolic dish.
Nuffield Radio Laboratory, Manchester University	Jodrell Bank, England	76 m (250 ft) steerable dish Lovell Telescope.
Mullard Radio Astronomy Observatory	Cambridge, England	Various interferometers (Ryle Telescope), source of the various 'C' catalogues
Radio Observatory, Nançay (Observatory of Paris)	Nançay, France	305 m (1000 ft) partially steerable meridian transit array.
ATNF Parkes Observatory	NSW, Australia	64 m (210 ft) independent steerable dish, also networked to form VLBI.
Dwingeloo Telescope	The Netherlands	25 m dish.

Figure 21.19. The 1000 ft radar-radio telescope (Cornell University) at Arecibo, Puerto Rico. (By courtesy of the NAIC-Arecibo Observatory, a facility of the NSF.)

Figure 21.20. The Ryle Telescope in compact configuration. (Photograph by Keith Papworth, reproduced courtesy of Mullard Radio Astronomy Observatory.)

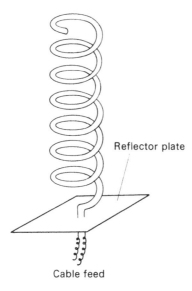

Reflector plate

Cable feed

Figure 21.21. A simple helical antenna.

the progress of the occultation, the diffraction pattern is effectively scanned across the aperture of the telescope and measurement of the variations of the strength of the signal with time allow the size and position of the source to be determined with good precision.

According to equation (20.3), the typical scale of a fringe on the Earth's surface is given by

$$x = (d_{\mathbb{C}} \, \lambda)^{\frac{1}{2}}.$$

Because of the wavelength values in the radio region, the timescales for the passage of the passage of the pattern over a telescope are of the order of several minutes relative to the millisec times for measurements undertaken by optical telescopes (see section 20.3.4). This method of pinpointing a source is, however, limited to the band of sky through which the Moon appears to travel.

21.8 Polarization

The ordinary half-wave dipole is sensitive to polarization which may be present in radio waves. By altering the orientation of the dipole in the beam and recording the receiver's response, the degree of linear polarization and its preferred direction of vibration can be determined. It is because of an antenna's sensitivity to polarization that the values of gain and effective area are related to a particular polarization form (see earlier discussion). If a telescope is defined to have a particular gain, G, in response to linearly polarized radiation in the preferred vibration, then the gain of the system to unpolarized radiation is $G/2$, since the antenna is only able to respond to half the energy incident on the telescope.

Circular and elliptical polarization can be detected by altering the relative phases of resolved components prior to the antenna and recording any changes in the recorder response as the orientation of the dipole is varied. Some antennas are designed to be in the form of a helix and this type is already suitable for accepting and detecting circular and elliptical polarization. The helical antenna with a given diameter of cylinder has a larger effective collecting area than an ordinary dipole of the same dimension. It also has the advantage of being less frequency-selective and is capable of being used over a wider range of wavelengths. Helical antennas are particularly useful for forming the basis of a two-dimensional array as there is much less interaction between the antenna elements than in the case for the dipole array. An elemental helical antenna is depicted in figure 21.21.

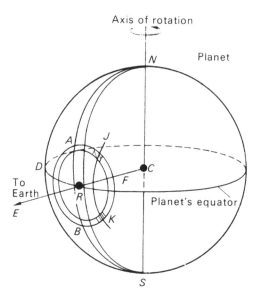

Figure 21.22. The production of a radar map by measuring the echo time and the wavelength spread of a radio pulse.

21.9 Radar observations

As well as receiving radiation generated by astronomical sources, radio telescopes are used frequently in radar studies of the various bodies of the Solar System. Pulses of radio energy are directed towards the body and then their faint echo is received after a delay depending on the distance of the object.

In planetary radar studies, observation of both the time delay between transmitting and receiving the radio pulse and any Doppler shifts in the wavelengths making up the pulse may be combined in building up a radar map of the planetary surface.

In the simple case depicted in figure 21.22, it is assumed that the axis of rotation of the planet is perpendicular to the direction to the observer, this direction CRE cutting the planet's surface at R. Then the ring $AFBD$ defines a zone on the planetary surface giving the particular time delay since, within that zone, the planetary surface is at the same distance from the observer. If the planet is rotating on its axis, the reflected pulse will exhibit Doppler shifts. Again there are zones on the planetary surface providing particular values of Doppler shift. For example, the zone $NARBS$ obviously defines a region which is neither approaching nor receding from the observer and so will produce zero Doppler shift. Areas J and K, however, lie on a zone all points of which are producing the same positive Doppler shift.

Although the real case is more complicated than the simple case depicted in the diagram, it can be seen that by examining the reflected pulse very carefully, and selecting particular time delays and Doppler shifts, identifiable points on the surface can be picked out, thus providing the basis of a radar map.

Finally brief comment may be made on the role of radio astronomy in the field of space probes. Without the application of the large radio receivers to pick up the weak signals of data or pictures transmitted by interplanetary probes, their voyaging out into the Solar System would have been futile.

Problems—Chapter 21

1. Determine the smallest angle on the sky that can be resolved by a radio dish of 100 m diameter working at a wavelength of 10 cm.

   How does this compare with a 5 m optical telescope operating at wavelengths around 500 nm?

2. A radio dish of 25 m diameter observes a point source providing a flux density of 140 m(illi)Jy. Calculate the power received if the bandwidth is 20 MHz.

3. A small radio dish with a narrow beam (less than $\frac{1}{2}°$) is directed to a plain wall of a building with a surface temperature of 27 °C (300 K). It is then directed to the Sun (6000 K). Estimate the ratio of the signal strengths for the two signals.

4. A dish type antenna has a diameter D but is used with an aperture efficiency of η. Show that the beam solid angle may be written as

$$\Omega_A = \frac{4\lambda^2}{\pi D^2 \eta}$$

   where λ is the operating wavelength.

   Suppose a telescope of diameter 25 m has an aperture efficiency of 0·6 and is used at a wavelength of 3 cm, determine the solid angle of the beam.

   If the beam comprises a simple cone, determine the diameter of the circular patch that is resolved on the sky.

Chapter 22

Telescope mountings

22.1 Optical telescopes

In order to be able to point any telescope to particular directions in the sky, it is necessary to mount the collector on a 'platform' which can be rotated about *two* axes. In the case of a refractor, the objective is normally held in a tube which is then attached to the platform. For a large reflector, the optics are usually supported in an open tubular frame in order to reduce the overall weight. The open frame, however, is more susceptible to air currents which give deterioration in the telescope's optical performance. Baffles may be included in the open frame system to cut out any extraneous sky light. It is very important that the tube or frame is free from flexure while the platform is oriented over its range of positions.

The simplest of mounting designs provides motion about vertical and horizontal axes and this is known as **alt-azimuth mounting** (see section 8.2). Rotation about the vertical axis allows a telescope to be set in azimuth and rotation about the horizontal axis allows the telescope to be set in altitude. A simplified version of an alt-azimuth mounting is depicted in figure 22.1.

This system is not very convenient for astronomical purposes as the telescope needs to be driven simultaneously about both axes if an object is to be tracked during the course of a night: the drive rates constantly change through the night and also depend upon the position of the object on the celestial sphere.

By altering the orientation of the head containing the bearings for the two axes, so that one of the axes is set parallel to the direction of the Earth's rotational pole, a more convenient form of mounting is obtained. This system is known as an **equatorial mounting** and its basic features are illustrated in figure 22.2. The axis which is parallel to the direction of the Earth's pole is known as the **polar axis** and the axis at right angles to this is called the **declination axis**. In directing the telescope to any particular point on the sky, the required angular rotations about the axes are directly related to the equatorial coordinates of that point (see section 8.3).

Consideration of the design of the equatorial mounting shows that the polar axis points in the direction given by $\pm90°$ declination. Rotation about the polar axis scans out a circle of constant declination on the celestial sphere whereas rotation about the declination axis scans out a circle of constant RA. It is standard practice to fit setting circles with encoders to the axes and, for larger telescopes, the settings are also displayed on a console. By using the setting circles, perhaps by control from the console, a particular point on the celestial sphere can be observed by setting the telescope to the correct hour angle—knowing the required RA and the local sidereal time—and to the correct declination. A preliminary setting can sometimes be achieved by fast slewing motors and final adjustment is then made by guidance motors. The slewing motors allow the telescope to be moved from object to object with the greatest possible speed.

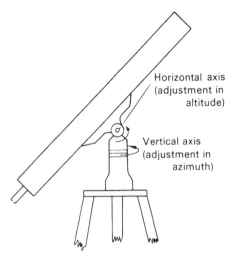

Figure 22.1. A simple alt-azimuth mounting.

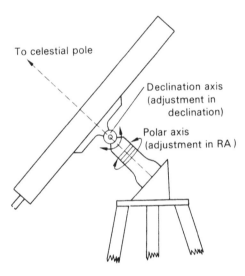

Figure 22.2. A simple equatorial mounting.

Once an equatorially-mounted telescope has been directed to a particular point on the celestial sphere, the point can be tracked or followed by applying rotation to the polar axis at an equal rate but in the opposite sense to the rotational speed of the Earth. Any errors in the **following**, such as those caused by flexure of the telescope or effects of atmospheric refraction, are compensated by the observer using supplementary guidance motors which are controlled by push-button. In some elaborate systems, exact following is achieved by image-position-sensing devices providing a servo-system with error signals which are made to control the guidance motors automatically.

The drive to a telescope is usually applied by an electric synchronous motor, via a gear train, to a worm and large wheel. A simple look at the accuracy required of a drive shows that, in general, some electronic elaboration is required.

Suppose that for an observation, it is hoped that the following of an image will match the resolution of the telescope. For a good telescope, the resolution may be of the order of 0·1 seconds of arc and this is equivalent to a movement of the sky during a period given by $0\cdot1/15$ seconds of time. If the observation takes 1 hr to make, the following must be good to

$$\frac{0\cdot1}{15} \text{ parts in } 60 \times 60 = 1 \text{ part in } \frac{60 \times 60 \times 15}{0\cdot1}$$

$$= 1 \text{ part in } 5 \times 10^5.$$

A synchronous motor driven directly from the mains frequency will normally be insufficient to provide such a high accuracy of following, as the variation in the mains frequency is usually greater than this.

There are several different methods which are used for controlling the speed of rotation of the telescope drive worm shaft. Perhaps the most common is the servo-system method whereby the speed of the following motor is constantly referred to a standard stabilized frequency generated by a quartz crystal. The guidance motors are usually connected to the worm shaft by means of mechanical differentials. Figure 22.3 shows the basic elements of a telescope drive and guidance system.

In some systems, the guidance motors can be dispensed with if, by the push-button control, the reference frequency can be instantaneously re-set to a value which is either greater or smaller than that used for following, depending on the sense of the adjustment required. In another system, the stabilized frequency, after appropriate division, can be made to drive a stepping motor directly. By using the push-button control to alter the dividing factor, the following rate can be altered and, hence, guidance can be achieved in this way.

22.2   Equatorial mountings

22.2.1   The German mounting

There are several designs of equatorial mounting, perhaps the most commonly used for refractors being the **German mounting** which is illustrated in figure 22.4. The declination axis is in the form of a beam across the top of the polar axis. The telescope is fixed to the end of the declination axis and the mounting requires a counterweight at the opposite end of the axis so that the system is balanced over the whole range of hour angles. Although the German mounting gives rise to larger stresses in the bearings and shafts than other designs, it has advantages in its ease of operation over the whole of the sky. It can easily be seen that for every point in the sky, the telescope can be set in one of two positions which are separated in hour angle by $180°$.

22.2.2   The fork mounting

Another mounting, known as the **fork mounting**, has a polar axis much like that of the German mounting but on its end is a fork through which the telescope can be swung about the declination axis. The system is illustrated in figure 22.5. This design eliminates the heavy declination yoke and counterweight of the German mounting. However, the fork mounting prevents easy access to the telescope's focal plane when it is directed to regions near to the pole and is unsuitable for refractors and the normal Cassegrain systems.

22.2.3   The English mounting

In the **English mounting**, the telescope is suspended inside the bearings of both the declination and polar axes. It is illustrated in figure 22.6. The telescope is able to swing between the frame which forms part of the polar axis. In this design, the suspended weight is kept down to the basic minimum of the telescope itself and is used for large telescopes. It has a disadvantage in that, in this form, it cannot be used to observe the region of the sky around the pole. This is no real disadvantage for observatories

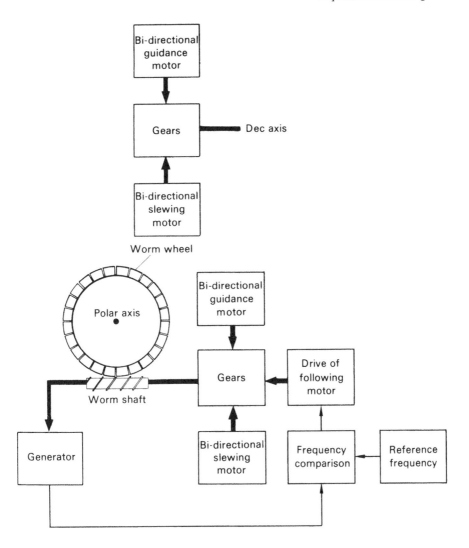

Figure 22.3. A telescope drive and guidance system.

which are located in low latitudes. A modified version of the English mounting, which allows polar observations, puts the telescope outside the frame of the polar axis, on the end of the declination axis but this then requires a counterweight to be applied on the opposite end of this axis.

22.2.4 The coudé system

In the case of some types of measurement, the analysing equipment is too cumbersome and heavy for attachment at the focal plane of the telescope, especially when the telescope needs to be oriented over a large range of positions. Such an instrument might be a spectrometer with a small RLD with a large camera. One way around this problem is to arrange the optical system of the telescope to provide an image at a fixed position in space, independent of the telescope's orientation, and to fix permanently the analysing equipment so that it accepts the energy contained in the image. The design of the equatorial mounting lends itself to such an optical system, whereby the converging light rays are directed along or through the polar axis. The system is known as a **coudé telescope** and an example of this system

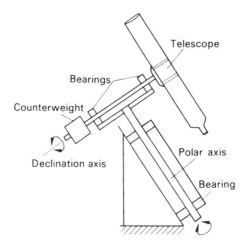

Figure 22.4. The principle of the German mounting.

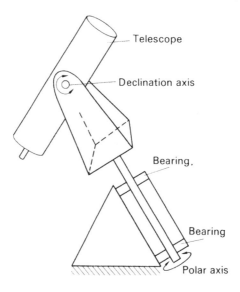

Figure 22.5. The principle of the fork mounting.

is illustrated in figure 22.7. It consists of a design similar to the Cassegrain but an extra flat mirror is placed with its reflecting surface on the declination axis so that the light is made to pass along the polar axis. The flat mirror is mechanically mounted to the telescope so that it rotates at half the rate of the telescope around the declination axis. As the reflection efficiency of a mirror depends on the angle of incidence of the incoming light, the transmission efficiency of the coudé system will depend slightly on the declination of the object which is observed. A coudé telescope has a high focal ratio, usually in the range $f/30$–$f/60$.

22.3   Telescope domes

The conventional method of housing an optical telescope is to provide it with a cylindrical or polygonal building which is covered by a hemispherical dome. Access to the light coming from particular

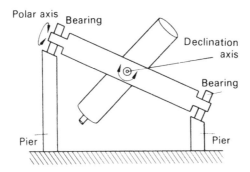

Figure 22.6. The principle of the English mounting.

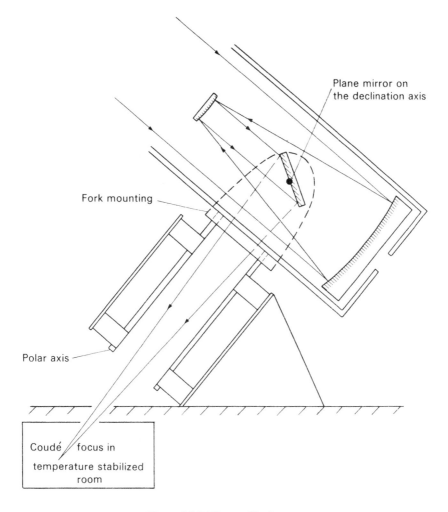

Figure 22.7. The coudé telescope.

directions in the sky is achieved by providing an aperture, usually in the form of the slit, which can be opened whenever the weather conditions are suitable. All aspects of the sky can be covered by allowing the dome to be rotated.

The design of any dome requires that the structure should not contribute to the disturbances which are apparent in all ground-based telescope images. Efforts are made to prevent any flow of air across the telescope aperture. Extremes of air temperature are to be avoided during the daily cycle and the design should ensure that the building does not act like a greenhouse.

As an example of design, the dome of the 200-in (5·08 m) Hale telescope at Mt Palomar, USA, approximates to a hemisphere of 41·5 m diameter. It consists of two 'skins' with a gap which is filled with crumpled aluminium foil. Air circulation through the gap, provided by vents at the top and bottom of the dome, takes away much of the heat which is absorbed from the sunlight during the day. Thus, by insulation and air circulation, the inflow of heat during the day is kept to a minimum; indeed, the rise in temperature of a telescope mirror during the day rarely exceeds the temperature of the preceding night by more than 2 °C.

With some elegant designs, as in the case of the 200-in, a model of the telescope is provided to act as an analogue computer. Its orientation is made to match that of the large telescope at all times. By a system of sensing devices and servos, it is made to drive the dome, for example, so that the telescope aperture is never obscured by the limits of the dome's slit. A safety mechanism is also provided so that the main telescope is never allowed to point below an altitude of 5°.

22.4   Radio telescopes

Extremely large constructions are required for the mountings of radio telescopes and, consequently, not every telescope is given adjustment to allow all points of the sky to be monitored. In some cases, however, it has been found possible to design the telescope so that it can be mounted equatorially. The mechanical problems for this type of mount are more severe than for alt-azimuth mountings and the larger telescopes with dish collectors are more often mounted by this latter method. The following of a celestial radio source is sometimes achieved by using a model which is mounted equatorially, providing an analogue computer. The model provides driving currents to the motors attached to the axes of the alt-azimuth mounted telescope and the system is servo-controlled so that the model telescope and the real telescope are simultaneously pointing to the same direction on the celestial sphere. An example using this system is the 250 ft (76 m) Lovell radio telescope at Jodrell Bank, England. In passing, it may be mentioned that the mechanical problems of mounting the 6 m optical telescope of the Russian Academy of Sciences in the Caucasus Mountains, have been solved in the same way. This telescope is mounted on alt-azimuth axes and is controlled by an analogue computer.

A few radio telescope dishes have been designed so that they remain in a fixed position and the skies are scanned by allowing the diurnal motion to carry a strip of the celestial sphere through the focus of the telescope. By tilting the antenna supports, so that the sensor is not at the exact focus of the dish, a range of strips can be scanned. In general, however, the area of sky that is available for observation is greatly reduced by using this type of fixed system. Nevertheless, the fixed-dish design allows extremely large collecting areas to be used. By choosing the observing station carefully, it is possible to allow the natural contour of the land to be used in providing the basic shape and support of the telescope dish. This idea has been used, for example, for the 1000 ft (305 m) steel mesh dish at Arecibo, Puerto Rico. By having an extendible boom to carry the antennas, this particular telescope allows a range of approximately 45° in observable declinations.

Observations in radio astronomy are not as weather-dependent as they are for the optical region and telescopes are protected from the weather only by paint and grease. It is usually unnecessary to provide them with a retractable cover in the same way that an optical telescope is protected by a dome. However, because of their large size and exposure, it is important that they are designed to be stable in high winds. According to their design, a limit is usually set to the wind speed for which the telescope can be operated.

Chapter 23

High energy instruments and other detectors

23.1 Introduction

The description of astronomical equipment and its physical principles has concentrated on the optical and radio spectral regions. This is fairly natural, following as it does the historical development of the subject. Optical and radio telescopes are also familiar instruments with observatories housing small visual telescopes usually available locally. Demonstrations of the principles of much of the subsidiary equipment are also readily found in educational laboratories.

However, astrophysical experimentation now covers much wider fields and, in this chapter, other observational equipment and techniques will be briefly described. The subject areas cover x-ray and γ-ray astronomy, ultraviolet observations, infrared, mm wavelengths, cosmical neutrino detection and the search for gravity waves and dark matter.

23.2 X-ray astronomy

23.2.1 X-ray energies

X-ray energy emitted by celestial sources is totally absorbed by the Earth's atmosphere and observations can only be made by special equipment launched by rockets. The first detections of cosmic x-rays were made from rocket flights in the early 1960s. Since then, there have been several satellite platforms carrying x-ray instruments for survey work and spectrometry with progress being made both in terms of sensitivity and spatial resolution.

In the spectral range of wavelengths shorter than the ultraviolet, it is usual to describe the radiation in terms of the **photon energy** rather than its wavelength or frequency, the unit being the **electron volt**. In the optical region, photon energies are typically of the order of 2 eV, while x-ray photons have energies of the order of a few keV. This can be checked by considering the wavelength (frequency) to photon energy conversion as follows.

In the optical region $\lambda \sim 500$ nm (5000 Å) so that

$$\nu = \frac{c}{\lambda} = \frac{3 \times 10^8}{5 \times 10^{-7}} = 6 \times 10^{14} \text{ Hz.}$$

The energy of a photon is $h\nu$ and for the optical region its value is

$$= 6 \cdot 6 \times 10^{-34} \times 6 \times 10^{14}$$
$$= 3 \cdot 96 \times 10^{-19} \text{ J.}$$

Section of
collimator

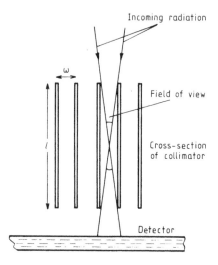

Incoming radiation

ω

Field of view

l

Cross-section
of collimator

Detector

Figure 23.1. The principle of an x-ray collimator telescope.

Converting the energy units from joules to electron volts gives a photon energy of

$$\frac{3 \cdot 96 \times 10^{-19}}{1 \cdot 6 \times 10^{-19}} \text{ eV}$$
$$\sim 2 \text{ eV}.$$

For x-rays, it has already been mentioned that the energies are a few keV and higher and the reverse calculation shows that wavelengths of $\sim 0 \cdot 5$ nm or 5 Å are covered. With such extremely short wavelengths, it will be readily appreciated that the principles used for optical telescopes cannot be extended to x-rays.

23.2.2   X-ray telescopes

Imaging of high-energy photons is difficult because of their extreme penetrating power. If the principles of optical design were extended to the x-ray spectrum, reflection surfaces could not be machined to the tolerances of the fraction of the wavelength necessary to obtain sharp images in the focal plane and, in any case, the x-rays would pass through the mirror! For energies under a few keV, special forms of reflector telescope can be designed. However, for higher energies, the only options for obtaining spatial isolation are **occultation, collimation** and **coincidence detection**.

A collimator is a device which physically restricts the field of view. One simple means of doing this is to use a series of baffles in a network of tubes as illustrated in figure 23.1. The field of view of such a collimator is given by $2w/l$, where w is the width of a tube aperture and l the tube length. With this kind of system, typical angular resolution is a few arc minutes. The energy range of this kind

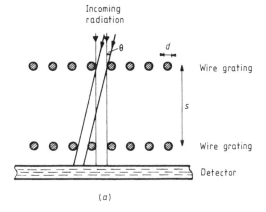

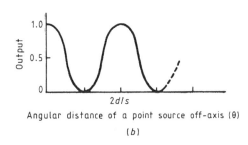

Figure 23.2. (*a*) The principle of an x-ray modulation telescope, (*b*) the form of the output signal for a point source as the telescope is panned.

of collimator is limited. For low energies, the x-rays may be reflected off the walls of the tubes, so accepting radiation from a wider range of angles on the sky. For high energies, the radiation can pass through the walls of the tube and any sense of directivity is completely lost.

A more sophisticated system is in the form of the **modulation collimator**. This comprises two (or more) parallel gratings separated by a small gap. A schematic diagram of the device is depicted in figure 23.2(*a*). Full intensity is passed when the x-ray source is on axis and also when it is at directions θ given by $2d/s, 4d/s, \ldots$, etc. If the collimator is scanned across the sky, then a given source will provide a sine-wave modulation as illustrated in figure 23.2(*b*). Other sources will provide similar signals but with a phase according to their position in the sky. The addition of several gratings to the system introduces modulations at other frequencies. By investigating the amplitudes and phases of the various frequencies within the combined signal by Fourier analysis, the spatial image and positions of the sources can be mapped. Orthogonal sets of gratings provide a means of determining the two-dimensional image distribution.

Direct imaging can also be achieved by using metallic reflection at grazing angles of incidence. The cross section of a simple design is shown in figure 23.3(*a*). For 2 keV x-rays, reflection efficiencies of the metal mirrors are the order of 50%. In order to prevent the image plane receiving radiations from objects not on axis, a central occulting disc is required and a larger collection area can be achieved by nesting a series of basic mirror systems as depicted in figure 23.3(*b*), the design being referred to as a Woltjer I system.

Following the first detection of celestial x-ray sources by transient rocket platforms, major steps were achieved by the first sky scanning instruments such as UHURU and Ariel V. With the launching of the Einstein Observatory in 1978 with a grazing mirror telescope, the sensitivity was sufficient to

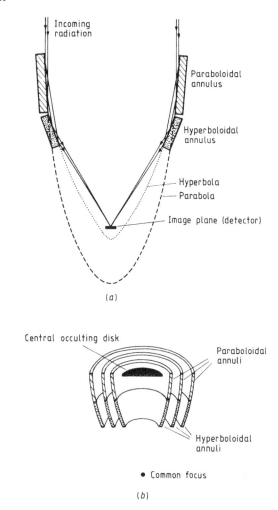

Figure 23.3. (*a*) A simple arrangement for an imaging x-ray telescope, (*b*) the simple imaging principle employed in a nested x-ray telescope.

allow investigation of a significant sample of normal stars and galaxies. The instrument was able to tackle the problem concerning the nature of the diffuse hard x-ray background with the result that it was probably due to a large number of very faint, distant sources, the conclusion being a little more definite for low-energy x-rays.

In the following two decades, satellite missions included EXOSAT, GINGA, ROSAT, ASCA and BeppoSAX but with barely any improvement on spatial resolution. During this period, any imaging was substantially worse by factors between 5 and 50 relative to most other wavelength domains. The situation has now been redressed by Chandra[W 23.1] launched by NASA in July 1999 and XMM-Newton[W 23.2] launched by ESA in December 1999. Chandra has provided x-ray images at sub-arc second resolution putting them on a par with pictures by ground-based optical and infrared telescopes. XMM-Newton (see figure 23.4) with its large collection area and multiple nested mirrors is particularly sensitive. Both systems carry grating spectrometers with specially developed CCD chips as detectors. Individual platforms such as HESSI (High Energy Solar Spectrometry and Imaging) have

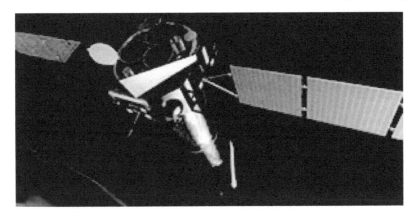

Figure 23.4. An artist's impression of the XMM-Newton x-ray telescope launched in December 1999 in prospect of a 10-year mission. (By courtesy of ESA.)

been also been recently launched for extended high-energy x-ray observations of the Sun with high spatial resolution.

As with all other kinds of telescope development, the game is to design instruments with larger collection apertures for better flux gathering and improved angular resolution. It may be noted that the current XEUS design project by ESA plans to increase the effective collection area over the early mission of ROSAT by factors of ~ 100 and ~ 1000. In order to achieve this, a strategy needs to be developed to assemble the telescopes in orbit. To establish the instruments, the XEUS system will comprise two freeflyers—a mirror spacecraft and a detector spacecraft, separated so that the detector is at the focus of the mirror. The plan is to launch the two vehicles together into a low Earth orbit and to position them by an active orbit control and alignment system.

As the proposed aperture exceeds the capacity of available launch vehicles, the mirror spacecraft will be designed to visit the International Space Station (ISS) to build it up with additional modules that are launched separately. To achieve this, the detector vehicle will have the ability to dock with the mirror spacecraft, after which the mirror spacecraft can dock with the ISS for refurbishment and for the mirror aperture to be further increased, thus converting XEUS1 to XEUS2 with an additional collection aperture some $\times 10$ larger. The detector vehicle will be replaceable at any time by launching a new version.

23.2.3 X-ray spectrometry

X-ray spectrometry or energy isolation can also be performed by using the crystal planes of materials to act in the same way as a diffraction grating for the optical spectrum, the principle being that of the **Bragg spectrometer** as sketched in figure 23.5. Interference is achieved according to the dimensions of the lattice and the wavelengths of the incident x-rays.

Some of the incident radiation is scattered by the first layer of the crystal lattice—radiation penetrating the crystal is scattered by successive layers. It can be readily seen from figure 23.5 that the path difference for the radiation emerging from two adjacent layers is $2d \sin \theta_g$ where d is the crystal plane separation and θ_g is the grazing angle.

For constructive interference to occur, the path lengths need to be equal to an integer number of wavelengths, that is

$$m\lambda = 2d \sin \theta_g$$

where m is an integer. The spectrum can, therefore, be readily scanned by altering the angle of θ_g by

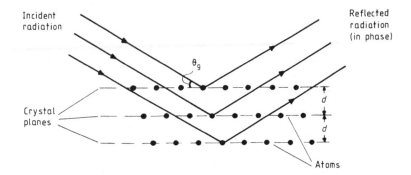

Figure 23.5. The principle of the Bragg spectrometer for x-rays.

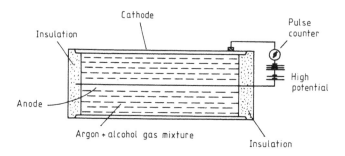

Figure 23.6. A simple schematic diagram of a Geiger counter.

rotation of the crystal surface. Materials used as 'gratings' are lithium fluoride, tungsten disulphide and graphite.

As will emerge in the following subsection, some x-ray detectors themselves allow the energy of each photon to be determined.

23.2.4 X-ray detectors

The first observation of a cosmic source other than the Sun was made in 1962 when soft x-rays were recorded and the object became known as Sco X-1. This experiment was made using a large-area **Geiger counter** flown on a sounding rocket.

The layout of a Geiger tube is shown in figure 23.6. Two electrodes within the enclosure are held at a high potential—but just below the discharge voltage of the gas medium. When ionizing radiation enters the chamber, the discharge breakdown is triggered with a subsequent pulse of current which is measured or counted. The amplification factor is typically 10^8 electrons. Saturation occurs so that the original photon energy cannot be elucidated and it is for this reason that they are generally no longer used, the **proportional counter** being preferred. Geiger counters also have a disadvantage in that following a pulse they suffer a dead-time, of the order of 200 μs, in which a second photon cannot be detected.

The proportional counter is closely related to the Geiger counter. A low voltage is applied across the tube so as to prevent saturation. The gain is of the order of 10^4–10^5 which is sufficient for detection by more conventional amplifiers. Provided all the energy of the x-ray photon is absorbed within the detection chamber, the size of the pulse relates to the photon energy. Thus,

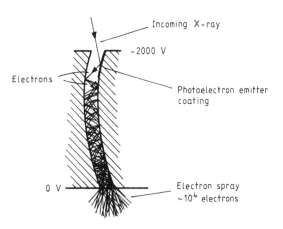

Figure 23.7. The principle of the microchannel plate.

one 30 eV photon produces 1 electron–ion pair,
one 1 keV photon produces 36 electrons and
one 10 keV photon produces 360 electrons.

For low photon energies, a window is required for the radiation to enter the chamber. This might be in the form of a thin plastic sheet, mica or beryllium. The window allows diffusion of gas from the chamber and this must be replenished continuously for extended use of the detector. The position of the x-ray photon path can be obtained by measuring the pulse from both ends of a resistive anode and comparing the pulse strengths and shapes.

For photon energies above 30 keV, gas-filled detectors have limited stopping power and devices such as the proportional counter become inefficient. In this higher frequency domain, a scintillation detector might be used. This comprises a crystal such as sodium iodide doped with thallium placed before a photomultiplier tube. Any incoming x-ray generates a flash of light which is detected photoelectrically. The intensities of the light pulses are proportional to the x-ray energies.

In the extended region of the ultraviolet and for low-energy x-rays, both amplification and image recording can be achieved by a **microchannel plate**, the thin plate pierced with rectangular holes of the order of 25 μm in diameter (see figure 23.7). The top of the plate is coated with a photoelectron emitter for the x-ray energies of interest. The curved tubes have a large potential across their length, so that the emitted electrons produce an avalanche by secondary emission on being accelerated down the tube. The electron sprays and the positions of origin corresponding to the original x-ray image are then recorded by a subsidiary detector system (for example, phosphor and TV camera or CCD).

Table 23.1. γ-ray energies.

Energy range	Energy (MeV)	Observational detection
Low	1–10	Satellite—scintillation counter
Medium	10–30	Satellite—spark chamber
High	30–1000	Satellite—spark chamber
Very high	10^5–10^7	Ground-based Čerenkov detector
Ultra high	10^9 up	Ground-based cosmic rays

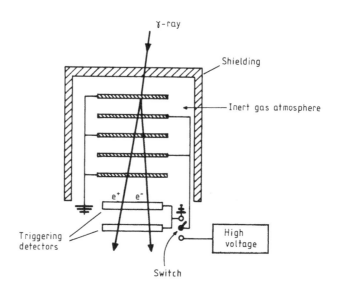

Figure 23.8. The basic components of a spark chamber.

23.3 γ-ray astronomy

23.3.1 Detectors and satellites

The γ-ray spectrum lies beyond the short wavelengths of hard x-rays and it is convenient to set the boundary of the two regions at photon energies of 1 MeV. Sources generating low-energy γ-rays require satellite platforms for their observation but at the extreme high-energy end of the spectrum, their arrival from cosmic sources can be recorded by the Earth's atmosphere contributing to the detection system. The wavelengths of γ-rays are less than 0·1 Å, telling us immediately that very energetic processes are involved in their generation with some of them being nuclear and certainly not being associated with electronic transitions in the shells of atoms. Table 23.1 gives a broad classification of γ-ray energies and their methods and detection.

The scintillation counter already described under x-ray astronomy is used to detect low-energy γ-rays. The medium-to-high energy γ-ray photons can be detected by a **spark chamber** whose operation is similar to that of a cloud chamber. Figure 23.8 outlines the principal components.

For a γ-ray to be detected, its impact on the first plate in the chamber produces an electron/positron pair ($e^- e^+$). Alternate plates in the chamber are charged to a high voltage and the strong electric fields cause breakdown of the gas along the paths of the charged particles. The trails of sparks can be recorded (TV or CCD cameras) with the results being telemetered to the Earth.

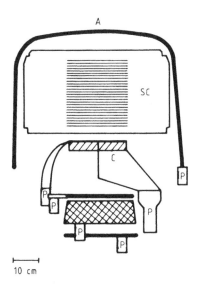

Figure 23.9. A sectional view of the COS B *γ*-ray detector equipment with a spark chamber (SC), a Čerenkov detector (C) and other photomultipliers (P) attached to scintillation systems. The enclosing dome (A) provides an anti-coincidence detection system.

Several early satellites have carried *γ*-ray detectors, notably Explorer II, OSO III, SAS II and COS B. The detection system on board SAS II, which was operational for six months in 1972/73 was centred on the 100 MeV region. Some 8000 photons were recorded during its operation. The COS B satellite was launched in August 1975 and the system was operational for about six and a half years before the gas supply ran out. A cross section of the COS B *γ*-ray detector is depicted in figure 23.9. Overall, some 200 000 *γ*-ray photons were detected with an angular precision of a few degrees. It is of interest to note that in many optical measurements the same number of photons might be recorded in one second! COS B detected about 25 sources, some of which have been identified optically.

A giant step forward occurred in April 1991 with the launch of the Compton *γ*-ray observatory. Unlike earlier missions, Compton covered a broader band of energies from 30 keV to 3×10^4 MeV with 10 times higher sensitivity and greatly improved angular resolutions and timing capabilities. One of the on-board packages, BATSE, designed to investigate the *γ*-ray burster detected a new source virtually every day, the 2000th event being recorded by the end of 1997, revealing a random distribution over the sky.

One of the aims of such telescopes is to be able to obtain a fix of any burster with sufficient accuracy to allow immediate follow-up of observations in other wavebands. In 1997, the BeppoSAX telescope detected a *γ*-ray burst that was identified optically within hours with a spectrum obtained by the KECK telescope suggesting that the event involved the most energetic phenomena at the limiting edges of the Universe.

23.3.2   *γ*-ray spectral lines

γ-rays may be generated by a variety of mechanisms such as matter–antimatter annihilation, radioactive decay, energetic particle collisions, synchrotron radiation (high-energy electrons being deviated by interstellar particles), Bremsstrahlung (high-energy electrons being deviated by a magnetic field) and by hot plasmas. Some other mechanisms involve the production of *γ*-rays with discrete energies so allowing positive identification of the process.

The most famous *γ*-ray line emission is that at 0.511 MeV resulting from an e^-e^+ collision at

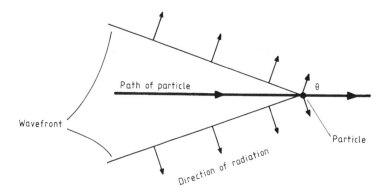

Figure 23.10. The generation of Čerenkov radiation.

rest, denoted by

$$e^+ + e^- \rightarrow \gamma + \gamma.$$

This line has been detected by the HEAO-3 satellite in the direction of the galactic centre. A possible explanation involves the production of positrons as secondaries in the interactions of cosmic rays with the interstellar gas—subsequent annihilation produces the 0.511 MeV line. However, the strength of the line appears to be too great in relation to the cosmic-ray intensity at the galactic centre. An alternative explanation involves the accretion of gas by a black hole with a mass several million times that of the Sun. Hot spots within the disc emit e^+e^- pairs. The positrons are then annihilated to produce the 0.511 MeV γ-ray line.

23.3.3 Čerenkov radiation and detection

When high-energy γ-rays pass through a medium, their production (e^+e^-) may take place with the resulting particles travelling faster than the local speed of light $(c/\mu,\ \mu$ being the refractive index of the medium). As a result of the abrupt changes in the electric field as the particle passes the atoms, Čerenkov radiation is emitted.

The radiation is concentrated into a cone spreading outward from the direction of motion of the particle (see figure 23.10). The half-angle of the cone is given by

$$\theta = \tan^{-1} \left[\left(\mu_\nu^2 \frac{V^2}{c^2} - 1 \right)^{1/2} \right]$$

where μ_ν is the refractive index at frequency ν and V is the velocity of the particle $(V > c/\mu_\nu)$. The form of the spectrum of the radiation is given by

$$I_\nu = \frac{e^2 \nu}{2\varepsilon_0 c^2} \left(1 - \frac{c^2}{\mu_\nu^2 V^2} \right)$$

where ε_0 is the permittivity of free space. The radiation may then be detected by a photoelectric device. Čerenkov detectors are employed on satellites to record γ-rays.

The principle of the detection of Čerenkov radiation is also used with the Earth's atmosphere providing the medium for the production of the light flashes. The form of the Čerenkov spectrum given here indicates that optical flashes are produced by γ-rays in the energy range 10^5–10^7 MeV resulting from pair production in the upper atmosphere. Two or more ground-based 'light buckets' need to be used to eliminate non-Čerenkov events by anti-coincidence discrimination. Several sources have been detected in this way and a 4^h8 periodicity in the famous object Cygnus X-3 has been monitored.

Several large experimental facilities are currently being established to record Čerenkov events. A facility with the acronym HESS (High Energy Stereoscopic System) is being constructed in the Khomas Highland of Namibia to investigate γ-ray skies at energies of several TeV (10^{12} eV). The observatory is due for completion in 2004 but the first elements may be operational before then. The final arrangement will comprise four identical telescopes each of 12 m diameter and comprising 400 individual elements. The energy of any incident γ-ray will be determined from the intensity of the Čerenkov image and its direction of arrival from the image shape and orientation, so being able to discriminate between air showers caused by γ-rays and those caused by cosmic-rays. Similar experiments are ongoing at the CANGAROO Observatory sited in Australia.

As with any optical telescope, these Čerenkov radiation detector systems cannot be used in daylight or when the weather conditions are unsuitable. An alternative method of ground-based detection of γ-rays is being developed at the Yangbajing High Altitude Cosmic Ray Laboratory in Tibet. The 'telescope' will occupy an area of 6500 m^2 and will be operational over the full 24 hour period. The 18 000 resistive plate detectors each measuring 56 cm × 60 cm are made of Bakelite. Each counter is covered by a 5 mm thick sheet of lead to convert the γ-ray photons into charged particles, thereby improving the performance of the device. The joint Italian–Chinese venture should be fully active by the end of 2003.

23.4   Ultraviolet astronomy

The ultraviolet electromagnetic spectrum is blocked by the Earth's atmosphere and, in order to make observations at these wavelengths, it is necessary to use rockets and satellite platforms. This domain contains a wealth of spectral lines suited to astrophysical interpretation. For example, the Lyman series (see section 15.7.3) is well within the ultraviolet region with Lα at 1215 Å; some of the spectral lines associated with ionized atoms are particularly useful for investigating stellar winds and calculating mass loss by this mechanism.

By far the most successful enterprise in ultraviolet astronomy has been the International Ultraviolet Explorer (IUE) spacecraft which was operational for some 20 years.

The 'IUE Observatory' comprised the flight system and the ground control. The flight system included the spacecraft and means of attitude control, the telescope and analysing equipment and the data transmission circuitry. The ground stations were centred at the Goddard Space Flight Centre (NASA), Greenbelt, Maryland, USA and at the Villafranca Satellite Tracking Station (VILSPA) near Madrid, Spain, operated by the European Space Agency (ESA). Data were obtained on a routine basis—8 hr per day at VILSPA and 16 hr a day at Goddard.

The 45 cm $f/15$ Cassegrain telescope was operated directly in real time from the ground by a control computer linked to the on-board computer. Ultraviolet spectrometry was undertaken from 1150 to 3200 Å with SEC Vidicon cameras to record the data. The observer was able to monitor the measurements in real time and adjust the observational routine to enhance the scientific value of the work in hand.

23.5   Infrared astronomy

For many years, photographic plates have been able to record images for radiation beyond the red sensitivity of the eye and, quite legitimately, observations made by these means were referred to as infrared astronomy, the wavelength region covered being from about 7500 to 12 000 Å. More recently, other detectors have been developed which are more efficient for infrared measurements and the spectral range has been extended to the tens of micrometres, there being several windows for which

Figure 23.11. The 3.8 m United Kingdom Infrared Telescope (UKIRT) at Mauna Kea, Hawaii. (By permission of the Royal Observatory, Edinburgh.)

molecular absorption by the atmosphere is sufficiently low for ground-based observations to be made. Water vapour is one of the chief absorbers and it is for this reason that the best sites for infrared observation are high, the station at Mauna Kea in Hawaii at 4200 m being a good example.

The modern age of infrared astronomy was given impetus by experiments launched on balloon platforms but recent emphasis has been placed on the design of special ground-based infrared telescopes and satellite observations.

The main atmospheric windows in the infrared are between 1 and 5 μm, 7 and 13 μm and around 20 μm. Passbands with labels J (1·25 μm), K (2·2 μm), L (3·4 μm), M (5·0 μm) and N (10·2 μm) have been developed for performing standard photometry.

Application of Wien's displacement law (section 15.5.3) for a temperature of 290 K shows that the peak emission from a black body at that temperature is at 10 μm. Thus, many measurements in the infrared relate to cool bodies, for example, solar system objects and cool stars. The same calculation also gives an immediate hint to some of the problems and difficulties of making infrared observations. The telescope itself will radiate strongly in the infrared and special care must be given to its design and structure. For optical telescopes, it is common practice to surround the secondary mirror by a baffle tube to prevent stray light from the sky entering the system. In the infrared, such an arrangement is an embarrassing source of radiation and is generally removed, particularly for observations at 5 μm and beyond. The optical components within the subsidiary analysing equipment are invariably refrigerated and the detector itself cooled to liquid helium temperatures. Again, by virtue of its temperature, the atmosphere between the incoming radiation and the telescope also radiates strongly and the infrared brightness is continually changing. This problem is dealt with by chopping frequently between the source and a nearby sky position and integrating the difference signal. The chopping may be performed by having a secondary mirror which 'wobbles' between two positions. A specially designed infrared telescope known as UKIRT (United Kingdom Infrared Telescope) is depicted in figure 23.11.

Modern infrared detectors include 'thermal' devices and solid state photo-conductive and photovoltaic cells. Perhaps the most widely used detector is the **germanium bolometer**, operated at liquid helium temperatures. Essentially, it is a gallium-doped germanium chip with a blackened (for absorption and increased efficiency) sensing area of less than 0·1 mm^2. The sensing depends on the change in electrical resistance of the chip as it absorbs the infrared radiation and increases its temperature.

The sensitivity of the infrared detector is normally characterized by its **noise equivalent power (NEP)**, this being defined as the power required of any incident radiation such that it would produce an rms (root mean square) value of the noise. In other words, the NEP is the incident power required to obtain a signal-to-noise value of unity. A good value for the NEP of a germanium bolometer is of the order of 7×10^{-15} W Hz$^{-1/2}$.

A well-known success has been the IRAS (Infrared Astronomical Satellite—see figure 23.12) which, in 1983, carried a refrigerated 57 cm telescope into orbit, surveying virtually the whole of the sky at four wavelength bands centred at 12, 25, 60 and 100 μm. The telescope optics and the main system were maintained below 5 K for the 11 months of the survey. The focal plane of the telescope held a template with a series of 62 slots providing different fields of view and followed by detectors operating at 1·8 K. The detector pattern was designed so that any source crossing the field was monitored by at least two of the detectors in the same wavelength band. Analysis of the thousands of recorded infrared sources has provided a new reference catalogue and has opened up new areas of astrophysical research.

23.6   Millimetre astronomy

At high sites where the amount of precipitable water vapour is particularly low, several atmospheric windows open in the sub-mm range of the electromagnetic spectrum. Radiations from space at these wavelengths are related to the spectral emissions of molecules within interstellar dust clouds (see section 15.7.4). For this spectral domain, it is more usual to express the radiation in terms of frequency rather than wavelength. The frequency range for this newly open window is from about 25 to 1000 GHz. One particular spectral line of interest in the mm range is that emitted by the CO molecule at 230 GHz. Figure 23.13 indicates the various windows with the frequencies of some molecular lines clearly available for observation. Special telescopes and techniques are required and, recently, several research groups have developed the necessary facilities.

The problem of telescope design is classical but new concepts have been incorporated because of the working wavelengths. A large aperture is required to collect the low fluxes. To obtain a good angular resolution, the mirror needs to be smooth to a fraction of a wavelength, that is a fraction of a mm. The James Clerk Maxwell Telescope (JMCT) (see figure 23.14) commissioned in 1987 and

Figure 23.12. The international Infrared Astronomical Satellite (IRAS) in preparation. (By permission of the Rutherford Appleton Laboratory.)

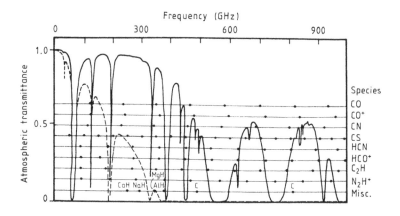

Figure 23.13. The atmospheric windows for mm-astronomy. The dots indicate the frequencies corresponding to emissions from various molecular species. Full curve, Mauna Kea (predicted); broken curve, sea level.

stationed on Mauna Kea mountain in Hawaii provides a 15 m dish made up of 276 separate panels of aluminium. Lightness of structure was achieved by pressing them from a honeycomb of lightweight aluminium and covering them with stretched aluminium sheets. Each panel is fixed to three motorized supports which can be controlled in positions to within ± 3 μm so that the complete mirror is adjustable to the required parabolic shape. The collection aperture can be considered as an example of 'active optics'.

A special dome has been designed with a large slot to allow the collection dish an uninterrupted

Figure 23.14. The 15 m dish of the James Clerk Maxwell Telescope (JCMT) at Mauna Kea, Hawaii, without the protective membrane which normally fits between the large slit of the observing turret. (By permission of the Royal Observatory, Edinburgh.)

view. However, the dish aperture is normally protected from the wind and from solar radiation by a large PTFE (polytetrafluoroethylene) membrane across the dome slit. This plastic sheet has a transmittance of 98% at mm wavelengths but absorbs 80% of the general solar radiation and allows observations to be made up to wind speeds of the order of 70 km h^{-1}. When the weather is calm, the shield can be removed to allow the most sensitive measurements which are degraded by the small amount of scattering that the membrane introduces.

One of the chief instruments developed at the Royal Observatory, Edinburgh, and attached to the JMCT is SCUBA (Submillimetre Common-User Bolometer Array) which allows images to be obtained. The array of bolometers is cooled to 0·06 K making it the world's most powerful submillimetre camera. It has had great success in mapping the centre of our galaxy, the galactic plane and other active galaxies. Because of the relatively small field of view, integration times are quite long and in order to piece the images together to produce a map, very many telescope hours are required. An improved version (SCUBA-2) is being developed with the prospect of reducing current observational times by a factor ~ 100.

23.7   Neutrino astronomy

23.7.1   Introduction

The neutrino (ν) is an elusive particle and even at this stage of the development of the theory of nuclear processes, its true nature is not perfectly understood. It has taken 70 years since Pauli proposed its existence in 1930 to conclude that the particle has a rest mass.

Pauli's postulation was made to allow the principle of conservation of mass and energy to be maintained in nuclear reactions by removing energy in some β-decay processes. Confirmation that the particle actually existed took some 25 years, the experimental problem being the very small **cross section** for absorption by any atom in its path. For example, one interaction which is used for its detection involves its absorption by chlorine atoms and the production of argon according to the interaction expressed as

$$\nu + {}^{37}_{17}\text{Cl} \rightarrow {}^{37}_{18}\text{Ar} + \text{e}^-.$$

The cross section (σ) of a chlorine atom for this absorption is only of order 10^{-46} m^2. The **mean free path** of a particle (i.e. the distance that the particle is likely to travel before undergoing an interaction) may be expressed as

$$\lambda = \frac{1}{\sigma N}$$

where N is the number density of atoms along the particle's path.

With such a small cross section, the mean free path is enormous. Thus, if it were possible to produce a column of pure liquid chlorine in which the neutrino could travel, its length would have to be greater than *one parsec* to match the mean free path. Possible sources of neutrinos are the thermonuclear processes at the centre of the Sun, and it is instructive to note that in the passage along the Sun's 700 000 km radius, there is only a one in 10^{10} chance of any neutrinos interacting with any other particle. This appreciation illustrates the difficulties of setting up a 'neutrino observatory'.

In addition to the detection of solar neutrinos, the 1987 supernova event in the Large Magellanic Cloud has accelerated interest in cosmic neutrinos and their detection. At that event, it has been estimated that about 10^9 neutrinos passed through the cross section of each human being but the total number of particles detected by the various neutrino telescopes around the world was of the order of 10.

23.7.2   Neutrino telescopes

The most publicized neutrino detection experiment has been operating since 1968 in a disused gold mine at a depth of 1·5 km. The detector is in the form of a large tank (over 600 tonnes) of tetrachloroethylene (C_2Cl_4) commonly known as cleaning fluid. One in four of the chlorine atoms is the isotope ${}^{37}_{17}\text{Cl}$ and hence, on average, one of the molecules contains one atom of the necessary isotope. Within the tank, there are of the order of 2×10^{30} atoms of ${}^{37}_{17}\text{Cl}$.

The argon (${}^{37}_{18}\text{Ar}$) that is produced by neutrino capture has a half-life of 35 days, decaying back to ${}^{37}_{17}\text{Cl}$ by capturing one of its inner orbital electrons, at the same time ejecting a 2·8 keV electron.

The detection process involves sweeping out the argon with helium, separating the gases through a charcoal trap and counting the number of 2·8 keV electron events.

Placing the tank some 1·5 km below ground shields it from cosmic rays and from natural and artificial radioactive sources. A thick water jacket surrounding the tank acts as a shield to absorb neutrinos. The efficiency of extracting the argon is monitored by introducing known quantities of $^{37}_{18}$Ar. The collected gas containing any $^{37}_{18}$Ar is shielded during the monitoring of its decays and anti-coincidence techniques are applied.

As with five other different observatory experiments established since the early 1970s using three different detector techniques, significantly fewer neutrinos are detected from the Sun than predicted by the theory associated with the thermonuclear generation of energy within its core. The 'missing solar neutrino problem' has been very significant in developing our ideas about the nature of the neutrino itself. The question as to whether the particle carries mass is a very fundamental one, it not being supported by the Standard Model of Particle Physics.

Neutrinos are known to carry one of three flavours, there being 'electron', 'muon' and 'tau' varieties. The Sun's thermonuclear reactions are sufficiently energetic to produce electron neutrinos but not muon or tau ones. A mooted possibility is that neutrinos, once produced, can oscillate from one flavour to another as a result of a tiny but non-zero mass. To explore the issue, a detector that can differentiate between the neutrino flavours is required. Such a capability has been established at the Sudbury Neutrino Observatory (SNO) in Canada.

The detector core comprises a chamber 34 m high and 22 m wide of 1000 tonnes of heavy water (D_2O) some 2 km below ground in the Creighton nickel mine near Sudbury, Ontario. The volume is surrounded by 10 000 photomultiplier tubes to detect Čerenkov radiation events (see earlier). The system has greater accuracy than all previous measurements and provides approximately one event per hour.

By using D_2O, the detector responds to three types of neutrino interaction referred to as charged current (CC), neutral current (NC) and elastic scattering (ES). In the CC interaction, only electron neutrinos are detected with the production of two protons and an electron; the NC interaction is equally sensitive to all flavours with the production of a proton and a neutron. The ES process is $\times 6 \cdot 5$ more sensitive to electron neutrinos than other neutrino flavours. By detecting the different decay products, it is possible to measure both the electron neutrino flux and the total neutrino flux. By looking at the ratio of reactions specific to ES and NC, precise information can be obtained as to whether neutrinos change flavour on their way between the Sun and Earth.

The first phase of the experiments, in collaboration with a measurement of the ES rate, as obtained by the Superkamiokande Observatory in Japan, has provided strong evidence that electron neutrinos disappear but are replaced by neutrinos of alternative flavour. Further evidence that electron neutrinos produced in the Sun oscillate into other neutrino flavours on their way to Earth and, hence, have mass has now emerged from differences in the signals during day and night. The SNO results suggest that, although the Standard Model is not destroyed, it is somewhat tarnished and requires modification.

23.8   Gravitational radiation

The investigation of the detection of gravitational radiation offers the last frontier in experimental observation of the Universe.

Gravitational radiation is predicted by Einstein's field equations. If an object with mass changes its position, then its gravitational effect on another object will change and information relating to change will propagate in the form of waves. Very large changes are required to release power in the form of gravity radiation and only cosmic phenomena beyond the laboratory can be considered to generate sufficient energy for detection. The kinds of event which might be detected by the current developing facilities are related to coalescing black hole binary systems, rotating neutron stars and stellar collapses. The frequency of the waves depends on the kind of event: frequencies of a few kHz will ensue from collapsing or exploding objects whereas frequencies of a few μHz might be expected to be generated within binary star systems.

The amount of gravitational radiation liberated by a source can be calculated according to the

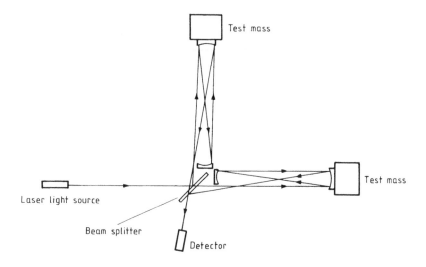

Figure 23.15. A simple layout for a possible gravity wave detector.

theory of general relativity. An estimate of the expected flux can be made by taking the simple case of a stellar binary system. For such a source, the energy L_G released may be expressed as

$$L_G \approx (2 \times 10^{-63}) \frac{M_1^2 M_2^2}{(M_1 + M_2)^{2/3}} \frac{(1 + 30e^2)}{P^{10/3}}$$

where M_1, M_2 are the component masses, P is the orbital period and e is the orbital eccentricity. In a typical dwarf nova system, $M_1 = M_2 = 1.5 \times 10^{30}$ kg, $P = 10^4$ s and the simple case of a circular orbit gives $e = 0$. Substituting for these values gives

$$L_G = 2 \times 10^{24} \text{ W}.$$

If the system is at a typical stellar distance, say 250 pc, the flux arriving at the Earth is

$$F_G = 3 \times 10^{-15} \text{ W m}^{-2}.$$

When compared with other radiations received from celestial objects, this is by no means small. For example, limiting faintness for detection of a radio source is approximately 10^{-29} W m^{-2} Hz^{-1}. However, the means of detection is very different from that for radio waves.

The progress of gravitational waves through a medium can only be detected by its effect on the mass centres in the body. In other words, it is the deformity or strain within the detector that needs to be measured as the gravity waves pass through it. Translating the gravitational flux to the anticipated amount of movement within a test mass detector suggests that the possible changes in length are exceedingly small, that is of the order of 10^{-12} times the diameter of a hydrogen atom.

The challenge of the detection of gravitational radiation was taken up by Weber in the United States and, in 1969, claims were made that gravity waves were being observed from the centre of the Galaxy. However, the detection was not confirmed by other workers and since that time, the sensitivity of various detector systems has increased by very large factors—still without success. A possible working arrangement for a gravity way detector is depicted in figure 23.15.

Currently, there are several experimental search groups pioneering systems of gravitational wave detectors. These include instruments such as GEO600 (Germany/UK), LIGO (USA), VIRGO (Italy/France) and TAMA300 (Japan). Their arrangements are based on a pair of monolithic masses

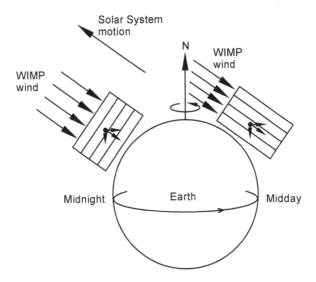

Figure 23.16. At midnight, any nuclear recoil from an interaction with the WIMP wind is essentially normal to the multiwires within DRIFT; twelve hours later, any recoil is more parallel to the multiwires.

with reflecting mirrors being suspended at the ends of two long light paths at right angles to each other. In the case of GEO600, the arm lengths are 600 m. Laser beams are despatched along evacuated tubes comprising the two arms and returned by reflection from the masses. At the intersection of the arms, an interference pattern is produced. Any change or oscillation in the path lengths alters the interference pattern which is continuously monitored. Key to the sensitivity of all the techniques is the engineering design whereby the masses are suspended and the stability of a laser system monitoring their separations as determined by the associated interferometer signal processing.

ESA is also preparing to launch the LISA (Laser Interferometer Space Antenna) mission comprising three identical spacecraft orbiting in a huge triangular formation with laser links to monitor the inter-platform distances. The passing of gravitational waves through the system will alter the relative phases of the connecting laser beams.

23.9 The missing mass problem

The gravitational behaviour of the Universe suggests that there is more matter present than is immediately detectable by the emanating radiation. It is estimated that 'Dark Matter' comprises about 90% of matter in the Universe in the form of weakly interacting massive particles (WIMPs). Such exotic particles have not been discovered but are predicted to exist by extensions of the Standard Model of Particle Physics used to describe matter and the forces in nature. It is also predicted that they might be detected directly by the low-energy recoils of atomic nuclei suffered on a WIMP collision.

A new kind of detector and observatory has recently been established with the aim of detecting WIMPs. The predicted rate of a nuclear recoil is less than 1 per 10 days kg^{-1} of detector material. The equipment used as the detector is known as the DRIFT (Directional Recoil Identification From Tracks). It comprises 1 m^3 of gas at low pressure with two multiwire proportional chambers (see earlier) to detect the tracks of ionization left by any nuclear recoil. At present, the readout from the system allows a two-dimensional track projection but future developments should provide a fully three-dimensional capability.

The principle behind the observations is that the directions of the recoils should change on a regular cycle following the Earth's rotation. The motion of the Solar System through the Galaxy effectively induces a WIMP wind from the direction of travel, this being at an angle ~42° to Earth's rotational pole. The wind vector, therefore, changes direction according to the rotational position of the sited detectors. For DRIFT at Boulby, UK (N hemisphere), the average recoil direction should vary from 'downwards' through the detector at midnight to 'southwards' at midday (see figure 23.16). Such a recorded signature will provide unique evidence for the existence of WIMPs.

The DRIFT system at Boulby is deep underground in an old coal mine to reduce interference from cosmic rays and other sources of noise. Although the observing location seems to be far away from 'Astronomy' this experimental research is most fundamental to our understanding of the Universe.

PART 4

EXPERIMENTAL WORK

Chapter 24

PROGRAMME: A scheme of practical projects and exercises is provided, forming a useful supplement to a first course in astronomy.

A programme of learning is more enjoyable if direct participation of the student is involved. One way to do this is through exercises which use data or through practical work.

Chapter 24

Practical projects

24.1 Introduction

The type of practical work that can be attempted obviously depends on the available equipment (telescopes, laboratory apparatus, planetarium, etc) or on the access to an astronomical library from which data can be obtained for exercises. It is, therefore, impossible to generalize and provide a set of experiments which will be completely suitable for everybody.

This chapter collects ideas for projects of a practical or exercises nature which can be used directly or adapted according to the resources of the student. Some of them have been or are used as parts of the course at the Glasgow University Observatory and are described here in relation to particular designs of equipment. Allowance for this must be made according to the equipment that might be to hand.

As the pieces of knowledge required for particular experiments may be scattered in several chapters of the book, it is not possible to present a series of exercises following the textual layout exactly. The scheme outlined here broadly follows the progress through the book.

The number of possible experiments and exercises that can be thought up is practically boundless. Those described here represent only a small selection but provide sufficient material for any course and are built from the chapters of this book. They have also been selected with the purpose of offering material which is more of the 'hands on' nature rather than on numerical projects related to astrophysics.

24.2 The Sun as a timekeeper

By using suitable geometries, it is fairly easy to convert the changing directions of shadows of fixed objects to the apparent movement of the Sun across the sky and, hence, to the passage of time. Such a specially designed device is a **sundial**.

Sundials abound in a variety of forms. Two of the most common types are described here together with the ideas which are necessary to make and use them properly: these are the **horizontal** and the **vertical** types. In addition, the use of a simple **noon-marker** is described. Before doing this, it is important to have an appreciation of two terms which effect that position of the shadow and its relationship to time. These are **longitude** and the **equation of time**.

(a) *Effect of longitude.* If two sundials are separated some miles apart in longitude, then the times at which the shadows fall on the noon-markers will be different: the further west the sundial, the later is the time for this event. Thus, if a time indicated by a sundial is to be converted to a civil time, allowance must be made for the sundial's longitude in relation to the reference longitude used to define the local time zone. In Britain, particularly in the winter period, the situation is very simple. For sundials west of the Greenwich meridian, a certain period of time according to the value of longitude must be added to give the civil time and for sundials east of Greenwich, a time correction must be subtracted. In summer time, an additional hour must be added to all sundial times.

As an example, consider a sundial in Glasgow where the longitude is $+4° 22'$ W, which is equivalent to 17 minutes 28 seconds of time. If t represents the time read from the sundial and T is the civil time, then

$$T = t + 17 \text{ min } 28 \text{ s} + 1 \text{ hr} \qquad \text{(if summer time is in operation)}.$$

(b) *The equation of time.* The best system of timekeeping is one in which time flows at an even rate, i.e. each unit of time as it passes should equal any other previous unit time interval. If the time between successive transits of the sun across a north–south meridian is measured by an accurate clock, it is found that this interval is not constant through the year. In some seasons, it is speeding up and in others it is slowing down. The effect is due to the variation of the Earth's speed as it revolves about the Sun in an elliptical orbit.

During those periods when the interval between successive transits is becoming shorter, it is obvious that the transits themselves must occur earlier each day and similarly when the transit interval is growing longer, the transits occur later each day. Relative to a noon defined by a regular clock, transits occur before that time in some seasons and after that time in others. The difference between the time indicated by the dial (corrected for longitude) and the civil time is known as the **equation of time**, i.e.

$$\text{Equation of time } (\mathcal{E}) = \begin{array}{c} \text{Time on the dial} \\ \text{(corrected for longitude)} \end{array} - \begin{array}{c} \text{Time by the civil clock} \\ \text{(corrected to UT)} \end{array}$$

Thus,

$$\text{Civil time} = \text{Time by sundial} - \mathcal{E}.$$

Knowing the equation of time for the particular day allows a time given by a sundial to be converted to civil time.

The value of the equation of time for any particular day may be taken from figure 9.5.

For some of the dials which may be found in your locality, the equation of time may also be tabulated on it. The equation of time is also tabulated in various books including, for example, *Norton's Star Atlas* in which values are given rounded to the nearest minute. For more accurate values of the equation of time, daily values for any year may be determined from *The Astronomical Almanac* by noting the Terrestrial Dynamical Time (TDT) of the solar ephemeris transit and using the relation

$$\mathcal{E} \text{ at } 12^\text{h} = 12^\text{h} - \text{tabulated value of TDT of ephemeris transit}.$$

24.2.1 A horizontal sundial

As the name implies, a horizontal sundial is designed so that the shadow cast by the style falls on to a horizontal surface (see figure 24.1). In order that the shadow line should be distinct, the style should be thin, of the order of 1 mm. It should be cut so that the angle which it makes with a horizontal plane is equal to the latitude of the location. When positioned correctly on the meridian the line of the style should point to the north celestial pole.

The markings to be inscribed on the plane to correspond to the hour lines may be drawn by applying the equation

$$\tan \gamma = \sin \phi \tan h$$

where γ is the angle between the noon-line and the edge of the shadow corresponding to time, h (converted to degrees), the angle by which the Sun is away from the meridian, and ϕ is the latitude. Obviously, at noon, h is equal to zero.

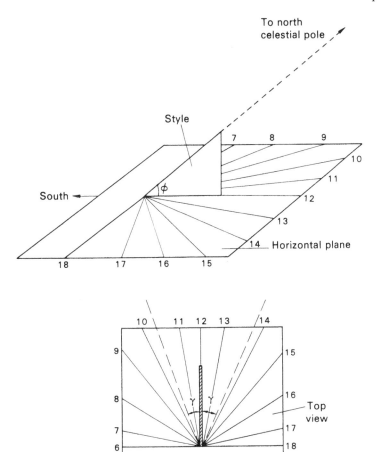

Figure 24.1. A horizontal sundial.

As an example, the value of γ is calculated for the times when the Sun is 2 hr ($h = 30°$) either side of the meridian, i.e. at 10 hr and 14 hr for a latitude $\phi = 55°\,53'$.

$$\begin{array}{ll} \sin\phi & 0\cdot8279 \\ \tan h & \times 0\cdot5773 \\ \tan\gamma & \overline{0\cdot4779} \\ \gamma & \underline{25°\,33'} \end{array}$$

For this situation, γ is equal to $25°\,33'$. To complete the sundial, values of γ should be calculated and drawn, say for each 10 minute interval.

In setting the dial, it is necessary to fix the marked plane horizontal and this can be done easily using a spirit level. The setting of the noon-line can be achieved by rotating the dial by small amounts each day until on any one day, the observed angle γ is equal on both sides of noon for equal time intervals either side of noon, say 2 hr.

24.2.2 A vertical sundial

The simplest form of vertical sundial makes use of a plane which passes through the east–west line as well as, of course, being vertical. The length of the style must point to the north celestial pole. Time

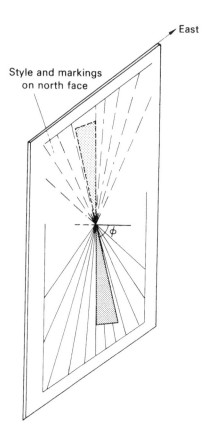

Figure 24.2. A vertical sundial.

lines should be drawn either side of the style, the angles γ of a line from the base of the style relative to the noon-line being given by the formula

$$\tan \gamma = \cos \phi \tan h$$

where ϕ and h are the same as for the horizontal dial.

In order to be able to record the early morning and late evening shadows of summer days when the Sun rises and sets north of the east–west line, the sundial may be continued on its reverse face if the style perforates the plane and is continued for projection of shadows on the north face (see figure 24.2).

24.2.3   A noon-marker

The simplest noon-marker can be achieved by using a vertical plane which is oriented also in the east–west direction. By fixing a disc containing a small central hole on an arm which projects at right angles from the plane, during the hours close to mid-day a patch of light surrounded by a shadow will appear on the plane. It is an easy matter to mark on the plane the centre of the bright patch, corresponding to an image of the Sun.

If during the course of a year, at each noon the Sun's position is noted, as expected the mark will rise and fall according to the noon altitude during the winter and summer. In addition, however, the marks move first eastwards and then westwards, repeating the excursions twice in the year, so producing a figure-of-eight pattern—the **analemma**. These departures from a vertical line correspond to the equation of time. On those days when the equation of time is zero, the Sun's image at local noon

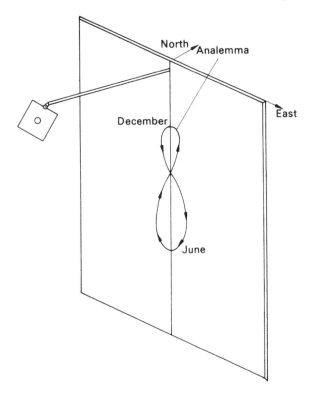

Figure 24.3. A simple noon-marker.

falls on the vertical line through the centre of the figure, whereas on those days when the departure from a line is greatest then the equation of time has its maximum value. This curve is an alternative way of displaying the equation of time. Its form is sometimes to be found on world maps or on globes.

Using a simple device similar to that depicted in figure 24.3, and after orienting it correctly in a similar manner as for the horizontal sundial, the analemma may be drawn during the course of a year and the variations of the equation of time deduced. Conversely, if the analemma has previously been drawn on to the vertical plane, the device may be used for announcing the passage of the local noon. Comparison with the civil time noon may be made after applying the correction for longitude.

24.3 The Sun as a position finder

At any particular instant, the apparent position of the Sun in the sky depends on the longitude and latitude of the observer. If positional measures of the Sun are used to determine the location with some degree of accuracy, an observer needs to use an optical instrument such as a theodolite or sextant and corrections such as those for refraction effects of the Earth's atmosphere must be applied. Nowadays, an observer's location on the Earth can be obtained to extreme accuracy by using GPS devices. Previous to the advent of such satellite technology, location fixes were derived from measurements by optical instruments.

It is of interest to note just how much GPS technology has improved the accuracy in respect of the location determination. Consider a simple situation of measuring the altitude of the Sun, $\alpha_\odot$, when it is exactly on the meridian. The determination of the latitude of the observer based on this altitude measurement above would involve the relation

$$\phi = 90° + \delta_\odot - \alpha_\odot.$$

Any error made in the measurement is directly translated into a location error along the line of longitude, i.e. the determined value of ϕ will be displaced either N or S relative to the true position. Suppose that optical measurements carry measurement uncertainties $\delta\theta = \pm10$ arc sec—this being typical of what can be achieved using simple optical equipment. In terms of the error, δd, in distance along the meridian, this corresponds to

$$\delta d = \pm10 \times \frac{2\pi R_\oplus}{360 \times 60 \times 60}$$

where $R_\oplus$ is the radius of the Earth. By substituting the appropriate value, $\delta d \approx 0\cdot3$ km.

By similar reasoning, an error in the knowledge of $\delta_\odot$ or in the calculation of the refraction correction to $\alpha_\odot$ of $\pm1''$ introduces uncertainties in position of approximately $\pm0\cdot03$ km or ±30 m.

Currently available GPS devices provide positional determinations to accuracies more than 10 times better than this last figure, i.e. positional fixes to ±3 m are readily achieved, and it is very obvious why location by optical instruments has been abandoned. Such weather-dependent systems with their associated labour of subsequent numerical calculations are things of the past.

It is, however, very instructive to use old optical devices to obtain data on the Sun's position in the sky and on its apparent movement. In addition to providing data and gaining familiarity with various reduction procedures, their application gives some feel as to the accuracy to which simple hand-held optical instruments provide positional fixes of the Sun and how these are translated to a determination of the observer's location on the Earth.

If such exercises are now attempted, it will be noted that *The Astronomical Almanac* or *AA* has evolved in ways more related to modern positional astronomy. Certain kinds of information are now presented differently relative to times past when positional determinations were regularly obtained from optical measurements. For example, the positions of the Sun (RA and δ) were formerly given for each day at 00^h UT rather than 00^h TDT as they are done currently. The examples provided here are from measurements obtained in recent times. Consequently, the presented numerical correction procedures are related to the present forms of the data tables and are slightly different than in the previous era.

24.3.1 Simple determination of latitude

Set up a simple sharp vertical gnomon to cast a shadow on to a horizontal plane having a surface on which the shadow tip can be marked. At intervals of about ten minutes, over a period of about one hour either side of the local noon, mark the position of the shadow tip. After the recordings are completed, measure the minimum shadow length from the base of the gnomon to the shadow tip. This corresponds to the time when the Sun is on the meridian. Knowing the height of the gnomon above the horizontal plane allows the maximum altitude to be determined using the formula

$$\tan\alpha_\odot = \frac{H}{L} \qquad \text{(see figure 24.4(a))}.$$

Knowing also the Sun's declination for that day, the latitude of the observer may be determined from

$$\phi = 90° + (\delta_\odot - \alpha_\odot) \qquad \text{(see figure 24.4(b))}.$$

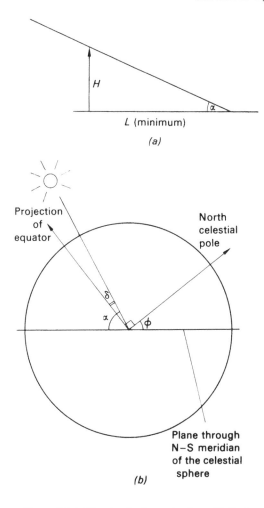

Figure 24.4. The simple determination of latitude.

24.3.2 Theodolite observations

This exercise involves the use of a theodolite to obtain the azimuth of the Sun at a particular time. Knowing the hour angle of the Sun at the time of the observation, the angle between the Sun and the meridian can be calculated. The position of the meridian may, in turn, be related to the observer by means of the azimuth of a fixed reference object and it is then determined once and for all.

Check that the theodolite is level and adjust if necessary. *__Having taken the precaution of checking that the filters are fitted to the telescope__*, direct the theodolite towards the Sun and record the following times:

- the moment when the limb of the Sun approaching the cross-wire makes contact with it; and
- the moment when the opposite limb breaks contact with the cross-wire, after the solar image has trailed across the field of view.

The mean of these two observations gives the time at which the centre of the Sun's disc was at the cross-wire. Note the angle on the horizontal circular scale of the theodolite at which the Sun was observed. Note also the angle of some distant fixed reference object.

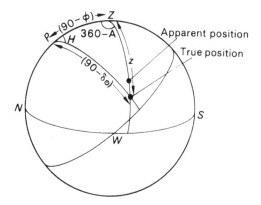

Figure 24.5. Determination of the Sun's azimuth

Now the situation is depicted in figure 24.5 showing that the apparent azimuth of a celestial object is not altered by the effect of refraction.

By using the sine formula, we have

$$\frac{\sin(90 - \delta_\odot)}{\sin(360 - A)} = \frac{\sin z}{\sin H}$$

giving

$$\sin z = \frac{-\cos \delta_\odot \sin H}{\sin A}.$$

Also, we have

$$\sin z \cos(360 - A) = \cos(90 - \delta_\odot) \sin(90 - \phi) - \sin(90 - \delta_\odot) \cos(90 - \phi) \cos H$$

giving

$$\sin z = \frac{\sin \delta_\odot \cos \phi - \cos \delta_\odot \sin \phi \cos H}{\cos A}.$$

Hence, elimination of $\sin z$ gives

$$\tan A = \frac{\sin H}{\cos H \sin \phi - \tan \delta_\odot \cos \phi}.$$

Example. On May 2nd 2000 the Sun's centre was at the cross-wire of a theodolite at $10^h 42^m 10^s$ UT. The reading on the horizontal circle was $222° 12'$; the reading corresponding to the reference object was $108° 27'$. The coordinates of the observing site are known to be $\lambda = 4° 18'.9$ W and $\phi = 55° 54'.3$ N.

Sun's declination

According to *The Astronomical Almanac (AA)* for May 2nd 2000 at 00^h (TDT),

$$\delta_\odot = + 15° 24' 10''.1$$
$$= +15° 24'.17.$$

By noting the declination for May 3rd 2000,

The daily increase in $\delta_\odot = +17'.75.$

Assuming a linear interpolation over the 24^h period

$$\text{Increment for TDT to UT (66 s)} = +0\!.'01$$
$$\text{Increment for } 10^h = +7\!.'40$$
$$\text{Increment for } 42^m\,12^s = \underline{+0\!.'52}$$
$$\text{Total increment} = +7\!.'93$$
$$\delta_\odot \text{ at time of observation} = +15°\,32\!.'10.$$

Sun's hour angle

For this calculation, the true hour angle will be determined based on the apparent LST, this in turn being related to GAST, i.e.

$$\text{HA}\odot = \text{LST} - \text{RA}\odot$$

and

$$\text{LST} = \text{GAST} - \lambda.$$

From the *AA* for May 2nd 2000 at 00^h (TDT),

$$\text{RA}\odot = \underline{2^h\,37^m\,49^s\!.54}$$

From the values for May 3rd 2000, the daily increase in RA$\odot = 3^m\,50^s\!.2 = \underline{230^s\!.20}$

$$\text{Increment from TDT to UT } (66^s) = +\qquad 0^s\!.2$$
$$\text{Increment for } 10^h = +1^m\quad 35^s\!.9$$
$$\text{Increment for } 42^m\,12^s = +\qquad 6^s\!.7$$
$$\text{Total increment} = \underline{+1^m\quad 42^s\!.8}$$

$$\text{RA}\odot \text{ at time of observation} = \underline{2^h\,39^m\,32^s\!.3}$$

$$\begin{aligned}
\text{Again from the } AA, \text{ at } 00^h \text{ (UT) GAST} \quad & 14^h\,40^m\,51^s\!.0 \\
\text{Longitude } 4°\,18\!.'9 \text{ W} = - & 17^m\,15^s\!.6 \\
\text{LST at } 00^h = \ & 14^h\,23^m\,35^s\!.4 \\
\text{Time of observation} = \ & 10^h\,42^m\,10^s\!.0 \\
\text{Conversion of } 10^h\,42^m\,10^s = \ & \underline{1^m\,45^s\!.5} \\
\text{LST at } 10^h\,42^m\,10^s = \ & 1^h\,07^m\,30^s\!.9 \\
\text{RA}\odot = \ & \underline{2^h\,39^m\,32^s\!.3} \\
\text{HA}\odot = \ & \underline{-1^h\,32^m\,01^s\!.4}
\end{aligned}$$

$$\text{HA}\odot = H = \underline{-23°\,0\!.'3}.$$

Now,

$$\tan A_\odot = \sin H / (\cos H \sin\phi - \tan\delta_\odot \cos\phi)$$

i.e.

$$\tan A_\odot = -\sin 23°\,0\!.'3 / (\cos 23°\,0\!.'3 \sin 55°\,54\!.'3 - \tan 15°\,32\!.'1 \cos 55°\,54\!.'3)$$
$$= -\sin 23°\!.005 / (\cos 23°\!.005 \sin 55°\!.905 - \tan 15°\!.535 \cos 55°\!.905)$$

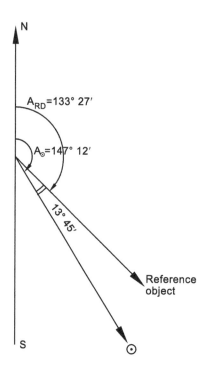

Figure 24.6. The Sun's azimuth relative to a fixed reference object.

giving, on reduction

$$\tan A_\odot = -0.644\,501.$$

Hence, $A_\odot = 147°198 = 147°\,12'$ E of N, to the nearest minute of arc.

Now the corresponding reading on the theodolite is $222°\,12'$ and the reading for the reference object is $108°\,27'$, the difference being $13°\,45'$. Thus, the reference object has an azimuth which is $13°\,45'$ less than that of the Sun at the time of the observation, giving it a fixed azimuth of $133°\,27'$ E of N. The situation is summarized in figure 24.6.

24.3.3 Sextant observations

This exercise involves the use of a sextant to make measurements of the Sun's altitude over a short period of time in order to determine the position of the observer. The times of these observations are noted using a Greenwich mean time chronometer of known error and rate.

The basic principles behind the finding of an observer's position on the Earth's surface are first of all described briefly.

(i) *The position circle.* In figure 24.7, it is seen that the latitude ϕ and longitude λ of a point, L (the **sub-solar point**), on the Earth's surface directly below the Sun, S, are related to the Sun's Greenwich hour angle (GHA$\odot$) by the relations

$$\phi = \delta_\odot$$
$$\lambda\,\mathrm{W} = \mathrm{GHA}\odot \qquad (\mathrm{GHA}\odot \leq 12^\mathrm{h})$$

or

$$\lambda\,\mathrm{E} = 24^\mathrm{h} - \mathrm{GHA}\odot \qquad (\mathrm{GHA}\odot \geq 12^\mathrm{h}).$$

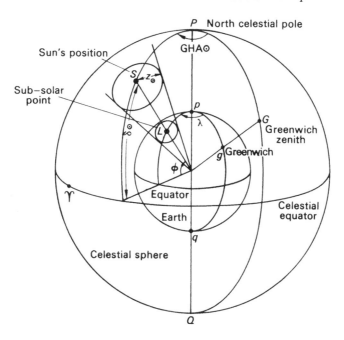

Figure 24.7. The position of the sub-solar point.

The Sun's geocentric zenith distance, $z_\odot$, is found at this time by applying corrections for sextant errors, refraction, semi-diameter and geocentric parallax to the observed altitude of the Sun as measured by the sextant. Then it is readily seen that the observer must have been on the small circle centred at L, found by drawing a circle of radius $z_\odot$, centre S and allowing the cone it makes with the Earth's centre to intersect the Earth's surface in a circle. This small circle is called a **position circle**.

Let us suppose that the observer's position (i.e. latitude and longitude) does not change with time. Some hours later, a second observation of the Sun's altitude will provide a second position circle. The observer must be at one of the two points of intersection of these circles when they are drawn on a map. Since the points are usually hundreds of nautical miles apart, the observer usually has no doubt as to which is the correct position.

The required information of the Sun's Greenwich hour angle and declination can be derived from *The Astronomical Almanac*, using the corrected chronometer times of the observations.

(ii) *The position line.* In practice, the observer makes use of the knowledge of the latitude and longitude, ϕ_D and λ_D respectively, of the approximate position D (the so-called **dead reckoning position**). This position will lie in the vicinity of the position circle. The line from D to the centre, L, of the position circle will, therefore, cut it at some point V (see figure 24.8).

In the *vicinity* of V, we can use a straight line as an approximation to the position circle. Then the observer will be on the straight line FJ tangential to the position circle at V and at a distance DV from the dead reckoning position. If the length DV and the value of the azimuth, $\angle pDV$, measured east of north, can be found, then on a chart the **position line**, as FJ is called, can be plotted. The distance DV is called the **intercept** (see figure 24.9). The angle pDV is obviously equal to $\angle PZS$. Also

$$\angle PZS = 360° - A$$

where $\angle PZS$ is within $\triangle PZS$. Angle SPZ is the Sun's hour angle H measured from the observer's meridian for the dead reckoning position D.

$$DV = ZW = SZ - SW.$$

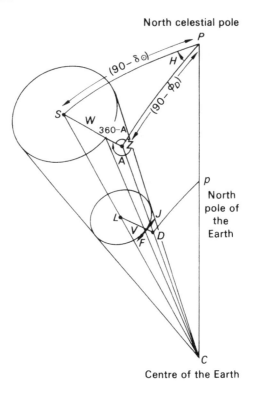

Figure 24.8. The position circle.

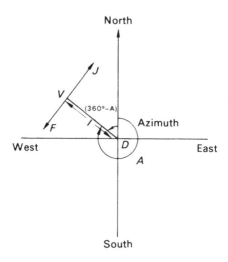

Figure 24.9. Construction of the intercept and the position line.

In $\triangle PZS$, SZ is the **calculated zenith distance**, z, i.e. the zenith distance of the Sun as it would have been measured by an observer at D and subsequently corrected in the usual way for refraction, etc. The arc SW, of course, is the **observed zenith distance**, $z_\odot$, of the Sun.

Then the intercept DV is given by

$$DV = \text{calculated zenith distance} - \text{observed zenith distance}$$

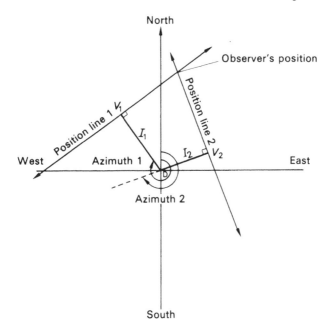

Figure 24.10. The intersection of two position lines allows the observer's position to be determined.

or

$$I = z - z_\odot.$$

In $\triangle PZS$, we then have,

$$\cos z = \sin \phi_D \sin \delta_\odot + \cos \phi_D \cos \delta_\odot \cos H \qquad (24.1)$$

giving z and

$$\sin z \cos A = \sin \delta_\odot \cos \phi_D - \cos \delta_\odot \sin \phi_D \cos H \qquad (24.2)$$

giving A.

(iii) *The plotting chart.* On the plotting chart a suitable scale is chosen, for example, one nautical mile being represented by one centimetre, remembering that one minute of arc is equivalent to one nautical mile. The north–south axis is a latitude axis but the east–west axis must be measured as departure. To convert to difference of longitude, the usual relation is used, namely

$$\text{departure} = \text{difference of longitude} \times \cos(\text{latitude}).$$

In order to find the observer's position, two position lines must be obtained from observations made a few hours apart. The intersection of the two position lines (on both of which the observer lies) will then give the increments in latitude and longitude (corrected from departure) to be applied to the dead reckoning coordinates ϕ_D and λ_D.

It should be noted that if the dead reckoning position D happens to lie within the position circle (in figure 24.8), the intercept DV will be negative since the observed zenith distance, z, is greater than the calculated in the distance, $z_\odot$. In this case, the intercept must be drawn in the opposite direction to the azimuth, i.e. in a direction of azimuth 180° less than the calculated one. Thus, in figure 24.10, we have a positive intercept, I_1, and a negative one, I_2.

The following scheme of instructions and data are taken from an exercise formerly carried out by first-year students in Astronomy at Glasgow University. In the United Kingdom at that time, British

Summer Time was in operation, where

$$\text{British Summer Time (BST)} = \text{Universal Time (UT)} + 1 \text{ hour.}$$

(iv) *Procedure*. Data concerning the Sun's position are tabulated in *The Astronomical Almanac* (*AA*)—(see pages C4–C19 therein)—and are referred to TDT with

$$\text{UT} = \text{TDT} - \Delta T.$$

To a good approximation for the summer of 2000, $\Delta T = 66$ s. Values of the numerical correction for other years can also be found in the *AA*—pages K8–K9.

The daily solar data (C4–C19) refer to the solar crossing of the meridian as the 'ephemeris transit', this corresponding to the TDT at which the Sun crosses an ephemeris meridian at longitude $1.002\,738 \times \Delta T$ east of the Greenwich meridian. Its occurrence is, thus, ΔT earlier on the TDT time scale than the transit of the Greenwich meridian. To all intents and purposes, to the accuracy that can be achieved with this experiment, the time of the Greenwich transit in UT can be taken as the ephemeris transit in TDT.

An approximate position of Glasgow University Observatory to be used as the dead reckoning position (DRP) can be taken as

$$\text{Longitude } (\lambda_D) = 4° \, 22' \text{ W}$$
$$\text{Latitude } (\phi_D) = 55° \, 53' \text{ N.}$$

Two sets of observations are required—**A** and **B**(i) or **B**(ii).

One set, **A**, is to be taken at the local meridian transit, when the Sun's altitude is changing slowly with time. Determination of the Sun's maximum altitude at the transit allows a position line to be determined. (The time at which the transit occurs cannot be measured accurately and only an approximate position of the observatory could be obtained if this information were to be used.) The advantage of a meridian set of observations is that the reduction of the observations is particularly simple.

Observations should be made at approximately three minute intervals, and cover a period from *at least* twenty minutes before to *at least* twenty minutes after meridian transit.

The second set, **B**, is to be taken over a period when the Sun's altitude is changing quite quickly with time. Using the curve which depicts this variation, a second position line can be obtained. Make a set of observations *EITHER* according to the scheme **B**(i) *OR* to scheme **B**(ii).

For **B**(i), obtain a series of measurements of the altitude of the Sun's centre during a period in the *morning*. Observations should be made *at least* 2 hr before meridian transit, at approximately two minute intervals, for a run of *at least* 10 minutes.

For **B**(ii), obtain a series of measurements of the altitude of the Sun's centre during a period in the *afternoon*. Observations should be made *at least* 2 hr after meridian transit, at approximately two minute intervals, for a run of *at least* 10 minutes.

IMPORTANT REMINDER
NEVER OBSERVE THE SUN **WITHOUT** THE SHADES
COVERING **BOTH MIRRORS**.

(a) A sextant, an artificial horizon and a chronometer are required. The artificial horizon, a trough of water, is required for sextant observations made on land. (In some old books, a trough of mercury is recommended for the artificial horizon. This is no longer accepted practice because of mercury's toxicity.) At sea a horizon is available. Any chronometer error must be checked *before* and *after* the series of observations.

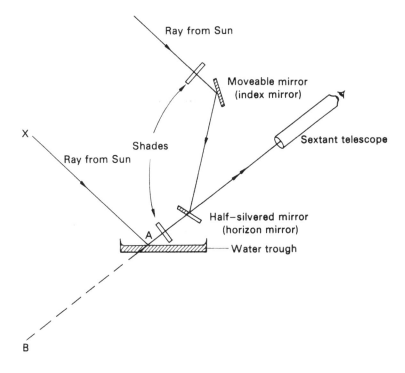

Figure 24.11. The sextant used with an artificial horizon.

(b) Insert the telescope in the sextant and focus it on a distant object.

(c) The index error (ΔR) of the sextant must be determined. Set the pointer on the zero of the scale and view the Sun through the telescope. Two images should be seen in the field of view. The image obtained by the direct view is known as the 'horizon image' (H-image), while the other is obtained by two reflections and is known as the 'index image' (I-image). Try to arrange the combination of dark shades so that the images are of equal intensity.

By using the micrometer adjustment, set the two images so that they are just in contact, with the H-image above the I-image. Note the reading (R_1) on the scale. Reverse the positions of images so that the I-image is above the H-image and just in contact with it. Note the reading (R_2) on the scale.

$$\Delta R = 360° - (R_1 + R_2)/2.$$

(R_1 and R_2 have values which should both be close to 360°.)

Repeat the index error observations several times to obtain a mean value.

Check the accuracy of the values of R_1 and R_2 by comparing the Sun's semi-diameter $SD_\odot$ as given in the *AA* with the value calculated from the observations:

$$SD_\odot = (R_1 - R_2)/4.$$

(d) Set up a water horizon on a rigid observing pillar or platform. Protect the surface from the wind by means of a glass roof cover. (The cover is not shown in figure 24.11.) Close one eye and stand in such a position that, with the other, the Sun can be seen reflected in the centre of the water. Without moving your head, interpose the sextant telescope between your eye and the water. The image of the Sun reflected in the water (i.e. the H-image) will be seen. Distortion of this image, e.g. into roughly rectangular shape, usually indicates that the ray is being reflected from the curved part of the water surface near its edge. The remedy is to move your head slightly, keeping the image in the field, until

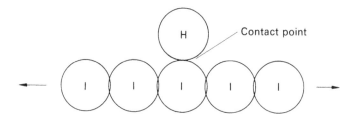

Figure 24.12. Sweeping the index image to check contact with the horizon image.

the telescope is pointing directly at the centre of the water, when the image will become round and sharp.

(e) The I-image will appear in the field of view if two conditions are satisfied:

1. The scale must be twice the Sun's altitude, i.e. it must be equal to $\angle XAB$ in figure 24.11.
2. The plane of the sextant must be vertical.

To find the I-image, set the scale to read twice the estimated altitude of the Sun. With the H-image in the field of view, move the index arm slowly to and fro over this part of the scale and, at the same time, rotate the entire instrument slowly through a few degrees about the telescope axis, so that the sextant swings backwards and forwards through the vertical plane. During these operations, the I-image will be seen to move sideways through the field. When this occurs, stop the movement of the index arm and bring the sextant into the vertical plane.

(f) Set the two images in contact, the H-image above the I-image, by means of the adjustment screw and observe the relative movement of the two—whether they tend to separate or overlap. For the forenoon observations, **B**(i), the altitude is increasing and the direction of movement is from the lower towards the upper limb. If, therefore, the images are separating, it is the lower limb which is being used to make the contact of images. For the afternoon observations, **B**(ii), this rule is reversed.

The *lower* limb should be used for the contact observations.

For the forenoon observations, make the images overlap by means of the adjustment screw and observe the chronometer time, to the nearest half-second, at which the last trace of overlapping disappears. For afternoon observations, set the images slightly apart and observe the time at which the images just come into contact. For observations near to the meridian, the Sun's altitude is changing slowly. In this case, put the images in contact by means of the adjustment screw and note the time when they are in contact.

The sextant vernier or micrometer should, in all cases, be read very carefully to the nearest 10 seconds of arc.

(g) For all observations, the plane of the sextant must be vertical. To ensure that this is so, rock the instrument to rotate round the telescope as axis. The effect of this is to make the I-image move to and fro across the field of view. 'Contact' occurs when this image, in passing across the field, just grazes the H-image (see figure 24.12).

(h) *Reduction of observations.* Before any calculations can be made using the measured altitudes, all the observations must be corrected for

(1) the index error,
(2) any scale error (see notes in lid of sextant box) and
(3) the effect of refraction by the Earth's atmosphere.

The method for making these corrections is illustrated in the sample set of observations.

(i) *Calculations.*

(1) Using the reduced observations from **A**, obtain a curve of the altitude of the Sun's centre against time, during the meridian transit. Determine the maximum altitude and the time when this occurs. Use this information to determine an approximate position of the Observatory. From the value of the maximum altitude, determine a position line.
(2) Using the reduced observations from either **B**(i) or **B**(ii), obtain a curve of the altitude of the Sun's centre against time. From a chosen point on this graph, determine a position line.
(3) By using the two position lines obtained from (1) and (2), determine an accurate position for the Observatory.

The methods for making these calculations are illustrated in the sample set of calculations.

For plotting the curves, it will be found more convenient to convert seconds of time to decimals of a minute of time and seconds of arc to decimals of a minute of arc.

(j) *Sample set of observations and calculations.*

Determination of approximate time of meridian transit: March 29th 2000: The *AA* provides the TDT time of the Sun's Ephemeris Transit which, to a first order, is the time of the Greenwich Transit.

	h	m	s	
Transit due at Greenwich:	12	04	39·3	UT
+	1	00	00	
	13	04	39·3	BST

Approximate longitude of the Observatory
$= 4°22'$ W $= 17^{m}28^{s}$ 17   28

Transit due at the Observatory: 13   22   07·3   BST

Determination of index error:

$$
\begin{array}{rclrrr}
R_1 & = & 360° & 31' & 10'' \\
R_2 & = & 359 & 27 & 00 \\
R_1 + R_2 & = & 719 & 58 & 10 \\
\div 2 & = & 359 & 59 & 5 \\
\therefore \Delta R & = & + & 0' & 55''
\end{array}
$$

Check on the accuracy of the readings:

$$
\begin{array}{rclrrr}
R_1 - R_2 & = & 1° & 4' & 10'' \\
\div 4 & = & & 16' & 02'' \\
\text{Measured diameter of the Sun} & = & & 16' & 02'' \\
\text{From the } AA, \text{ Sun's diameter} & = & & 16' & 01''
\end{array}
$$

Determination of an approximate position—observations **A**: After applying the corrections to the observations (see the worked example for **B**(i)), a curve can be drawn of the corrected solar altitude against corrected time (UT). Measurement of the Sun's maximum corrected altitude allows an accurate latitude position to be determined. The estimation of the time when the Sun's centre has this maximum altitude is not accurate but it is sufficient to allow an approximate longitude to be determined.

The longitude (λ) and latitude (ϕ) of the Observatory are given by

$$\lambda = (\text{UT of transit of } \odot) - (\text{UT of transit of } \odot \text{ at Greenwich})$$
$$\phi = 90° - (\text{true altitude of } \odot \text{ at transit}) + \delta_{\odot}$$

where $\delta_{\odot}$ is the declination of the Sun, at the time of its transit.

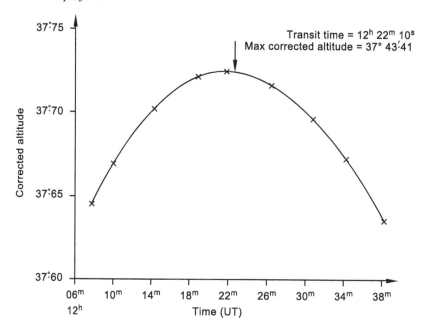

Figure 24.13. The variation of the solar corrected altitude with time over the period of meridian transit.

Example. From the curve obtained for March 29th 2000 (see figure 24.13),

Transit time:	12^h	22^m	10^s
Greenwich transit:	12	04	39·3
Longitude of Observatory (λ) :		17	30·7 West
Longitude of Observatory $=$		$4°$	$23'$ W

The maximum corrected altitude was determined to be $37°\ 43'\!.41$.

Sun's declination

According to the *AA* for March 29th 2000, $\delta_\odot$ at 00^h (TDT) is

$$+3°\ 24'\ 40''\!.1 = +3°\ 24'\!.67.$$

The daily increase in $\delta_\odot = +23'\!.33$. By making a linear interpolation,

$$\delta_\odot \text{ at time of transit} = +3°\ 37'\!.11$$
$$\phi = 90° - 37°\ 43'\!.41 + 3°\ 37'\!.11 = 55°\ 53'\!.7.$$
$$\therefore \text{ Latitude of Observatory} = 55°\ 53'\!.7$$

This information can now be used to draw the first position line.
Determination of a position line—observations **B**:

$$
\begin{aligned}
R_1 &= 360° & 31' & 36'' \\
R_2 &= 359° & 26' & 50'' \\
\therefore \Delta R &= & +0' & 47''
\end{aligned}
$$

For readings close to $60°$, the scale error (Δr) $= -20''$

$$\therefore \text{ Total correction to the readings} = +0'\ 27''$$

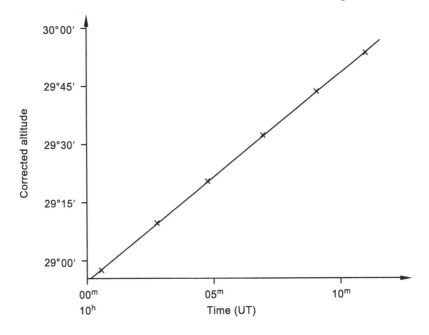

Figure 24.14. The variation of the solar corrected altitude with time well away from the meridian.

Observations taken on March 25th 2000: During the measurements, the altitude of the Sun ~29°

$$\begin{aligned} \text{Refraction} \quad &= \quad 58'' \cot(\text{alt}) \\ &= \quad 58'' (1 \cdot 80) \\ &= \quad \underline{1' \, 40''} \end{aligned}$$

Semi-diameter of Sun (from *AA*) = $16' \, 02''$.

$$\begin{aligned} \text{Correction for obtaining true altitude of Sun's centre} \quad &= \quad 16' \, 02'' - 1' \, 40'' \\ &= \quad \underline{14' \, 42''} \end{aligned}$$

Corrected time (UT)	Sextant reading (R)	Corrected Sextant reading $(R + \Delta R + \Delta r)$	Uncorrected altitude	Corrected altitude
$10^h \, 00^m \, 31^s$	$57° \, 25' \, 53''$	$57° \, 26' \, 20''$	$28° \, 43' \, 10''$	$28° \, 57' \, 32''$
$10^h \, 02^m \, 45^s$	$57° \, 50' \, 39''$	$57° \, 51' \, 06''$	$28° \, 55' \, 33''$	$29° \, 09' \, 55''$
$10^h \, 04^m \, 41^s$	$58° \, 11' \, 59''$	$58° \, 12' \, 26''$	$29° \, 06' \, 13''$	$29° \, 20' \, 35''$
$10^h \, 06^m \, 53^s$	$58° \, 35' \, 57''$	$58° \, 36' \, 24''$	$29° \, 18' \, 12''$	$29° \, 32' \, 34''$
$10^h \, 09^m \, 02^s$	$58° \, 59' \, 05''$	$58° \, 59' \, 32''$	$29° \, 29' \, 46''$	$29° \, 44' \, 08''$
$10^h \, 11^m \, 01^s$	$58° \, 20' \, 21''$	$58° \, 20' \, 48''$	$29° \, 40' \, 24''$	$29° \, 54' \, 46''$

Corrected Time (UT)	Corrected Altitude
$10^h \, 00^m 52$	$28° \, 57'\!53$
$10^h \, 02^m 75$	$29° \, 09'\!92$
$10^h \, 04^m 68$	$29° \, 20'\!58$
$10^h \, 06^m 88$	$29° \, 32'\!57$
$10^h \, 09^m 03$	$29° \, 44'\!13$
$10^h \, 11^m 02$	$29° \, 54'\!77$

From the curve of the corrected altitude of the Sun against corrected time (see figure 24.14), a point was chosen where the altitude was $29° 22'.34$ at $10^h 05^m.0$ UT.

Thus, the observed zenith distance: $\underline{(OZD) = 60° 37'.66}$.

By taking the values of RA$\odot$, $\delta_\odot$ from the *AA* and extrapolating, at the time of observation ($10^h 05^m$), the solar position is given by

$$RA\odot = 00^h 18^m 34^s.38$$
$$\delta_\odot = +2° 00' 35''.8$$

Sun's hour angle:

$$HA\odot = LST - RA\odot$$
and
$$LST = GAST - \lambda_D.$$

From the *AA*, at 00^h (UT), GAST = 12^h 11^m $01^s.99$
Approx longitude ($4° 22'$) = $\underline{ 17^m 28^s}$

$\therefore$ LST at 00^h (UT) 11^h 53^m $34^s.0$
Time of observation 10^h 05^m $00^s.0$
Conversion of $10^h 05^m$ to sidereal time $\underline{ 1^m 39^s.4}$

$\therefore$ LST at $10^h 05^m$ 22^h 00^m $13^s.4$
RA$\odot$ $\underline{00^h 18^m 34^s.4}$

HA$\odot$ -2 18^m $21^s.0$
$$\underline{HA\odot = H = -34° 35' 15'' = -34° 35'.25}$$

Calculation of intercept: The calculated zenith distance, z, is obtained by application of equation (24.1), namely

$$\cos z = \sin \phi \sin \delta_\odot + \cos \phi \cos \delta_\odot \cos H$$

or

$$\cos z = \sin(55° 53'.7) \sin(2° 00'.60)$$
$$+ \cos(55° 53'.7) \cos(2° 00'.60) \cos(-34° 35'.25)$$
$$= \sin(55°.895) \sin(2°.0099) + \cos(55°.895) \cos(2°.0099) \cos(-34°.5875)$$

giving, on reduction, $z = 60° 38'.11$.

$\therefore$ Calculated Zenith Distance (CZD) $60°$ $38'.11$
(OZD) $60°$ $37'.66$
Intercept $\underline{ 0'.45}$

As CZD > OZD, the intercept is marked off in the direction 'Towards' the Sun.
The azimuth of the Sun can be determined by application of equation (24.2), namely

$$\sin z \cos A = \sin \delta_\odot \cos \phi_D - \cos \delta_\odot \sin \phi_D \cos H.$$

For the previous example, $A = 139° 23'$ E of N. The value of A can then be checked by using

$$\cos \delta_\odot \sin H = -\sin z \sin A.$$

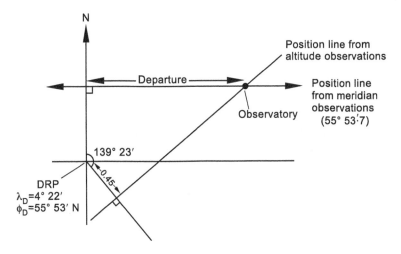

Figure 24.15. Position lines obtained from a morning observation and a transit of the Sun, allowing the observer's position to be determined.

The intercept is to be drawn in the direction 139° 23′ east of north, from the DRP, and its length is 0·45 nautical miles (see figure 24.15).

The intersection of the position line obtained earlier with the position line obtained from the observations of the meridian passage gives the position of the observatory in terms of latitude and departure from the DRP.

From figure 24.15, the latitude is given by 55° 53′.7 and the departure is equal to 1′.51. Now the difference in longitude is given by

$$
\begin{aligned}
\text{difference in longitude} \quad &= \quad \text{departure} \times \sec \phi \\
&= \quad 1'.51 \times 1.78 \\
&= \quad \underline{2'.69.}
\end{aligned}
$$

$$
\begin{aligned}
\text{Hence, the longitude is} \qquad & 4° 22' - 2'.69 \\
&= \quad \underline{4° 19'.3 \text{ W.}}
\end{aligned}
$$

$$
\begin{aligned}
\text{Summarizing:} \quad &\underline{\lambda = 4° 19'.3 \text{ W}} \\
&\underline{\phi = 55° 53'.7 \text{ N.}}
\end{aligned}
$$

This might be checked by using GPS.

24.4   Observational radio astronomy

Although most radio celestial objects require sophisticated instrumentation for their detection, it is possible to undertake a couple of simple exercises with a satellite dish and a power meter. The detector system is the kind used by professional satellite dish installers for orienting the collector to the best direction in the sky. The equipment is also capable of detecting the solar output at 10 GHz.

The receiving dish should be set in a cradle in the form of an alt-azimuth mount with scales marked in degrees. The supports can be made of wood with a large horizontal base carrying the azimuth scale marked out in degrees. The size of the circular scale should be sufficient to allow the azimuth setting to be read to an accuracy ~1/2°. The scale should be fixed with N = 0°, E = 90°, S = 180° and W = 270°. When setting out the equipment, the N direction may be established using a compass, making allowance for the local magnetic deviation. Control of the altitude setting can be arranged by

Figure 24.16. A simple satellite receiver dish mounted on an alt-azimuth frame with the output displayed on a power meter.

fitting a metal bar to the rear of the dish and by providing two uprights attached to the horizontal plate with notches in their tops in which the rod is seated. A segment of a large circular scale needs to be attached to the rod axis so that the altitude of the pointing of the dish can be read after fixing its position with an adjustable stop. A photograph of a working arrangement is shown in figure 24.16.

The telescope consists of a 55 cm dish that reflects the radio signals into a horn at its focus. The horn is connected to a 'low noise block' (LNB) which shifts the signals to lower frequency ($\sim$1 GHz) before sending them down a cable to a signal detector. In the power meter box, the signal is amplified again and then rectified with a diode to give a voltage dependent on the strength of the received radiation. The output voltage is displayed on a dial and the sensitivity of the equipment can be adjusted by a control knob.

Note that the dish is offset in that it points in the direction of the arm supporting the horn rather than the direction perpendicular to the wire mesh. (Allowance needs to be made for this in setting the 'zero' angle on the altitude scale.) Point the dish to the sky. After switching on the power meter, adjust the sensitivity knob until the meter reads mid-scale. A check can be made to see if the system is working by placing a hand over the cap protecting the horn. The system should give a deflection as your hand is much warmer than the sky. At 10 GHz, an empty patch of sky has an effective temperature of about 10 K, this being made up of about 7 K from the water vapour in the atmosphere and 3 K from the Big Bang cosmic background radiation—in comparison your hand should have a temperature slightly above 300 K.

24.4.1 Observing the Sun

The microwave output from the Sun consists of broad-band noise (or 'hiss' when converted to an audio signal) generated by the thermal motions of electrons in the Sun's atmosphere. At microwave frequencies, most of the radiation results from the long wavelength tail of the Sun's black body spectrum.

The telescope has a beam (see section 21.2) relating to the angular extent of the patch on the sky that it sees at any one time. To estimate the beamwidth, consider it corresponding to the diffraction limit given by the aperture of the telescope. If we take $D = 55$ cm and the wavelength, λ, to be 3 cm, then the beamwidth is

$$\theta = \frac{\lambda}{D} = 0{\cdot}054 \text{ rad} = 3°.$$

Following from section 21.2, any black body object that fills the beam will produce a signal which depends *only* on its temperature (not on its composition, distance, size, etc). This fact can be used to estimate the temperature of the Sun.

Any object that occupies only a fraction of the beamwidth will appear to have a temperature that is reduced by that same fraction. The Sun, with an angular diameter of about $1/2°$, does not fill the beamwidth of the telescope and the temperature derived from the immediate output signal needs to be scaled. For the equipment used here, the scaling factor is given by

$$\left(\frac{3}{1/2}\right)^2 = 36.$$

Microwaves pass through cloud quite well and, consequently, the experiment can be undertaken in overcast conditions, even when the Sun is obscured. According to the local time, an estimate of the azimuth of the Sun may be calculated making it easier to locate its direction.

Adjust the pointing of the dish and the sensitivity of the power meter until a maximum output is achieved. From this position, any slight movement of the dish *in any direction* will cause the signal to drop. Record the azimuth and altitude of the Sun and note the time at which the measurement was made.

Drop the altitude of the telescope to $< 0°$ (i.e. aim at the ground). The meter deflection should be similar to that corresponding to the Sun. Although the ground is only about 290 K, it 'fills the beam'. Using the scaling factor as outlined earlier, estimate the temperature of the Sun.

24.4.2 Observing geostationary satellites

Although the electronics associated with the telescope are very sensitive, the dish has insufficient collecting area to detect other celestial radio sources. With its wide beamwidth, the signal from any point source is very much diluted by the cold blank sky filling most of the beam. Obviously it is designed with sufficient collecting area to detect man-made transmissions generated by communication satellites orbiting the Earth.

Nearly all communication satellites are in geostationary orbits. That is to say, they orbit the Earth at the same rate as it spins, so that they appear to hover above the Earth's equator. For this condition, the orbital radius, R_s, measured from the Earth's centre, relative to the radius of the Earth, $R_\oplus$, is readily calculated as follows. For a body of mass, m, on the surface of the Earth, it experiences a gravitational force by the mass of the Earth, $M_\oplus$, and is matched by the surface acceleration, g. Thus,

$$\frac{GM_\oplus m}{R_\oplus^2} = mg \tag{24.3}$$

where G is the universal constant of gravitation.

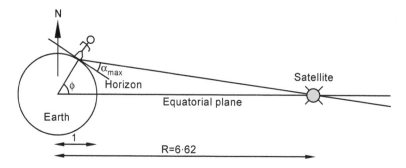

Figure 24.17. The apparent position of a geostationary satellite affected by parallax with its maximum altitude, α_{max}, being dependent on the observer's latitude, ϕ.

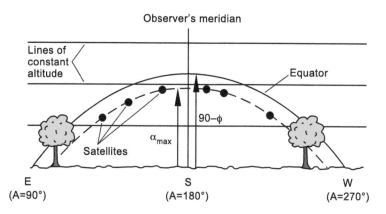

Figure 24.18. The altitudes of geostationary satellites at different latitudes showing the parallactic displacement of the Clarke belt.

For a body in orbit, the gravitational force is matched by the centripetal force. Hence,

$$\frac{GM_\oplus m}{R_s^2} = m\omega^2 R_s \tag{24.4}$$

where ω is the angular frequency (rad s^{-1}) of the satellite's motion.

By combining equations (24.3) and (24.4),

$$\frac{R_s}{R_\oplus} = R = \left(\frac{g}{\omega^2 R_\oplus}\right)^{\frac{1}{3}}. \tag{24.5}$$

By using the determined value of ω and with $g = 9.8$ m s^{-2} and $R_\oplus = 6371$ km, $R \approx 6.62$.

All geostationary satellites lie directly above the Earth's equator and are characterized by their **longitude**, λ_s, east or west of the prime (Greenwich) meridian. Their apparent positions, however, suffer from parallax because of their proximity and they appear to be below the celestial equator. With reference to figure 24.17, the altitude of the celestial equator is given by $90° - \phi$, where ϕ is the observer's latitude, whereas the maximum altitude for a satellite on the meridian is depicted as α_{max}. Looking at the sky (figure 24.18), we may note that the satellites occupying the Clarke belt are close to but slightly below the celestial equator. By plotting the observed altitudes of a number of satellites on this belt, we can determine α_{max} and, therefore, obtain a value for latitude of the observing station via the formula

$$\cos\phi = \frac{\cos^2\alpha_{max} + (R^2 - \cos^2\alpha_{max})^{1/2}\sin\alpha_{max}}{R}. \tag{24.6}$$

It will also be appreciated that for a satellite to be on the observer's meridian (i.e. with azimuth $A = 180°$), its longitude value must match that of the observer's longitude.

Thus, from measurements of the azimuth and altitude of a number of satellites, it is possible to determine the latitude and longitude of the observing station. In order to achieve this with reasonable accuracy, the measurements of the satellite positions should be made to an accuracy of about $\pm 1/2°$.

Table 24.1 gives a list of some of the geostationary satellites presently in orbit according to their longitudes about the Greenwich meridian, with a guide as to their relative signal strengths.

Table 24.1. Longitudes of geostationary communication satellites visible from Glasgow (λ_s − ve is E, λ_s + ve is W).

Name	Strength	λ_s	Name	Strength	λ_s
Eutelsat II f1	**	−48·0	Intelsat 707	*	+1·0
Eutelsat Sesat W4	**	−36·0	Telecom 2B	*	+5·0
Astra 2A II f1	****	−28·2	Telecom 2A	**	+8·0
Kopernikus 3	**	−23·5	Eutelsat II f2	**	+12·5
Eutelsat II f3	**	−21·5	Telstar 12	***	+15·0
ASTRA	****	−19·2	Intelsat 705	*	+18·0
Eutelsat W2	**	−16·0	NSS K	**	+21·5
Hot Bird	***	−13·0	Intelsat 605	**	+27·5
Eutelsat II f4	**	−10·0	Hispasat	**	+30·0
Eutelsat W3	**	−7·0	Intelsat 801	*	+31·5
Sirius	**	−5·0	Intelsat 601	*	+34·5
Telecom 2C II f1	**	−3·0	Telstar 11	**	+37·5
Thor	**	+0·8	Panamasat 3R	**	+43·0

Having set up the simple radio telescope, commence by finding the brightest satellite in the sky—ASTRA. It should be found at an azimuth of about 150°. Keep adjusting the sensitivity setting of the power meter to read mid-scale and 'home-in' on the satellite position in both altitude and azimuth. It should be possible to estimate the measured angles to about 0·2°. Note the two values.

Next, adjust the telescope to search for another satellite, Thor. This should be found at an azimuth of about 150°. Again home-in on its position and record its altitude and azimuth.

The altitude of the celestial equator changes only slowly with azimuth at a station with high latitude, so that the altitude of all the satellites should be fairly similar (within 10°). Similarly, the azimuthal angle between two satellites should be similar to their difference in longitude. Work your way onto other satellites, noting down their positions. Try to find at least six satellites.

With the data, construct a graph of satellite altitude, a, against satellite azimuth, A. Draw a smooth curve through the points and determine a value for α_{max} corresponding to an azimuth of 180°. Estimate how well the value has been determined. Together with the value for R as calculated in equation (24.5), insert the value of α_{max} in equation (24.6) to determine the latitude of the observing station. Make an estimate for the uncertainty in the determination by trial adjustments of the value of α_{max} according to the estimate of its accuracy.

Plot also the satellite longitudes against the measured satellite azimuths. From this graph, estimate the longitude a satellite would have if it were on the site's meridian (i.e. if it had an azimuth of 180°) and, hence, determine the longitude of the observing station. Again, make an estimate for the uncertainty of this determination.

24.5   Solar disc phenomena: practical exercises

Occasionally, when there happens to be a large sunspot on the Sun's disc and the Sun is seen through a thick haze, it is possible to see the spot with the unaided eye. However, for a more serious study of the solar surface phenomena, optical equipment is required and, in the first instance, a basic telescope is essential. Useful work may be done with telescopes of 50 mm diameter or larger. Both a refractor and a reflector may be used although it is generally accepted that thermal distortions of the optics when the instrument is directed to the Sun are less severe for a refractor so giving better images than a reflector.

24.5.1   Visual observations

Danger to Eyesight
It cannot be emphasized too strongly that the Sun should
NEVER
be viewed using instruments without special precautions.

Ordinary telescope eyepieces sometimes carry a screw thread to allow absorbing filters to be attached to them. This system is not suitable for use with the Sun as either the filter may not be sufficiently absorbing or the solar heat may cause it to splinter suddenly.

Direct telescopic views of the Sun should only be attempted with special equipment designed for the purpose. An old-fashioned device is the solar diagonal whereby the converging beam from the telescope is split into two very unequal components. Only about 10% of the collected light is used for viewing through an eyepiece, which also requires the addition of an absorbing filter; and the remaining 90%, which is in the second beam, is passed out of the system. An alternative is to fix an absorbing filter securely over the telescope collection aperture. These are commercially available in the form of coated mylar sheets stretched on a ring which fits over the end of the telescope tube.

24.5.2   Recording sunspots by drawing

A more convenient way of looking at the features on the solar disc is by the projection method. If smooth white paper is held at a distance behind the eyepiece of a telescope which is directed to the Sun, a large image may be focused on to the paper, this being very suitable for viewing by groups of people. The best eyepiece to use is one of low power, preferably a Huyghens type in which the component lenses are not cemented together. (Cemented lenses are prone to damage by the heat from the Sun.)

By making a frame which can be attached to the telescope so that the paper may be clipped to a board, it is then an easy matter to draw the positions and details of the solar markings. The contrast of the projected solar image may be improved by also fitting a shadow board around the telescope tube. A schematic projection frame is depicted in figure 24.19.

The projection technique also allows considerable detail to be seen and drawn. For example, it should be noticed that the solar disc is not uniformly bright and that it fades towards its edge (limb darkening). Sunspots and sunspot groups should be seen with structure—the spots showing a strong core (umbra) surrounded by a less black region (penumbra) which may show structure such as filaments radiating outwards. Small bright patches (faculae) may be evident particularly near to the spots or towards the edge of the Sun where the limb-darkening effect produces a background with lower illumination. The detection of faculae may be made easier by gently shaking the frame which holds the paper.

When drawing the solar markings by the projection method, it is important to provide axes to which specific points may be referred and the north–south, east–west lines may be generated for this purpose. As convention, the north, south, east and west limbs of the Sun are defined according to the

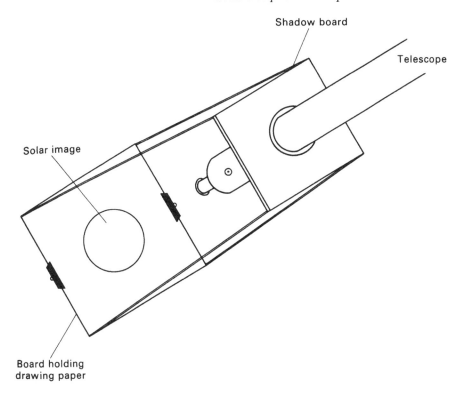

Figure 24.19. A simple projection board for drawing solar phenomena.

directions on the sky—to the unaided eye in the northern hemisphere, north is to the top and east the left.

Keeping the telescope fixed, allow the solar image to drift across the projection board and rotate the paper on the board until the upper edge of the Sun moves along or parallel to the upper edge of the paper. This edge of the paper now corresponds to the east–west direction on the sky. By noting the direction in which the image moves when the telescope is kept fixed, the image drifting because of the diurnal motion of the Sun across the sky, the west direction may be noted (with astronomical eyepieces, the left-hand side of the paper, when viewed from the telescope, is the west side of the Sun). A line at right angles to the upper edge of the drawing paper corresponds to the north–south axis and the north direction may be marked easily by seeing which way the image moves when the telescope's altitude angle is raised slightly. (If the image moves downwards, north is to the top.) Re-centre the image on the projection board and mark the position of a Sun's limb. After ensuring that the image is kept to the markings corresponding to the limb, lightly draw over the details of the structures of the sunspots and mark any faculae.

If a suitable focusing camera is available, very reasonable pictures of the solar markings may be taken without too much distortion if the projected image is photographed from a position very close to the telescope tube, looking as directly as possible at the solar image.

24.5.3 Determination of the solar rotation period

Drawings made daily of the Sun show that the features undergo slow evolutionary changes. In addition, it will be noted that there is a general drift of the sunspots from east to west as a result of the Sun's rotation. Determination of the angular movement of a feature over a measured period of time obviously allows the Sun's rotation period to be determined. Close examination of the movements of sunspots at different latitudes shows that the Sun does not rotate as a solid body but that different latitudes rotate at different rates. Before the motion of spots across the image can be converted to a rotation period, allowance must be made for foreshortening effects of seeing the spots on a curved surface. In order that the progress of the spots across the solid disc may be noted in terms of a shift in apparent longitude, it is important to know the relationship between the marked reference of the north–south, east–west axes of the sky and heliographic (solar) latitude and longitude.

The Sun's rotational axis is at an angle of about $7°25$ to the perpendicular of the plane of the ecliptic, this plane being defined by the Earth's orbit. Consequently, there will be times of the year when the north pole of the Sun will be slightly inclined towards the Earth and visible and other times when it is inclined away from the Earth on the other side of the Sun. In early December and June, the Earth is in a position so that the Sun's pole is at right angles to line of sight, passing through the projected limb of the Sun. Because the Earth's pole is also inclined by about $23°5$ to the ecliptic, the projection point of the Sun's north pole does not correspond to the sky's north–south axis but is inclined to it. However, at these times in December and June, the projection of latitude lines on the solar disc produces a series of straight lines parallel to the solar equator. In general, the motion of a sunspot follows a path of constant latitude and at these times of the year, they may be seen to move across the Sun's disc in straight lines.

At other times of the year, the motions of the spots across the disc follow elliptical paths, each path being the orthographic projection of a solar latitude as seen from the Earth. Since, at these times, the solar pole is inclined to the line of sight, it follows that the centre of the projected image of the Sun does not lie on the Sun's equator. In determining the Sun's period of rotation from the movement of sunspots, allowance must be made for this effect on the projection of the image. It is necessary to determine the heliographic coordinates of any spot from the drawings.

The heliographic latitude, B_0, of the centre of the image is tabulated; and it may be found for each day in the *AA* of the appropriate year. Also tabulated is the position angle, P, which the projection of the solar axis makes with the north–south direction and it is measured positively from the north point of the solar image towards east. For any particular day and time the values of B_0, P may require interpolation.

The formulas from spherical trigonometry which allow the heliographic latitude, B, the apparent longitude, L, of a spot to be determined are:

$$\sin B = \sin B_0 \cos \rho + \cos B_0 \sin \rho \cos(P - \theta)$$
$$\sin L = \sin \rho \sin(P - \theta) \sec B$$

where θ is the position angle of the sunspot on the disc, again measured from the north point of the disc towards the east, and ρ is the angle formed by the direction of the sunspot, the centre of the Sun and the direction of the Earth. The angle, ρ, is given by

$$\sin(\rho + \rho_1) = r/r_0$$

where r is the distance of the spot from the centre of the solar image and r_0 is the radius of the solar image. The angle ρ_1, in turn, is given by

$$\rho_1 = D_\odot r/2r_0$$

where $D_\odot$ is the angular diameter of the Sun, the values also being tabulated in the *AA*.

For a determined latitude, B, the time taken for a spot to move from one longitude to another may be used to calculate the time necessary for the spot to complete one rotation over the solar disc.

As an example of a heliographic coordinate determination, consider the drawing made for 2000

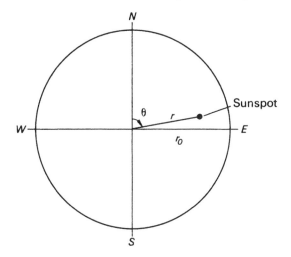

Figure 24.20. The coordinates of a sunspot.

January 2nd, 12–10 UT (see figure 24.20). From the drawing:

$$r_0 = 75 \text{ mm}$$
$$r = 50 \text{ mm}$$
$$\theta = 79°.$$

From the *AA*, we have interpolated values for B_0, P and $SD_\odot$ $(= D_\odot/2)$ as follows:

$$B_0 = -3°\!13$$
$$P = +1°\!65$$
$$SD_\odot = 16'\,15''\!92.$$

Hence,

$$P - \theta = -77°\!35$$
$$\rho_1 = (16'\,15''\!92)(50/75)$$
$$= 0°\!18$$
$$\sin(\rho + 0°\!18) = 50/75$$
$$\rho + 0°\!18 = 41°\!81$$
$$\therefore \; \rho = \underline{41°\!63}.$$
$$\sin B = \sin(3°\!13)\cos(41°\!63) + \cos(-3°\!13)\sin(41°\!63)\cos(-77°\!35)$$
$$\therefore \; B = \underline{10°\!52}.$$
$$\sin L = \sin(41°\!63)\sin(-77°\!35)\sec(10°\!52)$$
$$\therefore \; L = \underline{-41°\!28}.$$

On the following day, this same sunspot was determined to have an apparent longitude of $-26°\!51$. Thus, in one day its apparent change of longitude is $13°\!77$. In order to have an apparent change of $360°$, the time required is $360/13°\!77$ days, i.e. $26\cdot14$ days.

By following a sunspot over a longer period, a more accurate value for the Sun's rotation will ensue and, if spots are measured at different solar latitudes, it should be possible to determine the differential rotation effect.

24.5.4   The eccentricity of the Earth's orbit

Inspection of the *AA* shows that the apparent solar diameter changes from

a maximum of $16' \, 15''9$ on January 4th to

a minimum of $15' \, 43''1$ on July 4th

this being caused by the variation in the distance of the Earth from the Sun according to the eccentricity of its orbit.

If care is taken in making solar drawings with the same equipment throughout the year, with the image board holding the paper kept at a constant distance from the telescope, the variation in the size of the solar disc can be recorded and a value obtained for the eccentricity of the Earth's orbit. Investigation of the figures just given shows that the variation in the apparent size of the solar disc is about 3·5%. In order to detect this and measure it reasonably well over the year, the diameter of the projected image must be of the order of 100 mm giving a variation of about 3·5 mm.

24.5.5   Use of a pinhole camera

The simplest means of taking photographs of the Sun is with a pinhole camera. In this system, a tiny hole acts as a lens and the image that it produces can be recorded on film. A very convenient arrangement has the pinhole at one end of a long tube, blackened on its inside, with the body of a single lens reflex camera, without its lens, at the other end.

For the pinhole camera to give reasonable images, the focal ratio must be $f/1000$ or higher. Thus, a 1 mm hole attached to a 1 m tube is sufficient. Since the Sun subtends about $1/2°$, its image with a 1 m tube is equal to

$$\frac{1000}{2} \times \frac{\pi}{180} \text{ mm} = 8 \cdot 7 \text{ mm.}$$

Even with such a focal ratio, the illumination of the Sun's image is high and a typical exposure of 1/1000th of a second is required using Pan X film. The demand for an extremely high shutter speed may be relaxed by using either a higher focal ratio, filters over the pinhole or a slower film or even a photosensitive photographic paper rather than celluloid film.

A pinhole camera under good conditions is capable of allowing pictures of sunspots to be taken and is useful for recording the shape of the Sun, say at the time of a partial eclipse or when the Sun is just rising or setting.

When a celestial object is on the horizon, refraction alters its true altitude by over half a degree (see section 10.2). For an extended object such as the Sun, refraction has a differential effect so that the lower limb is more refracted than the upper limb. Thus, when the Sun is close to the horizon, it has an elliptical shape with the lower and upper limbs closer together than the east and west limbs. The distortion is easily seen with the unaided eye at a location where there is an uninterrupted view of the horizon. It may be photographed using a pinhole camera. (An ordinary camera is generally insufficient because of its short focal length.) From measurements of the images, a value for the differential refraction may be determined.

24.5.6   Atmospheric extinction

A pinhole camera arrangement is also a suitable means for measuring the apparent solar brightness. If, at the bottom of the light-type tube as described earlier, a solid state photodiode is placed with means of measuring the ensuing current, the response to solar radiation may be recorded as the zenith distance changes.

The size of the photosensitive area with respect to the solar image is not too important. When making a measurement, the tube should be directed towards the Sun and the maximum reading taken

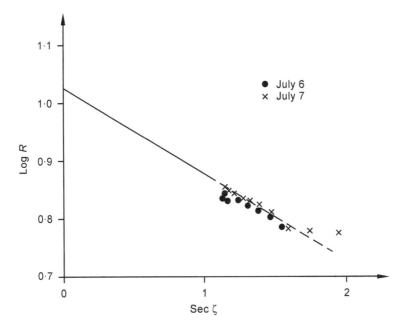

Figure 24.21. Data of the apparent solar brightness obtained by a pinhole photometer revealing the sec ζ dependence from which the zenith extinction is calculated.

as the instrument is adjusted in its pointing direction. When the best value from the detector is achieved, make a note of it and the time. From data obtained in this way, say at twenty minute intervals, it is then possible to explore Bouger's law (section 19.7.2) and determine the local zenith extinction for the wavelength passband associated with the sensitivity of the detector. This is done by plotting the logarithm of the readings against the secant of the solar zenith distance calculated from equation (24.1), namely

$$\cos z = \sin \phi \sin \delta_{\odot} + \cos \phi \cos \delta_{\odot} \cos H. \tag{24.1}$$

An example of data obtained from such an exercise is shown in figure 24.21. Equation (19.11) describes the behaviour of the apparent magnitude (arbitrary system), equivalent to the logarithm of the signal, with the secant of the zenith as

$$m(\zeta) = m_0 + \Delta m \sec \zeta.$$

Extrapolation of the measurements corresponding to $\zeta = 1$ provides a value for $m(0)$ which may also be expressed as

$$m(0) = m_0 + \Delta m.$$

Further extrapolation of the measurements to $\zeta = 0$ provides a value for m_0. Rearranging the previous equation gives

$$\Delta m = m(0) - m_0.$$

For the behaviour of the data displayed in figure 24.21, $m(0) = 0.88$ and $m_0 = 1.02$, giving a value for $\Delta m = -0.14$. It may be noted that such a value for extinction is fairly low with respect to optical measurements but it must be appreciated that the peak sensitivity of silicon diodes is in the far red of the spectrum.

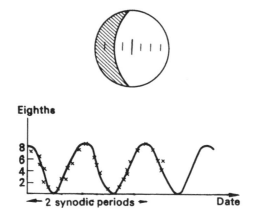

Figure 24.22. Estimation of the fraction of the lunar disc that is illuminated and the variation of this fraction with time.

24.6   The Moon's orbit: practical projects

It is possible, without having accurate astronomical instruments, to take observations over a few months that enable values of the Moon's orbital elements to be deduced. Observations are made of the Moon's phase, its sidereal position and its angular size.

The phase measurements are the easiest to make. If the diameter of the Moon's disc is divided into eighths, then it is possible with the naked eye to obtain fairly accurately the number of eighths of the diameter at right angles to the line joining the cusps that are illuminated. For example, in figure 24.22 the phase would be noted as six. The date and time of the observation is also noted. If observations are made in this fashion over some months and plotted against time, an accurate value for the synodic period, S, can be found.

By using a sky map such as *Norton's Star Atlas*, the Moon's sidereal positions on those nights when stars and Moon are visible may be plotted on the map. The Moon's right ascension and declination (read off from the star charts) and the time and date are recorded. As the months pass, it will be seen that not only does the Moon make circuits of the heavens, crossing from one side of the ecliptic to the other but the ascending and the descending nodes regress. The nodes are the points on the ecliptic at which the Moon is seen to cross that plane, the ascending and descending nodes being, respectively, where the Moon crosses the ecliptic from south to north and from north to south. After six months, the nodes will have regressed by almost 10 degrees. From a track on the map, or from a graph of right ascension against date and time, the value for the Moon's sidereal period, T, may be found.

Changes in the Moon's angular size are more difficult to detect by the naked eye. The Moon's orbital eccentricity has a value of the order of 1/20, so that the ratio of minimum to maximum angular semi-diameter is about 9/10. With some relatively simple manufactured equipment, however, the variations of the apparent angular diameter of the Moon can be measured with sufficient accuracy to allow determination of the Moon's orbital eccentricity.

The type of equipment required is a long, graduated rod (AB in figure 24.23). Two metre sticks attached end-to-end are suitable. A small piece of wood that can be slid along the rod is also required. A pencil stub attached to the metre sticks with an elastic band is good enough.

The rod is pointed in the direction of the Moon. By trial and error, a position is found for the piece of wood where its angular size equals that of the Moon. Let this position be distance, d, from the eye.

Then if $\alpha_{\leftmoon}$ and $R_{\leftmoon}$ are the semi-diameter and radius of the Moon respectively, while $2r$ is the

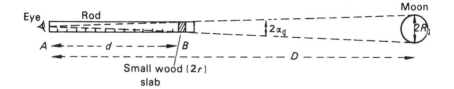

Figure 24.23. A simple device for measuring the angular diameter of the Moon.

length of the piece of wood as shown in figure 24.23,

$$\alpha_{\mathbb{C}} = r/d = R_{\mathbb{C}}/D \qquad (24.7)$$

where D is the Moon's distance from the observer.

Let D_P, D_A be the Moon's distances at perigee and apogee. Then,

$$D_P = a(1-e) \qquad D_A = a(1+e)$$

where a is the semi-major axis and e the eccentricity of the orbit.

Hence,

$$\frac{D_P}{D_A} = \frac{1-e}{1+e}$$

giving

$$e = \frac{D_A - D_P}{D_A + D_P}. \qquad (24.8)$$

But by equation (24.7) d is proportional to D, so that we may rewrite equation (24.8) as

$$e = \frac{d_A - d_P}{d_A + d_P}. \qquad (24.9)$$

A large number of readings over many months are taken. A graph of d against time is desired in order to obtain the maximum, d_A, and the minimum, d_P, values. In practice, however, it is found that the intrinsic errors in this crude method make it difficult to draw a suitable curve among the points as plotted (see figure 24.24(a)). Fortunately we know that the variation in distance is periodic and we can use this to enable us to abstract the required information.

The period in question is the anomalistic period, the time it takes the Moon to go from perigee to perigee. The anomalistic month is not at all that different in value from the length of a sidereal month, T, a value of which should have been obtained by this time.

If the x-axis or time-axis is then divided into such periods and the readings within each period lifted and re-plotted into the first sidereal period, the trend of the graph is usually obvious as shown in figure 24.24(b). This process of compacting the data is referred to as 'phase folding'.

From the graph (see figure 24.24(b)), values of d_A and d_P can be found, enabling a value for the eccentricity of the Moon's orbit to be calculated from equation (24.9).

Having obtained the eccentricity, e, we can find a date for the time of perigee passage from the same graph. It is the date at which the Moon's distance D was a minimum and, therefore, the date, τ, in figure 24.24(b).

From the track of the Moon's orbit plotted on the star atlas, it is easy to find a value for the longitude, Ω, of the ascending node, N, of the orbit on the ecliptic by estimating the angular distance along the ecliptic from ♈ to N.

The inclination, i, is best found by estimating the angular distance between the ecliptic and the orbit 90° from the nodes (see figure 24.25).

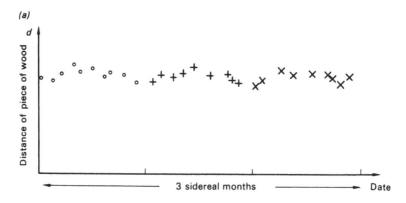

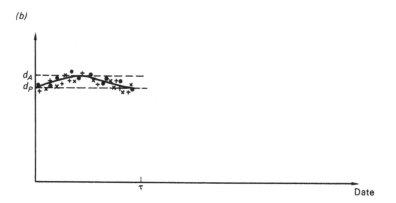

Figure 24.24. The variation of lunar distance with time: (*a*) over three months, (*b*) by superimposing the results into one sidereal month.

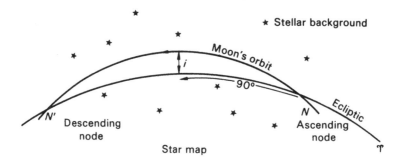

Figure 24.25. Estimation of the inclination to the ecliptic of the lunar orbit.

The argument of perigee, ω, that is the angular distance between the direction of perigee and the ascending node, N, is found by noting on the star chart the position of perigee. This can be done from our knowledge of the time of perigee passage and from the observations of the Moon's sidereal position at known times of observation.

The five elements thus obtained, namely e, i, Ω, ω, τ, give a scale model of the Moon's orbit. To obtain the scale, we require a knowledge of a, the semi-major axis. It is possible to obtain this also from simple observations carried out by one person with a star atlas.

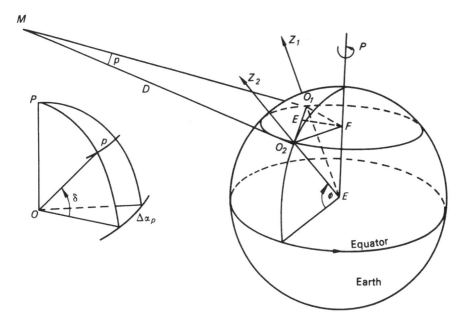

Figure 24.26. Determination of the Moon's distance by measuring the effect of diurnal parallax on the Moon's sidereal position.

24.6.1 Measuring the Moon's distance

Suppose that at full moon, an observer notes carefully the position of the Moon on a star atlas every hour, say from four hours before midnight to four hours after midnight. The difficulty of seeing stars relatively close to the bright Moon can be solved by holding a small disc at arm's length over the Moon to eclipse most of its light. Then the Moon's observed shift in sidereal position will be caused by (i) the Moon's motion in its orbit and (ii) the movement of the observer due to the Earth's rotation.

(a) The parallactic shift

In figure 24.26 the observer in latitude $\phi°$ N is moved from O_1 to O_2 during the night's observations, of duration t hours. Forgetting for the moment the Moon's orbital movement, let it be at M while observations are being made. Then, if δ is the Moon's declination, and p is the parallactic angle moved through by the Moon during t hours, we may write

$$p = \Delta\alpha_p \cos\delta \qquad (24.10)$$

where $\Delta\alpha_p$ is the parallactic displacement in right ascension. It may be noted that, as the observer moves along a small circle parallel to the equator, there is no corresponding parallactic shift in declination.

Let the Moon be on the meridian at a time halfway through the period of observation. Then, since p is a small angle,

$$p = O_1O_2/D \qquad (24.11)$$

D being the Moon's distance from the observer.

By equations (24.10) and (24.11),

$$D = \frac{O_1O_2}{\Delta\alpha_p \cos\delta}. \qquad (24.12)$$

Now in t hours, the angle $O_1 F O_2$ rotated through by the observer is θ, where

$$\theta = \frac{t}{24} \times 360°.$$

If $R_\oplus$ is the radius of the Earth,

$$O_2 F = O_2 E \cos\phi = R_\oplus \cos\phi.$$

Also

$$O_1 O_2 = 2 \times O_2 E = 2 \times O_2 F \times \sin(\theta/2).$$

Hence,

$$O_1 O_2 = 2 R_\oplus \cos\phi \sin\psi \qquad (24.13)$$

where $\psi = 7{\cdot}5t^\mathrm{h}$ and is measured in degrees.

Substituting in equation (24.12), we obtain

$$\frac{D}{R_\oplus} = \frac{2\cos\phi\sin\psi}{\Delta\alpha_p \cos\delta} \qquad (24.14)$$

giving the Moon's distance in Earth radii.

If, for example, observations were made over a six-hour period, the value of ψ is 45° and by equation (24.13), $O_1 O_2 \approx R_\oplus$. Since the Moon's distance is about 60 Earth radii, the parallactic angle is approximately one degree and, for an observer near 60° north latitude, $\Delta\alpha_p$ is almost two degrees.

(b) The Moon's orbital motion

During the night, the Moon will have moved in its orbit. Let us suppose we are hoping to measure the Moon's distance to an accuracy of about 10%. We remember that the Moon's mean motion is n, where $n = 360°/27\frac{1}{3}$ days $\approx \frac{1}{2}°$ h^{-1}. In six hours, therefore, the Moon's true anomaly, v, will increase by about three degrees. But by Kepler's second law, the angular velocity varies because of the eccentricity of the Moon's orbit. We must take this variation into account, at least to the first order.

It is known that to the first power of the eccentricity, the small increase, Δv, in the Moon's true anomaly, v, in a small time interval, Δt, is given by

$$\Delta v = n\Delta t (1 + 2e\cos v). \qquad (24.15)$$

If we take e to be of the order of 0·055, then

$$\Delta v = n\Delta t (1 + 0{\cdot}11\cos v)$$

showing that the shift, $n\,\Delta t$, in the Moon's true anomaly could be different from its mean value by an amount of the order of ±10% of its mean value, depending upon which part of its elliptical path it is in.

Equation (24.15), however, enables us to calculate Δv accurately enough. The eccentricity, e, will have been measured. n and Δt ($= t^\mathrm{h}$) are known and v, the angle between the direction of perigee and the Moon's direction, can be found from the observer's graphs.

It remains to compute what change in the Moon's right ascension, $\Delta\alpha_0$ say, will result from the Moon's orbital movement leading to an increase in its true anomaly of amount Δv.

In figure 24.27, the Moon's orbital plane, its orbit and the plane of the Earth's equator are shown.

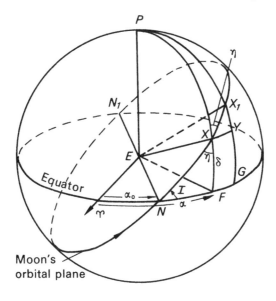

Figure 24.27. The celestial sphere depicting the Moon's orbital plane.

Now the angular shift $\Delta \nu$ is so small so that we may consider triangle XX_1Y to be plane, XY being the arc of a small circle parallel to the equator. If $\angle PXX_1$ is taken to be of size η, we may write

$$XY = XX_1 \sin \eta \qquad (24.16)$$

or

$$XY = \Delta \nu \sin \eta.$$

In $\triangle XNF$, $\angle XNF$ is the inclination I of the Moon's orbital plane *to the plane of the equator*, $\angle NXF = \eta$, $\angle XFN = 90°$.

If α and α_0 are the right ascensions of the Moon and the ascending node N of the Moon's orbit (again with respect to the equator),

$$NF = \alpha - \alpha_0.$$

Then by the sine formula in $\triangle NFX$,

$$\sin \eta = \frac{\sin(\alpha - \alpha_0) \sin I}{\sin \delta}. \qquad (24.17)$$

Equation (24.16) becomes

$$XY = \Delta \nu \frac{\sin(\alpha - \alpha_0) \sin I}{\sin \delta}.$$

But $XY = \Delta \alpha_0 \cos \delta$ and so

$$\Delta \alpha_0 = \frac{\Delta \nu \sin(\alpha - \alpha_0) \sin I}{\sin \delta \cos \delta} \qquad (24.18)$$

from which $\Delta \alpha_0$, the shift in the Moon's right ascension due to its orbital motion, can be found. It may be noted in passing that, as in finding i, the most accurate way of measuring I is to measure on the star chart the angular distance between the graph of the Moon's orbit and equator at points 90° from the nodes (N and N_1 in figure 24.27).

Table 24.2. Observations of Saturn, August 2000–January 2001.

Date	α			δ			λ
	h	m	s	°	′	″	
August 1	3	50	45	17	59	10	61°28
September 1	3	56	45	18	11	50	60°86
October 1	3	56	03	18	04	21	60°67
November 1	3	49	03	17	39	37	58°95
December 1	3	39	17	17	09	04	56°57
January 1	3	31	09	16	47	04	54°59

Now the observed shift in right ascension, $\Delta\alpha$, will be made up of the parallactic shift, $\Delta\alpha_p$, and the orbital shift, $\Delta\alpha_0$. Thus,

$$\Delta\alpha = \Delta\alpha_p + \Delta\alpha_0$$

giving

$$\Delta\alpha_p = \Delta\alpha - \Delta\alpha_0.$$

Care should be taken with regard to the algebra, since $\Delta\alpha_p$ is essentially a negative quantity while $\Delta\alpha_0$ is positive.

Having found $\Delta\alpha_p$, equation (24.14) gives the Moon's distance in Earth radii. If use is made of the polar equation for the ellipse, namely

$$r = \frac{a(1 - e^2)}{1 + e\cos\nu}$$

a value for a can then be calculated.

24.7 Planetary orbits

Positional data of the planets can be obtained by making regular observations over periods of weeks to months as they execute their orbits. From the data, estimates of the planets' distances can be made.

The same observations may also be made using a planetarium. Some machines provide sufficient accuracy for reasonable quantitative results to be obtained. A planetarium provides a useful means of concentrating into a few minutes a planet's annual movements against the stellar background. Differences between the inner and outer planets are quickly noted. All the terms such as elongation, phase angle, direct and retrograde motion, conjunctions, quadrature, stationary points, etc, can be demonstrated and appreciated in the course of an observing session.

24.7.1 The outer planets

The distances of the outer planets, Jupiter and Saturn, can be determined fairly easily by noting their positions, say once per month, over the period of opposition. This may be done either by a planetarium demonstration, by plotting the positions on a star map from direct observations or by obtaining the celestial coordinates from the *AA*.

A scheme for an exercise with Saturn and a set of results is given in table 24.2.

From the values of right ascension and declination (α, δ) the corresponding value for the ecliptic longitude, λ, may be calculated by using equation (8.10), namely

$$\tan\lambda = \frac{\sin\alpha\cos\varepsilon + \tan\delta\sin\varepsilon}{\cos\alpha}. \tag{24.19}$$

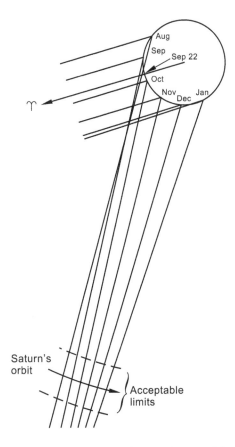

Figure 24.28. The construction for determining the distance of Saturn from the Sun.

For the graphical representation of the orbits of the Earth and Saturn, it may be assumed that they are coplanar. To a sufficient accuracy, it is only necessary to record the celestial longitude of Saturn. It can be taken that the Earth moves through $360/365°$ per day. Its positions in an assumed circular orbit can be plotted from reference to the angle between the Earth, the Sun and the first point of Aries, Υ, for the dates when the position of Saturn is observed. For a starting point, on Sep 22nd (when the Sun is at RA 12^h), the Sun, Earth and Υ are in line, the Sun being opposite to the direction of Υ. Using a scale of 1 AU = 25 mm, plot the positions of the Earth in its orbit according to the number of elapsed days from September 22nd. From those positions, draw the observed directions of Saturn in relation to Υ, these corresponding to values of λ as determined from equation (24.19).

Kepler's second law states that the radius vector joining the Sun to a planet sweeps out equal areas in equal times. If Saturn's orbit is assumed to be circular, then, in this case, the planet will sweep out equal arcs in equal times. There is, therefore, one particular circular orbit on which Saturn lies that cuts the Earth–Saturn direction lines, giving equal arcs along its orbit.

By trial and error, draw this circle on the construction and, hence, determine the distance of Saturn from the Sun.

Plots of the Earth orbit and the direction lines to Saturn are illustrated in figure 24.28. From the construction, Saturn's orbit might be placed at 8·2 AU from the Sun, an answer which is good to 10%. (NB: The exercise is suitable only for the orbits of Jupiter and Saturn—extremely accurate observations are required and allowance must be made for eccentricity if the orbit of Mars is to be constructed in this way.)

24.7.2   The inner planets

Distances for the inner planets can be obtained by measuring their elongation angles over a period and determining a value for maximum elongation, η_{max}. At this angle, the line joining the Earth to the planet is a tangent to the planet's orbit and, hence, the distance of the planet from the Sun relative to the Sun–Earth distance is given by

$$a = \sin \eta_{max}. \tag{24.20}$$

In order to obtain a value of η_{max} for Venus, say, measure the values of the right ascension of Venus and the Sun at approximately ten-day intervals. Again, this can be done conveniently in a planetarium. Either using real sky observations or those from a planetarium, positions of Venus may be plotted on a star map and the appropriate ecliptic longitude determined. For the exercise it is assumed that the orbits of Venus and the Earth are coplanar and that Venus lies exactly on the ecliptic. (If the ecliptic is not depicted on the map, the right ascension may be noted and converted to ecliptic longitude by the four-parts formula given below). Corresponding to each measurement of Venus, the right ascension of the Sun should be taken from the *AA*. This value can be converted to ecliptic longitude using the four-parts formula, namely

$$\cot \lambda = \cot \alpha \cos \varepsilon.$$

From a graph of elongation ($\lambda_V - \lambda_\odot$) against time, a value for η_{max} will be obtained for substitution into equation (24.20). It may be noted that if the data are taken around the time of the spring equinox, a sudden jump may appear in the calculated variation of elongation of Venus. This can result if either Venus or the Sun pass through the first point of Aries, e.g. $\alpha_\odot$ moves from $23^h 59^m$ to $00^h 00^m$. The problem can be rectified by adding 24^h to the later right ascension values.

24.7.3   Kepler's second law

The validity of Kepler's second law may be considered by examining the motion of a planet with an eccentric orbit, such as Mercury. Heliocentric data for the planets are tabulated in the *AA* and a sample sufficient for this exercise is shown in table 24.3, with the radius vector listed with the scale appropriate for plotting, together with the heliocentric longitudes. The data are for Mercury and are tabulated at seven-day intervals commencing at 2000 Jan 0.

Using a scale of 20 mm = 1 AU, plot the orbit of Mercury following the convention that the motion is counter-clockwise. Estimate and mark the position of the perihelion. Measure the eccentricity of Mercury's orbit from the construction.

To a sufficient accuracy, the area swept out by the radius vector in a seven-day interval may be expressed as

$$a = C r_1 r_2 \theta$$

where r_1 and r_2 are the radius vectors at the beginning and end of each seven-day period respectively, θ is the angle between the radius vectors and C is a constant. As the units of the measuring areas are arbitrary, let $C = 1$ for simplicity. Determine the value of each area which is swept out by the radius vector in a seven-day interval. Using 2000 Jan 0 as a starting point evaluate the total area swept out by the radius vector at the end of each seven-day period. Plot this running total against time.

On the same graph, plot the longitude of Mercury for each tabulated time. Choose a scale so that the two plots are close to each other. From this second graph, it is immediately apparent that the change in longitude with time is not constant. For the first graph, however, the plot should be a straight line, demonstrating that the rate of sweeping out areas by the radius vector is constant, thus verifying Kepler's second law.

Table 24.3. Mercury's orbit.

Date 2000	Longitude	Radius vector (mm)	r_1	r_2	θ	Area	Running total of area
January 0	249°.7	93·2					
January 7	268°.9	92·8					
January 14	289°.0	89·8					
January 21	311°.2	84·4					
January 28	337°.2	77·0					
February 4	8°.9	69·0					
February 11	47°.7	62·8					
February 18	91°.4	61·8					
February 25	133°.0	66·4					
March 3	167°.5	74·0					
March 10	195°.1	82·0					
March 17	218°.2	88·2					
March 24	238°.6	92·0					

24.8   The telescope

Only a small-sized telescope is necessary to demonstrate the result of increasing the collecting aperture over the eye's pupil and of utilizing the effect of magnifying power. Binoculars may also be used. For either kind of optics, the best appreciation is obtained if the instruments can be fixed to a device such as a tripod rather than just being handheld. Any telescope with a diameter of a few centimetres or more will reveal, with the aid of an eyepiece, craters on the Moon, the disks of planets, satellites of Jupiter and stars that cannot be seen by the unaided eye. Some stars which appear single to the unaided eye are seen as pairs when looked at with the aid of the telescope.

Objects of interest are usually listed with star maps and they will not be discussed here. Just to serve as an example, however, the improvement over the eye that the telescope affords can be appreciated by looking at the star ϵ *Lyrae*. For an observer possessing exceptionally keen sight, this star may be seen with the unaided eye as two fifth magnitude stars, ϵ^1 and ϵ^2, separated by 208 arc sec. Not only are the stars seen more distinctly when viewed with the telescope, each may be seen as doubles.

The area of the sky that can be investigated at any time depends on the field of view of the telescope. This may be determined by directing the telescope to the meridian, keeping it fixed and letting the stars drift across the field, timing them as they pass from one edge of the field to the other. The recorded time must first be converted to angular measure and then multiplied by the secant of the declination of the star.

The timing of the drift of stars may also be used to determine the calibration of a micrometer eyepiece. Once this has been done, comparison of the separation of a range of double stars can be made with a star catalogue.

Some aspects of the visual use of a telescope may be demonstrated in the laboratory. For these experiments (the investigation of resolving power and measurements of magnifying power), a small telescope needs to be fitted with a large iris diaphragm or provided with a set of masks with circular holes to be fitted over its aperture. The required source consists of two closely separated, illuminated pinholes (see figure 24.29). The pinholes, or 'artificial double stars', should be set at a distance so that they cannot be resolved as a pair by eye or with the telescope when the aperture is reduced but readily

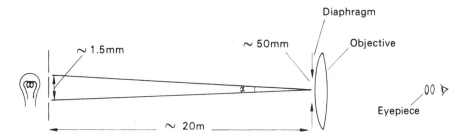

Figure 24.29. An experiment to investigate the resolution of a telescope when viewing an artificial double star.

resolved at full aperture. If possible, provide means for the insertion of red, green and blue colour filters.

24.8.1   Resolving power

Starting with the diaphragm at full aperture, slowly close it down until it is no longer possible to resolve the stars as being separate points of light. Measure the diameter of the aperture with the selected diaphragm at this condition. Repeat the procedure and take the mean from five results. By assuming $\lambda = 5500$ Å, determine the resolving power for the setting of the diaphragm (use equation (16.14)).

Repeat the experiment but this time commence with minimum aperture and slowly enlarge the diaphragm until the stars are resolved. Measure the diameter of the aperture required to resolve the stars under this operation.

Take five resolving power measurements with the blue, green and then the red filters in the beam. According to the effective wavelengths of the filter–eye combination, evaluate the resolving power of the telescope and see how it changes with wavelength.

Compare the observed values of resolving power with the actual angular separation of the double star.

These experiments serve to illustrate the difference between the resolving power as established by diffraction theory, using some chosen criterion, and the actual resolving power as determined with a telescope in combination with the eye.

In practice, when real double stars are being observed, the resolving power depends on the observing conditions, on the relative brightnesses of the stars and on their relative colours. The amount of scattered or background light may also affect the value of the resolving power. Apparent resolution can sometimes be achieved by partial knowledge of the objects such as the position angle of the line joining the two stars.

Some people are better than others at being able to resolve star pairs and may be able to resolve stars which are, in fact, closer than the Rayleigh limit.

24.8.2   Magnifying power

Direct the telescope to the day sky and place a thin piece of paper over the eyepiece. Measure the diameter of the exit pupil, d, for a range of values of the entrance aperture, D. From a plot of d against D, determine the gradient. The magnifying power is defined by the ratio D/d (see equation (17.5)) and, hence, the measured gradient provides a value for the magnifying power of the telescope and eyepiece combination.

Change the eyepiece for one with a different focal length and repeat the experiment.

24.9   Photography of star fields

Point-like star images on photographs can only be obtained with a system that has a drive and guidance. Interesting results can be obtained by attaching a 35 mm camera, with its ordinary or telephoto lens, to a telescope mounting which has a drive. A series of photographs taken with increasing exposure times reveals that fainter and fainter stars are recorded. Photographs taken by this means provide material for star identification, determination of plate scale, etc but generally allow only qualitative measurements of stellar magnitudes. Colour film may be used giving a rough sense of the colour of the stars but black and white emulsions can also be used.

In order to be able to compare stellar magnitudes with some accuracy, the photographs need to be taken with a telescope camera, providing a reasonable plate scale. The sizes of the star images on the plate need to be measured in turn by using some kind of microscope with a graticule or adjustable cross-wire. Depending on the plate, the images may sometimes allow projection on to a screen for measurement. Note that for black and white films, the star images comprise a collection of blackened grains. Plot catalogued magnitudes against the measured diameter, the square root of the diameter and the logarithm of the diameter. See which of the following equations gives the best fit:

$$m = a - bD,$$
$$m = a - bD^{\frac{1}{2}}$$
$$m = a - b\log_{10} D$$

and determine the constants a and b. Having obtained the magnitude–diameter relationship for a particular plate, use it to see how closely determined magnitudes agree with catalogued values. (NB: Unless the equivalent wavelength of the sensitivity of the chosen plate is close to the one corresponding to the listed magnitudes values, it may be necessary to limit the range of stars investigated for a magnitude–diameter relationship to one spectral type.)

Any measured diameter of an image depends on the separation of two points that the observer considers to be the extremities of a circular package whose perimeter fades into the fog of the plate. The criterion for deciding the extent of any image is, therefore, set by the individual observer. If magnitude–diameter relationships are obtained by different observers using the same plate and identical equipment, it is generally found that the constants a and b vary significantly from observer to observer. Once a relationship has been established by an observer using standard stars, it can be used only by that person for determining the magnitudes of other stars on the plate. Obviously, a set of magnitudes determined by various observers should agree, providing that they have used their own individual magnitude–diameter relationship.

As an alternative to performing magnitude determinations essentially following the principles used in professional astronomy, the measurement scheme might be modified in a way that perhaps makes the data reduction process easier. If the telescope aperture is fitted with a obscurator such as a pencil or a thick wire across a diameter then the stellar images are in the form of the diffraction pattern in the form of a line. The maximum strength of the record is at the image mid-point and tapers away into the fog at some distance along the line in both directions. The brighter the star, the longer is the perceived image. Thus, measurements of image lengths are related to the original brightnesses of the stars. It is a useful exercise to make an exposure of a star field and determine a magnitude/length calibration. From the scatter in the data providing the relationship, it is an easy matter to estimate the accuracy to which stellar magnitudes can be made with this diffraction arrangement.

24.10   Spectra

Good spectra of the Sun, perhaps using colour film, may be obtained by using a 35 mm camera with a replica transmission grating placed over its lens. (Such gratings may be purchased very cheaply.)

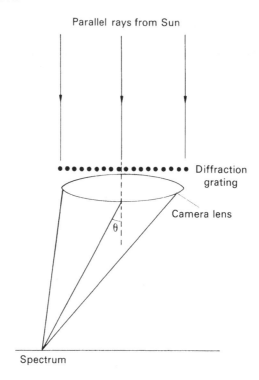

Figure 24.30. The production of a spectrum by using a replica transmission grating.

The optical system is similar to that used for objective prism spectra but differs in that the objective prism is replaced by an objective grating. With the usual commercially available gratings, the spacing of the ruling is such that the angular deviation of the spectrum is too great for it to be accepted by a conventional telescope. The field of view of a 35 mm camera normally allows the spectrum to be recorded. Consider a typical replica rating (see figure 24.30).

Such a grating may have 500 lines per mm, giving a ruling space of 1/500 mm. By equation (19.13), the deviation of the first-order spectrum is given by

$$\sin \theta = \frac{\lambda}{d}.$$

Thus, at a wavelength of 5000 Å, the deviation is given by

$$\theta = \sin^{-1}(5000 \times 10^{-7} \times 500)$$
$$= \sin^{-1}(0{\cdot}25)$$
$$= 15°.$$

A camera providing a field of a little wider than 30° would, therefore, be able to record most of the visible first-order solar spectra on either side of normal incidence. If the camera has a smaller field, it should be set at an angle to the Sun so that the angle of incidence on the grating is not normal.

Now the resolving power of a diffraction grating is given by

$$R = Nm$$

where N is the number of lines used on the grating and m is the order of interference. By assuming that the grating used is 20 mm long and that the first-order spectrum is recorded, the resulting power is

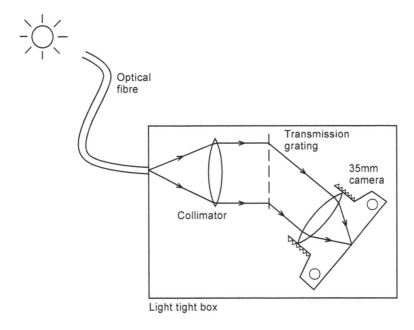

Figure 24.31. A simple solar spectrometer with a fibre optic feed.

given by

$$R = 500 \times 20$$
$$= 10^4$$

suggesting that it should be possible to resolve the spectral lines in the solar and stellar spectra.

In the case of the Sun, if the whole of the disc is used, the spread of half a degree in the rays incident on the grating causes a blurring of recorded detail in the spectrum. Consequently, it is unlikely that spectral lines will be identifiable on a photographed spectrum taken by this simple means.

In order to obtain a solar spectrum for an investigation of spectral lines, subsidiary objects must be used. First an image of the Sun must be obtained, allowing a small part of the disc to be isolated by a slit. The light passing through the slit must be collimated and the resulting beam must fall onto the objective grating camera.

An alternative to this system is to use a fibre optic to collect the light from the Sun simply by pointing its end in the solar direction, the emerging end acting as a slit of the spectrometer system. The fibre should make entry into a light tight box housing the spectrometer optics. The light cone leaving the fibre is $f/1$ and the collimating lens needs to have a matching or closely matching focal ratio. A redundant 35 mm camera lens may be used for this. Again, depending on the grating space and the field of view of the objective grating camera, the latter system may need to be set at an angle to the original collimated beam. The grating need not be attached to the camera which itself might be offset to the desired angle. In adjustment, the collimator should be at its focal length from the end of the fibre so producing a parallel beam and the camera should be focused for infinity. The diameter of the fibre can be taken as the width of an equivalent slit—the height of the recorded spectrum depends on this same diameter with a magnifying factor given by the ratio of the focal lengths of the camera and collimator lenses. A schematic layout of the required equipment is depicted in figure 24.31.

For stellar sources, the grating and the 35 mm camera can be used directly but it must be made to follow the particular star or star field, the exposure perhaps lasting a few tens of minutes. In order to give the recorded spectrum some height, it is necessary to arrange the ruled lines to be parallel either to right ascension or declination with tracking of the camera which drifts in declination or right ascension respectively. If a sufficiently fine grain film is used and care is taken, some stellar spectral lines will be resolved.

The spectra will appear on both sides of the undeviated and undispersed images, the latter being used to identify the stars in the field.

When colour film is used to record stellar spectra in this way, blue stars (A and B types) generally provide three brightness peaks in the blue, yellow/green and red, hinting as to how the colour recording process is effected. Bright red stars (K type) give only a weak contribution to the blue part of the spectrum and the stars can be easily spotted on the film.

24.11   Michelson's stellar interferometer

The largest known apparent angular diameters of stars are of the order of 10^{-7} radians and for the condition of minimum visibility of fringes to occur with the Michelson stellar interferometer, the separation of the apertures needs to be of the order of a few metres. An exact simulation of this experiment would be very difficult to achieve in the laboratory but the principle can be demonstrated by using a variation known as Anderson's method. Simple apparatus can be used to measure the angle subtended by an illuminated pinhole.

For the laboratory method,[1] the apertures are in the form of slits shaped like cats' eyes, at a fixed separation. They can be made simply by cutting three pieces of brass shim and glueing them together. The slits and the layout of the equipment are illustrated in figure 24.32.

The double-slit arrangement needs to be placed behind the 'telescope' at a distance within the focal length of the lens. The effective separation of the slits can be altered by moving the slit system along the optic axis. Suppose that without the double slit in the beam, the image of the artificial star is at a distance, S', from the objective. If the slits are set at a distance, D, from the image position and the separation of the slits is a, the effective slit separation, a', as seen from the star, is given by

$$a' = \frac{aF}{D}$$

where F is the focal length of the objective. The effective slit separation may, therefore, be controlled by varying the distance, D.

Set up a travelling microscope so that it lies on the axis of the optical bench at its end. Without the double slit in the beam, focus the microscope on the image of the artificial star formed by the objective. Adjust the tilt of the objective by means of the three screws until the best image is seen with minimum flare.

Check that by sliding the double slit along the optical bench, the slit face can be brought into focus without readjustment of the microscope.

Place the double slit on the optical bench close to the objective. When viewed through the microscope, the artificial star should now appear as a band of light with a fringe pattern across the band, the fringes appearing much the same as Michelson would have seen them with his interferometer. Adjust the position of the double slit in the holder until the sharpest fringe pattern is observed.

On moving the slits progressively towards the microscope, the visibility of the fringes will decrease until they disappear. Continuation of the travel causes the fringes to reappear but with less separation. After a little practice in controlling the fringe visibility, find the position for minimum visibility, corresponding to an effective separation of $a'_{\min}$ and note the position of $D_{\min}$ of the double slit on the optical bench. Alter the double-slit position and repeat this part of the experiment several times to obtain a mean value of the slit position which gives minimum visibility. Determine this

---

[1] The idea for this experiment has been taken from Palmer C H 1962 *Experiments and Demonstrations* (Baltimore, MD: Johns Hopkins Press).

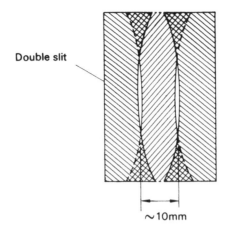

Double slit

~10mm

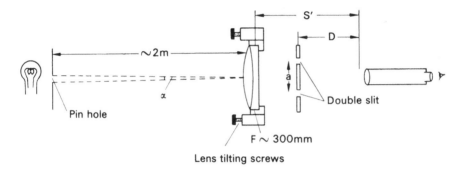

S'

D

~2m

Pin hole

α

a

Double slit

F ~ 300mm

Lens tilting screws

Figure 24.32. Laboratory equipment for simulating Michelson's stellar interferometer.

position relative to the image plane of the objective by measuring the position of the double slit when it is viewed to be in focus in the microscope.

Using the travelling microscope, obtain a mean value for the separation of the slits. By assuming the mean wavelength of the light to be 5500 Å and knowing the focal length of the objective, calculate the angular size of the artificial star using the formula

$$\alpha = \frac{1 \cdot 22\lambda}{a'_{\min}} = \frac{1 \cdot 22\lambda D_{\min}}{aF}.$$

As a check, calculate the actual angle subtended by measuring the physical size of the hole constituting the artificial star and the distance of the star from the objective.

24.12 Digital photography

The advent of two-dimensional solid state detectors has revolutionized 'photography'. The use of digital cameras for everyday use is now fairly common. Such equipment used for landscapes and portraits automatically provides colour pictures. Some of the latest models (SLR) have detachable lenses and could, in principle, be attached at the focus of a telescope without the camera lens being attached. However, there is no facility to make long exposures and, without cooling, the recorded pictures would be very subject to thermal noise. The formatting of the files is also generally not amenable to obtaining numerical information on brightness values. A 'landscape' digital camera could be used in a situation where the light level is sufficient as perhaps in the recording of solar spectra using the fibre optic spectrometer described earlier.

Digital cameras can, however, be applied to photography of the Moon and bright planets. In order to obtain images of planetary disks with a reasonable size, the method of eyepiece projection might be investigated. If the camera lens is detachable, then the arrangement is essentially the same as for drawing solar images as described in section 24.5.2, with the detector face replacing the paper on the projection board. (It is not advisable to experiment with solar images as damage to the camera might ensue.)

In the first place, the size of the planetary image in the focal plane in the telescope should be calculated (see equation (16.2)—the angular semi-diameter of the planet is tabulated in the *AA* and the focal length of the telescope should be known). This can be compared with the physical dimensions of the camera detector chip to estimate the size of the basic image relative to the picture format. According to the desired magnification, M, the projection distance, P, to the chip from the eyepiece is given by

$$P = (M + 1)F_e$$

where F_e is the focal length of the eyepiece.

If the camera lens is not detachable, then magnified images can also be obtained by using an eyepiece with the camera attached immediately behind. The eyepiece position needs to be set for normal viewing and the camera focused at infinity. Because the field of view is small making it difficult to home in on the desired object, the observations might be eased by placing a flip mirror and viewing system between the eyepiece and the camera. Because of seeing effects, not all of the images will be perfectly sharp. If a large number of identical exposures is made, then they might be subjectively graded according to their sharpness. A probability distribution of image quality can then be obtained and the predictions given by equation (19.12) checked out.

If an 'astronomical' CCD camera is available, then a whole range of observational schemes can be attempted. These cameras are normally available simply with a tube for attachment to a telescope and do not themselves carry any foreoptics. They also have built-in cooling so allowing long exposures to be made when attached to a telescope. Their operation requires a PC for their control with supplied software. The recorded images are stored in files with some standard format giving access for quantitative reductions or image processing. Investigation of their application is left to the student.

Further ideas for experiments and exercises can be found in some of the sources given in the bibliography.

Web sites

The first collection of web sites correspond to those which are referenced within the main text, the second block have not been referenced directly but provide other useful background information. There are many hundreds of addresses related to astronomy that are of interest but which have not been listed here. It is left to the student to explore these perhaps from links in the list below or by using the regular search engines.

- W 1.1—www.spaceweather.com
  Provides information and warnings of forthcoming aurora events according to observations of recent solar phenomena.
  An alternative site—www.sel.noaa.gov/SWN/index.html—with the title *Space Weather Now*—gives similar information with the current solar image indicating its level of activity.
- W 1.2—www.heavens-above.com
  Provides predictions of the visibility of the *International Space Station* according to the location of the observer.
- W 6.1—astro4.ast.vill.edu/
  'Click on' *Skyglobe*—the software is a 'planetarium' simulation of the sky and can be downloaded—it is distributed as shareware.
- W 6.2—www.seds.org/billa/astrosoftware.html
  This site provides a comprehensive list of planetarium software—demonstrations for some can be downloaded—with a view to purchase of the complete package.
- W 16.1—www.seds.org/billa/bigeyes.html
  The World's largest optical telescopes.
- W 16.2—www.ast.cam.ac.uk/astroweb/yp_telescope.html
  The site provides links to 'telescope' facilities around the world for the full range of spectral domains.
- W 18.1—www.aao.gov.au/
  This is the Anglo-Australian site—'click on' 'Images', then 'Images by Telescope—Anglo-Australian'; AAT 19 is a good example of an image processed by the unsharp masking technique.
- W 20.1—www2.keck.hawaii.edu
  Details are provided about the operation of the twin 10 m KECK optical telescopes.
- W 20.2—www.jach.hawaii.edu/JACpublic/JCMT/
  The site provides information about the James Clerk Maxwell Telescope designed for observational millimetre astronomy.
- W 20.3—linmax.sao.arizona.edu/help/FLWO/whipple.html
  The site of the Fred Lawrence Whipple Observatory on Mt Hopkins, near Amado, Arizona. As well as running the 6·5 m MMT, the observatory also operates a 10 m gamma-ray telescope.
- W 20.4—www.gemini.edu/
  The site of The Gemini Observatory involving the use of twin 8·1 m telescopes covering both the northern and southern skies. The telescopes operate with adaptive optics.

- W 20.5—www.mrao.cam.ac.uk/telescopes/coast/
  Information about the Cambridge Optical Aperture Synthesis Telescope is provided.
- W 20.6—www.aao.gov.au/
  The site of the Anglo-Australian Observatory—includes the 3.9 m AAT and the 1.2 m UK Schmidt.
- W 20.7—www.ast.cam.ac.uk/~dwe/SRF/camc.html
  A route into the site of The Carlsberg Meridian Telescope.
- W 21.1—www.atnf.csiro.au/SKA/
  This site gives details of the plans for the development of the world's most sensitive radio telescope—The Square Kilometre Array.
- W 21.2—info.gb.nrao.edu/GBT/
  The site of the National Radio Observatory (USA) with descriptions of the new Green Bank Telescope in West Virginia.
- W 21.3—www.mrao.cam.ac.uk/telescopes/ryle/
  This site describes the operation of the Ryle Telescope (RT) in Cambridge.
- W 21.4—www.stsci.edu/astroweb/cat-radio.html
  Most of the world's radio telescope facilities are listed with their links together with other useful sites related to radio wavelength observations.
- W 23.1—chandra.harvard.edu/
  The site of the Chandra X-ray Observatory Center.
- W 23.2—sci.esa.int/xmm/
  The European Space Agency (ESA) site describing the XMM-Newton x-ray telescope.

- —www.ing.iac.es/
  The site for the Isaac Newton Group of Telescopes—includes the 4.2 m WHT, the 2.5 m INT and the 1.0 m JKT.
- —www.jach/hawaii.edu/
  Joint Astronomy Center with telescopes on the summit of Mauna Kea on Hawaii.
  Adding 'JACpublic/UKIRT' to the address provides information about the United Kingdom Infra-Red Telescope, the World's largest telescope devoted to IR astronomy.
- —www.eso.org/
  The site of The European Southern Observatory operating telescopes in the southern hemisphere.
  Adding 'projects/vlt/' to the address provides information about ESO's four 8·2 m telescopes operating in concert.
  Adding 'projects/owl/' to the address provides information on the OverWhelmingly Large Telescope or OWL Project.
- —www.subaru.naoj.org/
  The site of Japan's 8·2 m optical–infrared telescope.
- —www.astro.lu.se/~torben/
  The Swedish Lund Observatory site. Clicking on 'Euro 50' provides information on Sweden's proposed 50 m telescope.
- —celt.ucolick.org/
  The site of the University of California Lick Observatory providing information on CELT (The California Extremely Large Telescope).
- www.aura-nio.noao.edu/
  AURA—New Initiatives Office—details of the 30 m Giant Segmented Telescope are provided.
- www.noao.edu/
  US National Optical Astronomy Observatory—an excellent site for pictures of telescopes and celestial objects.

- www.stsci.edu/
  The site of the Space Telescope Science Institute—Information about the Hubble Space Telescope is available, together with recent images of astronomical objects.

Appendix: Astronomical and related constants

A.1 Physical constants

Velocity of light $c = 2.997\,925 \times 10^8$ m s^{-1}
Constant of gravitation $G = 6.668 \times 10^{-11}$ N m^2 kg^{-2}
Planck constant $h = 6.625 \times 10^{-34}$ J s
Boltzmann constant $k = 1.380 \times 10^{-23}$ J K^{-1}
Stefan–Boltzmann constant $\sigma = 5.669 \times 10^{-8}$ W m^{-2} K^{-4}
Wien displacement law constant $\lambda_{max} T = 2.90 \times 10^{-3}$ m K
Mass of electron $m_e = 9.108 \times 10^{-31}$ kg
Electron charge $e = 1.602\,06 \times 10^{-19}$ C
Solar parallax $P_\odot = 8''794\,05$
Constant of aberration $= 20''49$
Constant of nutation $= 9''207$
Obliquity of ecliptic at any epoch t: $\varepsilon = 23° 26' 21''45 - 46''815T$
    where $T = (t - 2000.0)/100$
Constant of general precession $= 50''290\,966$ per Julian year
    (at standard epoch 2000·0)
Permittivity of free space $\epsilon_0 = 8.854\,24 \times 10^{-12}$ F m^{-1}

A.2 Time

Length of the second

1 dynamical second $= 1/31\,556\,925.975$ of length of tropical year at 1900·0
1 mean solar second (smoothed) $= 1$ dynamical second to about 1 part in 10^8
1 mean second $= 1.002\,737\,909\,3$ mean sidereal seconds

Length of the day

1 mean solar day	$=$	$1.002\,737\,909\,3$ sidereal days
	$=$	$24^h\,03^m\,56^s5554$ mean sidereal time
	$=$	$86\,636.5554$ mean sidereal seconds
1 sidereal day	$=$	$0.997\,269\,566\,4$ mean sidereal days
	$=$	$23^h\,56^m\,04^s0905$ mean solar time seconds
	$=$	$86\,164.0905$ mean solar seconds

Length of the month

Synodic	$29^d 530\,59$	29^d	12^h	44^m	03^s
Sidereal	$27\cdot321\,66$	27	07	43	12
Anomalistic	$27\cdot554\,55$	27	13	18	33
Nodical	$27\cdot212\,22$	27	05	05	36
Tropical	$27\cdot321\,58$	27	07	43	05

Length of the year

Tropical (Υ to Υ)	$365^d 242\,20$	365^d	05^h	48^m	46^s
Sidereal (fixed star to fixed star)	$365\cdot256\,36$	365	06	09	10
Anomalistic (perigee to perigee)	$365\cdot259\,64$	365	06	13	53
Julian	$365\cdot25$	365	06	00	00

A.3   Mathematical constants, systems of units and conversion factors

1 radian $= 57°295\,78 = 3437'\!747 = 206\,264''\!8$

Square degrees in a steradian $= 3282\cdot806$

A sphere subtends 4π steradians at its centre

$\pi = 3\cdot141\,59$

$e = 2\cdot718\,28$

$\log_{10} e = 0\cdot434\,29$

$\log_e 10 = 2\cdot302\,58$

1 km $= 0\cdot621\,37$ miles $= 0\cdot539\,96$ nautical miles $= 3280\cdot8$ ft

1 mile $= 1\cdot6093$ km

1 international nautical mile $= 1\cdot8250$ km $= 6076\cdot11$ ft

1 Å$= 10^{-7}$ mm

1 AU $= 149\,600\,000$ km $= 92\,960\,000$ miles

1 parsec $= 206\,264\cdot8$ AU

1 km s$^{-1} = 2236\cdot9$ miles per hour

1 foot per second $= 0\cdot304\,80$ metres per second

1 AU per year $= 2\cdot9456$ miles per second $= 4\cdot7404$ km s^{-1}

1 kilogram $= 2\cdot204\,622$ lb

1 lb $= 0\cdot453\,592$ kg

1 electron volt (eV) $= 1\cdot602\,06 \times 10^{-19}$ J

Basic SI units

Physical quantity	Name of unit	Symbol
length	metre	m
mass	kilogram	kg
time	second	s
electric current	ampere	A
thermodynamic temperature	degree Kelvin	K

Derived SI units for special quantities

Physical quantity	Name of unit	Symbol	Dimensions
energy	joule	J	$\text{kg m}^2 \text{ s}^{-2}$
power	watt	W	$\text{kg m}^2 \text{ s}^{-3}$
force	newton	N	kg m s^{-2}
electric charge	coulomb	C	A s
magnetic flux density	tesla	T	$\text{kg s}^{-2} \text{ A}^{-1}$
frequency	hertz	Hz	s^{-1}

Conversion of cgs units to SI units

$$
\begin{aligned}
1 \text{ ångström} &= 10^{-10} \text{ m} \\
1 \text{ dyne} &= 10^{-5} \text{ N} \\
1 \text{ erg} &= 10^{-7} \text{ J} \\
1 \text{ millibar} &= 10^2 \text{ N m}^{-2} \\
1 \text{ gauss} &= 10^4 \text{ T} \\
1 \text{ erg s}^{-1} &= 10^{-7} \text{ W}
\end{aligned}
$$

SI prefixes

Factor	*Prefix*	*Symbol*	*Factor*	*Prefix*	*Symbol*
10^{-3}	milli	m	10^{3}	kilo	k
10^{-6}	micro	μ	10^{6}	mega	M
10^{-9}	nano	n	10^{9}	giga	G
10^{-12}	pico	p	10^{12}	tera	T
10^{-15}	femto	f	10^{15}	peta	P
10^{-18}	atto	a	10^{18}	exa	E

A.4   Basic formulas

Trigonometry

Using a basic right-angled triangle (see figure A.1)

$$
\sin \theta = \frac{\text{opposite}}{\text{hypotenuse}} \qquad \cos \theta = \frac{\text{adjacent}}{\text{hypotenuse}} \qquad \tan \theta = \frac{\text{opposite}}{\text{adjacent}}.
$$

$$
\sec \theta = \frac{1}{\cos \theta} \qquad \csc \theta = \frac{1}{\sin \theta} \qquad \cot \theta = \frac{1}{\tan \theta}.
$$

$$
\sec = \text{secant} \qquad \csc = \text{cosecant} \qquad \cot = \text{cotangent}.
$$

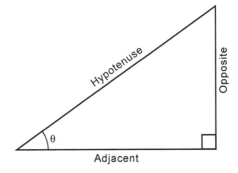

Figure A.1. A basic right-angled triangle.

Logarithms

$$y \log_{10} x = \log_{10}(x^y)$$

$$\log_{10}(x \times y) = \log_{10}(x) + \log_{10}(y) \qquad \log_{10}\left(\frac{x}{y}\right) = \log_{10}(x) - \log_{10}(y).$$

Bibliography

1 Source books

The Astronomical Almanac, Her Majesty's Stationary Office, published yearly

The Handbook of the British Astronomical Association, British Astronomical Association, Burlington House, Piccadilly, London, W1J 0DU, published yearly

Norton's Star Atlas, by A P Norton, 19th edition, ed Ian Ridpath, Longman Publishing Group, 1998

The Cambridge Star Atlas, by Wil Tirion, Cambridge University Press, 2001

Astronomy: A Handbook ed G D Roth, Springer, Berlin, 1975

Outline of Astronomy, two volumes, by H H Voigt, Noordhoof Publishers, Holland, 1974

The Cambridge Atlas of Astronomy, ed J Audouze and G Israël, Cambridge University Press, 1985

Colours of the Stars, by D Malin and P Murdin, Cambridge University Press, 1984

Philips Planisphere, There are various models—a small one of 130 mm diameter was published in 1994 by George Philip Ltd

Collins Dictionary of Astronomy, 2nd edn, ed V Illingworth, Harper Collins, 2000

Oxford Dictionary of Astronomy, ed I Ridpath, Oxford University Press, 1998

In addition, current discussion on the recent developments of the subject, on forthcoming, predicted events and on historical topics can be found in the excellent journal, *Sky & Telescope*. The magazine is also a good source for some experimental projects and old issues may be accessed from libraries, Sky Publishing Corporation, Cambridge, MA, USA

2 Practical projects

Further ideas for experiments and exercises can be found in:

Exercises in Astronomy, ed J Kleczek (a revised and extended edition of *Practical Work in Elementary Astronomy*, by M G J Minnaert), Reidel, Dordrecht, 1987

Astronomy: Observational Activities and Experiments, by M K Gainer, Allyn and Bacon Inc., Boston, 1974

Laboratory Exercises in Astronomy, 2nd edn, by J L Safko, Kendall/Hunt, 1978

Astronomy: Activities and Experiments, by L J Kelsey, D B Hoff and J S Neff, Kendall/Hunt, 1974

Activities in Astronomy, by D B Hoff, L J Kelsey and J S Neff, Kendall/Hunt, 1978

A description of the CCD camera and its application using small telescopes, together with a package for image processing, is given in:

The Handbook of Astronomical Image Processing, by R Berry and J Burnell, William Bell, Inc, 2000

A collection of data reduction exercises based on photographic material is available in:

> *Exercises in Practical Astronomy using Photography*, by M T Brück, Adam Hilger, 1990

3   Books on specific topics

Although the first four listed books are old, they are essentially benchmarks in their field.

> *The Sleepwalkers*, by A Koestler, The Macmillan Company, New York, 1969
> *A History of Astronomy*, by A Pannekeok, Interscience Publishers, New York, 1961
> *Megalithic Sites in Britain*, by A Thom, Oxford University Press, 1967
> *Megalithic Lunar Observatories*, by A Thom, Oxford University Press, 1969
> *Longitude*, by D Sobel, Fourth Estate Ltd, London, 1995
> *X-ray Astronomy*, by J L Culhane and P W Sanford, Faber and Faber, 1981
> *An Introduction to Radio Astronomy*, 2nd edn, by B F Burke and F Graham-Smith, Cambridge University Press, 2002

Answers to problems

Chapter 7

1. $32°, 78°, 259°, 20° 55', 62° 17', 28° 18', 234° 09'$
2. $10°, 53°, 17° 14', 164° 41', 77° 10'$
3. $600, 3180, 1034, 9881, 4630$
4. $472·1, 3405, 415·7, 1614$
5. (i) $c = 39° 59', B = 48° 19', A = 66° 53'$
   (ii) $a = 56° 22', B = 70° 15', C = 67° 46'$
   (iii) $b = 145° 29', A = 69° 19', C = 37° 07'$
   (iv) $a = 159° 00', A = 44° 22', B = 42° 24'$
6. (i) 5786 nautical miles   (ii) 5220 nautical miles
7. $1^d 19^h 77$
8. 5582 km; $53° 48'$ N
9. $72° 49'$

Chapter 8

1. Figures A8.1(a), A8.1(b)
2. Figure A8.2
3. Figure A8.3
4. Figure A8.4
5. $\lambda = 90°; \beta = 21° 33'$ N
6. $6° 34'$
8. $95° 55'$ W of N; $95° 55'$ E of N
9. $29° 46'$ N; $180° 01'$. Dec 21st; June 21st
   (*Hint*: The points of intersection of the galactic equator with the ecliptic must be at ♉ and ♋)
10. (*Hint*: $\angle PXZ = 90°$ when azimuth is a maximum for a circumpolar star)
11. $47° 42'$ ($3^h 10^m 48^s$); $\phi \geq 51° 16'$ N
12. $19° 46'$
14. $117° 0'14$ ($7^h 48^m 01^s$); $10^h 57^m 30^s$
15. $49° 57'$ E
16. $59° 47' 09''$; $72° 54' 25''$

RA X ~ 3ʰ
Dec X ~ 40°S

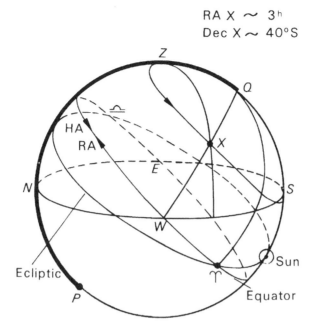

Figure A8.1(a). RA⊙ $= 21^{\text{h}}$. Therefore, date is March 21st $- 1\frac{1}{2}$ months $\approx$ February 6th.

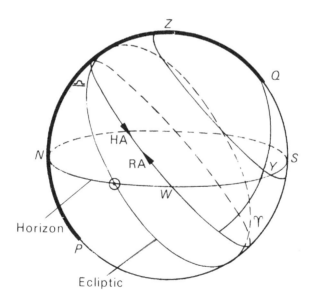

Figure A8.1(b). *Y* is the position of the star at sunset. Since RA⋆ $\sim 3^{\text{h}}$, ♈ is then as shown in the diagram. The ecliptic is then placed with the Sun ⊙ at sunset. Since the RA⊙ is then approximately $8\frac{1}{2}$ hours, the date is approximately July 29th.

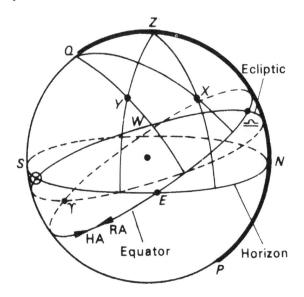

Figure A8.2. Hour angle of $X \approx 22^{\text{h}}$; declination of $X \approx 20°$ S, altitude of $Y \approx 40°$; azimuth of $Y \approx 80°$ E of S.

	Celestial longitude	Celestial latitude
X	$\approx 200°$	$\approx 15°$ S
Y	$\approx 250°$	$\approx 18°$ S

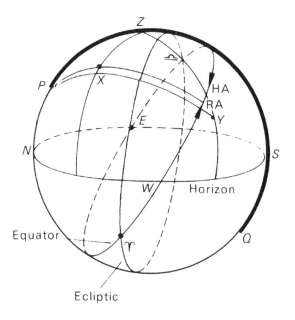

Figure A8.3. Azimuth of $X \approx 40°$ W; altitude of $X \approx 50°$.
Hour angle of $Y \approx 3\frac{1}{2}^{\text{h}}$; declination of $Y \approx 12°$ S.
The traveller is not stating the truth. The Sun's maximum northerly declination is $23°\ 26'$. The zenith lies on the $30°$ N declination circle.
Star X is circumpolar since $PX < PN$.

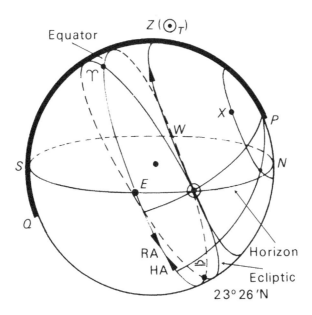

Figure A8.4. $\odot_T$ is the Sun's position when it transits; on June 21st this position at transit is Z in this latitude. Estimated LST is 6^h when Sun transits. Estimated altitude of X is $40°$. Estimated azimuth of X is $20°$ E of N at apparent noon. The star X is not circumpolar.

Chapter 9

1. $66° 34'$ north or south
2. $66° 34'$
3. $23^h 52^m$; $367\frac{1}{4}$
4. $4^h 46^m 21^s$; $14^h 04^m 40^s$; $17^h 15^m 46^s$; $20^h 14^m 29^s$
5. $6^h 47^m 27^s8$
6. $23^h 09^m 58^s7$
7. $22^h 46^m 20^s$
8. $56° N$
9. $2^h 04^m$
10. $17^h 17^m 7^s$
11. $20^h 47^m 32^s$ (*Note*: The approximate Greenwich Date removes the ambiguity about the chronometer dial being a 12^h or 24^h one)
12. $21^h 07^m 51^s$
13. $2^h 45^m 49^s$; $7^h 14^m 00^s$
14. $4^h 42^m 08^s$
15. June 21st: December 21st. (Sun's right ascension is $90°$ and $270°$, respectively, on these dates)
16. $3^h 57^m$. (*Hint*: Find RA$\odot$; find HA Venus at sunset, then at Venus set; convert interval of sidereal time to mean solar time)
17. $93^d 15^h66$
18. $93^d 14^h4$
19. Use equations (8.2), (8.12) and table 8.1 or 8.2

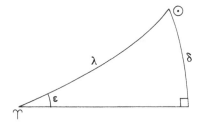

Figure A10.16.

Chapter 10

1. $25° 00' 28''1$
2. $50·49$ km
3. $143·6$ km
5. $42° 46'7$
6. $0·054$
7. Parallax $= \arctan (R_E/r_s) = 8°64$
8. (i) $0''04$; (ii) $0''035$
9. $13·51$; $21·28$; $7·97$
10. 1737 km; $55'0$
11. $9'16$; $19·92$ km
12. $695\,800$ km; $149·5 \times 10^6$ km
13.

| | Displacement | |
Date	RA	Dec
Mar 21st	$+1''26$	0
June 21st	0	$+0''378$
Sept 21st	$-1''26$	0
Dec 21st	0	$-0''378$

14. $\delta = 64° 51' 12'' \text{ N}; \phi = 47° 36' 21'' \text{ N}$
15. When $\lambda_\odot = \lambda \pm \pi/2$, the displacement is at right angles to the great circle from K to the star.
16. (a) $16·8$ days   (b) $30·1$ days. (*Hint*: in case (a) it is double the time it takes the Sun's declination to change from $23° 11'$ N to $23° 26'$ N). Use the triangle in figure A10.16 to find the Sun's longitudes at these times.
17. $1·76$ (*Hint*: In 12 hours the observer's position shifts by a distance $R_\oplus$, where $R_\oplus$ is the radius of the Earth.)
19. $62° 51'7$

Chapter 11

1. $-10''25$
2. (i) $359° 59' 39''51$;   (ii) $0°$;   (iii) $20''49$;   (iv) $0°$
3. $16''73$
4. $18°$

5.

	Maximum	Minimum
Right ascension	June 21st	December 21st
Declination N	September 21st	March 21st

7. Displacement in celestial longitude $= -21''81$; in celestial latitude $= 0$
8. 1448 years
9. The star lies on the great circle arc joining the zenith to the ecliptic point A which is $90°$ behind the Sun such that
$$k \tan z = \kappa \sin \theta$$
where θ is the angle between the star and A. $z = 18° 24'$
10. 6^{h}
11. $43°$ N; $47°$
12. 6859 years hence
13. $50''70$ per annum; $0·0143$ day
14. (i) $\lambda = 330°$; $b = 66\frac{1}{2}°$ N   (ii) $\alpha = 302° 14'$; $\delta = 49° 36'$ N
15. $48\frac{1}{2}°$ N; $90°$; 4189 years hence

Chapter 12

1. Superior
2. $115·9$ days
3. $46° 18'$
4. 4330 days
5. $23° 34'$
6. $5·483$
7. $0·8821$
8. $1·291$
9. $0·6395$ AU; $0·9457$
10. (i) $38° 46'$;   (ii) $97·3$ days
11. $120·6$ days
12. $0·0976$
13. (NB: 1968 is a Leap Year) Aug 28th; Mar 14th; 1978; Every 12th year

Chapter 13

1. 8 years
2. $0·6151$ year
3. $1·524$ AU
4. $1/1060$
5. $0·0316$ year
6. $341\,000$
7. 6200 km
8. 2078 km; $0·1251$; $79·1/1$
9. $2·081$ days (*Hint*: Use the first approximation to the binomial theorem)
10. (i) $0·382$ AU;   (ii) $1·205$;   (iii) $5·59$ AU year^{-1}
11. (i) $60·16$;   (ii) 3619

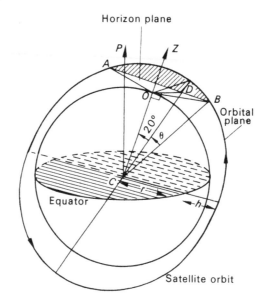

Figure A13.13. Triangle ODC is right-angled at O since plane $AOBD$ is the horizon plane. Triangle CDB is right-angled at D. Hence

$$\cos\theta = CD/CB = CD/(r+h)$$
$$\cos 20° = OC/CD = r/CD$$

Thus

$$\cos\theta = \left(\frac{r}{r+h}\right)\frac{1}{\cos 20°}.$$

Hence, time above horizon.

12. 1/7; 2/3
13. 52° 14′; 25·18 minutes; 21·82 minutes

Chapter 14

1. (i) 42 407 km;   (ii) 24 550 km;   0·7274;   5·272 hours
2. 141·88 minutes; 10 055 km; 0·2780; 80·66 minutes
3. (i) 195 639 km; 0·9648   (ii) 119·5 hours   (iii) 3·056 km s^{-1}
4. Julian Date 2 442 071·9; 1410 Julian Days

Chapter 15

1. 5, 3, 4, 2, 1, 6
2. 1·19
3. 15·8
4. +3·25
5. 8·3
6. 1·26 × 10^{-8} W
7. +7·75
8. 1·177

9. 5000 Å
10. Lα = 1215 Å—which is in the deep ultraviolet, well absorbed by the Earth's atmosphere
11. 1·0574 × 10^8 T
12. −42 km s^{-1}
13. 0·17 Å

Chapter 16

1. 43·6 mm
2. 34·4 arc sec mm^{-1}, 0.058 mm
3. Assuming that the 36 mm length is used → 137·5 mm
4. 1·9 mm
5. 2·91 × 10^{-2} mm
6. ≈3 × 10^7 s^{-1} (Assume λ = 550 nm)
7. 28·6 s^{-1}; 37·8:1
8. 5090 mm (*Hint*: Use lens-makers formula)
9. (i) 8000 mm (ii) $f/16$ (iii) −533 mm (iv) 1/25
   (*Hint*: Draw diagram and apply similar triangles)
10. 0·53%
11. 1·15%
12. 0·15 arc sec

Chapter 17

1. 150, Yes, $m > D/8$
2. $D = 6·7$ m
3. 1·34 × 10^{-6} rad; 1·34 × 10^{-2} mm (Airy disc = 2α × F)
4. Aberrated image size = 0·0025 (*Hint*: See example 16.2), Airy disc = 0·0013
5. Aberrated image size = 0·0015 (*Hint*: See example 16.3), Airy disc = 0·0016
6. 630 mm; 16^{m}2
7. 208 mm
8. τ = 17/10; Feb 1st 1575 (assuming no losses)
9. 13·4 arc sec

Chapter 18

1. (i) 13·9 (ii) 14·2 (both assuming no losses)
2. 100 s
3. 16^{m}3
4. ≈ 10^2
5. 0·26%

Chapter 19

1. 8·24
2. 4·9
3. 8·2, 86%
4. 22° 36′

5. 1·64
6. −0·545
7. 0·176 mag
8. 8^m5
9. 68·57 Å mm^{-1}
10. $R = 5 \times 10^4$; $V = 6$ km s^{-1}; $\Delta\lambda = 2\cdot5 \times 10^{-3}$ mm
11. RLD = 16·67 Å mm^{-1}; $R = 8\cdot4$; the seeing deteriorates the spectral resolution by a factor of ∼6.

Chapter 21

1. $4'.19$, $\alpha_{Opt}/\alpha_{Rad} = 10^4$
2. $1\cdot37 \times 10^{-17}$ W
3. $P_{\odot}/P_{Wall} = 20/1$
4. 6·8 arc min diameter

Index